HRW GEOMETRY

explore *communicate* APPLY

Integrating

MATHEMATICS
TECHNOLOGY
EXPLORATIONS
APPLICATIONS
ASSESSMENT

HOLT, RINEHART AND WINSTON
Harcourt Brace & Company
Austin • New York • Orlando • Atlanta • San Francisco • Boston • Dallas • Toronto • London

For permission to reprint copyrighted material, grateful acknowledgment is made to the following sources:

Discover Syndication, a division of Disney Magazine Publishing, Inc.:   Excerpt by Paul Hoffmann and five photographs by Annette Del Zoppo from "Egg Over Alberta" from *Discover,* May 1988, pp. 37, 38, 39, and 42. Copyright © 1988 by Discover Magazine and Annette Del Zoppo.

Dover Publications, Inc.:   "6 Accomplishments" and "31 Digits Are Symbols" from *My Best Puzzles in Logic and Reasoning* by Hubert Phillips. Copyright © 1961 by Dover Publications, Inc.

Griffith Institute, Ashmolean Museum: From Howard Carter's Notebook and sketch of "Band A."

Harcourt Brace & Company: From "Law of Sines: Case 1" on page 165 from *Trigonometry,* Revised Edition by Arthur F. Coxford. Copyright © 1981, 1987 by Harcourt Brace & Company.

Holt, Rinehart and Winston, Inc.:   From "Sun-Centered Coordinates of Major Planets for Aug. 17, 1990 in Astronomical Units" p. 690 from *HRW Algebra with Trigonometry.* Copyright © 1986, 1992 by Holt, Rinehart and Winston, Inc. Exercises 11–13 with drawing, p. 104; from "Critical Thinking Questions," pp. 110 and 159; drawing from p. 110, and graph from p. 159 from *HRW Geometry,* Annotated Teacher's Edition. Copyright © 1991, 1986 by Holt, Rinehart and Winston, Inc. "Brainteaser," on p. 622, from *Holt Algebra with Trigonometry.* Copyright © 1986, 1992 by Holt, Rinehart and Winston, Inc.

Knight-Ridder Tribune News Service: From " 'Parallax Conspiracy' Has Angry Umpires in a Tizzy" by Bill Conlin, as it appears in the *Albuquerque Journal,* October 21, 1993. Copyright © 1993 by Tribune Media Services.

NASA and Goddard Space Flight Center: From section 5.1.4 "Pointing Control Subsystem (PCS)" from *Hubble Space Telescope Media Reference Guide: 1st Serving Mission.*

National Council of Teachers of Mathematics: Adapted from "Activities: Spatial Visualization" by Glenda Lappan, Elizabeth A. Phillips, and Mary Jean Winter from *Mathematics Teacher,* vol. 77, no. 8, November 1984. Copyright © 1984 by the National Council of Teachers of Mathematics.

The New York Times Company: From "Gem Studded Relics in Egyptian Tomb Amaze Explorers" from *The New York Times,* December 1, 1922. Copyright © 1922 by The New York Times Company. From "Math Problem, Long Baffling, Slowly Yields" by Gina Kolata (with map "The Efficient Traveling Salesman") from *The New York Times,* March 12, 1991. Copyright © 1991 by The New York Times Company. From "Main Telescope Repairs are Completed" by John Noble Wilford from *The New York Times,* December 9, 1993. Copyright © 1993 by The New York Times Company. Headlines "Severe Earthquake Hits Los Angeles," "Collapsed Freeways Cripple City," "Scientists Say Unknown Fault Deep Within Earth Probably Caused Tremor" and map "Damage: First the Quake, Then Aftershocks and Fires" from *The New York Times,* January 18, 1994. Copyright © 1994 by The New York Times Company. From "Astronauts Snare Hubble Telescope for Vital Repairs" by John Noble Wilford from *The New York Times,* December 5, 1993. Copyright © 1993 by The New York Times Company.

Oxford University Press, London: "Fig. 2.5" on p. 36; "(h) prism" on p. 44; illustrations on p. 47; and "Fig. 3.7" on p. 71 from *Chinese Mathematics: A Concise History* by Lǐ Yǎn and Dù Shíràn, translated by John N. Crossley and Anthony W.-C. Lun. Translation copyright © 1987 by John N. Crossley and Anthony W.-C. Lun.

Science News, The Weekly Newsmagazine of Science: From "The Big Fix" by Ron Cowen from *Science News®,* vol. 144, no. 19, November 6, 1993, pp. 296-297. Copyright © 1993 by Science Service, Inc.

Greg Stec: From "Message of the Maya in Modern Translation" by Greg Stec (Retitled "Message of the Maya in Modern Times") from *The Christian Science Monitor,* June 22, 1989. Copyright © 1989 by Greg Stec.

United States Olympic Committee: "Olympic Rings" by The U.S. Olympic Committee.

Printed in the United States of America
 2 3 4 5 6 7 041 00 99 98 97 96

ISBN: 0-03-097775-4

AUTHORS

Kathleen A. Hollowell

Dr. Hollowell is widely respected for her keen understanding of what takes place in the mathematics classroom. Her impressive credentials feature extensive experience as a high school mathematics and computer science teacher, making her particularly well-versed in the special challenges associated with integrating math and technology. She currently serves as Associate Director of the Secondary Mathematics Inservice Program, Department of Mathematical Sciences, University of Delaware and is a past-president of the Delaware Council of Teachers of Mathematics.

James E. Schultz

Dr. Schultz is one of the math education community's most renowned mathematics educators and authors. He is especially well regarded for his inventive and skillful integration of mathematics and technology. He helped establish standards for mathematics instruction as a co-author of the NCTM "Curriculum and Evaluation Standards for Mathematics" and "A Core Curriculum, Making Mathematics Count for Everyone." Following over 25 years of successful experience teaching at the high school and college levels, his dynamic vision recently earned him the prestigious Robert L. Morton Mathematics Education Professorship at Ohio University.

Wade Ellis, Jr.

Professor Ellis has gained tremendous recognition for his reform-minded and visionary math publications. He has made invaluable contributions to teacher inservice training through a continual stream of hands-on workshops, practical tutorials, instructional videotapes, and a host of other insightful presentations, many focusing on how technology should be implemented in the classroom. He has been a member of the National Research Council's Mathematical Sciences Education Board, the MAA Committee on the Mathematical Education of teachers, and is a former Visiting Professor of Mathematics at West Point.

CONTRIBUTING AUTHORS

Larry Hatfield   Dr. Hatfield is Department Head and Professor of Mathematics Education at the University of Georgia. He is recipient of the Josiah T. Meigs Award for Excellence in Teaching, his university's highest recognition for teaching. He has served at the National Science Foundation and is Director of the NSF funded Project LITMUS.

Bonnie Litwiller   Professor of Mathematics, University of Northern Iowa, Cedar Falls, Iowa, Dr. Litwiller has been co-director of NSF institutes, project coordinator for the NCTM Addenda Project, and co-author of three books and 650 articles.

Martin Engelbrecht   A mathematics teacher at Culver Academy, Culver, Indiana, Mr. Engelbrecht also teaches statistics at Purdue University—North Central. An innovative teacher and writer, he integrates applied mathematics with technology to make mathematics more accessible to all students.

Kenneth Rutkowski   A mathematics teacher at James Bowie High School, Austin, Texas, Mr. Rutkowski is an innovative geometry teacher, who sponsors his school's mathematics honor society, serves on various professional committees and conducts creative teacher workshops.

• •

Editorial Director of Math
Richard Monnard

Executive Editor
Gary Standafer

Senior Editor
Ronald Huenerfauth

Project Editors
Charles McClelland
Joel Riemer
Michelle Dowell
Michael Funderburk

Design and Photo
Pun Nio
Diane Motz
Sam Dudgeon
Victoria Smith
Mavournea Hay
Cindy Verheyden
Michael Obershan
Alicia Sullivan
Lori Male
Katie Kwun
Monotype Editorial Services

Editorial Staff
Steve Oelenberger
Richard Zelade
Andrew Roberts
Pam Garner
Jane Gallion
Desktop Systems Assistant
Jill Lawson
Department Secretary

Production and Manufacturing
Donna Lewis
Amber Martin
Jenine Street
Shirley Cantrell

CONTENT CONSULTANT

Kenneth Rutkowski A mathematics teacher at James Bowie High School, Austin, Texas, Mr. Rutkowski is an innovative geometry teacher, who sponsors his school's mathematics honor society, serves on various professional committees and conducts creative teacher workshops.

MULTICULTURAL CONSULTANT

Beatrice Lumpkin A former high school teacher and associate professor of mathematics at Malcolm X College in Chicago, Professor Lumpkin is a consultant for many public schools for the enrichment of mathematics education through its multicultural connections. She served as a principal teacher-writer for the *Chicago Public Schools Algebra Framework* and has served as a contributing author to many other mathematics and science publications that include multicultural curriculum.

REVIEWERS

James A. Bade
Adlai Stevenson High School
Sterling Heights, Michigan

Tom Fitzgerald
Cocoa High School
Cocoa, Florida

Ona Lea Lentz
North High School
Minneapolis, Minnesota

Karen M. Lesko
Pacific High School
San Bernardino, California

Jean Mariner
St. Steven's Episcopal School
Austin, Texas

Gregory Massarelli
Watkins Memorial High School
Pataskala, Ohio

Susan May
Science Academy of Austin at LBJ
Austin, Texas

John S. Nesladek
Ozaukee High School
Fredonia, Wisconsin

Gary Nowitzke
Jefferson High School
Monroe, Michigan

Roger O'Brien
Polk County School District
Bartow, Florida

Ruth R. Price
Lee County High School
Sanford, North Carolina

Robert J. Russell
West Roxbury High School
West Roxbury, Massachusetts

Sandra Seymour
Science Academy of Austin at LBJ
Austin, Texas

Rosalind Taylor
W. C. Overfelt High School
San Jose, California

Jean D. Watson
High School of Commerce
Springfield, Massachusetts

Nanci Takagi White
University City High School
San Diego, California

CHAPTER 1 **Exploring Geometry** 2

1.0 Building Your Geometry Portfolio 4
1.1 Modeling the World With Geometry .9
1.2 *Exploring* Geometry Using Paper Folding 16
1.3 *Exploring* Geometry With a Computer 23
1.4 Measuring Length 29
1.5 Measuring Angles 37
1.6 Motions in Geometry 44
1.7 *Exploring* Motion in the Coordinate Plane 51
    Chapter Project *Origami Paper Folding* 58
    Chapter Review 60
    Chapter Assessment 63

MATH *Connections*

Algebra 15, 32, 34, 36, 41, 42, 50, 51, 67 - 70, 93, 100, 104
Coordinate Geometry 51-57
Statistics 35
Transformations 44-50, 51-57

CHAPTER 2 **Reasoning in Geometry** 64

2.1 *Exploring* Informal Proofs 66
2.2 Introduction to Logical Reasoning 72
    Eyewitness Math *Too Tough for Computers* 80
2.3 *Exploring* Definitions 82
2.4 Postulates 88
2.5 Linking Steps in a Proof 92
2.6 *Exploring* Conjectures that Lead to Proof 99
    Chapter Project *Solving Logic Puzzles* 106
    Chapter Review 108
    Chapter Assessment 111

CUMULATIVE ASSESSMENT **CHAPTERS 1–2** **112**

APPLICATIONS

Science
Archaeology 24
Chemistry 85
Geology 35, 85
Physics 28

Language Arts
Communicate 13, 20, 25, 33, 40, 48, 55, 68, 76, 85, 90, 96, 103
Eyewitness Math 80

Business and Economics
Construction 27

Life Skills
Navigation 42

Sports and Leisure
Aquarium 14
Nim 71, 98
Origami 19, 22
Scuba Diving 42

Visual Arts 50
Fine Arts 78

Other
Criminal Justice 78
Humor 77

CHAPTER **Parallels and Polygons** **114**

3.1 *Exploring* Symmetry in Polygons 116
3.2 *Exploring* Properties of Quadrilaterals 124
Eyewitness Math *Egg Over Alberta* 130
3.3 Parallel Lines and Transversals 132
3.4 Proving Lines Parallel 138
3.5 *Exploring* the Triangle Sum Theorem 143
3.6 *Exploring* Angles in Polygons 149
3.7 *Exploring* Midsegments of Triangles and Trapezoids 154
3.8 Analyzing Polygons Using Coordinates 160
Chapter Project *String Figures* 166
Chapter Review 168
Chapter Assessment 171

MATH
Connections

Algebra 122, 128, 137, 141, 153, 161, 164, 192, 206
Coordinate Geometry 160-164, 179, 220-224
Transformations 226-231

CHAPTER  **Congruent Triangles** **172**

4.1 Polygon Congruence 174
4.2 *Exploring* Triangle Congruence 180
4.3 Analyzing Triangle Congruence 186
4.4 Using Triangle Congruence 193
4.5 Proving Quadrilateral Properties 201
4.6 *Exploring* Conditions for Special Quadrilaterals 207
4.7 Compass and Straightedge Constructions 213
4.8 *Exploring* Congruence in the Coordinate Plane 220
4.9 *Exploring* the Construction of Transformations 226
Chapter Project *Flexagons* 232
Chapter Review 234
Chapter Assessment 239

CUMULATIVE ASSESSMENT **CHAPTERS 1-4** **240**

APPLICATIONS

Science
Archaeology 156
Engineering 120
Flood Control 159
Structural Engineering 158

Social Studies
Geography 123, 145
History 219

Language Arts
Communicate 120, 127,

136, 140, 146, 152, 157, 163, 177, 183, 190, 197, 203, 210, 215, 222, 229
Eyewitness Math 130

Business and Economics
Construction 164, 184, 192, 197, 202

Life Skills
Carpentry 136, 178, 184
Navigation 137, 187

Traffic Signs 200

Sports and Leisure
Baseball 206
Quilting 178, 184
Recreation 184

Other
Drafting 142
Fine Art 204
Interior Design 219
Surveying 199

CHAPTER **5** **Perimeter and Area** **242**

5.1 *Exploring* Perimeter, Circumference, and Area 244
Eyewitness Math *An Ancient Wonder* 252
5.2 *Exploring* Areas of Triangles, Parallelograms,
and Trapezoids 254
5.3 *Exploring* Circumferences and Areas of Circles 261
5.4 The "Pythagorean" Right-Triangle Theorem 267
5.5 Special Triangles, Areas of Regular Polygons 275
5.6 The Distance Formula, Quadrature of a Circle 283
5.7 Geometric Probability 289
Chapter Project *Area of a Polygon* 296
Chapter Review 298
Chapter Assessment 301

 MATH
Connections

Algebra 246, 247, 249,
251, 254-257, 259,
262, 264, 266, 269-
276, 278, 284, 285,
287, 288, 328, 331-
335
Coordinate Geometry
283-288, 326-330,
331-335
Maximum/Minimum
246, 250, 258, 265
Probability 290, 292

CHAPTER **6** **Shapes in Space** **302**

6.1 *Exploring* Solid Shapes 304
6.2 *Exploring* Spatial Relationships 311
6.3 An Introduction to Prisms 317
Eyewitness Math *Did the Camera Lie?* 323
6.4 Three-Dimensional Coordinates 326
6.5 Lines and Planes in Space 331
6.6 Perspective Drawing 336
Chapter Project *Building and Testing an A-Shaped Level* 344
Chapter Review 346
Chapter Assessment 349

CUMULATIVE ASSESSMENT **CHAPTERS 1–6** **350**

 **APPLICATIONS**

Science
Astronomy 330
Automobile Engineering 265
Chemistry 319
Civil Engineering 280
Environmental Protection 287
Meteorology 292

Social Studies
History 273

Language Arts
Communicate 248, 257, 263,
271, 279, 287, 292, 307,

314, 320, 328, 334, 340
Eyewitness Math
252, 323

Business and Economics
Agriculture 251
Construction 249
Farming 260
Irrigation 264
Landscaping 249

Life Skills
Decorating 259
House Painting 249
Navigation 284, 315, 335

Sports and Leisure
Fine Art 295, 342
Gardening 246
Recreation 309
Skydiving 293
Sports 266, 271
Theatre Arts 255

Other
Drafting 281
Measurement 277
Mechanical Drawing
304, 306
Solar Energy 250
Transportation 294

CHAPTER **Surface Area and Volume**   **352**

7.1 *Exploring* Surface Area and Volume 354
7.2 Surface Area and Volume of Prisms 359
7.3 Surface Area and Volume of Pyramids 366
7.4 Surface Area and Volume of Cylinders 373
7.5 Surface Area and Volume of Cones 379
7.6 Surface Area and Volume of Spheres 387
Eyewitness Math *Treasure of King Tut's Tomb* 394
7.7 *Exploring* Three-Dimensional Symmetry 396
Chapter Project 404
Chapter Review 406
Chapter Assessment 409

MATH
Connections

Algebra 354, 367, 374,
  381, 384, 388, 415,
  417, 422, 423, 425,
  428, 435, 444, 445,
  446, 447, 455, 459
Coordinate Geometry
  365, 369-402, 412-
  418
Maximum/Minimum
  355, 356, 357, 378,
  385, 386, 402
Probability 428
Transformations
  396-402, 412-218

CHAPTER **Similar Shapes**   **410**

8.1 *Exploring* Transformations and Scale Factors 412
8.2 *Exploring* Similar Polygons .419
8.3 *Exploring* Triangle Similarity Postulates 429
8.4 The Side-Splitting Theorem 436
8.5 Indirect Measurement, Additional Similarity Theorems 443
8.6 *Exploring* Area and Volume Ratios 452
Chapter Project Indirect Measurement and Scale Models . .460
Chapter Review 462
Chapter Assessment 465

CUMULATIVE ASSESSMENT **CHAPTERS 1–8** **466**

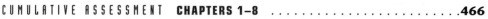

A P P L I C A T I O N S

Science
Aeronautical Engineering 358
Astronomy 448
Biology 357
Botany 357
Cartography 389
Civil Engineering 443
Earth Science 418
Geology 382
Marine Biology 361, 426
Optics 417
Physiology 357, 358
Wildlife Management 426

Language Arts
Communicate 356, 363, 370,
376, 383, 391, 400, 416, 424,
433, 440, 447, 456
Eyewitness Math 394

Business and Economics
Advertising 427
Architecture 365, 424

Construction 368, 372
Landscaping 427
Manufacturing 384
Small Business 385

Life Skills
Cooking 358
Food 392
Model Building 427

Sports and Leisure
Fine Art 427
Hot Air Ballooning 389, 391
Recreation 363, 385, 444
Scale Models 386
Sports 392, 458

Other
Consumer Awareness 377
Gasoline Storage 376
Pottery 401
Product Packaging 356,
364, 458
Surveying 435, 449

CHAPTER **Circles** 468

9.1	*Exploring* Chords and Arcs	470
9.2	*Exploring* Tangents to Circles	478
9.3	*Exploring* Inscribed Angles and Arcs	484
9.4	*Exploring* Circles and Special Angles	491
	Eyewitness Math *Point of Disaster*	500
9.5	*Exploring* Circles and Special Segments	502
9.6	Circles in the Coordinate Plane	509
	Chapter Project *The Olympic Symbol*	516
	Chapter Review	518
	Chapter Assessment	521

MATH
Connections

Algebra 473, 476, 477,
  479, 482, 483, 489,
  495, 496, 504, 506-
  509, 512-514, 526,
  527, 536, 537, 548,
  557, 565, 567, 568,
  573

Coordinate Geometry
  509-515, 548-554

Statistics 476

Transformations
  548-564

CHAPTER **Trigonometry** 522

10.1	*Exploring* Tangent Ratios	524
10.2	*Exploring* Sines and Cosines	533
10.3	*Exploring* the Unit Circle	541
10.4	*Exploring* Rotations With Trigonometry	548
10.5	The Law of Sines	555
	Eyewitness Math *Eyeglasses in Space*	562
10.6	The Law of Cosines	564
10.7	Vectors in Geometry	570
	Chapter Project *Plimpton 322 Revisited*	576
	Chapter Review	578
	Chapter Assessment	581

CUMULATIVE ASSESSMENT **CHAPTERS 1–10** 582

APPLICATIONS

Science
Astronomy 535, 540, 546, 547
Civil Engineering 475
Engineering 530, 531
Geology 505
Physics 570, 572
Space Flight 482
Wildlife Management 560

Language Arts
Communicate 474, 481, 489, 495, 506, 512, 528, 537, 545, 552, 558, 567, 574
Eyewitness Math 500, 562

Business and Economics
Architecture 559
Building Codes 539
Structural Design 514
Life Skills
Carpentry 488
Navigation 494

Sports and Leisure
Recreation 535, 538

Other
Communications 482, 497
Machining 507
Stained Glass 490
Surveying 520, 526, 531, 559

CHAPTER **11** Taxicabs, Fractals, and More   **584**

11.1 Golden Connections .... .586
11.2 Taxicab Geometry .... .594
11.3 Networks .... .600
11.4 Topology: Twisted Geometry .... .607
11.5 Euclid Unparalleled .... .614
11.6 *Exploring* Projective Geometry .... .621
11.7 Fractal Geometry .... .629
     Chapter Project *Random Process Games* .... .636
     Chapter Review .... .638
     Chapter Assessment .... .641

 **MATH** *Connections*
Algebra 588, 631, 632
Coordinate Geometry
594-606, 679-684

CHAPTER **12** A Closer Look at Proof and Logic   **642**

12.1 Truth and Validity in Logical Arguments .... .644
12.2 "And," "Or," and "Not" in Logic .... .650
     Eyewitness Math *The Ends of Time* .... .656
12.3 A Closer Look at If-Then Statements .... .658
12.4 Indirect Proof .... .665
12.5 Computer Logic .... .671
12.6 *Exploring* Proofs Using Coordinate Geometry .... .679
     Chapter Project *Two Famous Theorems* .... .685
     Chapter Review .... .688
     Chapter Assessment .... .691

CUMULATIVE ASSESSMENT **692–693**

Info Bank Table of Contents .... .695
Table of Squares, Cubes, Square and Cube Roots .... .696
Table of Trigonometric Ratios .... .697
Postulates, Theorems, and Definitions .... .698
Glossary .... .704
Index .... .714
Credits .... .725
Selected Answers .... .727

A P P L I C A T I O N S

Science
Environmental Science 678
Genetics 607

Social Studies
History 630

Language Arts
Communicate 590, 597,
604, 611, 618, 626, 633,
647, 653, 661, 668, 675,
683
Eyewitness Math 656

Business and Economics
Materials Handling 612

Sports and Leisure
Candy Making 635
Nine Coin Puzzle 628

Other
Communications 632
Klein Bottle 613
Law Enforcement 605
Transportation 595

CHAPTER 1

Exploring Geometry

LESSONS

1.0 Building Your Geometry Portfolio

1.1 Modeling the World With Geometry

1.2 *Exploring* Geometry Using Paper Folding

1.3 *Exploring* Geometry With a Computer

1.4 Measuring Length

1.5 Measuring Angles

1.6 Motions in Geometry

1.7 *Exploring* Motion in the Coordinate Plane

Chapter Project
Origami Paper Folding

Geometry is both ancient and modern. The traditional tools of the geometer, such as the compass and straightedge, and state-of-the-art computers and calculators are equally appropriate for the study of this fascinating subject.

This first chapter is for you to get acquainted with geometry. You will be introduced to a number of tools for exploring geometry, such as computer graphics and paper folding. In this way, you will discover what geometry is by *experiencing* it.

PORTFOLIO ACTIVITY

As you begin your study of geometry, set up your own portfolio. The first lesson of this book, Lesson 1.0, tells you how to do this. Throughout this book you will find many suggestions of things to include in your portfolio.

Start your portfolio right away. Look for examples of geometry that can be copied from or cut out of magazines or books. You can also use your own drawings or photographs.

3

LESSON 1.0 Building Your Geometry Portfolio

Why *Artists and other professionals often keep portfolios of their work. Although you are probably not yet a professional in any area, the work you do in school may help you decide on your future work and career.*

Building a portfolio will help you organize and display your work. Design it to show your work in a way that reflects your interests and your strengths. You should concentrate on the things you enjoy; these will probably be the things you do best. You might want to create geometric constructions at the computer, study the geometry of beehives and spider webs, or explore the geometry found in works of art.

Geometry in Nature

People have long been attracted to geometric figures in nature, such as the spiral shell of the chambered nautilus. The larger the shell grows, the more closely its proportions approach the value of the golden ratio, a very important number in mathematics. The underlying geometric principles of natural objects often seem to be the reason for their visual appeal. As you look around you, you will find many examples of geometric beauty in nature.

Geometry in Art

The artist Piet Mondrian (1872-1944) had a very deep interest in geometry. His desire to break art down to its purest forms led him to create solid shapes based solely on right angles which echo the rectangular shape of the canvas itself.

If you enjoy doing art, you should include works of your own in your portfolio. Even if you think you have little talent or interest in producing works of your own, try your hand at it. You might surprise yourself! The topics presented in this book will suggest different kinds of art with which to experiment.

Piet Mondrian. *Composition with Red, Blue and Yellow.* 1930. Oil on Canvas. 20″ x 20″.

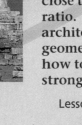

Geometry in Architecture

The dimensions of the Parthenon reveal the ancient Greek fascination with geometry ideas. The ratio of the height of the original structure to its width is very close to the exact value of the golden ratio. Geometry ideas are still in use in architecture. From the principles of geometry and physics, an architect knows how to design structures that will be both strong and beautiful.

Your Notebook and Journal

The photos above are details from two pages of Leonardo da Vinci's journals.

Mathematicians keep records of their theories and discoveries. You, too, should keep a notebook of your work, including tests, homework activities, and special activities such as research projects. Your teacher might also want you to keep a journal.

Putting Your Portfolio Together

Various containers can be used for your portfolio. File folders, accordion files, and even cereal boxes have been found to work very well. You might also have a number of activities such as string designs or physical models that will not fit into your actual portfolio container. These will still be considered a part of your portfolio.

On the following pages, there are six different things to do for your portfolio. You should begin right away and continue to add to your portfolio throughout the course—don't wait until the last minute.

You Can Begin Now . . .

1. Collect illustrations of geometry in nature, art, and engineering. Include drawings or photographs of your own if you wish.

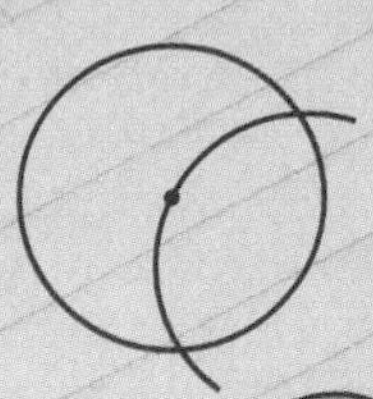

2. Study the "circle flower" design, which was begun with the circle in the center of the "flower." Use your compass to construct one or more circle flowers, adding shading to give them an attractive appearance.

You can continue adding circles to a circle flower to make a more elaborate design. Experiment by using ideas of your own.

3. Interesting designs can be created using only straight lines. One type of line design is made from string and is known as "string art." Make your own design, using either string or pencil, paper, and a straightedge.

Student project

4. Figures such as the ones shown on the right are known as **mandalas**, from the Sanskrit word for "circle" or "center." Try creating mandalas of your own. Write a report on mandalas and their history, including illustrations.

Aztec "calendar"

5. A special kind of art is the creation of "knot" designs such as those shown here. Experiment with your own designs. A good rule is to first create the overall pattern and then erase lines as necessary for the parts that go behind other parts in the figure.

6. Islamic culture, which at times has forbidden direct representations of real-world objects, became extremely rich in geometric and calligraphic art. Collect examples of Islamic or other geometric art for your portfolio, and perhaps try creating some designs in the style of Islamic art.

Modeling the World With Geometry

 After years of research, scientists have discovered the most fundamental elements of the physical world. The ideal world of geometry also has its fundamental elements.

Physicists recognize particles by their electronic signatures. In this computer-generated view, particle tracks emerge from the center of a collision.

Mathematically Perfect Figures

The points, lines, and planes of geometry can be used to create models of things in the physical world. However, geometric figures, unlike atoms and quarks, are not physical. **Lines** and **planes** in geometry have no thickness, and **points** have no size at all. Do you think such things can actually exist?

This book contains many drawings of geometric figures. But remember, the drawings themselves are not the same thing as the geometric figures they represent. Geometric figures exist "only in the mind."

Points Many familiar objects can be used to illustrate geometric points. When you look at the night sky and see the stars, the tiny dots of light seem like points. What are some other examples that could illustrate geometric points?.

Points are often shown as dots, but unlike physical dots, geometric points have no size. In geometry, points are named with capital letters such as *A* or *X*.

Lines A geometric **line** has no thickness. Unlike real-world "lines," it goes forever. A line is often named using two points on the line, such as line $\overleftrightarrow{AB}$, or just $\overleftrightarrow{AB}$. You can also name a line using a lowercase letter, such as line *m*.

line $\overleftrightarrow{AB}$, or $\overleftrightarrow{AB}$
line *m*, or *m*

Segments In drawings, geometric lines are usually represented by physical lines with arrowheads. The arrowheads are used to suggest that lines go on infinitely. **Segments,** on the other hand, have definite beginnings and endings. A segment is a portion of a line from one endpoint to another. It is named using its endpoints.

segment $\overline{AB}$, or $\overline{AB}$

ray $\overrightarrow{XY}$, or $\overrightarrow{XY}$

Rays A **ray** is a "half-line" that starts at a point and goes on forever. How is a geometric ray like a laser beam? A ray is named using its endpoint and one other point that lies on the ray. The endpoint is named first.

EXAMPLE 1

Name each of the figures below. Use the shorter form of the notation in each case.

 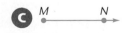 

Solution ➤

Angles An **angle** is formed by two rays that have the same endpoint. When there is no danger of confusion, an angle can be named by an angle symbol (∠) and just a single letter or number. Otherwise, an angle is named using three capital letters, with the **vertex** of the angle (the common endpoint of the rays) placed in the middle.

Planes A geometric **plane** extends infinitely in all directions. It is flat and has no thickness. You can think of any flat surface, such as the top of your desk or the front of this book, as representing a portion of a plane.

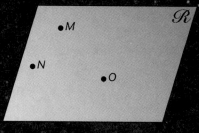

Plane *NMO*, or *R*

In the figure on the left, points *M*, *N*, and *O* lie on a flat surface. This flat surface represents a plane. A plane can be named by three points that lie on the plane (such as *M*, *N*, and *O*) or by a special script capital letter, such as *R*.

CRITICAL Thinking

In naming a plane, the three points you use must be **noncollinear**. That is, they must not lie in a straight line. (**Collinear** points lie on the same line.) If the points M, N, and O were collinear, could there be more than one plane that they would name? Make a sketch to illustrate your answer.

EXAMPLE 2

Name each of the figures below. Use the shorter form of the notation in each case.

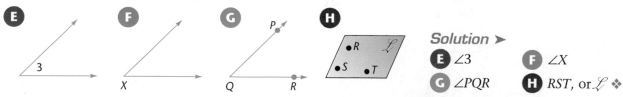

Solution ➤

E ∠3   F ∠X

G ∠PQR   H *RST*, or *L* ❖

Exploration · Discovering Geometry Ideas in a Model

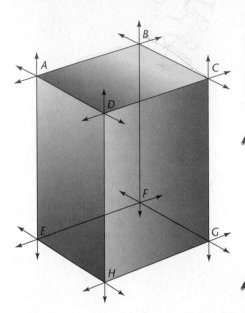

The drawing on the left illustrates how points, lines, and planes can be used to model a real-world object such as a box or perhaps the room you are in as you read this.

Each of the following questions represents an important mathematical idea. Record your answers to the questions in your notebook for future reference.

1 Examine the drawing. Identify the places where the lines pass through, or **intersect**, each other. What kind of geometric figure is suggested where lines intersect?

Complete the following statement: The intersection of two _?_ is a _?_ .

How many lines intersect at each corner of the drawing? Do you think there is a limit to the number of lines that can intersect at a single point?

2 Identify the places in the drawing where the planes pass through, or intersect, each other. What are the intersections called?

Complete the following statement: The intersection of two _?_ is a _?_ .

3 Look at points *A* and *B* in the drawing. How many lines pass through *both* of these points? Do you think it is possible for there to be another line, different from the one shown, that passes through both points *A* and *B*?

Complete the following statement: Through any two points _?_ .

4 Look at points *A*, *B*, and *C* in the drawing. How many planes pass through these three noncollinear points? Do you think it is possible for there to be another plane, different from the one shown, that passes through all three points *A*, *B*, and *C*?

Complete the following statement: Through any three noncollinear points _?_ .

5 Pick any plane in the drawing. Then pick two points that are on the plane. Name the line that passes through them. Is the line in the plane that you picked?

Complete the following statement: If two points are in a plane, then the line containing them _?_ . ❖

EXERCISES & PROBLEMS

Communicate

1. Explain how geometric figures are different from real-world objects.

2. Discuss why it is useful to have more than one way to name a line.

3. Tell how to count the number of segments that can be named in the figure below. Is there more than one way? Explain.

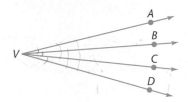

4. Tell how to count the number of angles that can be named in the figure below. Is there more than one way? Explain.

5. Tell how to count the number of rays that can be named in the figure on the right using just the given information. Why is the order of letters important in naming a ray?

6. Discuss why it is useful to have more than one way to name an angle.

How are geometric figures important in the design of this car?

Practice & Apply

In Exercises 7–10, refer to the triangle at right.

7. Name all the segments in the triangle.

8. Name each of the angles in the triangle using three different methods.

9. Name the rays that form each of the angles of the triangle.

10. Name the plane that contains the triangle.

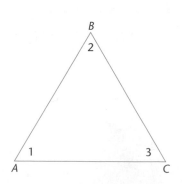

Aquarium In Exercises 11–16, identify each as being best modeled by a point, line, or plane.

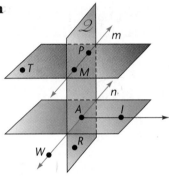

11. An edge of the aquarium
12. A grain of sand
13. A side of the aquarium
14. The bottom
15. A corner of the aquarium
16. Floating algae

In Exercises 17–23, tell whether each statement is true or false, and explain your reasoning.

17. Lines have endpoints.
18. Planes have edges.

19. Three lines that intersect in the same point must all be in the same plane.

20. Two planes may each intersect a third plane without intersecting each other.

21. Three planes may all intersect each other in exactly one point.

22. Any three points are contained in some plane.

23. Any four points are contained in some plane.

For Exercises 24–33, specify whether each statement is true in the figures that apply.

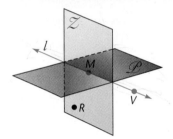

Figure 1

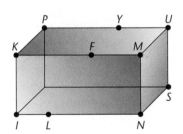

Figure 2

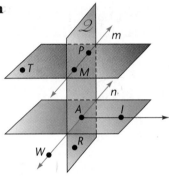

Figure 3

24. $\overline{KM}$ contains F.
25. Line n is the same as $\overline{WA}$.
26. $\overline{MU}$ and $\overline{MN}$ intersect at M.
27. Line l is on plane $\mathscr{P}$.
28. Line n intersects $\overline{AI}$ at point A.
29. Points I, N, and M are coplanar.
30. Points K, F, and M are collinear.
31. Points F, Y, and N are coplanar.
32. Plane $\mathscr{P}$ and plane $\mathscr{Z}$ intersect at $\overleftrightarrow{MV}$.
33. T, M, and P are coplanar.

Draw each of the figures for Exercises 34–36. Be sure to label each part you draw.

34. Plane $\mathscr{F}$ intersecting plane $\mathscr{L}$ in line p

35. Vertical planes $\mathscr{R}$ and $\mathscr{T}$ intersecting at line m

36. Three planes that never intersect each other

How many segments can be named in the figures below?

37.  A————————B

38. 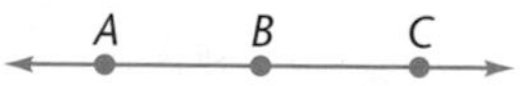 A———B———C

39.  A———B———C———D

40. 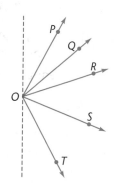 **Algebra** Write a general rule or a formula for finding the number of segments that are determined by a given number of points on a line. Can you explain why the rule works? Does your explanation prove that the rule will always work?

41. How many angles can be named in the figure. Write a general rule or a formula for finding the number of angles formed by a given number of rays radiating from a given point. (Assume all the rays lie to one side of a dotted line, as shown.)

Look Back

Complete each of the following. Draw a number line if you find it helpful.

42. $22 + (-6) =$

43. $7 + 15 =$

44. $11 - (-4) =$

45. $-81 - (-30) =$

46. $|-14 + (-35)| =$

47. $|13 - 10| =$

48. $-123 - 41 =$

49. $|21 + (-35)| =$

50. $|-54 + (-20)| =$

Look Beyond

51. 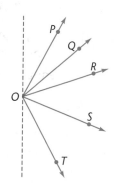 **Algebra** You can use geometry to help you answer the following question. Suppose that 4 people are exchanging cards. Each person gives 1 card to each of the other 3 people. How many exchanges are made? (**Hint:** Start by drawing 4 noncollinear points. These points represent the people. Now draw lines between the points to represent exchanges of cards. What is the total number of lines that you can draw?)

52. What if 5 people exchanged cards? How many exchanges would there be? Explain how to determine the number of exchanges when n people exchange cards.

Exploring

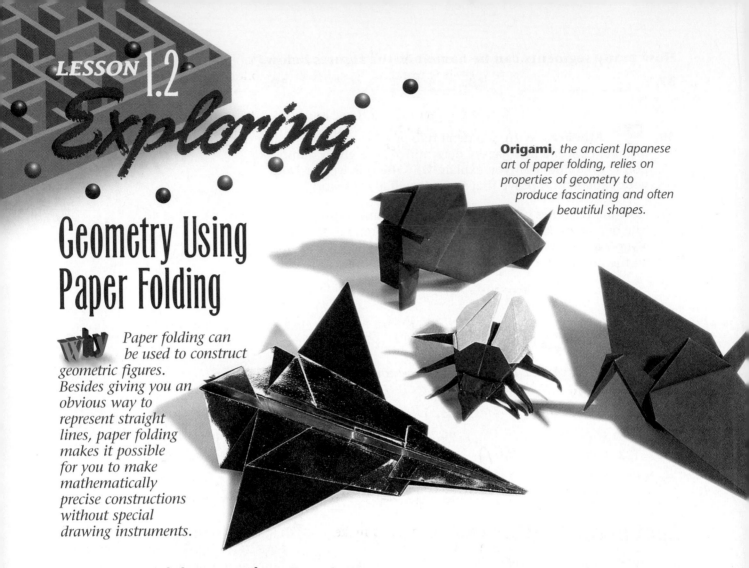

Origami, the ancient Japanese art of paper folding, relies on properties of geometry to produce fascinating and often beautiful shapes.

Geometry Using Paper Folding

Paper folding can be used to construct geometric figures. Besides giving you an obvious way to represent straight lines, paper folding makes it possible for you to make mathematically precise constructions without special drawing instruments.

Paper Folding: The Basics

Wax paper is especially good for folding geometric figures. For one thing, it is transparent, which allows you to match figures precisely when you fold. Another advantage of wax paper is that creases made in it are easy to see, especially against a dark background.

If you use wax paper, experiment with different tools for marking on paper. Also consider using "patty paper," which has the advantage of being easy to write on.

In the following explorations, work with a partner. One person can read the instructions while the other person does the paper folding. Discuss with your partner *why* you think the constructions work. The idea of knowing why things work is one of the most important things you should get from this course.

All of the terms in the explorations are probably already familiar to you. But to refresh your memory, consider the following three important geometry terms.

A **right angle** is a "square" angle, like the corner of a piece of paper. **Perpendicular lines** are lines that form right angles where they meet. **Parallel lines** are lines that are in the same plane but never meet, no matter how far they are extended.

Exploration 1 Perpendicular Lines and Parallel Lines

You will need
Folding paper
Marker that will write on folding paper

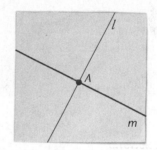

1. Fold the paper once to make a line. Label the line *l*.

2. Draw a point on line *l* and label it *A*. Fold the paper through point *A* so that line *l* matches with itself. Label the new line *m*.

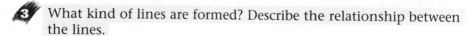
3. What kind of lines are formed? Describe the relationship between the lines.

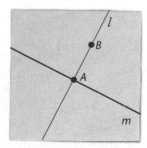

4. Start with perpendicular lines *l* and *m*. Mark a new point on line *l*, and label it as *B*.

5. Fold the paper through point *B* so that line *l* matches with itself as before. Label the new crease as line *n*.

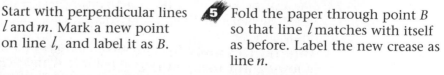
6. What is the relationship between lines *n* and *m*? ❖

Making Conjectures in Geometry

A conjecture is a statement we think is true. It is an "educated guess" based on observations.

Mathematical discoveries often start out as conjectures. In the explorations that follow, you will make conjectures about lines that divide angles and segments into congruent parts.

How would you measure the distance from the tree to the fence? Would you choose *a*, *b*, or *c?*

In geometry, the distance from a point to a line is the length of the perpendicular segment from the point to the line. Thus, the distance from the tree (point X) to the fence (line $\overleftrightarrow{AC}$) is the length of segment $\overline{XB}$, (m$\overline{XB}$, or XB).

CRITICAL Thinking Why do you think that the distance from a point to a line is defined along a perpendicular segment? What are the advantages of this definition?

Exploration 2 Segment and Angle Bisectors

You will need
Folding paper and marker
Ruler (for measuring only)

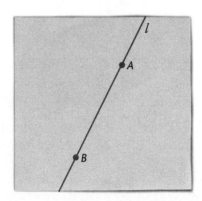

A **segment bisector** is a line that divides a segment into two equal parts. If that segment is perpendicular to the line, it is the **perpendicular bisector**. An **angle bisector** is a line that divides the angle into two equal angles.

1 Start with line *l* containing points *A* and *B*. Fold the line so *B* falls on *A*. Label the crease line *m*. Describe line *m* in relation to segment $\overline{AB}$.

2 Choose several points on line *m*. Measure the distance from each point to *A* and to *B*. Write a conjecture about the points on the perpendicular bisector of a segment.

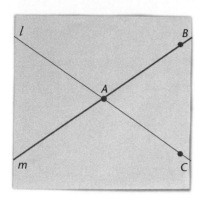

3 Fold two non-perpendicular lines *l* and *m*. Label their intersection point *A*. Label a point *B* on line *m* and a point *C* on line *l*. Fold the paper through *A* so that segment $\overline{AB}$ falls on segment $\overline{AC}$. Label the crease as line *n*. Describe line *n* in relation to ∠*BAC*.

4 Label point *X* on the part of line *n* that is in the interior of ∠*BAC*.

5 Fold the paper to make a line through *X* perpendicular to line *l*. Where the perpendicular intersects *l*, label the point *S*. Measure the segment $\overline{XS}$.

6 Fold the paper to make a line through *X* perpendicular to line *m*. Where the perpendicular intersects *m*, label the point *T*. Measure segment $\overline{XT}$.

7 Make a conjecture about the distance from a point in the angle bisector of an angle to the sides of the angle.

8 Test your conjecture with a point in each of the bisectors of the other angles formed by *l* and *m*. ❖

APPLICATION

Cultural Connection: Asia The *crane base* is the starting point for many of the traditional origami figures, including the popular paper crane. The folds below are one way of beginning a crane base.

1. Fold a square piece of paper into quarters. Then crease it along the diagonals as shown.

2. Fold the outside edges to align with the diagonal creases.

Now unfold the paper. If you have made your diagonal creases correctly, you should see a four-pointed star in the folds of your paper. How many perpendicular bisectors can you find? How many angle bisectors? ❖

EXERCISES & PROBLEMS

Communicate

1. When you fold line *l* onto itself in Exploration 1, which pairs of angles match up? What conclusion can you draw about the angles from the fact that they match up? In view of this, how can you define perpendicular lines?

2. When you construct parallel lines *m* and *n* in Exploration 1, how many right angles are formed? Make a conjecture about how you can determine if two lines are parallel.

3. When you fold *A* onto *B* in Exploration 2, how do you know (without measuring) that the new line divides $\overline{AB}$ into equal parts?

4. How can you use the results of Exploration 2 to determine if a given line is a perpendicular bisector of a segment? an angle bisector? How many measurements would you need to make? What would they be? Discuss.

Practice & Apply

Use pieces of unlined paper to do each of the following Exercises. Tracing paper or other paper that you can see through works best. You will also need a ruler.

5. Fold a piece of paper to make three lines, so that each line intersects the other two at different points. Label the lines *l, m,* and *n.* Use a marker or pencil to emphasize the three segments connecting the intersection points of the lines. What kind of geometric shape have you drawn?

6. Label the intersection of lines *l* and *m* as point *A,* of lines *l* and *n* as point *B,* and of lines *m* and *n* as point *C.* Then label the segment connecting *A* and *B* as *c,* the one connecting *A* and *C* as *b,* and the one connecting *B* and *C* as *a.*

   Describe the relationship of the names of the points to the names of the sides. How do you think this arrangement of names might be useful?

7. Recall how you constructed perpendicular lines in Exploration 1. Using a new piece of paper, construct a triangle with two sides perpendicular to each other. What kind of triangle have you formed?

8. Fold a new piece of paper and label two points on the fold line as A and B. Use these points to construct the perpendicular bisector of $\overline{AB}$. Select a point C on the perpendicular bisector. Draw $\overline{CA}$ and $\overline{CB}$. Using a conjecture you made in Exploration 2, what can you conclude about the lengths of $\overline{CA}$ and $\overline{CB}$?

9. Write a conjecture about triangles that have a vertex located on the perpendicular bisector of one of the sides.

10. Construct a rectangle using the techniques for folding perpendicular lines. (Recall that all four angles of a rectangle are right angles.) Do any of the sides of your rectangle seem to be equal? parallel? Write your own conjectures about the sides of a rectangle.

11. Fold a new piece of paper twice to make two intersecting lines. Construct the angle bisector of the smaller angle between the two lines. Label the angle bisector l. Construct the angle bisector of the larger angle between your original two lines. Label this angle bisector m. What do you observe about the lines l and m? Write a conjecture about the relationship between the bisectors of angles formed by intersecting lines.

12. Draw segment $\overline{AB}$ on a piece of waxed paper; then construct its perpendicular bisector. Trace over the crease. Next, fold the paper once through the intersection of the two lines so that A and B each align with the perpendicular bisector. Poke holes through the paper to mark the point where A lines up on the perpendicular as A' and the point where B lines up on the perpendicular as B'. Connect A, A', B, and B'. What kind of shape does this appear to be? Name the segments that form the *diagonals* of the shape.

13. Write your own conjectures about the diagonals of the geometric figure you constructed in Exercise 12.

 ## *Look Back*

Name the geometric figure that each item suggests. **[Lesson 1.1]**

14. The edge of a table

15. The wall of your classroom

16. The place where two walls meet

17. The place where two walls meet the ceiling

Name each figure, using more than one method when possible. **[Lesson 1.1]**

18.

19.

20.

21.

22.
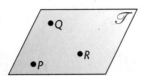

23. A

Evaluate each of the following expressions.

24. $13 - (-13)$

25. $-13 + (-13)$

26. $-13 - 13$

27. $|19 + 27|$

28. $|19 - 27|$

29. $|-19 - 27|$

Look Beyond

30. Cultural Connection: Asia Using a few simple folds, origami artists are able to suggest the shapes of real-world objects. The results, which range from humorous to elegant, are often quite striking.

There seems to be no limit to the powers of imagination of origami artists through the years. A recent creation is this sleek rendition of a fighter plane.

Experiment with creations of your own. You can find origami instruction books and special paper in hobby stores.

Exploring

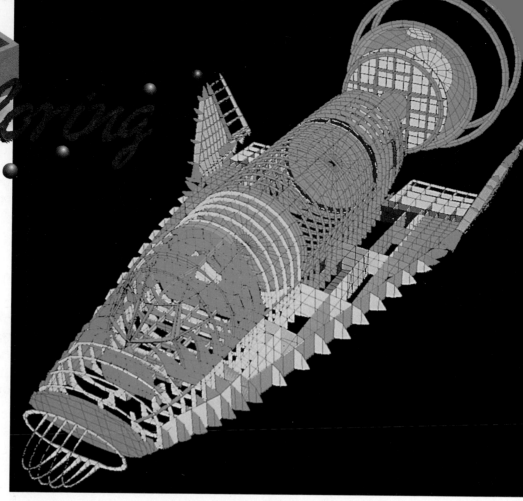

Geometry With a Computer

why *Powerful computer software has been created to solve problems in design. There is also computer software available to explore geometry ideas.*

If you do not have access to a computer graphics tool, you should complete the activities in this lesson using paper folding or pencil and paper. The ideas presented here will be used in later lessons.

Special Parts of Triangles

Before you begin the explorations, you will need to know some special parts of triangles.

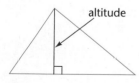

altitude

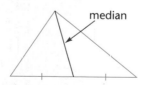

median

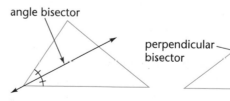

angle bisector

perpendicular bisector

An **altitude** is a line segment from a vertex drawn perpendicular to the line containing the opposite side.

A **median** is a line segment from a vertex to the midpoint of the opposite side.

An **angle bisector** is a line, a segment, or a ray that bisects an angle of the triangle.

A **perpendicular bisector** is a line or a segment that bisects and is perpendicular to a side of the triangle.

For any given triangle, how many different altitudes can you draw? How many medians? angle bisectors? perpendicular bisectors?

Exploration 1 *Some Special Points in Triangles*

Geometry Graphics

You will need
Geometry technology or
Folding paper and a compass

In each of the following activities, you should discover a special point related to the triangles you create. Save your figures for the second exploration. If you are using computer software, you may be able to "drag" each of the vertices of the triangles to explore different triangle shapes.

1 Draw or fold a triangle. Then construct the angle bisector of each of the angles.

2 Draw or fold a triangle. Then construct the perpendicular bisector of each of the sides.

3 Draw or fold a triangle. Then construct the three medians of a triangle.

4 Draw or fold a triangle. Then construct the three altitudes of the triangle.

5 Share your results with other members of your class. What do you notice? Write down your observations in the form of four separate conjectures. ❖

Special Circles Related to Triangles

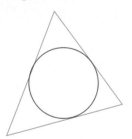

inscribed circle

circumscribed circle

For any triangle, you can draw an **inscribed circle** and a **circumscribed circle**. An inscribed circle, as the name suggests, is *inside* the triangle and touches its three sides. A circumscribed circle is *outside* a triangle (*circum* means "around") and contains each of its three vertices.

Before you do the following exploration, think: How could you find the center points for drawing the inscribed and circumscribed circles of a given triangle?

THE METROPOLITAN MUSEUM OF ART, ROGERS FUND, 1

Archaeology *An archaeologist wants to find the original diameter of a broken plate. She makes an outline of it, then draws a triangle with all three of its vertices on the circumference. The outer edge of the plate is thus part of a circle that circumscribes the triangle. If she can find the center of the circle she can determine the plate's original diameter.*

•Exploration 2 *Constructing Special Circles*

You will need
Geometry technology or
Folding paper and a compass

1 Use the triangles you created in Exploration 1. Draw circles with their centers at the special points you discovered in each activity. If you are using paper and pencil, vary the size of the circles by changing your compass settings. If you are using a computer, vary the size of the circles by using the pointer tool.

2 What do you discover about each special point? Write down your discoveries as conjectures about special points connected with triangles. ❖

CRITICAL
Thinking

Why do you think that points at the intersection of certain lines (or segments or rays) work as centers for circumscribed or inscribed circles? (**Hint:** Recall the conjecture you made about segment and angle bisectors in Lesson 1.2.)

EXERCISES & PROBLEMS

Communicate

1. The center of a circle inscribed in a triangle is called the **incenter** of a triangle. Can the incenter be outside the triangle? Explain why or why not.

2. The center of a circle circumscribed around a triangle is called the **circumcenter** of the triangle. Can the circumcenter be outside the triangle? Explain why or why not.

3. The intersection of the medians of a triangle is called the **centroid** or **center of mass** of the triangle. Can the centroid be outside the triangle? Explain why or why not.

4. The intersection of the altitudes of a triangle is called the **orthocenter** of the triangle. Can the orthocenter be outside the triangle? Explain why or why not.

5. Which of the special circles can you draw using the intersections of the perpendicular bisectors of a triangle? Explain why you can do this. To help you form an explanation, discuss the following questions.

Every point on the perpendicular bisector of $\overline{AB}$ is the same distance from what two points on the triangle? (Recall your conjectures from Lesson 1.2.) What can you say about the perpendicular bisectors of the other two sides?

If a point is on all three perpendicular bisectors of a triangle, what can you conclude about its distance from the vertices of the triangle? Explain why.

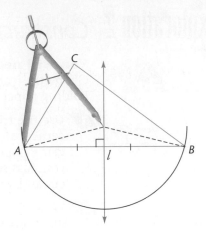

Try this with more than one point on l.

Practice & Apply

Technology For each of the Exercises below, you can use a computer software package such as *The Geometer's Sketchpad*™ or *Cabri*™. If you don't have access to geometry technology, you can use paper-and-pencil methods.

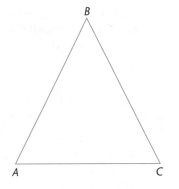

Draw a triangle like the one on the right. Draw altitude $\overline{AD}$, angle bisector $\overline{AE}$, and median $\overline{AF}$. Then draw a perpendicular bisector m of side $\overline{BC}$.

6. What is the order of points D, E, and F?

7. Experiment with different shapes of triangles. What do you observe about the order of points D, F, and E?

8. See if you can find a triangle in which $\overline{AF}$, $\overline{AE}$, $\overline{AD}$, and m are collinear. What seems to be special about the triangle?

9. If you drew all possible angle bisectors, medians, altitudes, and perpendicular bisectors of a triangle and found that they all intersected in a single point, what would be true of the triangle? Illustrate.

In Exercises 10–14 you will discover an important geometry idea.

10. Draw a triangle with angles all less than 90°. Find its circumcenter (the intersection of its perpendicular bisectors). Is the circumcenter inside or outside the triangle?

11. Draw a triangle that has one of its angles greater than a right angle. Find its circumcenter. Is the circumcenter inside or outside the triangle?

12. If the pattern of Exercises 10 and 11 holds, where do you think the intersection of the perpendicular bisectors would be for a right triangle—inside, outside, or somewhere else?

Draw a right triangle. Test your conjecture by finding the intersection of the perpendicular bisectors. Where is it located?

13. Use the intersection point in Exercise 12 (the circumcenter) to draw a circle through each of the vertices of the triangle.

How does the longest side of the triangle divide the circle? Are the parts of the circle equal or unequal?

14. Express your result from Exercise 13 as a conjecture about certain triangles and half-circles.

15. Construction A mechanical contractor is installing the air conditioning system in a large commercial building. He needs to run a round duct through a triangular opening in the structural steel above the ceiling of the top floor of the building. The dimensions of the opening are 38 inches, 68 inches, and 92 inches. What is the diameter of the largest duct that can pass through the opening? Use geometry software or paper folding to model the problem.

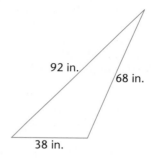

For Exercises 16 and 17, use triangles with all their angles smaller than right angles—acute triangles. You will discover three different results.

16. Find the midpoint of each side of an acute triangle. Then connect them to form another triangle. Your drawing should now contain four small triangles inside the larger one. Cut out the four triangles and compare them. What do you observe?

17. Construct a triangle *ABC*, with medians. For each median, measure the distance from the intersection or centroid to the side it touches, such as $\overline{OX}$, as well as to the vertex it touches, such as $\overline{OC}$. Record your measurements in a table like the one shown.

What do you notice about the relationship between the figures in the first row and those below in the second row? Express your answer in the form of a conjecture about the centroid of a triangle.

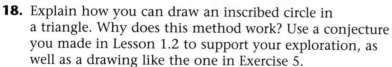

Triangle ABC. Triangles are named using their "corners," or vertices.

OA = ?	OB = ?	CO = ?
OY = ?	OZ = ?	OX = ?

18. Explain how you can draw an inscribed circle in a triangle. Why does this method work? Use a conjecture you made in Lesson 1.2 to support your exploration, as well as a drawing like the one in Exercise 5.

Euler Lines The following exercise is especially suited for computer software, but it can also be done by using paper-and-pencil methods.

19. In a single triangle, find the circumcenter, the orthocenter, and the centroid. Connect each of the points with segments. What do you notice? Repeat with different triangles, or compare your results with those of your classmates. Write a conjecture about these three points.

20. 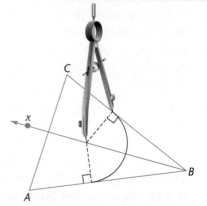 **Portfolio Activity** Collect samples of computer-generated art. Do some research to find out about different ways computers are used in art and publishing.

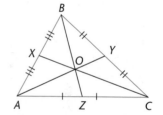

Try this with more than one point on $\overrightarrow{BX}$.

Choose any two different positive numbers.

21. Add them together. Do you get a positive or a negative number?

22. Subtract the smaller number from the larger number. What kind of number do you get? Now take the absolute value of this number.

23. Subtract the larger number from the smaller number. What kind of number do you get? Now take the absolute value of this number.

24. Compare your answers from Exercises 22 and 23 above. Explain why they are different or the same.

Choose any two different negative numbers.

25. Add them together. Do you get a positive or a negative number?

26. Subtract the smaller number from the larger number. What kind of number do you get? Now take the absolute value of this number.

27. Subtract the larger number from the smaller number. What kind of number do you get? Now take the absolute value of this number.

28. Compare your answers from Exercises 26 and 27 above. Explain why they are different or the same.

Look Beyond ~~~

29. Cut out a triangle from a stiff piece of cardboard. Draw a median of the triangle. See if you can balance the triangle along the line you drew. What can you conclude about the area of the triangle on each side of a median? Express your answer as a conjecture.

30. Draw the other two medians of your triangle. See if you can now balance the triangle on a single point placed at the centroid. In your own words, explain why the centroid is known as the center of mass.

31. **Physics** According to physics, a freely falling body should rotate around its center of mass. Test this theory by giving your triangle a toss like a Frisbee. What do you observe?

LESSON 1.4 Measuring Length

why We use rulers and other measuring devices all the time. We ask each other questions such as these: How tall are you? How far do you live from here? Rulers and other methods of measurement work because of some basic agreements and assumptions.

A ruler in the eyepiece of the microscope makes it possible to measure in units of one-millionth of a meter, which are known as microns.

The Measure of a Segment

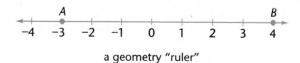

a geometry "ruler"

In defining the measure or length of a segment, we will use a ruler.

In geometry, a "ruler" is a **number line**—that is, a line that has been set up to correspond with the real numbers. The coordinate of a point on a number line is the real number that matches the point. In the illustration, -3 is the coordinate of A and 4 is the coordinate of B.

How would you find the distance between points A and B? Try $|-3 - 4|$. Also try $|4 - (-3)|$. What do you notice about the absolute values? This leads to the definition of the measure of a segment.

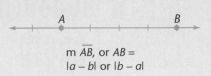

MEASURE OF SEGMENT $\overline{AB}$

Let A and B be points on a number line, with coordinates a and b. Then the measure of segment $\overline{AB}$, which is called the length, is $|a - b|$ or $|b - a|$.

The measure of segment $\overline{AB}$ is written as mAB or simply as AB.

m $\overline{AB}$, or $AB =$ $|a - b|$ or $|b - a|$

1.4.1

EXAMPLE 1

Find the measures (lengths) of $\overline{AB}$, $\overline{AX}$, and $\overline{XB}$ on the number line.

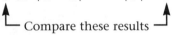

Solution ➤

AB (or m$\overline{AB}$) $= |-4 - 4| = |-8| = 8$ or $|4 - (-4)| = |8| = 8$
AX (or m$\overline{AX}$) $= |-4 - 1| = |-5| = 5$ or $|1 - (-4)| = |5| = 5$
XB (or m$\overline{XB}$) $= |1 - 4| = |-3| = 3$ or $|4 - 1| = |3| = 3$ ❖

└─ Compare these results ─┘

Try This   Find ST.

Constructing a Ruler

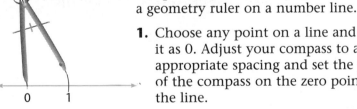

The geometry rulers that you will use in this book—with a few interesting exceptions—have evenly spaced divisions. A compass can be used to make a geometry ruler on a number line.

1. Choose any point on a line and label it as 0. Adjust your compass to an appropriate spacing and set the point of the compass on the zero point of the line.

   Use the pencil part of the compass to draw a short mark that crosses the number line on one side of zero. Label the point of intersection as 1.

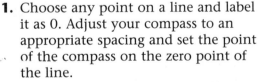

2. Set the point of the compass on the newly labeled point and draw another mark that crosses the line in a new place. Label the new intersection with an integer that is 1 larger than the coordinate of the previous point.

3. Repeat the previous step as many
times as desired. ❖

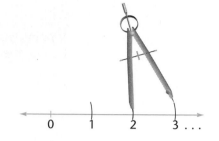

How would you construct the negative
numbers on the number line? Do you
think it matters on which side of the
zero point you place the negative
numbers?

Once the integers are placed on the
number line, a standard is usually
applied to name the **unit length**.
Two standards commonly used are
the inch and the meter.

Congruence

Segments that have the same
length are **congruent**. That is,
they match exactly. If you move
one of them onto the other, they
fit together perfectly.

The segments on the ruler that
was constructed above were all
congruent because the same
compass setting was used for
each one.

What do you think would happen if the divisions on a ruler were not
evenly spaced? Could you be sure that segments with the same measure
would be congruent, or that congruent segments would have equal
measures? Explain your answer.

In this book, "tick marks" show that
segments are congruent. Segments
that have a single tick mark are
congruent. Similarly, all segments
with two tick marks are congruent
to each other, and so on.

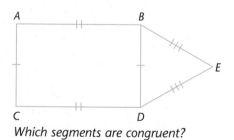

Which segments are congruent?

Segment Addition

Look again at the number line in Example 1. Notice that *X* is between *A*
and *B*. The relationship among the distances *AX*, *XB*, and *AB* leads to an
important assumption for geometry, known as the Segment Addition
Postulate.

SEGMENT ADDITION POSTULATE

If point *R* is between points *P* and *Q*, then *PR* + *RQ* = *PQ*.

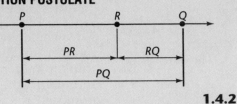

1.4.2

Are *segments* really being added in this postulate? or *measures* of segments? In this course, addition and other arithmetic operations are defined for numbers, not geometric figures.

Which ones make sense?

$$\begin{cases} AB + CD = 5 \\ \overline{AB} + \overline{CD} = 5 \\ m\overline{AB} + m\overline{CD} = 5 \end{cases}$$

The Segment Addition Postulate makes an important connection between ideas in algebra and geometry.

EXAMPLE 2

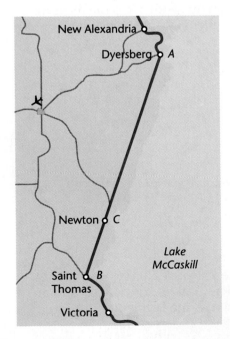

Algebra The towns of Dyersberg, Newton, and Saint Thomas are located along a straight portion of Ventura Highway. The distance from Dyersberg to Saint Thomas is 25 miles. The distance from Dyersberg to Newton is one mile more than three times the distance from Newton to Saint Thomas. Find the distances from Dyersberg to Newton and from Newton to Saint Thomas.

Solution ➤

First represent each town as a point on a line segment. Dyersberg = *A*, Saint Thomas = *B*, and Newton = *C*. Let *x* be the distance in miles from *B* to *C*. Then the distance from *A* to *C* will be $3x + 1$.

Since *C* is between *A* and *B*, *AC* + *CB* = *AB*. So,

$$(3x + 1) + x = 25$$
$$4x + 1 = 25$$
$$4x = 24$$
$$x = 6 = BC$$

So *AC* = $3x + 1 = 3(6) + 1 = 19$ miles. ❖

EXERCISES & PROBLEMS

Communicate

1. "Congruent segments are segments that can be moved so that they match exactly." Explain how this idea is used in paper folding to find the perpendicular bisector of a segment.

2. Explain why it is important for a ruler to have equally spaced divisions.

3. The distance from 0 to 1 on a ruler, known as the **unit length**, can be any desired size. Give as many examples as you can of units of different length.

4. If the centimeter were the only unit of measure for length, what problems would this create? Discuss why it is convenient to have different units of measure for measuring length.

Geometry graphics software often allows you to select your units of measure.

5. Explain why each of the following statements does or does not make sense.

   $\overline{MN} + \overline{OP} = 30$ cm $MN + OP = 30$ cm
   m$\overline{MN}$ + m$\overline{OP}$ = 30 cm

6. Once you have constructed the integers on your ruler, why would you want to subdivide the unit length into smaller divisions?

Practice & Apply

7. Find the measures of the segments determined by the points on the number line. Show that it does not matter which point you subtract from which.

In Exercises 8 and 9, name all the congruent segments.

8.

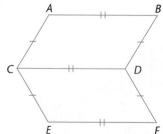

9.

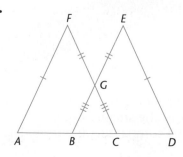

In Exercises 10–12 you are given that A is between points M and B. Sketch each figure and find the missing measures.

10. $MA = 30$ $AB = 15$ $MB = \underline{\ ?\ }$

11. $MA = 15$ $AB = \underline{\ ?\ }$ $MB = 100$

12. $MA = \underline{\ ?\ }$ $AB = 13.3$ $MB = 29.6$

Find the indicated value in Exercises 13–15.

13. $XZ = 25$ $x = \underline{\ ?\ }$

14. $XY = 25$ $XZ = \underline{\ ?\ }$

15. $YZ = 25$ $XY = \underline{\ ?\ }$

Algebra   Towns A, B, and C are situated along a straight highway. B is between A and C. The distance from A to C is 41 miles. The distance from B to C is 2 miles more than twice the distance from A to B.

16. Write an equation for the distances, and then solve the equation to find the distances between the towns.

17. Town X is between A and B, 6 miles from A. What is the distance from X to C.

For each of the statements in Exercises 18–23, write S for "Sense" and N for "Nonsense."

18. $XY = 5,000$ yd   **19.** $\overline{PQ} = 32$ in.   **20.** m$ST = 6$ cm

21. $XY + XZ = 32$ cm   **22.** m$\overline{PR} = 46$ cm   **23.** $XY - XZ = 12$ cm

24. Cultural Connection: Africa   The Egyptian Royal Cubit is subdivided into 28 units known as digits or fingers. From this basic unit, a number of others were created:

4 digits = 1 palm	12 digits = 1 small span
5 digits = 1 hand's breadth	14 digits = 1 great span
6 digits = 1 fist	16 digits = 1 foot (t'eser)
8 digits = 1 double palm	24 digits = 1 short cubit

How does the modern metric system of length measure resemble the ancient Egyptian system in subdividing larger units into smaller ones and in building up larger units of length from smaller ones? Give examples.

The length of the Royal Cubit, in modern units, is 20.67 inches. Calculate the length of the smaller units in inches. How does the Egyptian foot (t'eser) compare with our modern foot?

25. Think how you could use paper folding to subdivide a given unit into smaller parts. Suppose you first bisected the interval, then bisected each half, and so on. Write the next few terms of the sequence below, which gives the size of the units you could obtain by this process.

$$\frac{1}{2}, \frac{1}{4}, \cdots$$

To find a distance like three-fourths, you can add three intervals of a fourth together. Name at least three distances that you *cannot* find by this process.

There is more than one way of creating a measuring scale, and some of them might seem very strange. Do you think that "alternative rulers" might have real-world uses?

26. What can happen if the divisions of a ruler are not appropriately spaced? Draw your own "alternative ruler" to illustrate your answer. Show how segments of equal "length" might not be congruent.

27. Geology The **Richter scale** for indicating the intensity of earthquakes has an interesting feature. An earthquake of magnitude 2 has 10 times the ground movement of an earthquake of magnitude 1. Magnitude 3 is 10 times magnitude 2. In fact, each magnitude level has 10 times the ground movement of the next lower magnitude. What is the relationship between a magnitude 1 and a magnitude 3 earthquake? between a magnitude 1 and a magnitude 8 earthquake?

28.  **Statistics** On a standardized test, Susan scored higher than anyone else in the class. Her **raw score** (number of correct answers) was 1 answer higher than James', who came in second. Compare the raw scores and the **percentile** ratings of the four students below to see how the percentile measuring scale is like an alternative ruler. Discuss, using calculations to emphasize the point.

Name	Raw Score	Percentile Ranking
Susan	63	99.0
James	62	98.6
Stewart	43	57.3
Myrna	42	55.3

Two of the earth's continental plates meet at the San Andreas fault. When the plates slip, the resulting shock waves may be felt all over the world.

Look Back

In Exercises 29–32, give the name for the intersection points of each of the special objects related to a triangle. Tell whether each can be outside of the triangle or not. [Lesson 1.3]

29. the perpendicular bisectors   **30.** the angle bisectors

31. the medians   **32.** the altitudes

In Exercises 33 and 34, tell which of the above intersection points can be used to do the following. [Lesson 1.3]

33. Draw a circle "around" a triangle.   **34.** Draw a circle "inside" a triangle.

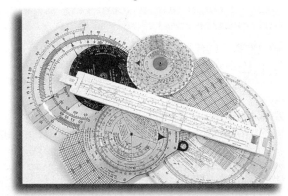

In Exercise 35, recall the famous result you found when exploring the perpendicular bisectors of a triangle. [Lesson 1.3]

35. State a conjecture about the relationship between a right angle and a part of a circle.

Look Beyond

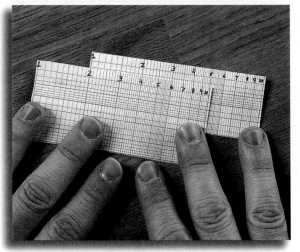

36.   **Algebra**   Slide rules, which were once used for doing approximate calculations, depended upon the properties of logarithms. You can make a simple slide rule from logarithm paper.

Cut out two strips of paper and place them together as shown. To multiply the numbers 2 and 3, move the left end of the top strip (the index) to the number 2 on the lower strip. Notice that the number 3 is just above the number 6 on the lower strip, which is the product of the 2 and 3.

Notice that a logarithmic scale is numbered from 1 to 10. By adjusting the decimal point you can approximate any number you like. For example, the number 3 on the scale can represent 0.003, 0.03, 0.3, 3, 30, 300, 3000, . . . If you keep track of your decimal points correctly, you can multiply any two numbers together—approximately.

Multiply at least five pairs of numbers of your own choosing using logarithmic scales. Compare the answer you get with the answers you get from your calculator. How accurate were your answers?

37. Explain the operation of a slide rule in terms of the following formula from algebra: $\text{Log } A + \text{Log } B = \text{Log } (A \times B)$

LESSON 1.5 Measuring Angles

why *Angle measure is used in many professions. For example, professional pilots use an angle measure known as the "heading" of a plane to navigate safely through the skies.*

Defining Angle Measure

A **protractor** is used to measure angles on a flat surface. As with a ruler for measuring length, the divisions of a protractor must be evenly spaced. Then you can be sure that if two angles have the same measure they are congruent, and *vice versa*.

To be sure you understand how a protractor is used, study the following example.

EXAMPLE 1

Use a protractor to find the measure of ∠CAB.

Solution ➤

1. Put the center of the protractor at the vertex.

2. Align the 0 point of the protractor scale with ray $\overrightarrow{AB}$.

3. Read the measure (in degrees) where ray $\overrightarrow{AC}$ intersects the protractor.

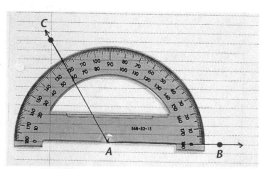

The measure of ∠CAB is 121°, or m∠CAB = 121°. ❖

Why isn't the measure of ∠CAB 59° instead of 121°?

A protractor may be thought of as another type of geometry ruler. This "ruler" is a half-circle with coordinates from 0 to 180. You can use it to define the measure of an angle such as ∠AVB.

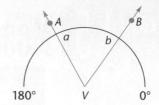

180°   V   0°

MEASURE OF ANGLE ∠AVB

Suppose the vertex *V* of ∠AVB is placed on the center point of a half-circle with coordinates from 0 to 180. Let *a* and *b* be the coordinates of the points where $\overrightarrow{VA}$ and $\overrightarrow{VB}$ cross the half-circle.

Then the **measure of ∠AVB,** written as m∠AVB, is $|a - b|$ or $|b - a|$.   **1.5.1**

Angles, like segments, are measured in standard units. The unit of angle measure is the degree. This is the measure of the angle that results when a half-circle is divided into 180 equal parts.

 **CRITICAL**
Thinking

In creating a protractor, does the size of the half-circle make a difference? Why or why not?

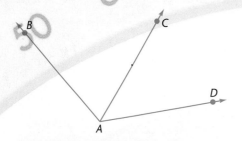

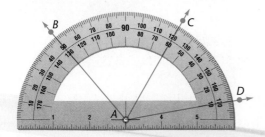

EXAMPLE 2

Use a protractor to find the measures of ∠BAC, ∠CAD, and ∠BAD. What is m∠BAC + m∠CAD?

Solution ➤

1. To measure ∠BAC, notice the points where $\overrightarrow{AB}$ and $\overrightarrow{AC}$ pass through the half-circle scale of the protractor (50 and 120). Using these coordinates, m∠BAC = $|50 - 120| = |-70| = 70°$.

2. The coordinate of $\overrightarrow{AD}$ is 170. So, m∠CAD = $|120 - 170| = |-50| = 50°$.

3. Similarly, m∠BAD = $|50 - 170| = |-120| = 120°$.

4. Using the answers from Steps 1 and 2, m∠BAC + m∠CAD = 70 + 50 = 120°. Notice that this agrees with the answer to Step 3. ❖

Congruent Angles

Angles, like segments, are congruent if one can be moved onto the other so that they match exactly. Tick marks are also used to show that angles are congruent.

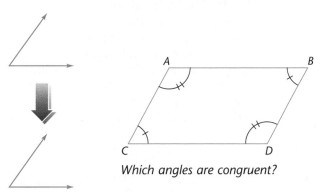

Which angles are congruent?

Pairs of Angles

The result in Step 4 of Example 2 suggests that m∠BAC + m∠CAD = m∠BAD. This leads to the following postulate.

ANGLE ADDITION POSTULATE

If point *S* is in the interior of ∠PQR, then
m∠PQS + m∠SQR = m∠PQR.

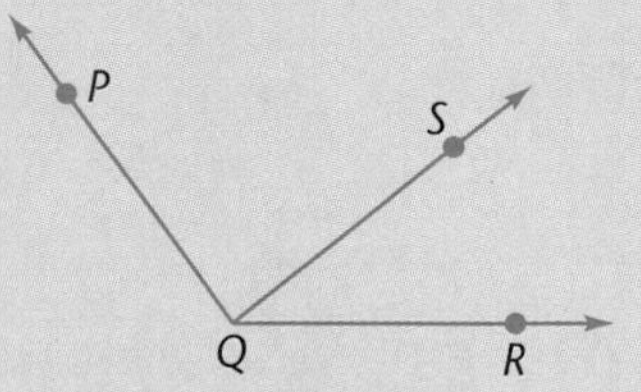

1.5.2

CRITICAL *Thinking*

As in the Segment Addition Postulate, you should ask yourself: What is really being added? Is it angles or measures of angles?

Which statement makes sense?
$$\begin{cases} \angle A + \angle B = 180° \\ m\angle A + m\angle B = 180° \end{cases}$$

SPECIAL ANGLE SUMS

If the sum of the measures of two angles is 90°, then the angles are **complementary.**

If the sum of the measures of two angles is 180°, then the angles are **supplementary.**

1.5.3

Linear Pairs

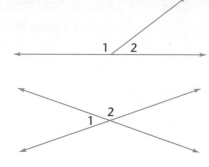

You will need
Protractor

When the endpoint of a ray intersects a line, two angles are formed. These angles are called a **linear pair**. In each figure, ∠1 and ∠2 are a linear pair. ∠1 is called the **supplement** of ∠2 and vice versa.

1 Draw three different linear pairs and measure each angle. Record your data in a table as shown at the right. The last row in the table asks you to make a **generalization**.

m∠1	m∠2	m∠1 + m∠2
?	?	?
?	?	?
?	?	?
x	?	?

2 Make a conjecture about linear pairs based on the information in your table.

3 If two angles form a linear pair, must they be supplementary? If two angles are supplementary, will they form a linear pair? Explain your reasoning.

4 Choose a point on a scale of your protractor and note the two different numbers indicated. What is the sum of these numbers? Explain why. ❖

EXERCISES & PROBLEMS

Communicate

1. When you travel on land, you do not ordinarily use compass directions, or headings, to find your way. What kinds of information do you use instead?

2. If an angle is 43°, what is its complement? What is its supplement?

3. How is angle measure like length measure? How is it different? Discuss.

4. Explain why one of the following statements makes sense and the other does not.

   m∠A + m∠2 = 190°
   ∠X + ∠Y = 150°

Name some other activities in which you would use a compass.

5. If two supplementary angles are congruent, what kind of angles are they? Why must each angle measure 90°?

6. Explain why folding a line onto itself produces perpendicular lines.

Practice & Apply

Find the measure of each angle in the diagram.

7. m∠AVB

8. m∠AVC

9. m∠AVD

10. m∠BVC

11. m∠DVB

12. m∠CVD

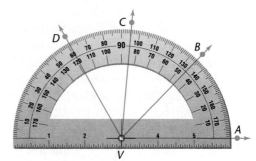

13. Name all the congruent angle pairs in the figure below.

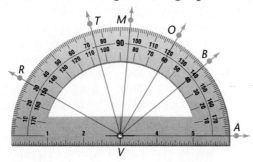

14. What is the angle between the hands of a clock if the time is 3:00?

15. Through what angle does the minute hand move between 3:00 and 3:10. Explain your answer.

16. In the diagram, m∠ALE = 31° and m∠SLE = 59°. What is the measure of ∠SLA? Which postulate allows you to draw this conclusion? What is the relationship between ∠SLE and ∠ALE?

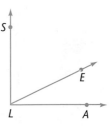

Find the missing measures in Exercises 17–19.

17. m∠SML = 80°, m∠LMI = 34°, m∠SMI = ?

18. m∠SML = 21°, m∠LMI = ?, m∠SMI = 43°

19. m∠SML = ?, m∠LMI = 34°, m∠SMI = 52°

20. Suppose that three rays each have their endpoint at the same point V but are not coplanar. Would the Angle Addition Postulate apply to the angles that are formed? Illustrate your answer with a sketch. (You can use rays starting from the corner of a box.)

Algebra   In the diagram, m∠LNO = 87°, m∠LNI = (2x − 8)°, and m∠INO = (x + 50)°.

21. What is the value of x?

22. What is m∠INO?

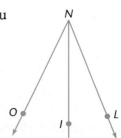

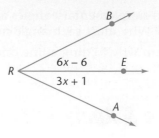

Algebra In the diagram, m∠ARB = 65 + x. Find the value of x, and then find each indicated angle measure.

23. m∠ARB

24. m∠BRE

25. m∠ERA

6x − 6

3x + 1

Navigation The heading of an airplane is defined in terms of a 360° circle, or "protractor." North is defined as 000, and the numbering is done from the 0 point along a circle in the clockwise direction. Find the heading for each of the following compass directions.

26. N   **27.** E   **28.** S

29. W   **30.** NE   **31.** NW

32. NNE   **33.** ENE   **34.** SSW

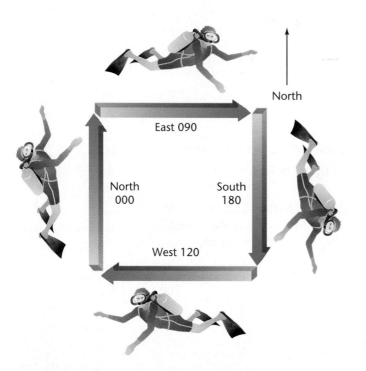

North

East 090

North 000   South 180

West 120

35. Scuba Diving In order to return to the same point, a diver may navigate a square pattern. If the diver plans to swim NE first, then SE, what compass headings must he or she use to navigate a square? Besides getting each angle correct, what else must a diver make sure of in order to move in a square?

● Chicago

355°

● Pilot   ● X

Navigation In Exercises 36 and 37, a pilot is flying to Chicago. In order to fly there in a straight line, he finds that he must keep his heading at 355.

36. Using the diagram, would the pilot's heading be greater or less than 355 if he started further east, say at point X? Explain.

37. On the way, the pilot encounters a thunderstorm along his planned route to Chicago. He finds he must adjust his heading to 15° to make the detour past the storm. Through how many degrees does the nose of his plane need to swing before the plane is on its new course?

38. In creating a ruler, the part of the ruler between 0 and 1 can be made any desired size. Can the part of a 180° protractor between 0 and 1 be made of any desired size? Explain your answer.

39. Angles may be measured using units other than degrees. One way of measuring angles is in gradians. There are 100 gradians in 90 degrees. Is 1 gradian smaller or larger than 1 degree? Why?

Look Back

Answer questions 40–43 using the diagram shown. [Lesson 1.1]

40. Name an angle in two different ways.

41. Identify a line that is coplanar to the angle.

42. Name all the lines that are intersections of two planes.

43. Identify three points that are collinear.

Find the measure of the segments below using two different methods.   [Lesson 1.4]

44.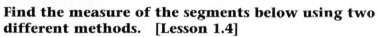

45.

46. If $AB = 27$, find AC and BC.

Look Beyond

47. Cultural Connection: Asia Our system of degrees comes from the Babylonians. The Babylonians considered a circle to be composed of a set of 360 congruent angles with each vertex at the center of a circle. Why do you think they chose 360? (Hint: How many numbers divide 360 evenly?)

48. Each degree measures a small angle; however, people who measure angles to navigate, to find angles between stars, or to build houses may need even more precise measurements. Often, instead of using decimals to divide the degree, they use minutes and seconds. There are 60 minutes in a degree and 60 seconds in a minute.

How many seconds are there in each degree?

How many seconds are there in 1.5 minutes?

How many minutes are there in 1.75 degrees?

The ancient Babylonians were sophisticated mathematicians.

Motions in Geometry

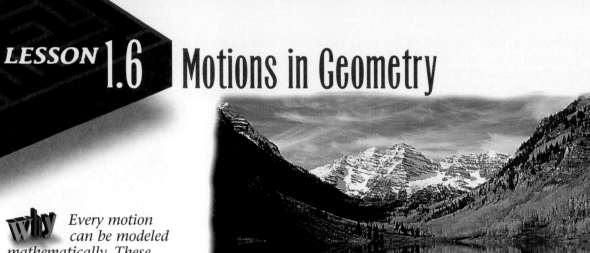

why *Every motion can be modeled mathematically. These three photos represent the three basic mathematical "motions," which are known as transformations.*

In mathematics, three basic **transformations** are used to model motion. They are **translations, rotations,** and **reflections**. Any real-world motion can be analyzed in terms of one or more of these transformations.

Translations

In geometry, where we deal with mathematical images, we speak of the **image** and **pre-image** of an object that undergoes a motion or transformation. Compare the two positions or "snapshots" of the same skier in motion.

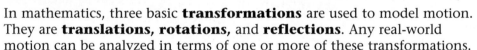
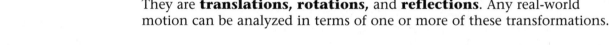

The pre-image of the skier has been **translated**. Notice that the pre-image and the image are exactly the same size and shape—that is, they are congruent. In geometry, a motion or transformation that does not change the size and shape of a figure is known as **rigid**.

EXAMPLE 1

Describe how the points have been translated in the figure. Identify the pre-image and the image. How far has point *A* moved? How far have you moved points *B* and *C*? What do you notice about the *direction* of the motion of points?

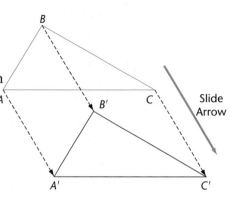

Solution ➤

The pre-image is △*ABC* and the image is △*A'B'C'*. Point *A* has moved 2 centimeters in the direction of the arrow. In fact, every point in △*A'B'C'* has moved 2 centimeters in the same direction. ❖

CRITICAL
Thinking

An arrow that shows the direction and motion of points in a translation is known as the slide arrow. Just one slide arrow is needed to describe the motion of the whole figure. When a **slide arrow** is drawn for a translation, does it matter where it is placed on the drawing?

Rotations

A **rotation** is a rigid motion that moves a geometric figure about a point known as the **turn center**. Probably the most obvious example of a rotation is a wheel turning in place. In this case the turn center is at the center of the rotating object. But this is not true of all rotations.

EXAMPLE 2

Describe the rotation in the figure. Identify the pre-image, the image, and the rotation center. Where is the rotation center of the figure? Describe the motion of the different points of the figure.

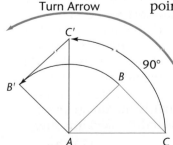

Solution ➤

The pre-image is △*ABC* and the image is △*AB'C'*. The rotation center is the point *A*. Every point in △*ABC* has been moved 90° in a counterclockwise direction about point *A*. ❖

Reflections

Hold a pencil in front of a mirror and focus on the pencil point P. The reflection P', called the image of the point, will appear to be on the opposite side of the mirror at the same distance from the mirror as P.

In a reflection, a line plays the role of the mirror, and geometric figures are "flipped" over the line. In the explorations that follow, paper folding is used to produce the appropriate motions.

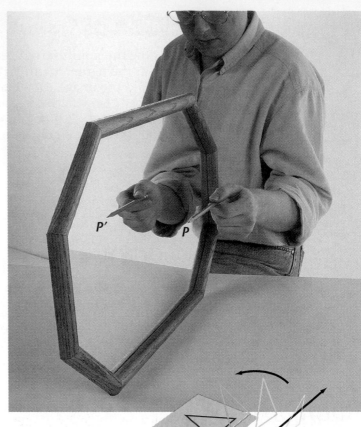

Image

"Mirror" Line   Pre-image

 Exploration 1 *The Reflection of a Point*

You will need
Pointed object
Ruler
Protractor

P

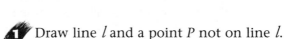
l

P
X
l Y
P'

1 Draw line l and a point P not on line l.

2 Reflect point P through l as follows. Fold your paper along l. Use a pointed object such as a sharp pencil or a compass point to punch a hole through the paper at point P. Label the new point P'.

3 Draw $\overline{PP'}$ and label the intersection of $\overline{PP'}$ and l as point X. Add a new point Y on l.

   a. Measure the length of $\overline{XP}$ and $\overline{XP'}$.

   b. Measure $\angle PXY$ and $\angle P'XY$.

4 Formulate a conjecture about the relationship between l and $\overline{PP'}$. Write your own definition of a reflection in terms of the relationship. ❖

 CRITICAL *Thinking*   What happens if you place point P on line l? What is the image of P? Make sure your conjecture in Step 4 is written so that it takes this into consideration.

•Exploration 2 *Reflections of Segments and Triangles*

You will need
Pointed object
Ruler
Protractor

1 Draw line *l* and segment $\overline{AB}$ that does not intersect *l*. Fold your paper along *l* and punch holes through *A* and *B* to obtain image points *A'* and *B'*. Connect the image points to form segment $\overline{A'B'}$.

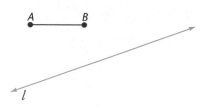

2 Measure $\overline{AB}$ and $\overline{A'B'}$.

Formulate a conjecture about a segment and its reflection through a line.

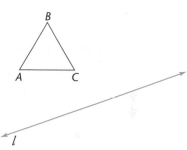

3 Draw line *l* with △*ABC* on one side.

Fold your paper along *l* and create the reflection △*A'B'C'* of △*ABC*.

4 Measure the angles and sides of each triangle. What do you observe? Are the triangles congruent? Will one fit over the other exactly? If you are not sure, cut them out and try it.

Formulate a conjecture about triangles and their reflection through a line. ❖

Try This Reflect each of the figures through the reflection line *l*. Does your conjecture about reflections of triangles and segments seem to hold when the figures touch the reflection line?

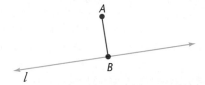

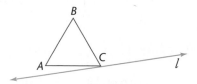

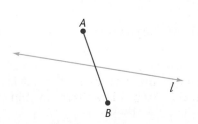

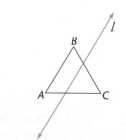

EXERCISES & PROBLEMS

Communicate

In Exercises 1–4, determine whether each of the following is a translation or a rotation. Explain why.

1. A canoe floating with the current

2. A canoe that capsizes in still water

3. The movement of the Earth's center about the sun in 6 months

4. Migrating birds flying in formation overhead

5. Explain how the pre-image and the image of a wheel rolling straight ahead is a combination of two different transformations.

6. Of the two figures below, which one best illustrates a reflection of a figure through a line?

7. When you project the image of a slide onto a screen, the pre-image (on the slide) is expanded to fill the screen for viewing. Explain how this transformation is different from those studied in this lesson.

Practice & Apply

In the diagram in Example 1 on page 45, point A shifted to A' by 2 cm in the direction shown.

8. Describe how point B moved to point B'.

9. Given any point of $\triangle ABC$, tell how to locate its image point in $\triangle A'B'C'$.

In the diagram in Example 2 on page 45, $\overline{AC}$ turned 90 degrees to coincide with $\overline{AC'}$.

10. $\overline{AB}$ turned how many degrees to coincide with $\overline{AB'}$?

11. $\overline{BC}$ turned how many degrees to coincide with $\overline{B'C'}$?

12. Which point traveled farther, B to B' or C to C'? Explain.

Refer to items 3 and 4 in Exploration 2 on page 47. Sketch the diagram.

13. Draw $\overline{BB'}$. What line appears to be equidistant from B and B'?

14. Repeat Exercise 13 for $\overline{CC'}$ and $\overline{AA'}$.

15. Given any point in △ABC, use the results of Exercises 13 and 14 to describe the location of its image point in △A'B'C'.

Copy the pictures shown and reflect them through the mirror line drawn. What is the result?

16.

The word *MOM* has some unique characteristics when flipped over the mirror lines. Show the result on your own paper.

17.

While visiting the beach, Theresa saw footprints like those shown.

18. Which pair of footprints is a reflection?

19. Which pair of footprints is a translation?

20. The bird footprints illustrate a fourth type of transformation known as a **glide reflection**. A glide reflection is a combination of a reflection and a translation. Using your own illustrations, explain how a glide reflection is a combination of these two transformations.

A.

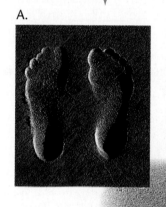

B.

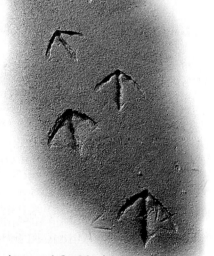

The shape below, a *net*, can be drawn, cut, and folded on the dotted lines to construct a cube. Use this net to help you answer Exercises 21–23.

21. How can you draw this net using one square and performing translations.

22. How can you draw this net using one square and performing reflections.

23. How can you draw this net using one square and performing rotations.

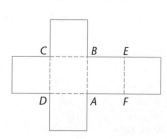

24. What transformation describes the relationship between the picture carved into the wood and the picture imprinted on the paper?

Wood block   *Print*

 Look Back

In Exercises 25–26, complete the sentences.   [Lesson 1.1]

25. Points that lie on the same line are said to be _____?_____ .

26. Points or lines that lie on the same plane are said to be _____?_____ .

In Exercises 27–29, explain your answer using illustrations when helpful.   [Lesson 1.1]

27. Is it possible for two points to be collinear? Why or why not?

28. Is it possible for three points to be noncoplanar? Why or why not?

29. Is it possible for two lines to be noncoplanar? Why or why not?

30. According to a conjecture you made in an earlier lesson, the diagonals of a square are _____?_____ and _____?_____ to each other. **[Lesson 1.2]**

 Algebra **For Exercises 31–34, find the indicated measures. WN = 48, ∠CAR ≅ ∠EAR.**

31. *IN* **[Lesson 1.4]**

32. *WI* **[Lesson 1.4]**

33. m∠CAR **[Lesson 1.5]**

34. m∠EAR **[Lesson 1.5]**

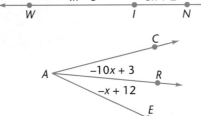

Look Beyond

35. **Portfolio** Collect photographs that illustrate the different kinds of motions you have studied in this lesson, or make drawings of your own. Label each drawing or photograph.

36. **Visual Arts** Trace the shape shown onto your paper. To the right side of your picture, draw two parallel lines about an inch and a half apart. Reflect the picture over the closest parallel line; then reflect the result over the other parallel line. How might you describe the final picture in terms of another transformation discussed in this lesson?

Exploring Motion in the Coordinate Plane

Why *A coordinate plane enables you to describe geometry ideas using algebra. For example, you can use algebra to find the transformation of a geometric figure without using special construction tools.*

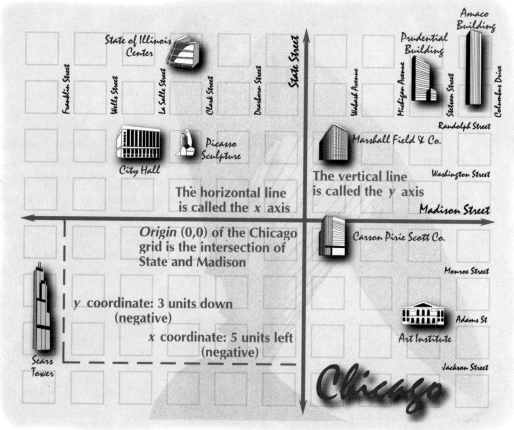

State of Illinois Center

Franklin Street · Wells Street · La Salle Street · Clark Street · Dearborn Street · State Street

Picasso Sculpture

City Hall

The horizontal line is called the x axis

Wabash Avenue · Michigan Avenue · Stetson Street · Columbus Drive

Amaco Building

Prudential Building

Randolph Street

Marshall Field & Co.

The vertical line is called the y axis

Washington Street

Madison Street

Origin (0,0) of the Chicago grid is the intersection of State and Madison

Carson Pirie Scott Co.

Monroe Street

y coordinate: 3 units down (negative)

x coordinate: 5 units left (negative)

Adams St

Art Institute

Jackson Street

Sears Tower

Chicago

The city of Chicago is laid out like a coordinate system. The Sears Tower, once the tallest building in the world, is located at (–5, –3).

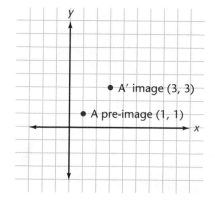

y

• A′ image (3, 3)

• A pre-image (1, 1)

x

Algebra By doing algebraic operations on the coordinates of a point, you can relocate it on the coordinate plane. For example, if you multiply the *x*- and *y*-coordinates of the point (1, 1) by 3, the result is the new point (3, 3):

Pre-image → Image

$$(1, 1) \quad \rightarrow \quad (3, 3)$$

This operation, which is known as a transformation, can be expressed as a rule using the following notation:

$$(x, y) \quad \rightarrow \quad (3x, 3y)$$

Exploration 1 *Translation*

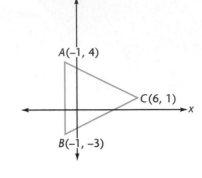

$A(-1, 4)$

$C(6, 1)$

$B(-1, -3)$

You will need
Graph paper
Ruler
Protractor

1 Draw triangle $\triangle ABC$ on graph paper.

2 Pick a number between -10 and $+10$ (but not zero). Then choose either x or y.

3 Depending on your choice, add the number you picked to either the x-or the y-coordinates of the points A, B, and C.

Plot each of your new points and connect each pair with a segment. Does your new figure seem to be congruent to the original one? Measure the sides and angles of each triangle if you are not convinced.

4 The original figure has been translated to a new position. In what direction has it moved, and by how much? Draw a slide arrow for the translation. Compare your results with those of your classmates.

5 Formulate a conjecture about how to translate a figure horizontally or vertically on the coordinate plane. What would you do to each point in the figure?

6 Express your conjecture in the form of rules for moving a given point h units horizontally or v units vertically. Assume that h and v can be either positive or negative.

	Pre-image	Image
Horizontal movement	$(x, y) \rightarrow$	$(\underline{\ ?\ }, \underline{\ ?\ })$
Vertical movement	$(x, y) \rightarrow$	$(\underline{\ ?\ }, \underline{\ ?\ })$ ❖

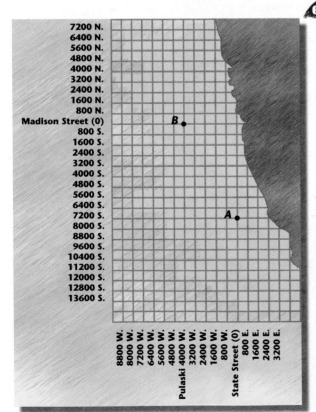

APPLICATION

On the map, each square is 8 blocks long on each side. Suppose you are driving from A at State and 72nd Street to B at Madison and Pulaski. The rule for the horizontal movement is $(x, y) \rightarrow (x - 40, y)$. The rule for the vertical movement is $(x, y) \rightarrow (x, y + 72)$. These rules can be combined to show both horizontal and vertical movement.

$(x, y) \rightarrow (x - 40, y + 72)$

What does the rule represent? ❖

•Exploration 2 Reflection Over the x- or y-Axis

You will need
Graph paper
Ruler
Protractor

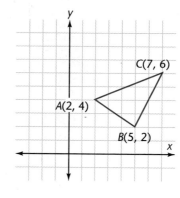

C(7, 6)
A(2, 4)
B(5, 2)

1 Draw the triangle on graph paper.

2 Multiply the x coordinate of each vertex by −1.

3 Plot each of the new points and draw a segment between each pair. Does the new triangle seem to be congruent to the original one? Measure each triangle.

4 Repeat Steps 2 and 3 using the y-coordinate instead.

5 The original triangle has been reflected in each case. What line is the axis of reflection, or "mirror", in each?

6 Formulate a conjecture about how to reflect a figure through the x- or y-axis. What would you do to each point in the figure?

7 Express your conjecture as a rule for reflecting a given point through the x- or y-axis.

	Pre-image	Image
Reflection through x-axis	(x, y) →	(_?_ , _?_)
Reflection through y-axis	(x, y) →	(_?_ , _?_) ❖

CRITICAL *Thinking*

Experiment with figures in other positions, such as the ones below. Do your conjectures seem to be true for all of them?

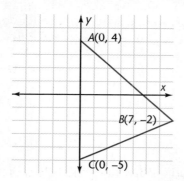

A(0, 4)
B(7, −2)
C(0, −5)

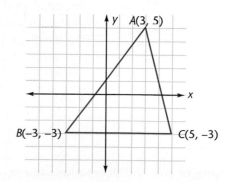

A(3, 5)
B(−3, −3)
C(5, −3)

Exploration 3 *180° Rotation About the Origin*

You will need
Graph paper
Ruler
Protractor

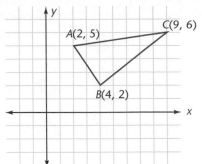

1. Draw the figure on graph paper.

2. Multiply the *x*- and *y*-coordinate of each vertex by −1.

3. Plot each of your new ordered pairs as points and draw a segment between each one. Does the new figure seem to be congruent to the original one? Measure each triangle.

4. The original figure has been rotated. What is the turn center of the rotation? How much has the figure been rotated? Draw a turn arrow for the rotation.

5. Formulate a conjecture for an 180° rotation about the origin. Express your conjecture for rotating a given point in the original figure as a rule.

Pre-image	**Image**

180° rotation about the origin (x, y) → (_?_ , _?_) ❖

In the figure on the right, the upper left panel has been reflected first through the *y*-axis and then through the *x*-axis.
Describe how a 180° rotation about the origin is a combination of these two reflections. Do you think this idea could be extended so that any rotation is a combination of two reflections?

EXERCISES & PROBLEMS

Communicate

1. Explain how you would find the *x*- and *y*- coordinates of the given points on the coordinate plane.

2. Explain how to plot a point on a coordinate plane from its *x*- and *y*-coordinates.

3. Compare the graphs of the points (1, 8) and (8, 1). Would you say that the order of the points mattered? Why do you think the coordinates of a point are known as *ordered pairs?* Discuss.

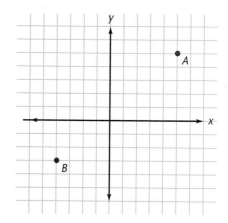

Practice & Apply

In Exercises 4–7, use the street map of Chicago to translate the shuttle bus as indicated. Write the rule you used in each case.

4. Translate the bus up 16 blocks.

5. Translate the bus down 32 blocks.

6. Translate the bus to the right 24 blocks.

7. Translate the bus to the left 8 blocks.

Each square represents 8 blocks.

In Exercises 8–12, use graph paper to draw the transformations of the figure as indicated. Write the rule you used below the transformed figure in each case.

8. Reflect the figure through the *x*-axis.

9. Reflect the figure through the *y*-axis.

10. Rotate the figure 180° about the origin.

11. Reflect the figure first through the *x*-axis and then through the *y*-axis. What is the result?

12. Reflect the figure first through the *y*-axis and then through the *x*-axis. What is the result? How does your result compare with the result in the previous exercise?

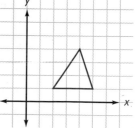

In Exercises 13–21, state the result of applying each rule to a figure on the standard coordinate plane.

13. $(x, y) \rightarrow (x + 7, y)$ **14.** $(x, y) \rightarrow (-x, y)$ **15.** $(x, y) \rightarrow (x - 6, y + 7)$

16. $(x, y) \rightarrow (x - 8, y)$ **17.** $(x, y) \rightarrow (x, y - 4)$ **18.** $(x, y) \rightarrow (x, y - 7)$

19. $(x, y) \rightarrow (-x, -y)$ **20.** $(x, y) \rightarrow (x - 7, y)$ **21.** $(x, y) \rightarrow (x, y + 2)$

22. Draw the figure on a piece of graph paper. Graph the transformation that results from applying the first rule to the figure. Then apply the second rule to the new figure, and so on, graphing each new figure.

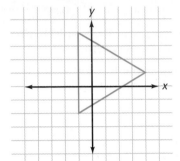

    Rule 1 $(x, y) \rightarrow (x, -y)$
    Rule 2 $(x, y) \rightarrow (-x, y)$
    Rule 3 $(x, y) \rightarrow (x, -y)$
    Rule 4 $(x, y) \rightarrow (-x, y)$

What is the final result of the series of transformations?

Algebra **Exercises 23–30 allow you to discover another rule for transforming an object.**

23. Fill in the table on the right, in which $y = x$ for each row.

24. Use the values in the table on the right to help you graph the line $y = x$.

25. What angle does the line form with the y-axis?

26. Plot each of the following points on your graph. Connect the points with segments.

    $A(2, 1)$ $B(5, 2)$ $C(6, 4)$

y	x
?	?
?	?
?	?

27. For each of the points in the previous exercise, reverse the order of the coordinates to get new points, A', B', and C'. Plot those points on your graph and connect them with segments.

28. What do you observe about the two figures you have drawn?

29. Write the rule for the transformation you have drawn. (**Hint:** Your rule will use just the letters x and y, no minus signs or numbers.)

30. What is true of the coordinates of a point that lies *below* the line $y = x$? of the coordinates of a point that lies *above* the line?

Look Back

Tell how you would find each of the following points of a triangle. **[Lesson 1.3]**

31. incenter   **32.** centroid   **33.** orthocenter   **34.** circumcenter

Express each of the following as a compass heading. **[Lesson 1.5]**

35. W   **36.** NW   **37.** SW   **38.** SSW

Describe each of the following transformations. **[Lesson 1.6]**

39. translation   **40.** rotation   **41.** reflection   **42.** glide reflection

Look Beyond

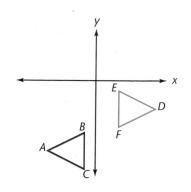

43. What two transformations would you apply to △ABC to get △DEF? Express these as a single rule.

44. Apply the *same rule* to the figure that results. What do you get? Repeat the process three or four times. Describe what happens.

45. Write a general rule for a glide transformation in which the axis of reflection is the *y*-axis and the vertical movement is *v*.

46. Write a general rule for a glide transformation in which the axis of reflection is the *x*-axis and the horizontal movement is *h*.

47. Cultural Connection: Africa An early Egyptian drawing dating from 2650 B.C.E. is a coordinate plane for the shape of a rounded vault. The numbers, which are the small vertical marks on the drawing, give the height of the curve at intervals of 1 cubit.

Make your own sketch of the curve using the dimensions on the drawing. Recall that a palm is equal to 4 fingers and a cubit is equal to 7 palms (or 28 fingers). Begin by filling in a table like the one below.

x-coordinate (in fingers)	y-coordinate (drawing inscription)	y-coordinate (in fingers)
0 cubit = 0 fingers	3 cubits, 3 palms, 2 fingers	98 fingers
1 cubit = ?	3 cubits, 2 palms, 3 fingers	?
2 cubits = ?	3 cubits	?
3 cubits = ?	2 cubits, 3 palms	?
4 cubits = ?	1 cubit, 3 palms, 1 finger	?
5 cubits = ?	?	?

Select an appropriate scale for a coordinate grid and plot the points from your table. Then draw a smooth curve through the points.

Origami Paper Folding

Origami, the ancient Japanese art of paper folding, produces intriguing figures from simple paper folds. In this project, you will use paper folding to create a paper crane. First, follow steps one through six to make the crane base. Then continue with the final folds to make your crane.

Use a 6- to 8-inch square of paper.

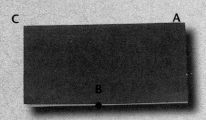

1. Fold the paper in half as shown.

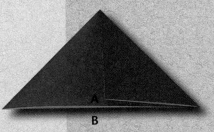

2. Fold point A forward and down to point B. Fold point C backward and down to point B.

3. Open the paper at point B, and lay it flat.

4. Crease the paper by folding points D and E inward to point F.

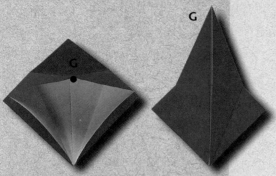

5. Open the figure and pull G upward, reversing the folds you made in Step 4.

6. Turn the paper over and repeat step 5.

You have now completed the crane base. Continue with the next steps to complete the crane.

7. From the crane base, fold points *H* and *I* upward into point *J*. Turn the paper over and repeat.

8. Pull points *K* and *L* outward and reverse the folds. These will become the tail and neck of the crane.

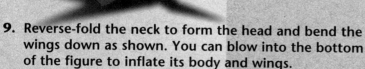

9. Reverse-fold the neck to form the head and bend the wings down as shown. You can blow into the bottom of the figure to inflate its body and wings.

Your crane is finished.

10. Unfolding a paper crane will reveal a pattern of creases. Draw over and label geometric figures formed by the creases. Are any of the figures congruent with one another?

11. The crane base can be used to create many origami figures. Can you figure out how to fold a frog or fish?

12. Find an origami book at your library and expand your collection of origami figures.

Origami Dragon

Chapter 1 Review

Vocabulary

altitude	23	linear pair	40	pre-image	44
angle	11	median	23	ray	10
angle bisector	18	noncollinear	11	reflection	44
circumscribed circle	24	number line	29	right angle	17
collinear	11	parallel lines	17	rigid	44
congruent	31	perpendicular bisector	18	rotation	44
image	44	perpendicular lines	17	segment	10
inscribed circle	24	plane	9	translation	44
line	9	point	9	turn center	45

Key Skills and Exercises

Lesson 1.1

➤ **Key Skills**

Identify and name parts of a figure.

In the figure, A and C are endpoints of the segment $\overline{AC}$, which is part of line $\overleftrightarrow{AC}$. $\overrightarrow{AB}$ is a ray; it starts at point A, passes through point B, and goes on forever. $\overrightarrow{AB}$ and $\overrightarrow{AC}$ form $\angle BAC$. Points A, B, and C lie on plane ABC.

➤ **Exercises**

1. Name the segments and angles in the triangle, and the plane that contains it.

2. Name all the rays and angles in the figure.

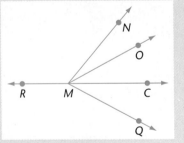

Lesson 1.2

➤ **Key Skills**

Use paper folding to make conjectures about lines and angles.

Using folding paper, construct a perpendicular bisector of a segment $\overline{AB}$. Locate a point X on the perpendicular bisector. From measuring the distances XA and XB you might write the following conjecture: "Every point on the perpendicular bisector of a segment is equidistant from the endpoints of the segment."

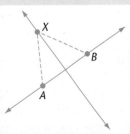

➤ **Exercise**

3. Using folding paper, construct the bisector of an angle. Measure the perpendicular distances from a point on the bisector to the rays that form the angle. Write a conjecture about points on the bisector of an angle.

Lesson 1.3

➤ Key Skill

Indentify special parts of triangles and their intersection points.

Find the intersection point of the perpendicular bisectors of a triangle. Set your compass at the intersection and draw a circumscribed circle.

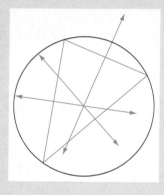

➤ Exercises

Use geometry technology to construct these figures.

4. A right triangle circumscribed by a circle.

5. An inscribed circle that touches each side of a triangle at the midpoint.

Lesson 1.4

➤ Key Skills

Determine if segments are congruent.
Are $\overline{CD}$, $\overline{DE}$, and $\overline{EF}$ congruent segments?

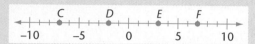

$mCD = \left| -7 - (-2) \right| = 5$

$mDE = \left| -2 - (-3) \right| = 5$

$mEF = \left| 3 - 7 \right| = 4$

$\overline{CD}$ and $\overline{DE}$ are congruent; $\overline{EF}$ is not congruent to the other segments.

Add measures of segments.

To add measures AB and BX, we first find the measure of each segment.

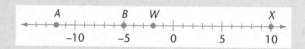

$mAB = \left| -12 - (-5) \right| = \left| -7 \right| = 7$

$mBX = \left| -5 - 10 \right| = \left| -15 \right| = 15$

$mAB + mBX = 7 + 15 = 22$

➤ Exercises

6. Which segments, if any, are congruent?

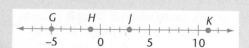

In Exercises 7–9, *A* is between points *R* and *P*. Sketch each figure and find the missing measure.

$-5 - 1 = 6$
$3 - 1 = 2$

7. $RA = 25$ $AP = 13$ $RP = \underline{\ ?\ }$ **8.** $RA = \underline{\ ?\ }$ $AP = 39.5$ $RP = 200$

9. $RA = 121$ $AP = \underline{\ ?\ }$ $RP = 350$

Lesson 1.5

➤ Key Skills

Find relationships between measures of angles.

In a parallelogram opposite angles are congruent; in other words, their measures are equal. In parallelogram *RSTU* the measure of $\angle URS$ and $\angle STU$ is twice the measure of $\angle TUR$ and $\angle RST$.

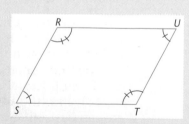

All the following relationships are true: $m\angle URS = m\angle STU = m\angle TUR + m\angle RST = 2m\angle TUR = 2m\angle RST$.

➤ Exercises

In Exercises 10–12 find the missing measures.

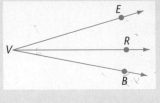

10. m∠EVR = 12° m∠RVB = 26° m∠EVB = __?__

11. m∠EVR = 63° m∠RVB = __?__ m∠EVB = 126°

12. m∠EVR = __?__ m∠RVB = 75° m∠EVB = 90°

13. For which exercises, if any, are ∠EVR and ∠RVB congruent in Exercises 9–11?

Lesson 1.6

➤ Key Skills

Identify transformations: translation, rotation, and reflection.

The figure below shows $\overline{AB}$ translated to $\overline{A'B'}$, reflected to $\overline{A''B''}$, and rotated to $\overline{AB'''}$.

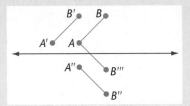

➤ Exercises

14. The letters *S*, *B*, and *W* can be thought of as transformations of the letters *C*, *P*, and *V*. Sketch these transformations and identify each as a reflection, translation, or rotation. Label the pre-image and the image.

Lesson 1.7

➤ Key Skills

Use the coordinate plane to transform geometric objects.

Using the rule $(x, y) \rightarrow (x + 3, y + 3)$, draw the transformed triangle. This represents a translation.

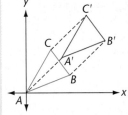

➤ Exercises

In Exercises 15–17 copy the figure and draw the transformation on graph paper. State what type of transformation it is.

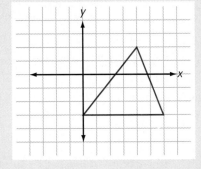

15. $(x, y) \rightarrow (x - 4, y - 4)$

16. $(x, y) \rightarrow (x, -y)$

17. $(x, y) \rightarrow (-x, -y)$

Applications

18. Describe how to find the center of gravity of a triangle. (Hint: Find the centroid.)

19. **Travel** Bastrop, Smithville, and LaGrange are towns on a highway that runs straight from Bastrop, through Smithville, to LaGrange. The distance from Bastrop to Smithville is 22 kilometers. The distance from Bastrop to LaGrange is 5 kilometers more than 3 times the distance from Bastrop to Smithville. Write an equation for the distances, and then solve to find the distance between Smithville and LaGrange and between Bastrop and LaGrange.

Chapter 1 Assessment

1. Name the segments and angles in the triangle, and the plane it lies on.

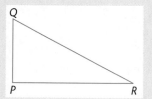

In the figure below, lines *l* and *m* are parallel.

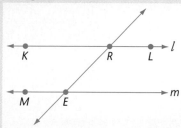

2. Write a conjecture about the relationship of ∠LRE to ∠MER.
3. How would you test your conjecture?

4. Complete the conjecture: *The center of a circle inscribed in a triangle is the same point as the intersection of the triangle's ___?___.*
   a. altitudes b. angle bisectors c. medians d. perpendicular bisectors

In Exercises 5–7, *N* is between points *A* and *D*. Sketch each figure and find the missing measure.

5. *AN* = 51 *ND* = 6 *AD* = __?__
6. *AN* = __?__ *ND* = 14.7 *AD* = 24
7. *AN* = 111 *ND* = __?__ *AD* = 180
8. For which exercises, if any, are $\overline{AN}$ and $\overline{ND}$ congruent in Exercises 5–7?

In Exercises 9–11 find the missing measures.

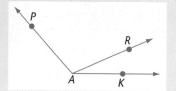

9. m∠*PAR* = 48 m∠*RAK* = 52 m∠*PAK* = __?__
10. m∠*PAR* = 89 m∠*RAK* = __?__ m∠*PAK* = 126
11. m∠*PAR* = __?__ m∠*RAK* = 45 m∠*PAK* = 135

12. A four-leaf clover can be thought of as transformations of a single leaf. Sketch the transformations and identify them as reflections, translations, or rotations. Label the pre-image and the image in each.

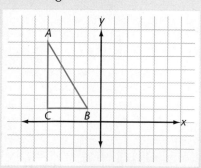

In Exercises 13–15 draw each transformation on graph paper and state what type of transformation it is. Use the pre-image shown.

13. $(x, y) \rightarrow (x + 5, y + 5)$
14. $(x, y) \rightarrow (-x, -y)$
15. $(x, y) \rightarrow (-x, y)$

CHAPTER 2

Reasoning in Geometry

LESSONS

2.1 *Exploring* Informal Proofs

2.2 Introduction to Logical Reasoning

2.3 *Exploring* Definitions

2.4 Postulates

2.5 Linking Steps in a Proof

2.6 *Exploring* Conjectures That Lead to Proof

Chapter Project
Solving Logic Puzzles

egardless of who you are or what you do, there are times when you must reason clearly. Whether you are scaling a cliff or playing a game of Nim with a friend, you must make sound decisions in order to be successful. In either case, the basic principles underlying your reasoning processes are similar.

Geometry presents a unique opportunity for studying reasoning processes. Since ancient times, geometry has been for many people the foremost example of a fully reasoned system.

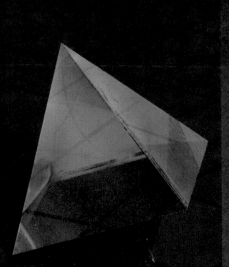

PORTFOLIO ACTIVITY

The game of Nim is played by two persons, who take turns. Twelve counters are arranged in rows of 3, 4, and 5 as shown. On a player's turn, he or she must remove one or more counters from any one of the three rows. The object of the game is to take the last remaining counter.

Try to figure out a strategy that will ensure you of winning if you have the first turn. Present your strategy with an explanation of why it works.

Exploring

Informal Proofs

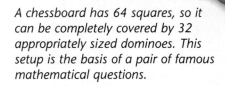

why Have you ever needed to prove something you said? If you used a logical argument, you were probably able to make your case. Logical arguments that make a definite point are **informal proofs.**

A chessboard has 64 squares, so it can be completely covered by 32 appropriately sized dominoes. This setup is the basis of a pair of famous mathematical questions.

If the opposite corners are cut from the chessboard, can the board be completely covered by 31 dominoes? If your answer is yes, can you offer a method for showing that it is possible? If your answer is no, can you explain why it cannot be done? In either case you would be giving an informal proof of what you say.

Here is an informal proof that the reduced chessboard cannot be covered by the dominoes:

"The squares that were cut off were of the *same color*, leaving more squares of one color than the other on the remaining board.

"Any arrangement you make with the dominoes must cover the same number of dark squares as light squares. This is because each domino will cover one dark square and one light square, no matter where you place it on the board. Therefore, it is not possible to cover the reduced board with the 31 dominoes."

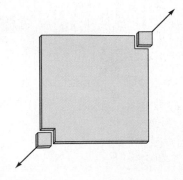

Exploration · Three Challenges

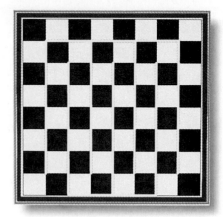

You will need
No special tools

1 Suppose you change the chessboard problem so that you cut off one dark square and one light square from anywhere on the board.

Now the remaining board can be covered with dominoes. Use the diagram on the right and your own explanation to prove this is true.

The pathway through the maze suggests a proof.

ALGEBRA
Connection

2 How could you find the sum of the first n odd counting numbers without actually adding them up? Fill in a table like the one below and see if you can discover the answer.

n	First n odd numbers	Sum of the first n odd numbers
1	1	1
2	1, 3	4
3	1, 3, 5	9
4	1, 3, 5, 7	16
5	?	?
6	?	?
n	?	?

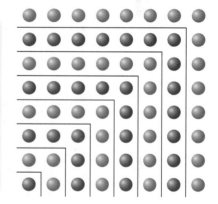

The diagram to the right of the table is a "proof without words" of the algebraic result you may have discovered. Explain in your own words how the diagram "proves" the result.

3 You are given the diagram on the right, built entirely of squares. It is known that the area of Square C is 64 and that the area of Square D is 81. Use this information to determine whether the overall figure is a square. (To be a square, all the sides must have the same length.) ❖

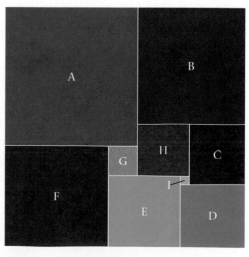

What Is a Proof?

ALGEBRA
Connection

A proof is a fully convincing argument that a certain claim is true. But before you allow yourself to be convinced by a supposed proof, you should make sure that it is sound.

There are many different styles of proofs in mathematics. For example, the calculations below prove that for the given equation, $x = 4$.

$$5x + 4 = 24 \qquad \text{Given}$$
$$5x = 20 \qquad \text{Subtraction Property of Equality}$$
$$x = 4 \qquad \text{Division Property of Equality}$$

Formal Proof

Proof

Statements	Reasons
1. $\triangle ABC$ is isos.	Given
2. $\overline{AB} \cong \overline{BC}$	Def. isos. $\triangle$
3. $\overline{BD} \perp \overline{AC}$	As constructed
4. $\overline{BD} \cong \overline{BD}$	Reflexive prop $\cong$
5. $\angle BDA$ is rt $\angle$ $\angle BDC$ is rt $\angle$	Def. $\perp$ segments

In geometry many proofs are formal in nature. Such proofs are carefully related to a *mathematical system* like the one you will be studying in this book. In such proofs, reasons or justifications are given for every statement that you make.

One kind of formal proof is known as a **two-column proof**. In proofs of this type, you write your statements in the left-hand column and your reasons in the right-hand column. For some proofs you may find this format to be the most convenient way to present your ideas in a simple, orderly fashion.

EXERCISES & PROBLEMS

Communicate

1. In your own words, describe what a proof is.

2. How many different proofs of one claim might there be?

3. Explain why 31 dominoes could not cover a chessboard if you removed 2 dark squares.

4. Explain in your own words how you proved that the overall shape of the diagram in Step 3 of the exploration was or was not a square. Do you think it might be possible for someone to find a "hole" in your argument?

5. In Step 2 of the exploration, a geometric solution is given to an algebraic problem. Discuss the difference between algebra and geometry using this exploration to help you.

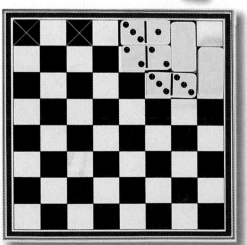

Practice & Apply

Use the sequence $\frac{1}{2}$, $\frac{1}{4}$, $\frac{1}{8}$, $\frac{1}{16}$, . . . for Exercises 6–12.

6. **Algebra** If the pattern of the sequence is to double the denominator to find the next term, then what are the next four terms in the sequence?

7. Are the numbers in the sequence getting larger or smaller? Explain your answer?

8. **Technology** Use your calculator to sum the first 8 terms of the sequence. Use a table like the one below to help you organize your sums.

Number of terms	Terms	Sum of terms
1	$\frac{1}{2}$	.5
2	$\frac{1}{2} + \frac{1}{4}$	.75
3	$\frac{1}{2} + \frac{1}{4} + \frac{1}{8}$	.875
4	$\frac{1}{2} + \frac{1}{4} + \frac{1}{8} + \frac{1}{16}$	.9375
5	_____	_____
6	_____	_____
7	_____	_____
8	_____	_____

9. What number do the sums seem to be getting closer and closer to?

10. You can demonstrate the infinite sum in Exercise 8 geometrically. Start by drawing a square with an area of 1 square unit. Label the lengths of the sides of the square.

11. Divide your square into the fractions of the sequence you identified in Exercise 6, starting with $\frac{1}{2}$. Use the diagram on the right as a guide.

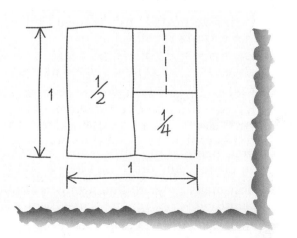

12. Explain how your illustration in Exercise 10 "proves" the sum of the infinite sequence. Is it a geometric or an algebraic proof?

Algebra   Use the table for Exercises 13–18.
The pattern of numbers is from the cubes of
integers (i.e., $1^3 = 1$, $2^3 = 8$, $3^3 = 27$, $4^3 = 64$, etc.).
The table continues infinitely.

A	B	C
1	8	27
64	125	216
•	•	•
•	•	•

13. In which column does the number 1 billion occur?
One method for finding the appropriate column is
to continue filling in the table until you reach 1 billion.
What is the disadvantage of using this method?

14. Begin to analyze the table. 1,000,000,000 is what number raised to
the third power? (Use the following to help you if you are having
trouble: $10^3 = 1000$, $100^3 = $ ___?___)

15. Continue to analyze the table. Notice that column A contains
entries for the numbers 1^3, 4^3, 7^3, . . . What number entries occur
in column B? in column C?

16. What is true of every entry in column C?

17. In what column does the entry for the number 999 occur?

18. How can you add to the above information to finish your proof
that 1 billion occurs in column A? Write the final step in the proof.

Algebra   Exercises 19–22 refer to the following conjecture:
$1 + 2 + 3 + 4 + . . . + n = \frac{n(n + 1)}{2}$ where n can be any positive integer.

19. Check to verify that the conjecture holds for $n = 5$ and $n = 6$.

20. Draw a rectangle of dots that is 6 across and 7 down.
How many dots are there on the whole rectangle?

21. If n represents the width of the rectangle and $n + 1$
represents the height, what is an expression for the
number of dots in the whole rectangle?

22. Divide the rectangle of dots into two equal parts by
circling exactly half of them in the lower right-hand
part of the rectangle, as shown. What is an
expression, in terms of n, for the number of dots
in the circled part of the rectangle?

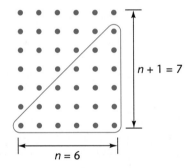

23. How does the diagram suggest a proof for the claim that the sum of
the integers from 1 to n is equal to $\frac{n(n + 1)}{2}$? You should be sure
that it works for all values of n.

Look Back

24. Draw plane $\mathcal{Q}$ and three points A, B, C on plane $\mathcal{Q}$. Draw point P not
on plane $\mathcal{Q}$ and draw lines $\overleftrightarrow{PA}$, $\overleftrightarrow{PB}$, and $\overleftrightarrow{PC}$. **[Lesson 1.1]**

25. How many different line segments can you
identify in the diagram? Name them.
[Lesson 1.1]

26. How many different angles can you identify in the diagram? **[Lesson 1.1]**

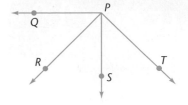

27. Use the diagram below to find the value of *x* and the measure of ∠KIY and ∠TIY.   **[Lesson 1.5]**

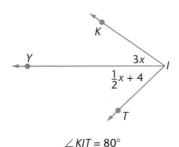

∠KIT = 80°

Look Beyond

28. The positive integers are written in a triangular array as shown. In what row is the number 1000?

$$1$$
$$2\ 3$$
$$4\ 5\ 6$$
$$7\ 8\ 9\ 10$$
$$11\ .\ .\ .$$

29. **Portfolio Activity**

You can also play the game of Nim, on page 65, by changing the object of the game. In a second version of the game, the winner is the person who forces his or her opponent to take the last counter.

A clever strategist will discover that the game of Nim can be won by the player who succeeds in leaving just two rows with two counters in each row. This is true for either version of the game. Explain why this combination will always lead to a win.

The player on the right is removing one penny. Which player will win?

Introduction to Logical Reasoning

WHY *A proof, whether formal or informal, requires* **logical reasoning.** *Logical reasoning assures you that the conclusions you reach in an argument are true— if the rest of the statements in the argument are true.*

Organisms can be classified according to their structure. For example, all of these flowers belong to the orchid family. Class organization, which also extends to man-made things, is the basis of logical reasoning.

Drawing Conclusions From Conditionals

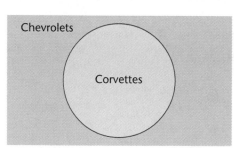

Chevrolets

Corvettes

The force of logic comes from the way information is structured. For example, all Corvettes are Chevrolets, a fact which can be represented by a Venn diagram like the one on the left.

From the Venn diagram, it is easy to see that the following statement is true:

> If a car is a Corvette, then it is a Chevrolet.

Statements like the one above are called **if-then**, or **conditional**, **statements**. In logical notation, conditional statements are usually written in the following form:

$$\text{If } \boldsymbol{P} \text{ then } \boldsymbol{Q}$$
$$\text{or } \boldsymbol{P} \Rightarrow \boldsymbol{Q} \text{ (read: "P implies Q")}$$

In conditional statements the P part—following the word "if"—is the hypothesis. The Q part—following the word "then"—is the conclusion.

If a car is a Corvette, then it is a Chevrolet.

$\underbrace{\qquad\qquad\qquad}_{\textit{hypothesis}}$ $\underbrace{\qquad\qquad\qquad}_{\textit{conclusion}}$

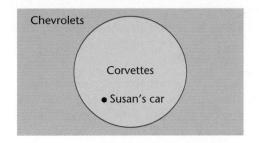

Chevrolets

Corvettes

• Susan's car

Now consider the following statement:

Susan's car is a Corvette.

By placing Susan's car into the Venn diagram, you can see that it is a Chevrolet. This result is a conclusion from a logical argument. Notice that the argument has three parts:

1. If a car is a Corvette, then it is a Chevrolet.

2. Susan's car is a Corvette.

3. Therefore, Susan's car is a Chevrolet.

The process of drawing conclusions using logical reasoning is known as **deductive reasoning** or **deduction**.

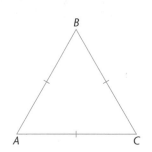

EXAMPLE 1

Use deduction to show that triangle *ABC* is an isosceles triangle. Illustrate your argument with a Venn diagram.

Solution ➤

An isosceles triangle has at least two congruent sides, and an equilateral triangle has three. From this information and the given drawing, you can write the following statements:

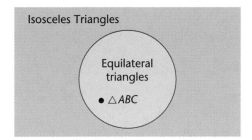

Isosceles Triangles

Equilateral triangles

• △*ABC*

1. If a triangle is equilateral, then it is isosceles.

2. Triangle *ABC* is equilateral.

Then you can write a third statement that is the logical conclusion of Statements 1 and 2. (The Venn diagram illustrates the logic.)

3. Triangle *ABC* is isosceles. ❖

What Is a "True" Conditional?

In the above example, Statements 1 and 2 lead to Statement 3, a *logical conclusion*. A conditional (like Statement 1) that leads only to true conclusions is said to be a **true conditional**. Such conditionals accurately reflect the reality they are supposed to represent. So, if Statement 2 is true, you can be sure that Statement 3 is also true.

Suppose someone says, "If a person performs classical music, then that person dislikes jazz." Can you think of a person who proves that this is not true?

Yo Yo Ma provides a **counterexample** that proves this conditional to be false. If you represent the conditional with a Venn diagram, you will see that there is no place for Yo Yo Ma to go—either inside or outside of the diagram.

Yo Yo Ma, a famous classical cellist, is also a lover of jazz.

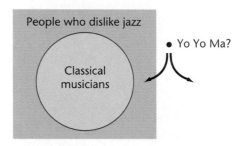

The diagram cannot be completed, because the conditional is false.

Reversing Conditionals

When you interchange the *if* and the *then* parts of a conditional statement, the new statement, which is also a conditional statement, is called the **converse** of the original statement.

> **Statement:** If a car is a Corvette, then it is a Chevrolet.
> **Converse:** If a car is a Chevrolet, then it is a Corvette.

The original statement is true. But what about its converse? If there is a Chevrolet that is not a Corvette—and there certainly is—then the converse is false.

EXAMPLE 2

Write a conditional with the conclusion *the triangle is isosceles* and the hypothesis *a triangle is equilateral*. Then write the converse of your statement.

Solution ➤

> **Statement:** If a triangle is equilateral, then it is isosceles.
> **Converse:** If a triangle is isosceles, then it is equilateral. ❖

Does your original statement seem to be true? Does the converse?

Try This Write a conditional with the hypothesis *a figure is a square* and the conclusion *the figure has four congruent sides*. Then write the converse. Does your original conditional seem to be true? Does its converse? If either seems to be false, find a counterexample.

Logical Chains

Conditionals can be linked together. The result is a **logical chain**. In the following example, three different conditionals are linked together to form a chain. (It does not matter whether the conditionals are actually true.)

EXAMPLE 3

Given: **1.** If cats freak, then mice frolic.

   2. If sirens shriek, then dogs howl.

   3. If dogs howl, then cats freak.

Prove: Show that the statement below follows logically from the given statements.

   If sirens shriek, then mice frolic.

Solution ➤

Identify the *if* clause of the statement you are trying to prove:

   If sirens shriek . . .

Look for a statement that begins with "If sirens shriek":

   If sirens shriek, then dogs howl.

Look for a statement that begins with "If dogs howl":

   If dogs howl, then cats freak.

Look for a statement that begins with "If cats freak":

   If cats freak, then mice frolic.

Finally, by linking the first *if* and the last *then* together, you can conclude:

   If sirens shriek, then mice frolic.

Notice that there is a zigzag pattern in the resulting steps of the argument as you set it up:

   If sirens shriek, then dogs howl.

   If dogs howl, then cats freak.

   If cats freak, then mice frolic. ❖

CRITICAL *Thinking*

Notice that we have not proven that mice are frolicking. What have we proven instead? What is lacking in the argument that would be necessary to prove that mice are actually frolicking?

In proving the conditional, "If sirens shriek, then mice frolic," you have actually relied on an idea which has been assumed to be true without mentioning it. What is that idea?

IF-THEN TRANSITIVE PROPERTY

Suppose you are given:
    If *A* then *B*
    If *B* then *C*

You can conclude:
    If *A* then *C*

2.2.1

CRITICAL *Thinking*   Explain how you use the same property repeatedly when you link long chains of conditionals together.

EXERCISES & PROBLEMS

Communicate

1. Look at the satellite photo of the United States. Can you conclude that it rained all day in Miami. Why or why not?

2. Draw a Venn diagram that illustrates the statement, "If it rains, then it is cloudy."

3. What is the converse of the given statement? Use your Venn diagram to illustrate whether or not the converse is true.

4. Explain how to write the converse of a given conditional statement.

5. Explain how to disprove a given conditional.

On March 21, it was cloudy all day in southern Florida.

Practice & Apply

For Exercises 6–9, refer to the following statement:

All people who live in Ohio live in the United States.

6. Rewrite the statement above as a conditional statement.

7. Identify the hypothesis and the conclusion of the statement.

8. Draw a Venn diagram that illustrates the statement.

9. Write the converse of the statement and construct its Venn diagram. If the converse is false, illustrate this with a counterexample.

In Exercises 10–12, use the information given to draw a conclusion about the individual named. Then draw a Venn diagram to illustrate the solution.

10. If an animal is a mouse, then it is a rodent. "Mikey" is a mouse.

11. If someone is a human being, then he or she is mortal. Socrates is a human being.

12. If a person files his or her income tax early, then he or she will get an early refund. Susan filed her income tax at 7:00 A.M.

If the conclusion about Susan is not actually true, what might be the reason(s) for this?

13. Humor Can logic be used to do the impossible? Consider the following argument.

Is this man disappearing?

The independent farmer is disappearing. That man is an independent farmer. Therefore, that man is disappearing.

How could you write the above argument using a conditional statement? How would you criticize the argument?

For Exercises 14–15, write a conditional statement with the given hypothesis and conclusion, and then write the converse of the statement. Is the original statement true? Is the converse? If either is false, give a counterexample to show that it is false. (You may use the fact that the sum of the measures of the angles of a triangle is 180°.)

14. Hypothesis: "m∠A + m∠B = 90°"

Conclusion: "m∠C = 90°"

15. Hypothesis: "m∠C = 90°"

Conclusion: "triangle *ABC* is a right triangle"

In Exercises 16–18, arrange the three statements to form a logical chain. Then write the conditional statement that the argument as a whole proves.

16. If it is cold, then birds fly south.
If the days are short, then it is cold.
If it is winter, then the days are short.

17. If quompies pawn, then rhomples gleer.
If druskers leer, then homblers fawn.
If homblers fawn, then quompies pawn.

Do you think an argument has to make sense to be logical?

18. If the police radar catches Tim speeding, then Tim gets a ticket.
If Tim drives a car, then Tim drives too fast.
If Tim drives too fast, then the police radar catches him speeding.

Should Tim's parents let him borrow the family car? Which statement(s) might Tim challenge to help him get the keys?

19. Criminal Justice A robber broke into an apartment. He left no fingerprints, but he did leave a mess behind, which included his personal comb. The police have a suspect in custody. You are the lawyer for the suspect, and upon meeting your client, you discover that he is completely bald. Using conditional statements, write a logical argument that you would present in court to prove the innocence of your client.

For Exercises 20–24, refer to the following statement:

If m∠A + m∠B < 90°, then the triangle is obtuse.

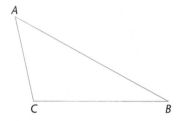

20. Identify the hypothesis.

21. Identify the conclusion.

22. Prove the statement.

23. Write the converse of the statement.

24. Prove or disprove the converse.

25. Fine Arts Choose one of the objects below. Suppose you were going to argue that the object is a work of art. Which of the following conditional statements would you use to lay the foundations for your argument? If you would like, supply your own conditional statement.

> *If an object displays form, beauty, and unusual perception on the part of its creator, then the object is a work of art.*

> *If an object displays creativity on the part of the creator, then the object is a work of art.*

Sol Lewitt, *Two Open Modular Cubes/Half Off*, 1972

Jean Miro, *Tre Donne*, 1935

Write an outline of the argument you would use. Or if you prefer, write an argument that the object you choose is *not* a work of art.

Look Back

26. A floor is modeled by what geometic figure? **[Lesson 1.1]**

27. "Perpendicular" is defined to mean what? Using folding paper, construct a segment and its perpendicular bisector. **[Lesson 1.2]**

28. What is an angle bisector? Construct one using folding paper. **[Lesson 1.2]**

In a triangle, which of the following intersect in a single point? [Lesson 1.3]

29. altitudes

30. medians

31. perpendicular bisectors

32. angle bisectors

33. Reflect the drawing shown with respect to the given line. **[Lesson 1.6]**

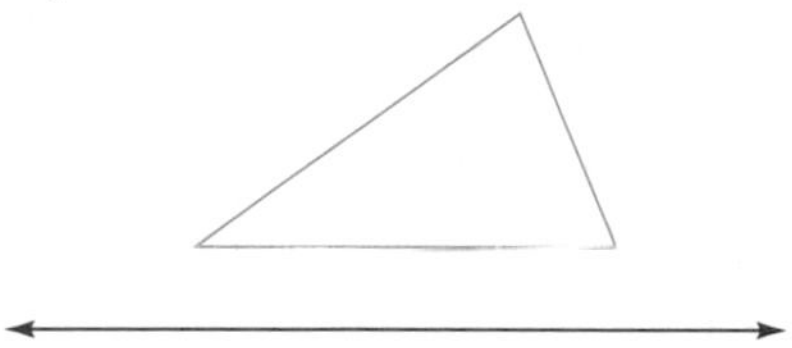

34. Name three types of transformations. **[Lesson 1.7]**

Look Beyond

Another kind of reasoning that is important in mathematics is called inductive reasoning. An inductive argument reaches conclusions based on the recognition of patterns. **Use inductive reasoning in Exercises 35–37 to find the next number in each sequence.**

35. 5, 8, 11, 14, _____

36. 20, 27, 36, 47, 60, _____

37. $3, 1, \frac{1}{3}, \frac{1}{9}$, _____

38. The **Fibonacci sequence** has the pattern shown below. Numbers that occur in the sequence are known as Fibonacci numbers. What are the next five numbers in the sequence?

1, 1, 2, 3, 5, _____

In artichokes and other plants, the number of spirals in each direction are Fibonacci numbers. This artichoke has 5 clockwise spirals and 8 counterclockwise spirals.

Too Tough for Computers

Communication News

Math Problem, Long Baffling, Slowly Yields

By Gina Kolata, *New York Times*

A century-old math problem of notorious difficulty has started to crumble. Even though an exact solution still defies mathematicians, researchers can now obtain answers that are good enough for most practical applications.

The traveling salesman problem asks for the shortest tour around a group of cities. It sounds simple—just try a few tours out and see which one is shortest. But it turns out to be impossible to try all possible tours around even a small number of cities.

Companies typically struggle with traveling salesmen problems involving tours of tens of thousands or even hundreds of thousands of points.

For example, such problems arise in the fabrication of circuit boards, where lasers must drill tens to hundreds of thousands of holes in a board. What happens is that the boards move and the laser stays still as it drills the holes. Deciding what order to drill these holes is a traveling salesman problem.

Very large integrated circuits can involve more than a million laser-drilled holes, leading to a traveling salesman problem of more than a million "cities."

In the late 1970s, investigators were elated to solve 50-city problems, using clever methods that allow them to forgo enumerating every possible route to find the best one. By 1980, they got so good that they could solve a 318-city problem, an impressive feat but not good enough for many purposes.

Dr. David Johnson and Dr. Jon Bentley of AT&T Bell Laboratories are recognized by computer scientists as the world champions in solving problems by getting approximate solutions. They began by working on problems involving about 100,000 cities. By running a fast computer for two days, they can get an answer that is guaranteed to be either the best possible tour or less than 1 percent longer than the best possible one.

In most practical situations, an approximate solution is good enough, Dr. Johnson said. By just getting to within about 2 percent of the perfect solution of a problem involving drilling holes in a circuit board, the time to drill the holes can usually be cut in half, he said.

The researchers break large problems into many smaller ones that can be attacked one by one, and give these fragments to fast computers that can give exact answers.

For example, Dr. Bentley said, "If I ask you to solve a traveling salesman problem for 1,000 cities in the U.S., you would do it as a local problem. You might go from New York toward Trenton and then move to Philadelphia," he explained. Then the researchers would repeat this process from other hubs, like Chicago, and combine the results. "We end up calculating only a few dozen instances per point," Dr. Bentley said. "If you have a million cities, you might do only 30 million calculations."

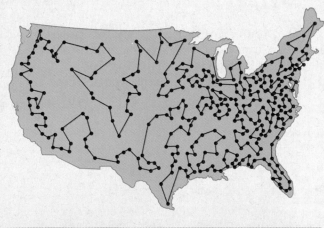

As of 1968, the computer solution for the shortest route connecting 532 cities with AT&T central offices looked like this. It was the biggest such problem calculated up to that time. Now, a route for 2,392 destinations has been computed, and mathematicians are working on a 3,038-city problem. (Source: New York University and Institute for Systems Analysis)

Mathematical Reasoning

The most obvious way of solving a "traveling-salesman" problem is to try all possible routes—which turns out to be very time consuming, even for the fastest computers. So there has been a search for methods of solving the problem that will reduce the number of steps. Finding shortcut methods involves conjecture and mathematical reasoning.

Cooperative Learning

Suppose that a computer can calculate a billion routes per second. How long do you think it would take the computer to calculate all possible routes for 50 cities? In Activities 1–3 you will find out.

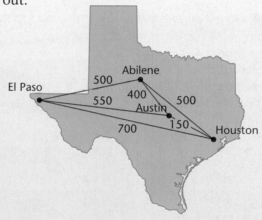

1. A music group based in Austin plans to tour 3 cities, Abilene, Houston, and El Paso. The map at the right shows the distances in miles between cities.

   a. What is the shortest possible route? (In these problems, a route means a path that starts and ends at one city and passes through every other city once and only once.)

   b. List all the possible routes. How do you know you have listed them all?

Number of cities	Number of routes
3	?
4	?
5	?
6	?

2. Now see what happens as you change the number of cities.

   a. Copy and complete the chart. Keep filling in rows until you identify a pattern that you are sure will continue.

   b. Describe the pattern you have found.

   c. If n is the number of cities, write a formula for R, the number of possible routes.

3. You can now use your formula to estimate the computer time required.

   a. How many routes are possible for 50 cities?

   b. Suppose a computer could calculate a billion route lengths every second. How many seconds would it take the computer to calculate the length of every route for 50 cities? How many years is that?

   c. Suppose thousands of computers worked at the same time. Do you think they could make all the calculations in a reasonable time? Why or why not?

LESSON 2.3

Exploring

Definitions

why *Have you ever disagreed with someone, only to find out that you and the other person had different definitions of the same terms? In mathematics it is especially important to know the definitions of the terms you are using.*

Which of the figures in Card 3 are floppers? Simply by observing the differences between Card 1 and Card 2 figures, it is possible to write a definition of a flopper.

A flopper is a figure with one eye and two tails.

Using this definition, you can see that figures (d) and (e) are floppers.

Definitions have a special property when they are written as conditional statements. For example,

If a figure is a flopper, then it has one eye and two tails.

You can also write the converse of the statement by interchanging the hypothesis and conclusion:

If a figure has one eye and two tails, then it is a flopper.

Notice that both the original statement and its converse are true. This is the case with all definitions! The two true statements can be combined into a compact form joining the hypothesis and the conclusion with the phrase "if and only if," represented by $p \Leftrightarrow q$.

$$p \text{ if and only if } q, \quad \text{or} \quad p \Leftrightarrow q$$

By combining the statement and the converse, you create the following definition of a flopper, expressed in logical terms.

A figure is a flopper if and only if it has one eye and two tails.

Venn diagrams can be used to represent two parts of the definition.

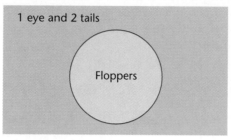

Statement

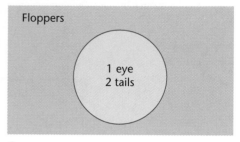

Converse

CRITICAL *Thinking*

Can you imagine a Venn diagram that would represent the fact that both the statement and its converse are true? How would the "inner" and "outer" parts of the diagram be related? Discuss.

•Exploration 1 *Capturing the "Essence of a Thing"*

You will need
No special tools

1 Look at the figure on the right. Make up your own name for the object. Then think: What must be true of a geometric figure in order for it to be a (your name for the object)?

2 According to your idea of a (your name for the object) which of the objects below would you consider to be one? There are no set rules for this. The conditions are up to you!

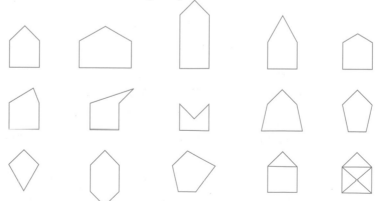

3 Write your own definition of a (your name for the object). Base your definition on your answers to Step 1, or any other decisions you might want to make about the object. Test your conditions for the object to be sure they actually provide a true definition. ❖

•Exploration 2 *Creating Your Own Objects*

You will need
No special tools

Create your own object; then give it a name and a definition. Your object need not be geometrical, or even mathematical, but it should be something you can draw. Test your definition to be sure it is a true one. Share your definitions with others. ❖

▶ EXTENSION

An important geometry concept is that of **adjacent angles**. By examining the figures in the first two boxes, you should be able to form an idea about what adjacent angles are—and also what they are not. This information will enable you to write your own definition.

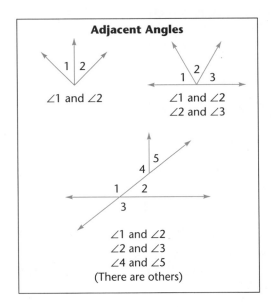

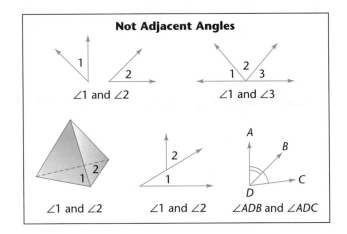

Notice that adjacent angles have a common vertex, a common ray, and do not overlap. So in the third box, angles 1 and 2 are adjacent angles, as are angles 3 and 4 and angles 4 and 5.

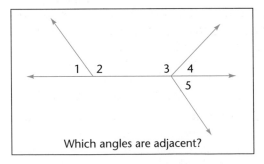

Which angles are adjacent?

A difficulty in defining adjacent angles is to clearly express what is meant by *non-overlapping angles*. The following definition illustrates one way of doing this:

> *Adjacent angles are two angles in a plane that have their vertices and one ray in common, but no interior points in common.*

EXERCISES & PROBLEMS

Communicate

1. Explain why a definition is a special form of if-then statement.
2. Why is it necessary to have undefined terms in geometry?
3. Explain why the following statement is not a definition:

   A tree is a plant with leaves.

4. These are blips.   These are not blips

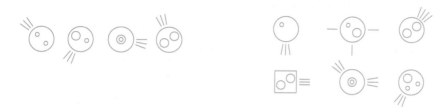

Find a definition for a blip and identify which of the following are blips.

a.   **b.**   **c.**   **d.**

Practice & Apply

In Exercises 5–9, use the following steps to determine whether the given sentence is a definition. First write the sentence as a conditional statement. Then write the converse of the conditional. Finally, write an if-and-only-if statement. Does your final statement seem to be true? Discuss your answer. What is your conclusion? (Exercise 5 is partially worked.)

5. A teenager is a person over 12 years old.
   Conditional statement: If a person is a teenager, then . . .
   Converse: If a person is over 12, then . . .
   If-and-only-if: A person is a teenager if and only if . . .

6. A teenager is a person from 13 to 19 years old.

7. Zero is the integer between -1 and 1.

8. **Geology**   Granite is a very hard crystalline rock.

9. **Music**   A sitar is a lutelike instrument of India.

10. **Chemistry**   Write a definition for propane as a conditional. Write the converse and then an if-and-only-if statement. Does your final statement seem to be true?

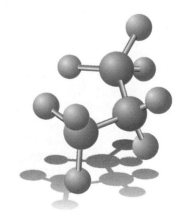

Propane is a colorless gas with the chemical formula C_3H_8.

11. The following are fliches:

The following are not fliches:

Which of the following are fliches?

a.   **b.**   **c.**   **d.**

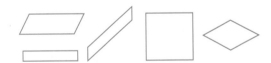

12. The following are zoobies:

The following are not zoobies:

Which of the following are zoobies?

a.   **b.**   **c.**   **d.**

13. The following are parallelograms:

The following are not parallelograms:

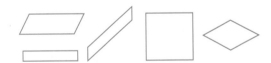

Which of the following are parallelograms?

a.   **b.**   **c.**   **d.**

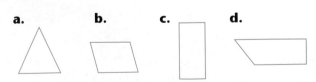

14. The following are regular polygons:

The following are not regular polygons:

Write a definition for a regular polygon.

15. Make up problems that are similar to Exercises 11–14.
Share them with your classmates.

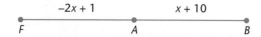

Look Back

16. Point A is the midpoint of $\overline{FB}$. Find x and $\overline{FA}$.
[Lesson 1.4]

$$F \quad \overset{-2x+1}{\longleftarrow} \quad A \quad \overset{x+10}{\longrightarrow} \quad B$$

For exercises 17–20, refer to the diagram.
[Lesson 1.6]

17. Reflect the triangle through the line of reflection drawn.

18. Draw line segments connecting corresponding parts.

19. What relationship appears to exist between the line segments you drew for Exercise 18?

20. What relationship appears to exist between the line segments you drew for Exercise 18 and the mirror of reflection?

Refer to the diagram for Exercises 21–23.   [Lesson 1.7]

21. On your own paper, slide the triangle five units to the right as shown by the arrow.

22. Draw line segments connecting corresponding parts of your drawing.

23. What relationship appears to exist between the segments you drew for Exercise 22?

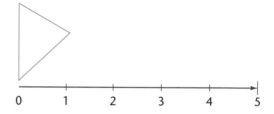

Look Beyond

24. Tonia, Julie, and Susanne are three exceptional students. Use the clues to answer the question below:

a. Two are exceptionally mathematical, two are exceptionally musical, two are exceptionally artistic, and two are exceptionally athletic.

b. Each has no more than three and no less than two exceptional characteristics.

c. If Tonia is exceptionally mathematical, then she is exceptionally athletic.

d. If Julie and Susanne are exceptionally musical, then they are exceptionally artistic.

e. If Tonia and Julie are exceptionally athletic, then they are exceptionally artistic.

Who is not exceptionally athletic? (Hint: Determine the students who are artistic first.)

LESSON 2.4 Postulates

Viewed from a distance, roads may appear as straight lines. Which of Euclid's five postulates does this photo suggest?

Why *The geometry you will study in this book is, for the most part, based on a "textbook" written by Euclid, who lived around 300 BCE. Euclid's geometry begins with five simple postulates.*

The importance of Euclid's work lies not so much in what he discovered, as in the way he organized the knowledge of geometry that already existed in his time.

Euclid began with a few basic **postulates**, or statements that are accepted as true. Then, reasoning logically from his postulates, he was able to prove a number of **theorems**, or statements that are proven to be true. These theorems were then used to prove other theorems, until a vast body of knowledge was built up. *What do you think are the advantages of such a system?*

In Lesson 1.1 you discovered a number of "obvious truths" about points, lines, and planes. The list below states them formally as postulates. This list, while not the same as Euclid's list, should give you the flavor of the foundation of a deductive geometry system.

These postulates might seem as obvious to you as 2 + 2 = 4, but that's precisely what makes the geometry system built up from them so powerful.

POSTULATES	
Through any two points, there is exactly one line.	
or	
Two points determine exactly one line.	**2.4.1**
Through any three noncollinear points, there is exactly one plane.	
or	
Three noncollinear points determine exactly one plane.	**2.4.2**
If two points are on a plane, then the line containing them is also on the plane.	**2.4.3**
The intersection of two lines is exactly one point.	**2.4.4**
The intersection of two planes is exactly one line.	**2.4.5**

EXAMPLE 1

Answer each question and state the postulate that justifies your answer in each case.

A Name two points that determine line *m*.

B Is there a unique line through points *C* and *D*?

C Name three points that determine plane ℛ.

D Name the intersection of line *l* and line *m*.

E Does plane ℛ contain $\overleftrightarrow{BE}$?

F What is the intersection of plane 𝒬 and plane ℛ?

G Do points *B*, *C*, and *F* determine a plane?

Solution ➤

A Points *B* and *C*. "Through any two points there is exactly one line." Points *X* and *C* or points *X* and *B* are also valid answers since they are also contained on the unique line *m*. Try to imagine another line besides *m* through any of these pairs of points if you are not convinced.

B Yes. "Two points determine exactly one line." Notice that the line does not have to be represented in the drawing for it to exist.

C Points *E*, *B*, and *C*. "Through any three noncollinear points, there is exactly one plane." Can you think of other valid answers?

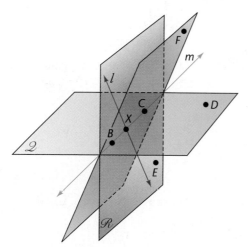

D Point *X*. "The intersection of two lines is exactly one point."

E Yes. "If two points are on a plane, then the line containing them is also on the plane." Which other postulate guarantees that there is a unique line through points *B* and *E*?

F $\overleftrightarrow{BC}$. "The intersection of two planes is exactly one line."

G Yes. "Three noncollinear points determine exactly one plane." The plane does not have to be represented in the drawing for it to exist mathematically. Here is how the drawing would look with the plane drawn. ❖

Try This   If two lines intersect, must there be a plane that contains them?

Photographers use three-legged stands known as **tripods** to hold their cameras steady. You may have also noticed surveyors using tripods, because it is very important that their instruments not move when they measure angles or distances. Tripods are also used to provide stable bases for portable telescopes.

CRITICAL
Thinking

Why do you think tripods provide more stability than four-legged tables? Why is this particularly important when the surface on which they rest is rough or irregular?

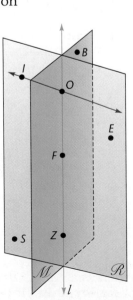

EXERCISES & PROBLEMS

Communicate

1. Why is it necessary to use postulates in geometry?
2. Explain the difference between a postulate and a theorem.
3. Draw a picture to illustrate each postulate on page 88.
4. Describe a space in which it is possible to draw more than two lines through two different points. Hint: Consider the lines of longitude on a globe.

Practice & Apply

For Exercises 5–11, use the illustration to the right to answer each question.

5. Name three points that determine plane $\mathcal{R}$.
6. Name two points that determine line l.
7. Name the intersection of plane $\mathcal{R}$ and plane $\mathcal{M}$.
8. Name the intersection of l and $\overleftrightarrow{IO}$.
9. How many planes contain points I, O, and E?
10. How many planes contain points O, F, and Z?
11. If line l is contained in plane $\mathcal{M}$, then what points must be contained in plane $\mathcal{M}$?

Each of your solutions to the previous exercises is justified by one of the postulates listed below. Match the exercise number (5–11) to the postulate given. You may have more than one answer.

12. Through any two points there is exactly one line.

13. Through any three noncollinear points there exists exactly one plane.

14. If two points are on a plane, then the line containing them must also lie on the plane.

15. Two lines intersect in a point.   **16.** Two planes intersect in a line.

For each of the following statements, draw one diagram for which the statement is true and draw another diagram for a counterexample.

17. Two lines intersect in a point.   **18.** Three points determine a plane.

19. Two planes intersect in a line.   **20.** Three planes intersect in a line.

21. A line and a point determine a plane.

Look Back

22. Solve for x and find the values of the measures of the angles. **[Lesson 1.5]**

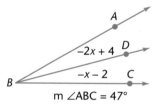

23. What do an angle bisector and the midpoint of a segment have in common?   **[Lesson 1.5]**

24. Draw a Venn diagram to illustrate the following argument:

Every member of the Culver High School football team goes to Culver High School. Brady is a member of the Culver High School football team. Therefore, Brady goes to Culver High School.   **[Lesson 2.2]**

Look Beyond

25. The following is a series of if-then statements. Construct a Venn diagram from these statements, and finish the concluding statement.

If you are taking geometry, then you are thinking hard.
If you are thinking hard, then you are developing good reasoning skills.
If you are developing good reasoning skills, then you will be able to succeed in many different careers.
If you are able to succeed in many different careers, then you will be able to choose your profession.
Therefore, if you are taking geometry, then . . .

26. Why do you think it is important to minimize the number of postulates in the Euclidean system of logic?

27. Any three noncollinear points determine a plane. The following box has 8 corners, each of which represents a point. Draw at least four other planes not shown in the diagram.

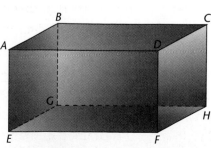

Linking Steps in a Proof

*In algebra, you used the **properties of equality** to link the steps of a proof together. Each link is a **justification** or **reason** that guarantees that each new statement you write in a proof will be true—if the statements you start with are true.*

Euclid's geometry starts with a list of twenty-three **definitions,** five **postulates**, and five other statements which Euclid called **Common Notions**. The Second Common Notion reads as follows:

IF EQUALS ARE ADDED TO EQUALS, THEN THE WHOLES ARE EQUAL.

What **algebraic property of equality** do you recognize in this statement?

The properties of equality are used to solve equations. For example, to solve $x - 3 = 5$, the Addition Property of Equality (Euclid's Second Common Notion) is used to add 3 to each side of the equation. That is,

$$x - 3 + 3 = 5 + 3$$
$$or \quad x = 8$$

In modern language, the Addition Property of Equality states:
For all real numbers a, b, and c,
if $a = b$ then $a + c = b + c$.

Recall the other properties of equality that you learned in algebra. State the properties that hold for subtraction, multiplication, and division.

The Algebraic Properties in Geometry

ALGEBRA
Connection

In the drawing, the lengths of the two outer segments are equal. What can you conclude about the lengths *AC* and *BD* of the two overlapping segments?

It's easy to see that
$$AC = 4 + 5 = 9$$
$$\text{and } BD = 5 + 4 = 9.$$

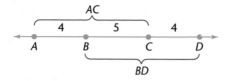

That is, *AC = BD*.

How does this conclusion illustrate the Addition Property of Equality? In the language of the Second Common Notion, what "equals" are added to what other "equals" in the two statements?

Linking Steps to Create a Proof

ALGEBRA
Connection

The following example illustrates how each of the steps in a detailed or "formal" proof can be linked, directly or indirectly, to the given information. The idea that is proved is the same as the one you have been working with. As you will see, one of the links or justifications in the proof is the Addition Property.

EXAMPLE 1

You are given that *AB = CD*. Prove that the lengths *AC* and *BD* are equal.

Solution ➤

The proof is given in two parts.

Part I
You are given that *AB = CD*. The Segment Addition Postulate can be used to write two equations about the lengths *AC* and *BD*.

$$AB + BC = AC \qquad BC + CD = BD$$

BC has been added to *AB* and *CD*. This gives a clue. Add the inner distance to each side of the given equation.

$AB = CD$	Given
$AB + BC = BC + CD$	Addition Property of Equality

$$\underbrace{AB + BC} \qquad \underbrace{BC + CD}$$
$$\quad AC \qquad\qquad BD$$

There are several other properties of equality. One of them is called the Substitution Property. It states that if you have another name for *a*, such as *b*, you may replace *a* with *b* anywhere it appears in an expression.

For all real numbers a and b,
if $a = b$, then either a or b may be replaced with the other in any expression.

Part II
Now you can substitute the distances AC and BD into the equation to get the conclusion you are looking for:

$AC = BD$ (Substitution Property) ❖

The result we have just proven, along with its converse, can be stated as a theorem. In your own future work you can use the theorem to justify a statement without going through the whole proof.

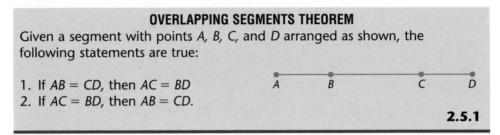

OVERLAPPING SEGMENTS THEOREM
Given a segment with points A, B, C, and D arranged as shown, the following statements are true:

1. If $AB = CD$, then $AC = BD$
2. If $AC = BD$, then $AB = CD$.

2.5.1

Try This   Use the pattern of the proof in the example to write the following proof on your own.

Given: $m\angle AOB = m\angle COD$

Prove: $m\angle AOC = m\angle BOD$

State this result, along with its converse, as a theorem called the **Overlapping Angles Theorem** (2.5.2).

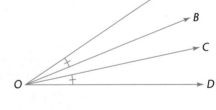

The Equivalence Properties

In addition to the algebraic properties of equality given above, there are three very important properties known as the **Equivalence Properties of Equality**. They are so "obvious" that you probably don't even think of them when you use them.

EQUIVALENCE PROPERTIES OF EQUALITY	
Reflexive Property of Equality	For all real numbers a, $a = a$. **2.5.3**
Symmetric Property of Equality	For all real numbers a and b, if $a = b$, then $b = a$. **2.5.4**
Transitive Property of Equality	For all real numbers a, b, and c, if $a = b$ and $b = c$, then $a = c$. **2.5.5**

Any relationship that satisfies the three properties of Reflexivity, Symmetry, and Transitivity is called an **equivalence relation**.

Try This   Write three statements and call them the Equivalence Properties of Congruence. Instead of the real numbers *a*, *b*, and *c*, write *Figure A*, *Figure B*, and *Figure C*. And instead of "=," write "≅."

The drawings to the right illustrate the Equivalence Properties of Congruence. Tell which property each figure illustrates. Does congruence seem to be an equivalence relation?

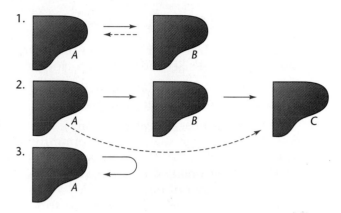

EXAMPLE 2

The first stamp pictured on the right measures 3 cm × 6 cm. The second stamp has the same measurements as the first. The third stamp is exactly the same as the second.

What can you conclude about the first and the third stamps from the given information? State your conclusions using geometry terms. What property discussed in this lesson justifies your conclusion?

Solution ➤

The first stamp is congruent to the third stamp. The justification for this conclusion (from the given information) is the transitive property of congruence. ❖

Russian stamps commemorating the Russian American Company

CRITICAL *Thinking*

In Lesson 1.4 you used a compass to draw congruent segments on a line. What property of congruence guarantees that the segments you draw by the compass method are actually congruent? (Hint: Think of the segment that exists between the points of the compass even though it is not shown.)

EXERCISES & PROBLEMS

Communicate

1. In the figure, $MO = NP$. How does the figure illustrate Euclid's Second Common Notion?

2. Discuss the following statement: Algebra can help us visualize geometry, and vice versa.

3. Fiona and Jada each have 22 mystery books which they do not share. The community library has 120 mystery books for everyone to share. Do Fiona and Jada have access to the same number of mystery books? What basic property enables you to answer this question without doing a calculation? Explain.

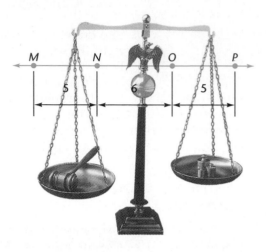

Explain how the balance scale illustrates Euclid's Common Notion that "if equals are added to equals, then the wholes are equal."

Practice & Apply

In Exercises 4–6, identify the properties of equality that justify the conclusion.

4.
$$x + 6 = 14 \qquad \text{Given}$$
$$x + 6 - 6 = 14 - 6 \qquad \text{(Property?)}$$
$$x = 8$$

5.
$$(a + b) = (c + d) \qquad \text{Given}$$
$$(c + d) = (e + f) \qquad \text{Given}$$
$$(a + b) = (e + f) \qquad \text{(Property?)}$$

6.
$$AB + CD = XY \qquad \text{Given}$$
$$CD + DE = XY \qquad \text{Given}$$
$$AB + CD = CD + DE \qquad \text{(Property?)}$$

In the diagram, $UT = AH$. Answer the questions in Exercises 7–12 to prove that $UA = HT$.

7. What two segments sum to UA?

8. What property allows you to answer 7?

9. What two segments sum to HT?

10. What property allows you to answer 9?

11. Write down the information that is given in the diagram about the lengths of segments UT and HA. Add TA to each side of your equation. What property allows you to add this length to both sides?

12. Use substitution and the results from Exercises 7 and 9 to get the conclusion you are looking for.

13. What theorem could you use to justify the result of Exercises 7–12 without doing the proof?

14. Draw a diagram that illustrates a conclusion you could draw from Exercises 7–12 using the second part of the Overlapping Segments Theorem on page 94.

In the diagram, m∠PLA = m∠SLC. Answer the questions in Exercises 15–20 to prove that m∠PLS = m∠CLA.

15. What two angles sum to m∠PLS?

16. What property allows you to answer Exercise 15?

17. What two angles sum to m∠CLA?

18. What property allows you to answer Exercise 17?

19. Write down the information that is given in the diagram about the measures of angles. Add m∠ALS to each side of your equation (remember that m∠ALS is the same angle as m∠SLA).

20. Now substitute angles to get the conclusion you are looking for (see your answers to questions 15 and 17).

21. What theorem could you use to justify the result of Exercises 15–20 without doing the proof?

22. Draw a diagram that illustrates a conclusion you could draw from Exercises 15–20 using the second part of the Overlapping Angles Theorem.

Use the three triangles drawn below for Exercises 23–25.

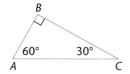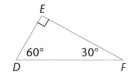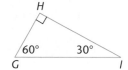

23. If *AB* = *DE* and *DE* = *GH*, what do you know about the relationship of *AB* and *GH*?

24. What property did you use in 23?

25. Illustrate the same property using the same triangles but using angles of the triangles instead of sides.

26. JoAnn wears the same size hat as April. April's hat is the same size as Lara's. Will Lara's hat fit JoAnn? What property of equality supports your answer?

In Exercises 27–29, write your own proofs using any form you like.

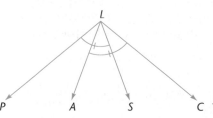

27. Given: m∠PLS = m∠ALC
Prove: m∠PLA = m∠SLC

28. Given: m∠CBD = m∠CDB
        m∠ABD = 90°
        m∠EDB = 90°

Prove: m∠ABC = m∠EDC

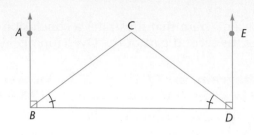

29. Given: m∠BAC + m∠ACB = 90°
        m∠DCE + m∠DEC = 90°
        m∠ACB = m∠DCE

Prove: m∠BAC = m∠DEC

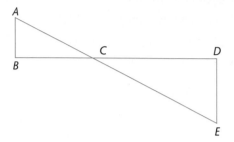

Look Back

30. Can three planes intersect in a point? If yes, draw an example.   **[Lesson 1.1]**

31. Use the midpoint definition and the figure shown to find the value of x and the length of the segments. **[Lesson 1.4]**

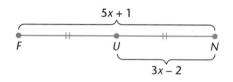

32. If you were to reflect a shape with respect to a line, would your original shape be congruent to the reflection?   **[Lesson 1.6]**

33. Write a true if-then statement whose converse is false. **[Lesson 2.2]**

34. In what special situation are the converse and the original statement both true?   **[Lesson 2.2]**

Look Beyond

35. **Portfolio Activity**   If you leave two rows with two counters each in a game of Nim, then you can win no matter what the other player does. The following strategies will allow you to achieve this winning combination—or an even quicker win, depending on what your opponent does.

a. On your turn, leave two rows with the same number of counters in each row (3 or 4 in each).

b. On your turn, leave three rows with 1, 2, and 3 counters, in any order.

Explain how these combinations can be converted to eventual wins, no matter what your opponent does, for either version of the game.

In either version of the game of Nim, the person who plays first can always win. Try to discover the first move that will allow you to always win. Explain.

These are winning combinations.

Exploring

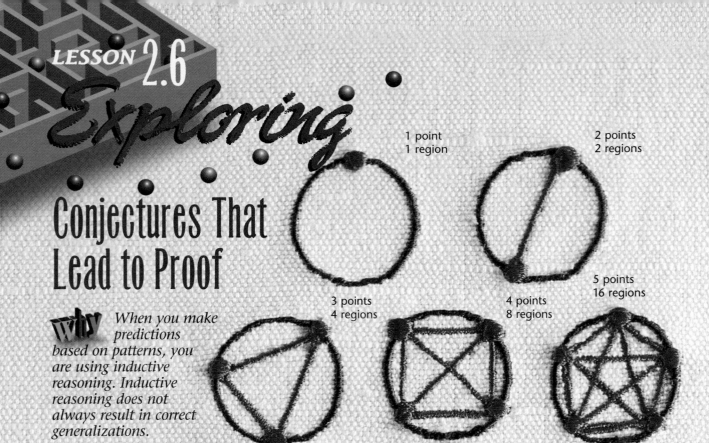

1 point
1 region

2 points
2 regions

3 points
4 regions

4 points
8 regions

5 points
16 regions

Conjectures That Lead to Proof

When you make predictions based on patterns, you are using inductive reasoning. Inductive reasoning does not always result in correct generalizations.

A Need for Proof

Most people who look at the pattern of circles shown above make the following generalization or conjecture:

> *The number of regions doubles each time a point is added.* True or false?

This conjecture can be tested by drawing a picture. What do you discover? Is the conjecture true? We will address this question later.

For a conjecture to be considered true in mathematics, it must be proven true using deductive reasoning. That is, it must be shown to follow logically from statements that are already known or assumed to be true. In the following exploration you will fill in a simple proof of a conjecture.

Exploration 1 *The Vertical Angles Conjecture*

Geometry Graphics

You will need
Geometry technology or
A ruler and protractor.

Vertical angles are the opposite angles formed wherever two lines intersect. You can think of a pair of scissors as forming vertical angles, with the blades making one of the angles and the handles the other.

1 Draw several pairs of vertical angles.

2 Measure each pair.

3 Make a conjecture about vertical angles.

ALGEBRA
Connection

4 What is the relationship of ∠3 to ∠1 and ∠2?

5 Complete: m∠1 + m∠3 = <u>?</u>
m∠2 + m∠3 = <u>?</u>

6 What property of equality lets you conclude that

m∠1 + m∠3 = m∠2 + m∠3?

7 What property of equality lets you conclude that

m∠1 = m∠2? ❖

Inductive and Deductive Reasoning

The steps in the exploration move from **inductive** to **deductive reasoning**. In Steps 1–3 you use inductive reasoning to make a conjecture. In Steps 4–7 you use deductive reasoning to complete an informal proof that all vertical angles are equal.

Throughout this book you will look for patterns to make generalizations and conjectures. As you gain experience in using the logic of deductive proofs, you will learn to write complete proofs of your own conjectures.

Here is an example of a formal proof of the theorem which states that all vertical angles have equal measure. A two-column format is used.

VERTICAL ANGLE THEOREM	
All vertical angles have equal measure.	**2.6.1**

Given: ∠1 and ∠2 are vertical angles.

Prove: m∠1 = m∠2

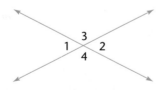

Proof

STATEMENTS	REASONS
1. ∠1 and ∠2 are vertical angles	Given
2. m∠1 + m∠3 = 180 m∠2 + m∠3 = 180	Angles that form linear pairs are supplementary
3. m∠1 + m∠3 = m∠2 + m∠3	Substitution Property
4. m∠1 = m∠2	Subtraction Property ❖

Exploration 2 Reflections Through Parallel Lines

You will need
Geometry technology or
Folding paper and a ruler and protractor

Geometry Graphics

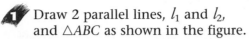

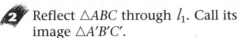

1 Draw 2 parallel lines, l_1 and l_2, and $\triangle ABC$ as shown in the figure.

2 Reflect $\triangle ABC$ through l_1. Call its image $\triangle A'B'C'$.

3 Reflect $\triangle A'B'C'$ through l_2. Call its image $\triangle A''B''C''$.

4 Study the relationship between the original triangle and the final triangle you drew. What single transformation would seem to produce the final image from the original image?

5 Measure the distance between each vertex in the original triangle and its image in the final triangle—that is, the distance AA'', BB'', and CC''. What do you discover?

6 Do all the points in the final triangle seem to have moved in the same direction? Explain your answer.

7 Measure the distance between l_1 and l_2. How does this distance compare with the distances you measured in Step 5.

8 Make a conjecture about the reflection of a figure through parallel lines. Include your results from Step 7 in your conjecture. **(Theorem 2.6.2)** ❖

CRITICAL *Thinking*

What would you need to prove about the reflection of a point through two parallel lines to show that your conjecture is true?

An Informal Proof

The diagram suggests an informal proof of the result you discovered in Exploration 2. See if you can follow it.

1. Which distances are indicated as being equal? Why?

2. What is the distance between the two lines?

3. What does the expression $2D_1 + 2D_2$ or $2(D_1 + D_2)$ represent in the drawing? How does it compare with the distance between l_1 and l_2

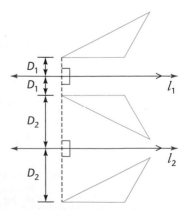

CRITICAL
Thinking

Do you think your result would hold true for any point you might choose on the figure? How could you prove that the direction of movement of each point is the same—that is, parallel?

•Exploration 3 Reflections Through Intersecting Lines

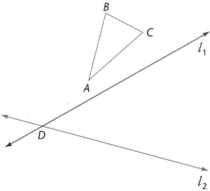

You will need
Geometry technology or
Folding paper, ruler and protractor

1 Draw two intersecting lines l_1 and l_2, and $\triangle ABC$ as shown in the figure.

2 Reflect $\triangle ABC$ through l_1. Call its image $\triangle A'B'C'$.

3 Reflect $\triangle A'B'C'$ through l_2. Call its image $\triangle A''B''C''$.

4 Study the relationship between the original triangle and the final triangle you drew. What single transformation would seem to produce the final transformation from the original image?

5 What seems to be special about the intersection point D of the lines l_1 and l_2 in the transformation you have created?

6 Make a conjecture about the reflection of a figure through two intersecting lines. Measure various angles with their vertices at point D, *such as* $\angle ADA'$ and $\angle A'DA''$. Include your results in your conjecture. (Theorem 2.6.3)❖

Try This Sketch a proof of your conjecture in the above exploration. It should be similar to the informal proof on the previous page. What is lacking from your present supply of theorems and postulates for a full proof?

The Amazing Thing About Proofs

Conjectures, like the prediction about the regions of the circle at the beginning of this lesson, may turn out to be false. This is because they are based on a limited number of observations or measurements. So, you should always ask whether the next case might prove your conjecture to be false.

Proofs are different. In the proof of the vertical angles conjecture, for example, it does not matter what the measures of the different angles on the drawing are. They could be *any size at all,* and the proof would still work. Therefore, a single proof covers *all possible cases.*

EXERCISES & PROBLEMS

Communicate

1. Discuss the advantages of deductive reasoning over inductive reasoning.

2. What is the value of inductive reasoning? Discuss, giving examples to support your position.

3. In the diagram, which pairs of angles are congruent? State a theorem that allows you to draw this conclusion without actually measuring the angles. Discuss.

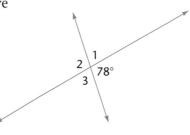

4. Explain how you could *translate* a given figure using only reflections.

Practice & Apply

5. Identify the pairs of congruent angles in the diagram to the right.

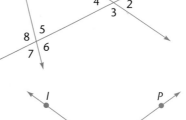

In the diagram to the right, which angle is congruent to the following:

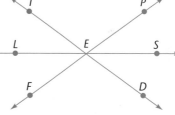

6. $\angle LEI$ 7. $\angle SEF$ 8. $\angle PED$

9. Find the measures of all the angles in the diagram.

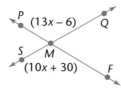

In Exercises 10–13, find the measure of $\angle ABC$.

10.

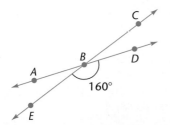

11.

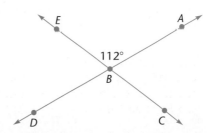

12.

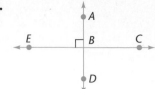

13.

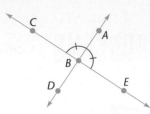

Algebra   In Exercises 14–17, find the value of *x*.

14.

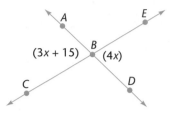

$(3x + 15)$ $(4x)$

15.

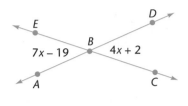

$7x - 19$ $4x + 2$

16.

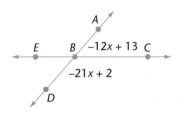

$-12x + 13$

$-21x + 2$

17.

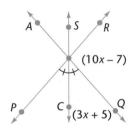

$(10x - 7)$

$(3x + 5)$

Use the results of the informal proof on page 101 for Exercises 18–20. If you reflect △*MNO* first through *m* and then through *n*, how far apart will the original figure and the final figure be if the distance between *m* and *n* is

18. 5 cm?   **19.** 10 cm?   **20.** *x* cm?

$m \parallel n$

For Exercises 21–23, refer to the figure on the right.

21. If you wanted to use reflections to translate the figure 10 cm in the direction of the arrow, how far apart should your parallel lines be?

22. How would you determine the direction of the parallel lines in Exercise 21?

23. Is there more than one placememt of parallel lines that will give the translation described in Exercise 21? Use diagrams to illustrate your answer.

Use the results from Exploration 3 on page 102 for Exercises 24 and 25.

24. What angle of intersection of the two lines is needed to produce a double reflection of a triangle that is equivalent to a rotation of the given number of degrees about the intersection?

    a. 190°   **b.** 90°   **c.** 2*x*°

25. For each angle of intersection of two lines, determine the degrees of an equivalent rotation about the intersection when a triangle is reflected twice.

    a. 30°   **b.** 50°   **c.** *y*°

The proof that follows is a proof of this theorem: *If two angles are supplements of congruent angles, then the two angles are congruent.*
Study the proof, then answer Exercises 26–28.

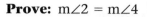

Given: m∠1 = m∠3, ∠1 and ∠2 are supplementary angles,
∠3 and ∠4 are supplementary angles.

Prove: m∠2 = m∠4

Proof

STATEMENTS	REASONS
1. m∠1 + m∠2 = 180°, m∠3 + m∠4 = 180°	Definition of Supplementary Angles
2. m∠1 + m∠2 = m∠3 + m∠4	__?__ Property of Equality
3. m∠1 = m∠3	__?__
4. m∠2 = m∠4	__?__ Property of Equality

26. Which property of equality is the justification for Step 2 of the proof?

27. What is the justification for Step 3 of the proof?

28. Which property of equality is the justification for Step 4 of the proof?

Look Back

29. A segment, line, ray, or plane that intersects a segment at its midpoint is called a segment __?__ .   **[Lesson 1.2]**

In Exercises 30 and 31, refer to the following statements:

- If I am well rested in the afternoon, then I am in a good mood.
- If I sleep until 8:00 A.M., then I am well rested during the afternoon.
- If it is Saturday, then I sleep until 8:00 A.M.

30. Arrange the statements in a logical order.
[Lesson 2.2]

31. Write the conditional statement that the three statements taken together prove.   **[Lesson 2.2]**

32. What three groups of statements does Euclid's geometry begin with?   **[Lesson 2.4]**

Look Beyond

33. The number of regions in the circle pattern on page 99 seem to double every time. Look for a possible deeper pattern in the diagram. Suppose the "real" pattern for the last row is 1, 2, 3, 4, . . . Use this pattern to fill in the numbers above the diagram. Do the numbers on the top row seem to match the actual regions in the circle figures? Test this by drawing circles with 6, 7, and 8 points. Place the points in an irregular way so that you get the maximum regions in each case.

34. If you found that the numbers in the diagram worked for a hundred, a thousand, or even more points on a circle, would you have proved that the pattern is correct? Explain why or why not.

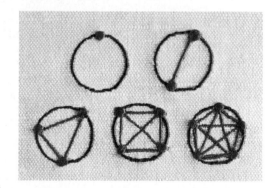

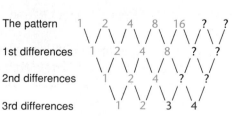

Is this the "real" underlying pattern?

SOLVING LOGIC PUZZLES

Geometry often involves using logical thinking to conclude what is possible or not for a given set of mathematical conditions. Logic puzzles provide a way to sharpen logic skills. Try the two logic puzzles by the British writer "Caliban."

Activity 1

Accomplishments

"My four granddaughters are all accomplished girls." Canon Chasuble was speaking with evident self-satisfaction. "Each of them, " he went on, "plays a different musical instrument and each speaks one European language as well as — if not better than — a native."

"What does Mary play?" asked someone.

"The cello."

"Who plays the violin?"

"D'you know," said Chasuble, "I've temporarily forgotten, alas! But I know it's the girl who speaks French."

The remainder of the facts which I elicited were of a negative character. I learned that the organist is not Valerie; that the girl who speaks German is not Lorna; and that Mary knows no Italian. Anthea doesn't play the violin; nor is she the girl who speaks Spanish. Valerie knows no French; Lorna doesn't play the harp; and the organist can't speak Italian.

What are Valerie's accomplishments?

Make a chart like the one below to determine the correct combinations. Mark an "x" for combinations which are not true and a "•" for combinations which are true. The first clue has been marked to help you get started.

	Cello	Violin	Organ	Harp	French	German	Italian	Spanish
Mary	•	x	x	x				
Valerie	x							
Lorna	x							
Anthea	x							
French								
German								
Italian								
Spanish								

After you have filled in the "x's" and "•'s" from the initial clues, you will have to deduce the remaining relationships. For example, suppose you know that Mary plays the violin and the violinist doesn't speak German. Then you can conclude that Mary doesn't speak German. But this is only an example. Find out for yourself!

Activity 2

In this logic problem, each letter represents a digit.

Digits Are Symbols

"Digits are only symbols, " said Miss Piminy to her class. "We could use other symbols instead. Suppose, for example, that SE is the square of E. Then STET might also represent a perfect square."

"I don't see that, Miss Piminy," said Troublesome Tess, the *enfant terrible* of the class.

"Don't you, Tess? Then use your wits," said Miss Piminy. "Your comment," she added "gives me an idea." She wrote on the board:

$$
\begin{array}{r}
S\ E\ E \\
T\ E\ S\ S \\
\hline
\bullet\ \bullet\ \bullet\ \bullet
\end{array}
$$

"There's an addition sum, students. S, E, and T have the values mentioned already. What's the answer to my sum? — in my own notation of course."

How quickly can you produce the answer?

Write a step-by-step procedure which explains how to find the answer.

"Caliban" is the pen name of a famous British puzzle writer named Hubert Phillips. Many of his puzzles appeared regularly in British newspapers. A number of his puzzle books have been published in the United States.

Two moves will solve this cube. What color will the bottom face be?

Computer games often require logical thinking.

Chapter 2 Review

Vocabulary

conditional statement	72	inductive reasoning	100	theorem	88
converse	74	logical chain	75	Transitive Property of	
counterexample	74	postulate	88	Equality	94
deductive reasoning	73	Reflexive Property of		true conditionals	73
Equivalence Properties of		Equality	94	two-column proof	68
Equality	94	Symmetric Property of		vertical angles	99
equivalence relation	95	Equality	94		

Key Skills and Exercises

Lesson 2.1

➤ Key Skills

Give an informal proof for a conjecture.

In the figure the points on a circle have been connected with segments. The table shows the number of points and the number of segments connecting the points.

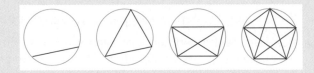

points	2	3	4	5
segments	1	3	6	10

Conjecture: *If there are n points on a circle, then the number of segments necessary to connect all the points is $\frac{n}{2}(n-1)$.* You can test this conjecture by sketching the next figure in the sequence, counting the number of segments, and comparing that to $\frac{6}{2}(6-1) = 15$.

➤ Exercises

The vertical line separates the S into four parts.

1. Make a conjecture about the number of parts that result from any number of vertical lines.

2. Write or draw an informal proof of your conjecture.

Lesson 2.2

➤ Key Skills

Draw a conclusion from a conditional.

What conclusion can you draw from these two statements?

If an animal is a cat, it has four legs.
Sandy is a cat.

Conclusion: *Sandy has four legs.*

State the converse of a conditional.

The converse of the previous conditional is "*If an animal has four legs, it is a cat.*" A counterexample proves this converse is false: *Peek the dog has four legs.*

➤ *Exercises*

In Exercises 3–4 refer to the statements:

If a "star" doesn't flicker, then it's really a planet.
The Evening Star doesn't flicker.

3. Draw a conclusion about the Evening Star.

4. Write the converse of the conditional statement. Is it a true conditional?

Lesson 2.3

➤ *Key Skills*

Write and test a definition.

Is this a definition? "*A right angle is an angle whose measure is 90°.*"

If the conditional form of a statement is true and its converse is also true, then the statement is a definition.

If an angle is a right angle, then its measure is 90°.   ← **Statement is true**
If an angle measures 90°, then it is a right angle.   ← **Converse is true**

Since both these conditionals are true, the original statement is a definition.

➤ *Exercises*

In Exercises 5–6 demonstrate whether the statement is a definition.

5. Parallel lines are lines that are always the same distance apart.

6. A goldenrod is a yellow wildflower.

Lesson 2.4

➤ *Key Skills*

Use postulates to justify a statement.

In the figure below, lines *l* and m intersect at point C. The postulate that justifies this statement is, "The intersection of two lines is exactly one point."

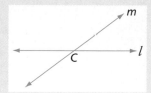

➤ *Exercises*

Use the figure to the right to answer each exercise. Then justify each answer with a postulate.

7. Name three points that determine plane $\mathscr{R}$.

8. Name the intersection of planes $\mathscr{R}$ and $\mathscr{M}$.

9. How many planes contain points *A*, *C*, and *D*?

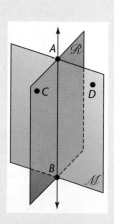

Lesson 2.5

> ### Key Skills

Use the properties of equality to justify statements.

If $\overline{AB} \cong \overline{CD}$, you know that the relationship between $\overline{AC}$ and $\overline{BD}$ is that they are congruent by the Addition Property of Equality.

> ### Exercises

Given ∠*EZN* ≅ ∠*PZO*.

10. Prove: ∠*EZO* ≅ ∠*PZN*

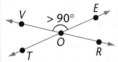

Lesson 2.6

Key Skills

Use deductive reasoning to prove a conjecture.

In the figure m∠*VOE* > 90°. Show that m∠*VOT* and m∠*ROE* are each less than 90°.

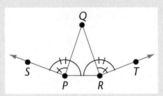

m∠*VOT* + m∠*VOE* = 180°, by the definition of supplementary angles. Because m∠*VOE* is greater than 90°, m∠*VOT* must be less than 90°, or the sum of the two angles would be greater than 180°. m∠*VOT* = m∠*ROE* because they are vertical angles, so m∠*ROE* must also be less than 90°.

> ### Exercises

Given that m∠*SPR* = m∠*TRP*, and m∠*QPS* = m∠*QRT*, prove that m∠*QPR* = m∠*QRP*.

By ___(11)___ , m∠*QPS* + m∠*QPR* = m∠*SPR* and m∠*QRT* + m∠*QRP* = m∠*TRP*. Since it is given that m∠*SPR* = m∠*TRP*, then m∠*QPS* + m∠*QPR* = m∠*QRT* + m∠*QRP* by ___(12)___ . It is also given that m∠*QPS* = m∠*QRT*. Therefore, m∠*QPR* = m∠*QRP* by ___(13)___ .

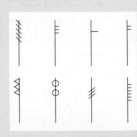

Applications

14. Computer Programming A programmer is trying to find the error in the logic of a program. In English the chain of logic is:

1. Is the first character on the screen a number or a letter?

2. If it is an E, then exit the program.

3. If it is any other letter, then go to the word-handling routine.

   Find the logical gap and write a conditional statement that could bridge the gap.

15. Archaeology The symbols in the top row are from the ogham alphabet, found in archaeological digs in southern Ireland.

The symbols in the bottom row are not from the ogham alphabet.

Write a definition of ogham symbols.

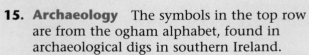

Chapter 2 Assessment

In Exercises 1–2 refer to the sequence below.

$$\frac{1}{2}, \frac{1}{4}, \frac{1}{8}, \frac{1}{16}, \cdots$$

1. Make a conjecture about the sequence's relation to zero.
2. Write or draw an informal proof of your conjecture.

In Exercises 3–4 refer to the statements.

If an animal has six legs, it is an insect.
A flea has six legs.

3. Draw a conclusion about a flea.
4. Write the converse of the conditional statement. Is the converse a true conditional?

5. Demonstrate whether the statement is a definition:
   A tuba is a brass musical instrument.

These objects are tori (singular, torus).

These objects are not tori.

6. Which of these objects are tori?
7. Write a definition of a torus.

In Exercises 8–9 use the figure to the right to answer each question. Then justify each answer with a postulate.

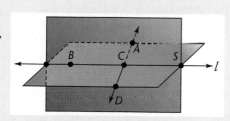

8. Name two points that determine line l.
9. Name the intersection of line l and $\overleftrightarrow{AD}$.

10. What property justifies this conclusion:
    $m\angle ABC = m\angle CBA$?
    a. Addition Property of Equality
    b. Reflexive Property of Equality
    c. Symmetric Property of Equality
    d. Transitive Property of Equality

Chapters 1–2 Cumulative Assessment

College Entrance Exam Practice

Quantitative Comparison. Exercises 1–4 consist of two quantities, one in Column A and one in Column B, which you are to compare as follows.

A. The quantity in Column A is greater.

B. The quantity in Column B is greater.

C. The two quantities are equal.

D. The relationship cannot be determined from the information given.

	Column A	Column B	Answers
1.	$N \quad B \quad\quad S$ $-4 \quad -1 \quad\quad 4$ NB	BS	(A) (B) (C) (D) **[Lesson 1.4]**
2.	$\overline{WZ}$ is an angle bisector. W X Z Y m∠XWZ	m∠YWZ	(A) (B) (C) (D) **[Lesson 1.5]**
3.	$\overline{EF}$ is congruent to $\overline{HG}$. E F H G EH	FG	(A) (B) (C) (D) **[Lesson 2.5]**
4.	1, 4, n, . . . 9	n	(A) (B) (C) (D) **[Lesson 2.1]**

In Exercises 5–9 identify each as a translation, rotation, or reflection. **[Lesson 1.7]**

5.

6.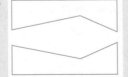

7. $(x, y) \rightarrow (-x, y)$

8. Which is a true definition? **[Lesson 2.3]**

   a. The midpoint of a segment divides the segment into two equal parts.

   b. An acute triangle has an angle less than 90°.

   c. Two points determine exactly one line.

   d. An equilateral triangle is an isosceles triangle.

9. m∠1 = 35° and m∠2 = 35°. What property justifies this conclusion: m∠1 = m∠2? **[Lesson 2.5]**

    a. Addition Property of Equality   **b.** Reflexive Property of Equality

    c. Subtraction Property of Equality   **d.** Substitution Property of Equality

10. Write a conjecture about ∠GYO and ∠RYE. **[Lesson 1.2]**

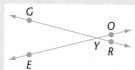

11. Construct an equilateral triangle and a circumscribed circle. **[Lesson 1.3]**

12. Copy the figure at the right and sketch the translation (x + 4) of △PQR. **[Lesson 1.7]**

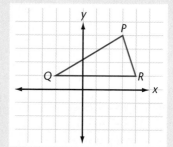

13. Write the converse of this statement: If you have a piece of cake, then you can eat it. Is the converse a true conditional? **[Lesson 2.2]**

14. Write a conclusion from these two statements: If the maple leaves are falling, then it must be October. It is October. **[Lesson 2.2]**

15. The figures on the right are splorts.

The figures on the right are not splorts.

Write a definition of a splort. **[Lesson 2.3]**

In Exercises 16–17 use the figure to answer each question. Then justify each answer with a postulate. [Lesson 2.4]

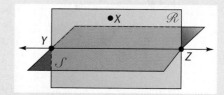

16. Name the intersection of planes 𝓡 and 𝓢.

17. How many planes contain points X, Y, and Z?

Free-Response Grid Exercises 18–20 may be answered using a free-response grid commonly used by standardized test services.

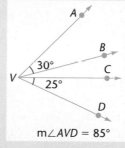

18. Hither, Thither, and Yon are towns on a highway that runs straight from Hither, through Thither, to Yon. The distance from Hither to Thither is 41 miles. The distance from Thither to Yon is 2 miles more than twice the distance from Hither to Thither. What is the distance between Thither and Yon? **[Lesson 1.4]**

19. What is m∠AVC? **[Lesson 1.5]**

m∠AVD = 85°

20. What is the measure of ∠MON? **[Lesson 1.5]**

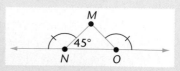

CHAPTER 3

Parallels and Polygons

Rectangles, triangles and hexagons are examples of polygons. The crisscross interplay of parallel beams and polygons in the photograph of the Thorncrown Chapel suggests the theme of this chapter.

LESSONS

3.1 *Exploring* Symmetry in Polygons

3.2 *Exploring* Properties of Quadrilaterals

3.3 Parallel Lines and Transversals

3.4 Proving Lines Parallel

3.5 *Exploring* the Triangle Sum Theorem

3.6 *Exploring* Angles in Polygons

3.7 *Exploring* Midsegments of Triangles and Trapezoids

3.8 Analyzing Polygons Using Coordinates

Chapter Project
String Figures

Thorncrown Chapel, Eureka Springs, Arkansas
Designed by E. Fay Jones

As you will learn, parallel lines and their properties provide the basis for classifying and exploring four-sided polygons known as quadrilaterals.

Patterns of polygons are often used for decorative purposes. The tiling pattern in the background is from the Alhambra, a famous Islamic fortress in Spain.

PORTFOLIO ACTIVITY

A repeating pattern that covers a surface is known as a **tessellation.** The word comes from the Latin word meaning "little square stones." Tessellations in the time of the Romans were mosaics made of small colored stones.

More recently, geometers have taken an interest in figures that cover a plane surface in ingenious ways, such as those displayed in the works of M. C. Escher. In this chapter you will examine some of the tessellations of Escher. You will also be invited to create tessellations of your own.

LESSON 3.1
Exploring Symmetry in Polygons

Symmetrical polygons give this American Indian blanket design an attractive appearance. They also have interesting mathematical properties.

Defining Polygons

In Lesson 2.3 you learned to use examples and nonexamples to formulate definitions. Use the following figures to define a **polygon**.

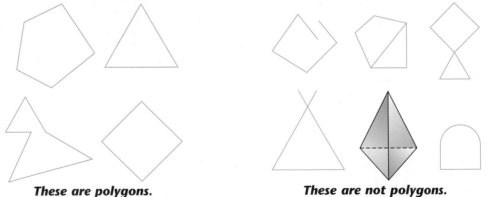

These are polygons.   **These are not polygons.**

Compare your definition of a polygon with the definition below. Does your definition have more requirements than it needs?

POLYGON

A polygon is a closed plane figure formed from three or more segments such that each segment intersects exactly two other segments, one at each endpoint.

3.1.1

A polygon is named according to the number of its sides. Familiarize yourself with the list below.

Polygons, With Number of Sides			
Triangle	3	Nonagon	9
Quadrilateral	4	Decagon	10
Pentagon	5	11-gon	11
Hexagon	6	Dodecagon	12
Heptagon	7	13-gon	13
Octagon	8	*n*-gon	*n*

Reflectional Symmetry

Is the cat's face symmetrical?

Hold your pencil in front of the photograph of the blanket on page 116 so that the part of the photo on one side of your pencil is the mirror image of the part on the other side. The line of your pencil is an **axis of symmetry**.

> **REFLECTIONAL SYMMETRY**
> A plane figure has **reflectional symmetry** if and only if its reflection image through a line coincides with the original figure. The line is called an **axis of symmetry**. 3.1.2

Imagine giving each of the following figures a complete flip over the dotted lines shown. As you can see, the resulting figure will coincide exactly with the original figure. Therefore, each figure has reflectional symmetry, and each dotted line is an axis of symmetry.

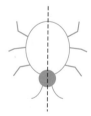 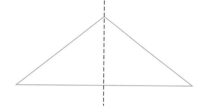 

•Exploration 1 Axes of Symmetry in Triangles

You will need
Folding paper

One way to name a triangle is by the number of its congruent sides:

Equilateral	3 congruent sides
Isosceles	at least 2 congruent sides
Scalene	0 congruent sides

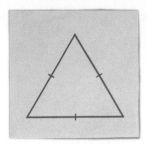

Equilateral

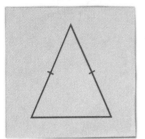

Isosceles

Scalene

As you do each of the following steps, recall geometry terms you studied in previous lessons, such as angles, bisectors, and altitudes of triangles. Make these terms part of your conjectures. *Write your conjectures in your notebook for future reference.*

1 Draw an example of each kind of triangle shown and find one or more axes of symmetry, where possible.

2 How many axes of symmetry does each triangle have?

3 Fold each figure through its axes of symmetry, if any. When you do, which angles coincide? Which segments coincide?

4 Make conjectures about angles that are congruent in the three different kinds of triangles.

5 How is each axis of symmetry related to the side of the triangle it passes through? Express your answers as conjectures.

6 How is each axis of symmetry related to the angle whose vertex it passes through? Express your answers as conjectures. ❖

Rotational Symmetry

> ### ROTATIONAL SYMMETRY
> A figure has rotational symmetry if and only if it has at least one rotation image that coincides with the original image. We say that a figure has a **rotation image of *n* degrees** if a rotation by *n* degrees about a fixed point results in an image that coincides with the original. Rotation images of 0° or multiples of 360° do not have rotational symmetry.
>
> **3.1.3**

This flower has approximate rotational symmetry. How many times will it match its original image if it rotates one complete turn about its center?

Exploration 2 *Rotational Symmetry in Regular Polygons*

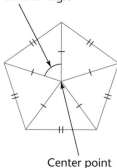

Central Angle

Center point

You will need
Tracing Paper

A **regular polygon** is a polygon in which all the sides and all the angles are congruent. A square, for example, is a regular polygon.

The **center** of a regular polygon is the point that is equidistant from all the vertices of the polygon.

A **central angle** of a polygon is an angle formed by two rays originating from the center and passing through adjacent vertices of the polygon.

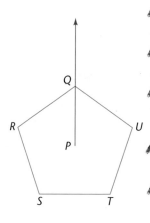

1 Sketch several different regular polygons. Do they seem to have rotational symmetry about their center points?

2 Do you think all regular polygons have rotational symmetry? Express your answer as a conjecture.

3 Draw or trace a regular pentagon such as the one in the illustration. Label the center P and the vertices Q, R, S, T, and U. Draw ray $\overrightarrow{PQ}$. Then copy the figure onto a sheet of tracing paper.

4 Use a pencil point to anchor one figure on top of the other at their centers so that the figures coincide.

5 Rotate the tracing paper counterclockwise until the trace of Q coincides with R. Continue rotating Q through each of the points S, T, U, and finally back to Q. How many moves did you make in all?

6 Notice that each of the turns rotates through one of the central angles of the pentagon. Do the central angles seem to be equal? Explain why or why not.

7 Use the information in Steps 1–6 to determine the measure of a central angle of a regular pentagon. Then measure a central angle with a protractor. How does your measure compare with your determined value?

8 Write a conjecture about the measure of a central angle of a regular polygon with n sides. Write a formula based on your conjecture. ❖

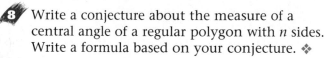

Engineering Jim is designing a special windmill for a wind-powered electrical generator. The windmill has eight blades, which will fit inside a regular octagon. What is the central angle of the blades? What is the rotational symmetry of the windmill?

Divide 360° by the number of sides of the octagon to find the central angle:

$$360° \div 8 = 45°$$

With each rotation through a central angle, the figure matches up with itself. So the figure has a rotational symmetry of 45°, 90°, 135°, 180°, 225°, 270°, 315°, 360°. At 360° the figure has returned to its original position. But you can keep going: 405°, 450°, 495°, . . . ❖

CRITICAL *Thinking*

Can a polygon have a rotation image of 0°? What about negative degrees, such as −45°, −90°, . . .

EXERCISES & PROBLEMS

Communicate

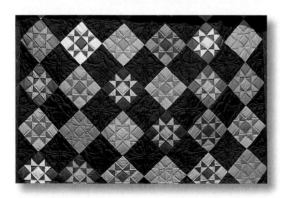

1. What kind of symmetry can you find in the quilt on the left? Identify the axis of symmetry for some of the patterns.

2. Explain how the quilt illustrates the definition of symmetry. Identify a point and its pre-image. How do the two points relate to the axis of symmetry of the photograph?

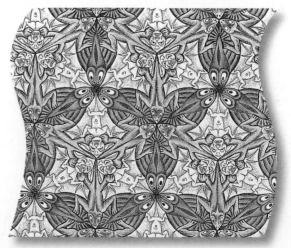

3. Identify regions of the Escher woodcut at right that have rotational symmetry. Where are the centers of rotation?

4. Discuss the symmetry of a regular hexagon. How many axes of symmetry does it have? What are its rotational symmetries?

5. Can you draw a hexagon with just two axes of symmetry? Would it have rotational symmetry? Discuss.

Practice & Apply

In Exercises 6–9, tell how many axes of symmetry, if any, each figure has. If the figure has rotational symmetry, make a list of its rotational symmetries.

6.

7.

8.

9.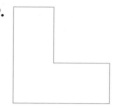

10. How many axes of symmetry does a circle have? Explain your answer.

Below are incomplete sketches of figures that have reflectional symmetry. **Copy and complete each figure.**

11.

12.

13.

Below are incomplete sketches of figures that have rotational symmetry. **Copy and complete each figure.** The center and the amount of rotation are given.

14.
60°
Clockwise

15.
180°

16.
180°

17. Which completed figures in Exercises 11–13 also have rotational symmetry? Explain your answer.

18. Which completed figures in Exercises 14–16 also have reflectional symmetry? Explain your answer.

In Exercises 19–21, use a protractor and ruler. Draw $\overline{AB}$ with $AB = 4$ cm. (→ means "coincides with")

19. Draw an axis of symmetry for $\overline{AB}$ that passes through $\overline{AB}$ at a single point.

20. When you reflect $\overline{AB}$ over its axis of symmetry:

$A \rightarrow$ __?__ $B \rightarrow$ __?__ $\overline{AB} \rightarrow$ __?__

A   B

21. What is the relationship between segment $\overline{AB}$ and its axis of symmetry?

22. Draw ∠ABC with m∠ABC = 90°, and AB = BC. Then draw an axis of symmetry for ∠ABC.

23. When you reflect ∠ABC over your axis of symmetry,

    a. $\overrightarrow{BA}$ → ___?___

    b. $\overrightarrow{BC}$ → ___?___

    c. ∠ABC → ___?___

24. What is the relationship between ∠ABC and its axis of symmetry?

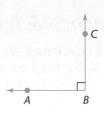

Describe how to draw an axis of symmetry for the following.

25. an 80° angle

26. a 130° angle

27. In general, describe how to draw an axis of symmetry for any given angle.

ABCD is a rhombus; that is, its four sides are of equal measure. **Determine whether each of the following is a result of a rotation, a reflection, or both.**

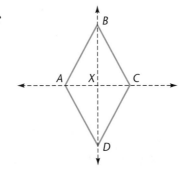

28. $\overline{BX}$ → $\overline{DX}$

29. $\overline{AD}$ → $\overline{AB}$

30. $\overline{AD}$ → $\overline{CB}$

31. ∠BCX → ∠BAX

32. ∠BCX → ∠DCX

33. ∠BCX → ∠DAX

34. ∠AXD → ∠CXD

Cultural Connection: Africa The designs below are taken from Egyptian bowls dating back to 3500 B.C.E. Describe the symmetries of each.

35.
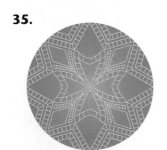

36.

37.

38.

 Algebra Use a graphics calculator or graph paper to draw each graph. Then write an equation for the axis of symmetry for each.

39. $y = (x - 1)^2 + 3$

40. $y = 2(x - 4)^2 + 3$

41. $y = -(x + 2)^2 + 3$

42. $y = -2(x + 5)^2 + 3$

43. $y = |x| + 3$

44. $y = |x + 3|$

Look Back

Use the diagram for Exercises 45–47.   [Lesson 1.1]

45. Name the intersection of $\overleftrightarrow{AB}$ and $\overleftrightarrow{MN}$.

46. Name three points that determine plane $\mathscr{T}$.

47. Name the intersection of planes $\mathscr{T}$ and $\mathscr{S}$.

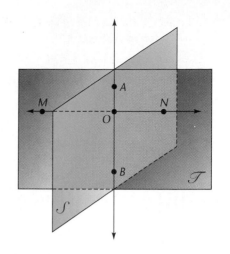

Find the missing measures in Exercises 48–50. [Lesson 1.5]

48. m∠ABD = 80°, m∠ABC = 30°, m∠CBD = ?

49. m∠ABC = 25°, m∠CBD = 60°, m∠ABD = ?

50. m∠ABC = m∠CBD, m∠ABD = 88°, m∠ABC = ?

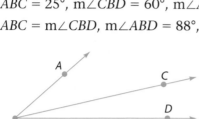

Look Beyond

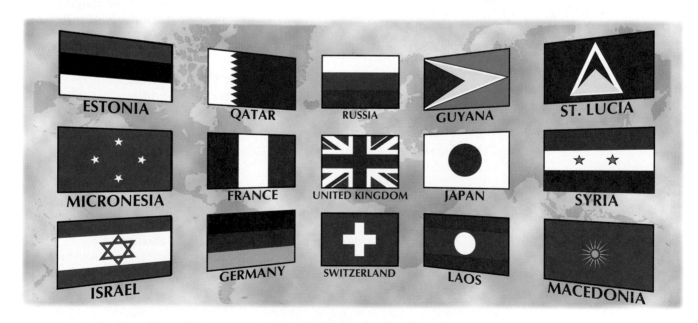

51. Geography   Name five countries whose flags have a horizontal axis of symmetry. Sketch each flag and identify the continent for each country.

52. Name five countries whose flags have a vertical axis of symmetry. Sketch each flag and identify the continent for each country.

53. Name five countries whose flags have rotation images of 180°. Sketch each flag and identify the continent for each country.

LESSON 3.2
Exploring

Properties of Quadrilaterals

why *Did you know that a baseball diamond is a special type of quadrilateral—namely, a rhombus? (It is also a square.) The length of each side is 90 feet, and the pitcher's mound lies on the diagonal from home plate to second base at a point about 6 feet closer to home plate than to second base.*

Any four-sided polygon is a quadrilateral. Quadrilaterals that meet certain conditions are called *special quadrilaterals*.

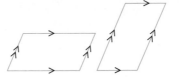

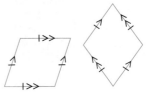

A **trapezoid** is a quadrilateral with one and only one pair of parallel sides.

A **parallelogram** is a quadrilateral with two pairs of parallel sides.

A **rhombus** is a parallelogram with four congruent sides.

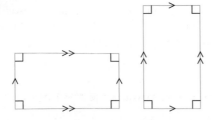

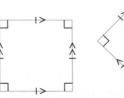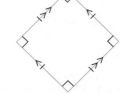

A **rectangle** is a parallelogram with four right angles.

A **square** is a parallelogram with four congruent sides and four right angles.

124   CHAPTER 3

CRITICAL *Thinking*

Which of the quadrilaterals on the previous page seem to have rotational symmetry? Which seem to have reflectional symmetry? What can you learn about quadrilaterals from the symmetry properties? Keep this in mind as you do the explorations that follow.

The opposite sides of a parallelogram are parallel, by definition. However, a parallelogram has many more properties. For example, it appears that the *opposite sides of a parallelogram are congruent.*

In the following explorations you will make conjectures about other properties of parallelograms. Record your conjectures in your notebook for future reference.

•Exploration 1 *Parallelograms*

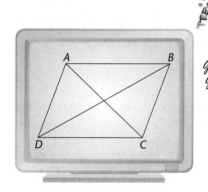

You will need
Geometry technology or
Ruler and protractor

Geometry Graphics

1 Draw a **parallelogram** that is not a rhombus, a rectangle, or a square. Measure the angles and the sides of the figure. What do you notice?

2 Draw diagonals connecting the opposite vertices. List any parts of your figure that appear to be congruent. You may want to measure these parts to verify that they are congruent. If you are using geometry technology, vary the shape of your figure by dragging one of the vertices.

3 What conjectures can you make about sides, angles, or triangles in the figure? (One was given just before this exploration. Try to find at least five conjectures in all.) ❖

•Exploration 2 *Rhombuses*

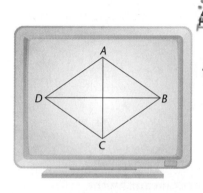

You will need
Geometry technology or
Ruler and protractor

Geometry Graphics

1 Draw a rhombus that is not a square. Draw diagonals connecting the opposite vertices. Make measurements as in the previous exploration.

2 Do the conjectures you made about parallelograms in Exploration 1 seem to be true of your rhombus? Discuss why they should or should not be true.

3 What new conjectures can you make about rhombuses that are not true of general parallelograms? (Try to find two new conjectures.) ❖

Exploration 3 *Rectangles*

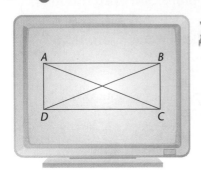

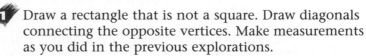

You will need

Geometry technology or
Ruler and protractor

Geometry Graphics

1 Draw a rectangle that is not a square. Draw diagonals connecting the opposite vertices. Make measurements as you did in the previous explorations.

2 Do the conjectures you made about parallelograms in Exploration 1 seem to be true for rectangles? Discuss why this should or should not be true.

3 What new conjecture can you make about rectangles that is not generally true of parallelograms? ❖

Exploration 4 *Squares*

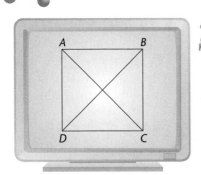

You will need

Geometry technology or
Ruler and protractor

Geometry Graphics

1 Draw a square with diagonals connecting the opposite vertices. Make measurements as you did in the previous explorations.

2 Which conjectures that you made in the previous explorations seem to be true of squares? Discuss. ❖

Summary

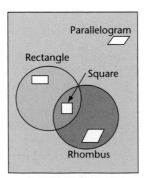

You may have noticed that there is a certain structure in the definitions and properties of quadrilaterals. This structure is revealed by the Venn diagram.

Is a square a rectangle? Is a square a rhombus? To answer these questions, use the Venn diagram of quadrilaterals. Do squares seem to have all the properties of rectangles? of rhombuses?

CRITICAL *Thinking*

Certain regions of the Venn diagram "inherit" properties from the larger regions in which they are located. Explain what is meant by the "inheritance" of properties.

EXERCISES & PROBLEMS

Communicate

1. What type of special quadrilateral is the "least specialized"? Explain your answer.

2. Why is every rhombus also a quadrilateral and a parallelogram?

3. Explain why every square is also a quadrilateral, a rectangle, a rhombus, and a parallelogram.

4. How many examples of quadrilaterals can you find in your classroom? List them.

Practice & Apply

Decide whether each statement is true or false, and explain your answer. Use the Venn diagram given on page 126.

5. If a figure is a parallelogram, then it cannot be a rectangle.

6. If a figure is a rhombus, then it is a parallelogram.

7. If a figure is not a parallelogram, then it cannot be a square.

8. If a figure is a square, then it is a rhombus.

9. If a figure is a rectangle, then it cannot be a rhombus.

10. If a figure is a rhombus, then it cannot be a rectangle.

PARL is a parallelogram whose diagonals meet at point *O*.
In Exercises 11–16, find the indicated measures given the following: m∠*PLR* = 70°, m∠*LAP* = 20°, *LO* = 6, *LR* = 10.
Give a justification for your answer using the parallelogram conjectures you made in this lesson.

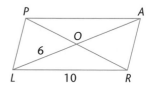

11. m∠*PAR*   12. m∠*LPA*   13. *AO*

14. *PA*   15. m∠*ALR*   16. m∠*PLO*

RECT is a rectangle, RHOM is a rhombus, SQUA is a square.
Copy each figure and fill in the missing angle measures.

17.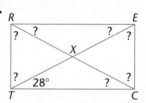

18.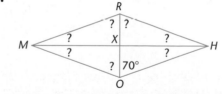

19.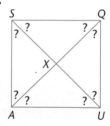

WXYZ is a rectangle. WX = 10 cm; WY = 12 cm.

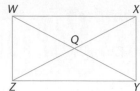

20. $ZY = ?$ **21.** $QY = ?$ **22.** $QZ = ?$

23. What type of triangle is $\triangle ZQY$?

24. Name two different angles that must be congruent to $\angle QZY$.

25. 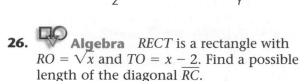 **Algebra** *DIAM* is a rhombus with $m\angle MIA = x$ and $m\angle ADI = 2x$. Find the measure of each angle of *DIAM*.

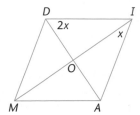

26. **Algebra** *RECT* is a rectangle with $RO = \sqrt{x}$ and $TO = x - 2$. Find a possible length of the diagonal $\overline{RC}$.

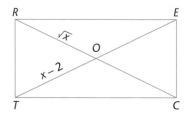

Look Back

Write each of the following statements as a conditional and then write the converse of the conditional. Is the original statement a definition? Explain your answer. **[Lesson 2.3]**

27. A hypsometer is a device used to measure atmospheric pressure and thus determine height above sea level.

28. An iris is a flower with sword-shaped leaves and three drooping sepals.

29. A whale is a mammal.

Which of the shapes in the objects below have rotational symmetry? reflectional symmetry? **[Lesson 3.1]**

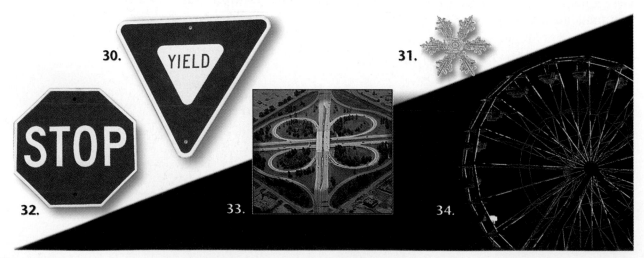

30. YIELD

31.

32.

33.

34.

Look Beyond

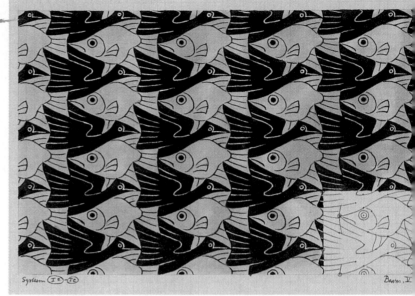

35. **Portfolio Activity**

One of Escher's tessellation patterns is known as a **translation tessellation** because each of the repeating figures can be seen as a translation of other figures in the design. You can make your own translation tessellation by following the steps below. You may have to keep adjusting your curves until you get something recognizable!

Draw your figures, as indicated below, on large-grid graph paper. The grid will help you copy the parts of your drawing. Or, if you like, you can use tracing paper. You can also do the steps using computer tessellation software.

a. Start with a square, rectangle, or other parallelogram. Replace one side of the parallelogram, such as $\overline{AB}$, by a broken line or curve.

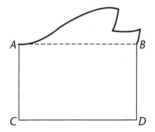

b. Translate the broken line or curve to the other side of the parallelogram. (If you are using a tracing of the curve, place it under your drawing and trace it again.)

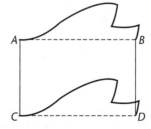

c. Repeat Steps A and B for the other two sides of your parallelogram.

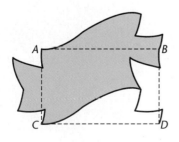

d. Your figure will now fit together with itself on all sides. Make repeated tracings of your figure in interlocking positions. You may want to add pictorial details.

Egg Over Alberta

By Paul Hoffman

The town leaders of Vegreville, Alberta, contacted Ronald Dale Resch, a computer science professor, for a special project—to build a 31-foot Easter egg.

The problem Resch faced was that no one other than a chicken had ever built a chicken egg. With no formal training in mathematics, Resch relied on his ability to play with geometric abstractions in his mind, then with his hands or a computer, to turn those abstractions into physical reality.

Resch assumed that someone had developed the mathematics of an ideal chicken egg. He soon found, however, that there was no formula for an ideal chicken egg.

After four months of contemplation and simulation, Resch realized that he could tile the egg with 2,208 equilateral triangles of identical size and three pointed stars (equilateral but non-regular hexagons) that varied slightly in width, depending on their position on the egg.

For six weeks, Resch led a team of volunteers in assembling the egg. Residents of Vegreville were afraid it might blow down. Long after the egg was finished, Resch used a computer to analyze the egg's structural integrity and found that it was ten times stronger than it needed to be.

Many Ukrainians still practice the ancient tradition of painting pysanki, eggs decorated with colorful geometric patterns.

Computer generated drawing of egg

Cutting the egg tiles from aluminum

Partially assembled egg

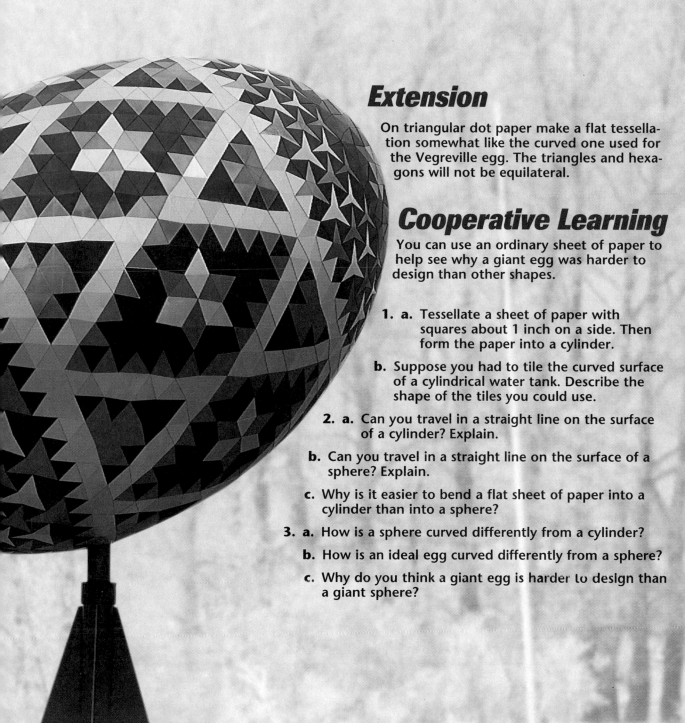

Extension

On triangular dot paper make a flat tessellation somewhat like the curved one used for the Vegreville egg. The triangles and hexagons will not be equilateral.

Cooperative Learning

You can use an ordinary sheet of paper to help see why a giant egg was harder to design than other shapes.

1. **a.** Tessellate a sheet of paper with squares about 1 inch on a side. Then form the paper into a cylinder.

   b. Suppose you had to tile the curved surface of a cylindrical water tank. Describe the shape of the tiles you could use.

2. **a.** Can you travel in a straight line on the surface of a cylinder? Explain.

   b. Can you travel in a straight line on the surface of a sphere? Explain.

   c. Why is it easier to bend a flat sheet of paper into a cylinder than into a sphere?

3. **a.** How is a sphere curved differently from a cylinder?

   b. How is an ideal egg curved differently from a sphere?

   c. Why do you think a giant egg is harder to design than a giant sphere?

Parallel Lines and Transversals

Why *For practical as well as artistic reasons, parallel lines are often a prominent feature in architecture and engineering. The striking appearance of parallel cables, which help support the suspended portion of a bridge, inspired Joseph Stella's painting of the Brooklyn Bridge.*

There's not much to go on.

Examine the parallel lines. What conjectures can you make? How many discoveries do you think you might make looking at just one pair of parallel lines?

Draw a line that intersects the parallel lines. This line is called a **transversal**. Now there are many discoveries to make.

The transversal changes the picture.

Transversals and Special Angles

TRANSVERSALS

A transverşal is a line, ray, or segment that intersects two or more coplanar lines, rays, or segments, each at a different point.   **3.3.1**

Notice that, according to the definition, the lines, rays, or segments that are cut by the transversal do not have to be parallel. This will be important in the next chapter.

Exploration · *Special Angle Relationships*

You will need
Geometry technology or
Lined paper and protractor

In each of the steps of this exploration, the terms interior and exterior will be used as shown.

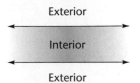

Exterior

Interior

Exterior

Start with two parallel lines. (If you are using lined paper, select two horizontal lines on the paper.) Then draw a third line that intersects both of them. Label each of the eight angles that are formed.

There are traditional names for certain special pairs of angles in the figure you drew. Measure each of the angles in the special pairs defined in the following steps:

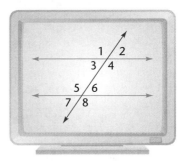

 1 Angles 3 and 6 are **alternate interior angles**, as are angles 4 and 5.

 2 Angles 1 and 8 are **alternate exterior angles**. Name another pair of alternate exterior angles.

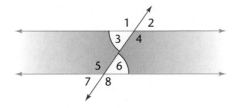

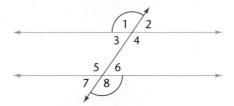

 3 Angles 3 and 5 are **consecutive interior angles**. Name another pair of consecutive interior angles.

 4 Angles 1 and 5 are **corresponding angles**. Name three other pairs of corresponding angles.

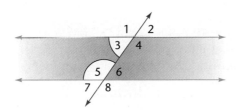

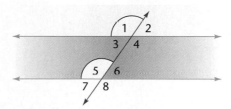

 5 Make a conjecture for each special pair of angles. ❖

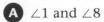
EXAMPLE

Indicate whether each is a pair of alternate interior, alternate exterior, consecutive interior, or corresponding angles:

A ∠1 and ∠8

B ∠7 and ∠3

C ∠5 and ∠4

D ∠3 and ∠5

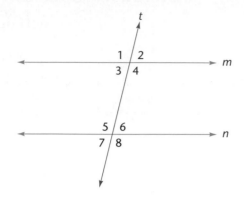

Solution ➤

A Alternate exterior angles

B Corresponding angles

C Alternate interior angles

D Consecutive interior angles ❖

Try This List three special pairs of angles not mentioned in the example.

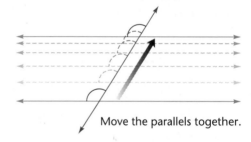

Move the parallels together.

In the exploration, one of the four conjectures you probably made is that corresponding angles are congruent.

Notice what happens if you slide (translate) one of the parallel lines closer to the other. Eventually, the highlighted corresponding angles will seem to come together as one. Do you think the corresponding angles will match exactly? Does this picture seem to support your conjecture? Explain.

One Postulate and Three Conjectures

We will not prove the Corresponding Angle Conjecture, but since it seems fully convincing, it will be a postulate in our system.

CORRESPONDING ANGLES POSTULATE
If two lines cut by a transversal are parallel, then corresponding angles are congruent.

3.3.2

There are three other conjectures that can be made in the exploration. Use the corresponding angles postulate as a model to state them formally.

Once the corresponding angles conjecture has been accepted as a postulate, it can be used to prove theorems. In particular, you can use it to prove the other three conjectures as theorems.

ALTERNATE INTERIOR ANGLES THEOREM

If two lines cut by a transversal are parallel, then alternate interior angles are congruent.   3.3.3

To prove the theorem, begin by drawing a diagram in which two parallel lines are cut by a transversal. Show that alternate interior angles, such as ∠1 and ∠2, are congruent.

Given: Line l ∥ line m
Line p is a transversal.

Prove: ∠1 ≅ ∠2

Plan: Study the diagram for ideas and plan your strategy.

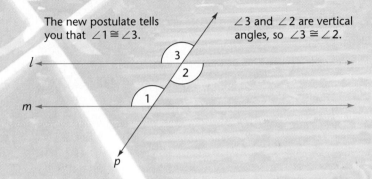

The new postulate tells you that ∠1 ≅ ∠3.

∠3 and ∠2 are vertical angles, so ∠3 ≅ ∠2.

Intersecting Railroad Tracks

Thus you have ∠1 ≅ ∠3 and ∠3 ≅ ∠2. So ∠1 ≅ ∠2 (the desired result).

What postulate or property allows you to draw the final conclusion? (Recall the properties of congruence.)

Now you can write out your proof. In proofs of this nature you may find it convenient to use a two-column format.

In writing proofs in geometry you may wish to develop abbreviations for use in stating reasons. For example, an abbreviation for the reason in Step 2 is

"∥s ⇒ Corr. ∠s ≅."

If you do use abbreviations, be sure you are consistent.

Proof

Statements	Reasons
1. Line l ∥ line m Line p is a transversal	Given
2. ∠1 ≅ ∠3	If parallel lines are cut by a transversal, then corresponding angles are congruent.
3. ∠3 ≅ ∠2	Vertical angles are congruent.
4. ∠1 ≅ ∠2	Transitive Property of Congruence.

EXERCISES & PROBLEMS

Communicate

1. What is a transversal?

2. Why do you think parallel lines are not part of the definition of a transversal?

3. Did you make any conjectures in the exploration that were not discussed later in the lesson? If so, what were they? Can you justify them?

4. Identify each transversal in the figure below, and identify the lines to which each is a transversal.

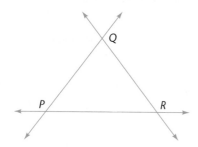

Can you identify transversals in the structure of the John Hancock building?

Practice & Apply

For Exercises 5–10, refer to the figure at right. In the figure, $p \parallel q$.

5. Line l is a transversal to lines ? .

6. Line k is a transversal to lines ? .

7. List all pairs of congruent alternate interior angles.

8. List all pairs of congruent alternate exterior angles.

9. List all pairs of supplementary consecutive interior angles.

10. List all pairs of congruent corresponding angles.

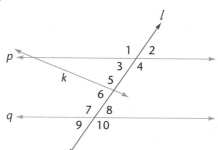

11. **Carpentry** In the drawing of the partial garage frame, the ceiling joist $\overline{PQ}$ and soleplate $\overline{RS}$ are parallel. How is the "let-in" corner brace $\overline{PT}$ related to $\overline{PQ}$ and $\overline{RS}$?

12. How are $\angle RTP$ and $\angle QPT$ related?

13. How is the corner brace $\overline{PT}$ related to the vertical studs it crosses? How are $\angle 1$ and $\angle 2$ related?

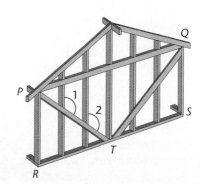

Navigation A periscope is an optical instrument used on submarines to view above the surface of the water. A periscope works by placing two parallel mirrors facing each other.

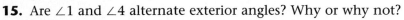

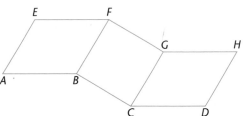

Periscope

14. Identify a transversal in the diagram. What does this line represent?

15. Are ∠1 and ∠4 alternate exterior angles? Why or why not?

16. Light rays reflect off the mirrors at congruent angles. For example, ∠3 ≅ ∠4. How can you prove that ∠1 ≅ ∠4?

17. If m ∠1 = 45°, find the measures of angles 2, 3, and 4.

Algebra Given: $l_1 \parallel l_2$, m∠2 = x, and m∠6 = 3x − 60. Find each of the angle measures.

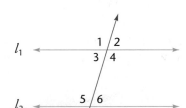

18. m∠1   19. m∠3

20. m∠4   21. m∠5

22. m∠7   23. m∠8

In Exercises 24–29, given △ABC with $\overline{DE} \parallel \overline{CB}$ and ∠ADE ≅ ∠AED, find the indicated angle measures.

24. m∠ADE   25. m∠AED   26. m∠DEB

27. m∠BDE   28. m∠CDB   29. m∠ABD

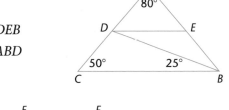

30. In the figure, AEFB and CGHD are parallelograms, and BFGC is a square. $\overline{AB}$ is parallel to $\overline{CD}$. If m∠A = 60°, proceed step by step to find m∠D.

31. Prove that if two lines cut by a transversal are parallel, then alternate exterior angles are congruent.

32. Prove that if two lines cut by a transversal are parallel, then consecutive interior angles are supplementary.

Look Back

33. Lines that intersect at right angles are said to be ? . [Lesson 1.2]

34. Describe what is meant when it is said that two lines are parallel. [Lesson 1.2]

35. In the figure, the measure of ∠2 is five times that of ∠1. Find m∠1 and m∠2. [Lesson 1.5]

36. Write this statement as a conditional: I have swimming practice on Tuesdays and Thursdays. [Lesson 2.2]

Look Beyond

Quadrilateral ABCD is a parallelogram.

37. Given: In quadrilateral ABCD, $\overrightarrow{CB} \parallel \overrightarrow{DA}$. Without drawing any additional lines, prove that the sum of the angles of the quadrilateral equals 360°.

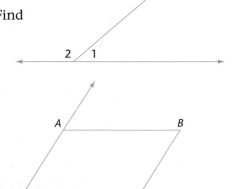

Proving Lines Parallel

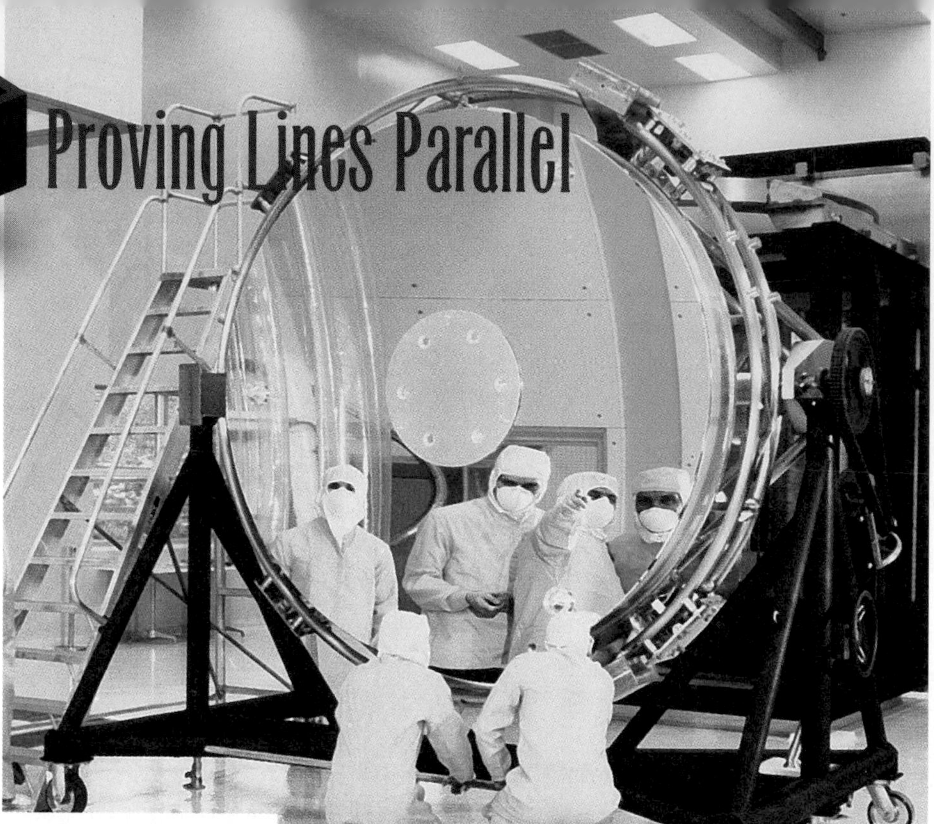

Suppose you needed to create a number of parallel lines, or you had to prove that certain lines were parallel. The converses of the transversal conjecture and theorems in Lesson 3.3 enable you to do these things.

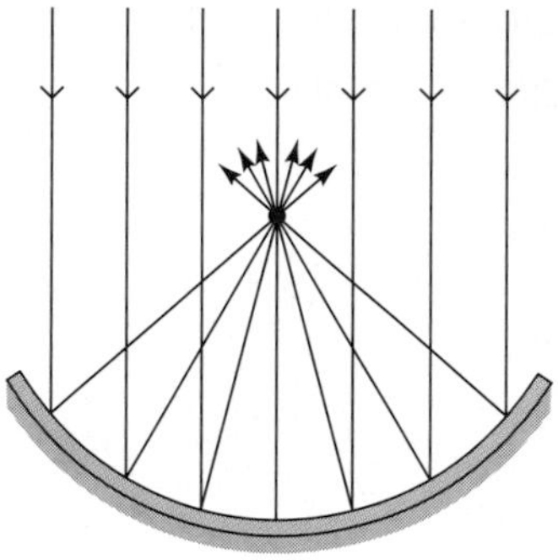

The parabolic mirrors of astronomical telescopes gather parallel light rays from distant stars and planets and direct them to a central point. In the photograph, technicians are inspecting the Hubble Space Telescope's mirror.

Forming the Converses

In Lesson 3.3 parallel lines were given in the hypotheses of the conjectures, and conclusions were made about certain special angles. What happens when special angles are given in the hypothesis, and conclusions are made about parallel lines? When a hypothesis and a conclusion are interchanged, the converse of the original statement is formed.

EXAMPLE 1

Write the converse of the Corresponding Angles Postulate.

Solution ➤

Identify the hypothesis and the conclusion of the Corresponding Angles Postulate. Then interchange the hypothesis and conclusion to form the converse.

Statement: If two lines cut by a transversal are parallel, then corresponding angles are congruent.

Converse: If corresponding angles are congruent, then two lines cut by a transversal are parallel. ❖

Another way of stating the converse is as follows:

THE CONVERSE OF THE CORRESPONDING ANGLES THEOREM
If two lines are cut by a transversal in such a way that corresponding angles are congruent, then the two lines are parallel.   **3.4.1**

Notice that the converse is labeled as a theorem. This is because you can prove it with the postulates and theorems you now know. However, the proof will be postponed because it involves a special form of reasoning.

Using the Converses

EXAMPLE 2

Suppose that m∠1 = 64° and m∠2 = 64° in the figure on the right. What can you conclude about lines *l* and *m*?

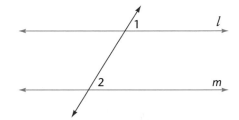

Solution ➤

∠1 and ∠2 are corresponding angles. The converse of the Corresponding Angles Postulate says that if such angles are congruent, then the lines are parallel. Conclusion: *l* ∥ *m*. ❖

EXAMPLE 3

Given a line *l* and a point *P* not on the line, draw a line through *P* parallel to *l*.

Solution ➤

Draw a line through point *P* and line *l*. Then measure ∠1.

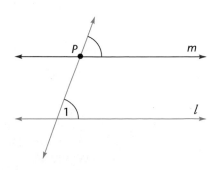

Using a protractor, draw a line *m* through *P* so that the new angle corresponds to ∠1 and has the same measure as ∠1. By the converse of the Corresponding Angles Theorem, the new line *m* is parallel to *l*. ❖

EXERCISES & PROBLEMS

Communicate

In Exercises 1 and 2, state the converse of each theorem.

1. If two parallel lines are cut by a transversal, then corresponding angles are congruent.

2. If two parallel lines are cut by a transversal, then consecutive interior angles are supplementary.

3. What can you conclude about two lines if a transversal forms corresponding angles that are *not* congruent? Explain your answer.

4. Complete the statement: If two lines that are not parallel are cut by a transversal, then alternate interior angles are __?__ .

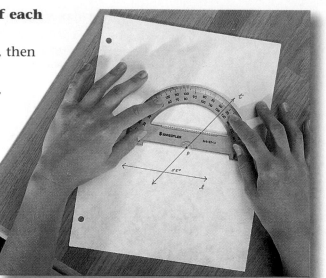

Practice & Apply

Copy the statements and supply the reasons in the proof of the theorem that follows. Refer to the diagram at right.

THEOREM
If two lines are cut by a transversal in such a way that consecutive interior angles are supplementary, then the two lines are parallel. **3.4.2**

Given: $\angle 1$ and $\angle 2$ are supplementary.

Prove: $l_1 \parallel l_2$

Proof:

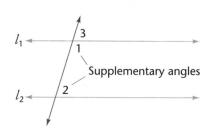

Supplementary angles

STATEMENTS	REASONS
$m\angle 1 + m\angle 2 = 180°$.	__(5)__
$\angle 1$ and $\angle 3$ are supplementary.	__(6)__
$m\angle 1 + m\angle 3 = 180°$	__(7)__
$m\angle 1 + m\angle 2 = m\angle 1 + m\angle 3$	__(8)__
$m\angle 2 = m\angle 3 \ (\angle 2 \cong \angle 3)$	__(9)__
$l_1 \parallel l_2$	__(10)__

Use the diagram for Exercises 11–14.

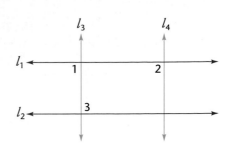

11. If m∠1 = m∠3, which pair of lines must be parallel? State a postulate or theorem to justify your conclusion.

12. If m∠1 = m∠2, which pair of lines must be parallel? State a postulate or theorem to justify your conclusion.

13. Redraw the given figure so that ∠1 ≅ ∠3 but ∠2 is not congruent to ∠1. Is $l_3 \parallel l_4$?

14. Redraw the given figure so that ∠1 ≅ ∠2 but ∠3 is not congruent to ∠1. Is $l_1 \parallel l_2$?

In Exercises 15–18, fill in the blanks. Use the figure at right.

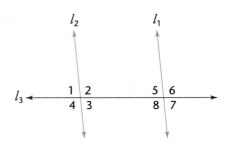

15. If ∠5 ≅ ∠3, then $l_1 \parallel l_2$ because _?_.

16. If ∠4 ≅ ∠8, then $l_1 \parallel l_2$ because _?_.

17. If ∠3 and ∠8 are supplementary, then $l_1 \parallel l_2$ because _?_.

18. If ∠5 ≅ ∠2 and $l_1 \parallel l_2$, then l_1 and l_2 are _?_ to l_3 because _?_.

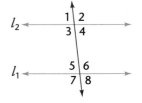

Algebra If m∠4 = 2x and m∠5 = x + 6, find the indicated angle measures in Exercises 19–26. l_1 is parallel to l_2.

19. m∠4   **20.** m∠5   **21.** m∠1   **22.** m∠2

23. m∠3   **24.** m∠6   **25.** m∠8   **26.** m∠7

In Exercises 27–36 you will complete proofs of two theorems you will use in future chapters.

THEOREM

If two lines are perpendicular to the same line, the two lines are parallel to each other.   **3.4.3**

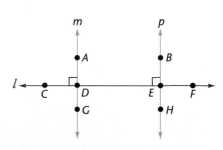

Given: lines l, m, and p with $m \perp l$ and $p \perp l$

Prove: $p \parallel m$

Proof:

Line l is a __27__ of m and p, by definition.
∠ADC ≅ ∠BED because perpendicular lines form __28__. Therefore, __29__ is parallel to __30__ by the converse of the __31__ Angles Postulate.

THEOREM

If two lines are parallel to the same line, then the two lines are parallel to each other.

3.4.4

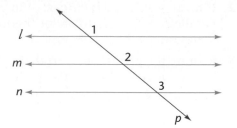

Given: lines *l*, *m*, and *n*, with transversal *p*; *l* ∥ *n*, and *m* ∥ *n*

Prove: *l* ∥ *m*

Proof:

Since *l* ∥ *n*, then ∠1 ≅ __32__ . Since *m* ∥ *n*, then ∠2 ≅ __33__ . So ∠1 ≅ __34__ by __35__ . Therefore, *l* ∥ *m* by __36__ .

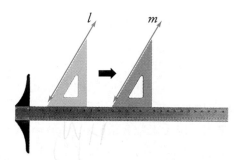

37. Drafting A T square and a triangle can be used to draw parallel lines. While holding the T square steady, slide the triangle along the edge of the T square. How can you prove that the resulting lines *l* and *m* are parallel?

Look Back

For Exercises 38–42, refer to the statement: Every rectangle is a parallelogram. [Lesson 2.2]

38. Rewrite the statement above as a conditional statement.

39. Identify the hypothesis and conclusion of the statement.

40. Draw a Venn diagram that illustrates the statement.

41. Write the converse of the statement and construct its Venn diagram.

42. Is the converse statement in Exercise 41 false? If so, illustrate this with a counterexample.

Look Beyond

Given: $l_1 \parallel l_2$; $\overrightarrow{PT}$ bisects ∠*RPQ*; $\overrightarrow{RT}$ bisects ∠*PRS*; m∠*RPQ* = 110°. **Find each of the following.**

43. m∠1

44. m∠2

45. m∠*PRS*

46. m∠3

47. m∠4

48. m∠5

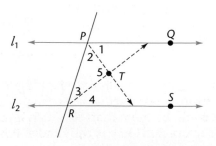

LESSON 3.5

Exploring The Triangle Sum Theorem

why *Euclid's geometry was founded on five postulates. One of them, the famous Parallel Postulate, has been thoroughly investigated by modern mathematicians. Entire systems of geometry depend on accepting or rejecting this postulate.*

The Parallel Postulate

The following postulate is a modern version of Euclid's Fifth Postulate.

THE PARALLEL POSTULATE

Given a line and a point not on the line, there is one and only one line that contains the given point and is parallel to the given line.　　**3.5.1**

In the previous lesson you constructed a line through a point parallel to a given line. You probably never questioned whether such a line actually existed, or whether there could be more than one such line. The assumption that there is exactly one such parallel line is the Parallel Postulate.

Lesson 3.5　Exploring the Triangle Sum Theorem　　**143**

The Parallel Postulate is used to prove a theorem about the sum of the measures of the angles of a triangle. Before you look at this proof, try the following exploration.

 **Exploration** *The Triangle Sum Theorem*

You will need
Scissors and paper

1 Cut out a triangle from the piece of paper.

2 Tear off two of the corners.

3 Position the two torn-off corners adjacent to the third angle as shown.

4 Make a conjecture about the sum of the measures of the angles of a triangle.

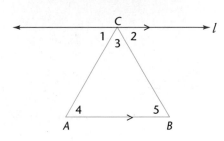

5 In the drawing on the right, line *l* has been drawn through a vertex of the triangle so that it is parallel to the opposite side. How does the Parallel Postulate relate to this figure?

6 Fill in the table for a figure like the one in the drawing. Use geometry theorems, not physical measurements.

m∠1	m∠2	m∠3	m∠4	m∠5	m∠3 + m∠4 + m∠5
40°	30°	?	?	?	?
20°	80°	?	?	?	?
30°	100°	?	?	?	?

7 Does the table support your conjecture? Explain. ❖

 **CRITICAL** *Thinking*

Why do the activities of the exploration fall short of proving this conjecture?

One conjecture from the exploration is stated below as a theorem.

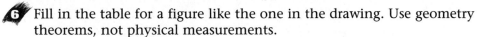

TRIANGLE SUM THEOREM
The sum of the measures of the angles of a triangle is 180°.   **3.5.2**

Proving the Triangle Sum Theorem

To prove the theorem, begin by drawing a line l through a vertex of $\triangle ABC$ so that it is parallel to the side opposite the vertex.

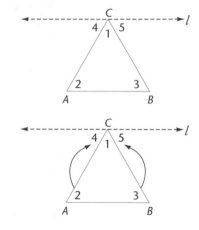

Given: $\triangle ABC$

Prove: $m\angle 1 + m\angle 2 + m\angle 3 = 180°$

Plan: Study the illustration, which is actually an informal proof of the theorem. You can use the ideas in it to write a more formal proof.

Proof:

STATEMENTS	REASONS
1. $m\angle 1 + m\angle 4 + m\angle 5 = 180°$	The angles fit together to form a straight line.
2. $l \parallel \overline{AB}$	As constructed (justification: the Parallel Postulate)
3. $\angle 2 \cong \angle 4$ $\angle 5 \cong \angle 3$	$\parallel$s $\Rightarrow$ alt int $\angle$s $\cong$
4. $m\angle 2 = m\angle 4$ $m\angle 3 = m\angle 5$	If $\angle$s are $\cong$, then their measures are $=$.
5. $m\angle 1 + m\angle 2 + m\angle 3 = 180°$	Substitution in Step 1.

Another Geometry

It is possible to create geometries in which the Parallel Postulate is not true. On the surface of a sphere, for example, where lines are defined differently from the way they are on flat surfaces, there are no parallel lines.

Geography On the surface of a sphere a "line" is defined as a **great circle**, which is a circle that lies on a plane that passes through the center of the sphere. The equator is a great circle on the surface of the earth. Lines of longitude, which run north and south, are also great circles. Notice that any two distinct "lines" (great circles) intersect in two points. Thus, there are no parallel lines on a sphere.

CRITICAL Thinking Discuss the following statement: "On the surface of a sphere, the shortest path between two points is not a straight line." What is the shortest path?

EXERCISES & PROBLEMS

Communicate

1. Explain how the paper-tearing exercise is like the proof of the Triangle Sum Theorem.

2. Explain why the paper-tearing exercise is not a proof of the Triangle Sum Theorem.

3. What role does the Parallel Postulate play in the proof of the Triangle Sum Theorem?

4. Why do you think airline pilots like to follow a great-circle route on long flights?

5. One alternative to the Parallel Postulate is that there is no line through a given point parallel to a given line. What are some others? On what kind of surface do you think this might be true?

Practice & Apply

In Exercises 6 and 7, find the missing angle measures.

6.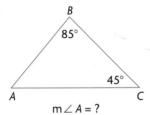

m∠A = ?

7.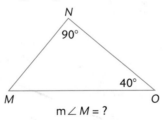

m∠M = ?

For Exercises 8 and 9, recall that an acute angle has a measure of less than 90° and an obtuse angle has a measure of greater than 90°.

8. Is it possible for a triangle to have more than one obtuse angle? Explain your reasoning.

9. If one of the angles of a triangle is a right angle, what must the other two angles be? Explain your reasoning.

Given: $\overline{DE} \parallel \overline{BC}$; $\overline{CF} \parallel \overline{BD}$; m∠ADE = 60°; m∠ACB = 50°. Find the following angle measures.

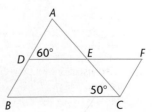

10. m∠A 11. m∠B 12. m∠EDB 13. m∠AED

14. m∠DEC 15. m∠FEC 16. m∠F 17. m∠ECF

For each of the triangles, find the indicated angle measures.

18.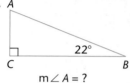

m∠ A = ?

19.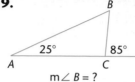

m∠ B = ?

20.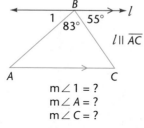

$l \parallel \overline{AC}$

m∠ 1 = ?
m∠ A = ?
m∠ C = ?

In Exercise 21 you will discover and prove an important geometry theorem. **Begin by copying and filling in the following table, which refers to the triangle to the right.**

21.

m∠1	m∠2	m∠3	m∠4	m∠1 + m∠2
?	50°	70°	?	?
60°	?	?	120°	?
?	30°	?	100°	?
?	45°	65°	?	?
31°	?	75°	?	?
—	—	x	?	?
—	—	?	x	?

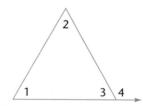

What do you notice about the values in the fourth and fifth columns of your table?

22. An angle like ∠4 is known as an **exterior angle** of a polygon. It is formed between one side of the polygon and the extended part of another side. How many exterior angles are possible at each vertex of a given polygon? Are their measures the same or different? Explain your answer.

23. In the triangle, angles ∠1 and ∠2 are known as the **remote interior angles** of ∠4, because they are the angles farthest from ∠4. State your observations from Exercise 21 as a theorem by completing the following statement.

EXTERIOR ANGLE THEOREM

The measure of an exterior angle of a triangle is equal to ___?___.   **3.5.3**

---

24. Explain how the last two lines of the table in Exercise 21 prove the theorem informally. Why do the first five lines fall short of proving the theorem?

The proof below is an incomplete formal proof of the theorem you stated in Exercise 23. **Complete the proof by filling in the missing reasons.**

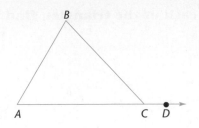

Given: △ABC with exterior angle ∠BCD and remote interior angles ∠A and ∠B

Prove: m∠BCD = m∠A + m∠B

Proof:

STATEMENTS	REASONS
△ABC with exterior angle ∠BCD	Given
m∠BCD + m∠BCA = 180°	__(25)__
m∠A + m∠B + m∠BCA = 180°	__(26)__
m∠BCD + m∠BCA = m∠A + m∠B + m∠BCA	__(27)__
m∠BCD = m∠A + m∠B	__(28)__

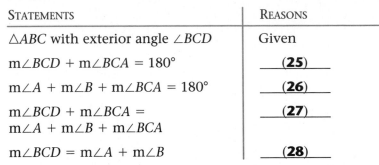

Look Back

29. The set of points that two geometric figures have in common is called their _?_. **[Lesson 1.1]**

30. _?_ points determine a line. **[Lesson 1.1]**

31. Adjacent supplementary angles form a _?_ pair. **[Lesson 1.5]**

Give reasons for the steps in the proof. **[Lesson 3.4]**

Given: $l_1 \parallel l_2$; ∠2 ≅ ∠3

Prove: $l_1 \parallel l_3$

Proof:

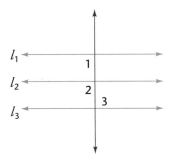

STATEMENTS	REASONS
$l_1 \parallel l_2$; ∠2 ≅ ∠3	__(32)__
∠1 ≅ ∠2	__(33)__
∠1 ≅ ∠3	__(34)__
$l_1 \parallel l_3$	__(35)__

Look Beyond

A quadrilateral has vertices P(2, 6), Q(2, 10), R(7, 10), and S(7, 6).

36. What kind of quadrilateral is PQRS? Explain your reasoning.

37. What is the perimeter of PQRS?

38. What is the area of PQRS?

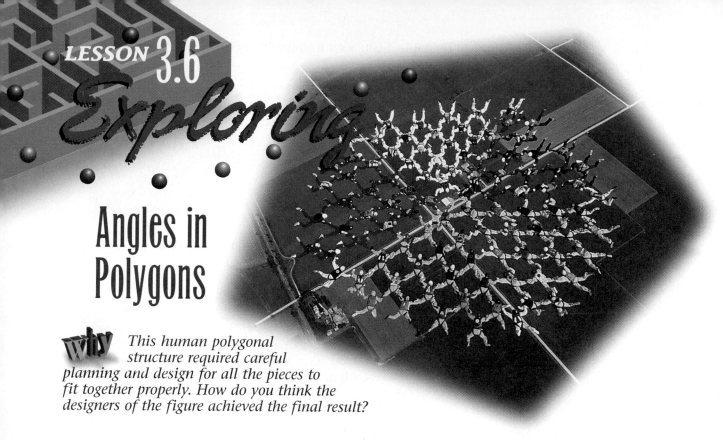

LESSON 3.6

Exploring

Angles in Polygons

Why *This human polygonal structure required careful planning and design for all the pieces to fit together properly. How do you think the designers of the figure achieved the final result?*

A **convex polygon** is one in which any line segment connecting two points of the polygon has no part outside the polygon.
For a **concave polygon** this is not true. In this book the word "polygon" will mean a convex polygon, unless stated otherwise.

convex polygon   concave polygon

•Exploration 1 *Angle Sums in Polygons*

You will need
Calculator (optional)

Pentagon *ABCDE* has been divided into three triangular regions by drawing all possible diagonals from one vertex.

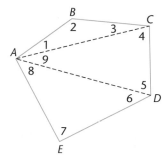

1 Find each of the following:

m∠1 + m∠2 + m∠3 = ?
m∠4 + m∠5 + m∠9 = ?
m∠6 + m∠7 + m∠8 = ?

2 Add the three expressions together.
(m∠1 + m∠2 + m∠3 + m∠4 + . . . + m∠8 = ?)

3 Use the diagram and the result from Step 2 to determine the sum of the angles of a pentagon. (m∠*EAB* + m∠*B* + m∠*BCD* + m∠*CDE* + m∠*E*)

Complete the table. Sketch each polygon and draw all possible diagonals from one vertex.

Polygon	Number of Sides	Number of Triangular Regions	Sum of Measures of Angles
Triangle	?	1	180°
Quadrilateral	?	?	?
Pentagon	?	3	540°
Hexagon	?	?	?
n-gon	?	?	?

5 Write a formula for the sum of the angles of a polygon in terms of the number of sides, *n*. ❖

EXTENSION

A **regular** polygon is one that has the measures of all its sides and angles equal. For instance, a square is a regular polygon. So is an equilateral triangle. In a square, each angle has a measure of 90°. In an equilateral triangle, each angle has a measure of 60°.

Fill in the chart below. Then write a formula for finding the measure of an angle of a regular *n*-gon.

Regular Polygon	Number of Sides	Sum of Measures of Angles	Measure of One Angle
Triangle	?	180°	?
Quadrilateral	?	?	90°
Pentagon	?	?	?
Hexagon	?	?	?
n-gon	?	?	?

The squares fit together at a point.

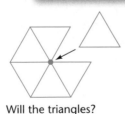

Will the triangles?

Some regular polygons fit together at a central point with no overlapping and no gaps. For example, notice how the squares fit together. Will the equilateral triangles fit together in the same way? Will regular pentagons? Will other *n*-gons?

For a regular *n*-gon to form a pattern like the one described above, the measure of one of its angles must be a factor of 360°. How many of the regular *n*-gons will form such a pattern? Make a list of them for future reference. ❖

CRITICAL *Thinking*

What kinds of symmetries are present in patterns of polygons that fit together and share a single point?

Exploration 2 Exterior Angle Sums in Polygons

You will need
Geometry technology or
Protractor, scissors

Geometry Software

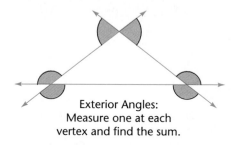

Exterior Angles:
Measure one at each
vertex and find the sum.

Interior
Angle

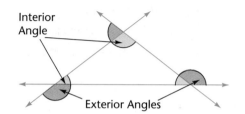

Exterior Angles

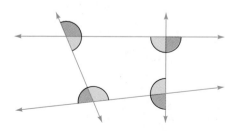

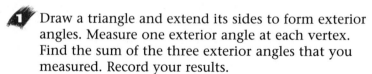

1 Draw a triangle and extend its sides to form exterior angles. Measure one exterior angle at each vertex. Find the sum of the three exterior angles that you measured. Record your results.

2 Cut out the exterior angles you measured and fit them together. Record your results.

3 Repeat Steps 1 and 2 for a quadrilateral.

4 Repeat Steps 1 and 2 for a pentagon.

5 Make a conjecture about the sum of the exterior angles of a polygon.

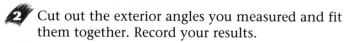

6 What is the sum of all the interior and exterior angles marked in the triangle?

7 What is the sum of all the interior and exterior angles marked in the quadrilateral?

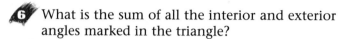

8 Write a formula for the sum of the interior and exterior angles of an *n*-gon as in Steps 6 and 7.

9 Complete the table.

Polygon	No. Sides	Sum (Ext. + Int. Angles)	Sum Interior Angles	Sum Exterior
Triangle	3	540°	180°	360°
Quadrilateral	?	?	?	?
Pentagon	?	?	?	?
Hexagon	?	?	?	?
n-gon	?	?	?	?

10 Use the formula from Exploration 1 and from Step 8 above to write an expression for the sum of the exterior angles of an *n*-gon. Use algebra to simplify the expression. ❖

EXERCISES & PROBLEMS

Communicate

1. What is the formula that gives the sum of the interior angle measures of an *n*-gon?

2. Describe an exterior angle of a polygon.

3. Is it possible to draw a quadrilateral such that three of its angles each have a measure of 60°? Give a reason for your answer.

4. Explain why a pentagon can have 5 obtuse angles.

Regular hexagons are the most efficient shape for bees to use when constructing a hive.

Practice & Apply

5. In the figure, find the values of *x*, *y*, and *z*.

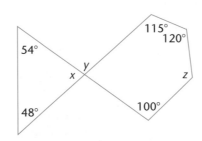

54°

115°
120°

y

x

z

48°

100°

Find the missing angle measures in Exercises 6–11.

6.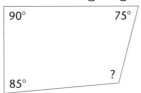

90°   75°

85°   ?

7.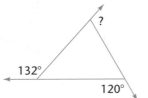

?

132°

120°

8.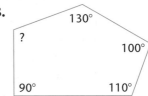

130°

?

100°

90°   110°

9.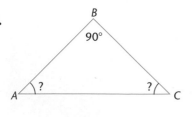

B

90°

A   ?   ?   C

10.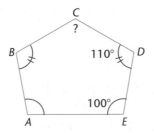

C

?

B   110°   D

100°

A   E

11.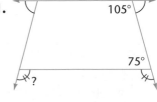

105°

75°

?

12. Make a sketch and determine the measure of an interior angle for each of the following: a regular quadrilateral, a regular pentagon, and a regular hexagon.

13. Determine the measure of an exterior angle for each of the figures in Exercise 12.

In Exercises 14 and 15, use your results from Exercises 12 and 13.

14. Write a formula for finding the measure of one interior angle of a regular *n*-gon.

15. Write a formula for finding the measure of one exterior angle of a regular *n*-gon.

For each regular polygon in Exercises 16–18, an interior angle measure is given. **Find the number of sides of the polygon. Then determine the sum of the measures of the exterior angles of each polygon.**

16. 135°   **17.** 150°   **18.** 165°

For each regular polygon in Exercises 19–21, an exterior angle measure is given. **Find the number of sides of each polygon and the sum of the measures of its interior angles.**

19. 60°   **20.** 36°   **21.** 24°

Algebra   Find each angle measure of quadrilateral *QUAD*.

22. ∠Q   **23.** ∠U

24. ∠A   **25.** ∠D

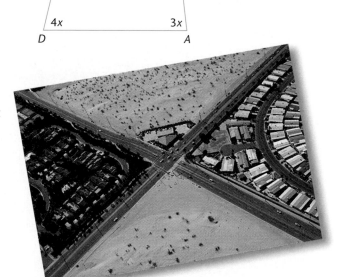

Look Back

26. Describe how the distance from a point to a line is determined. **[Lesson 1.4]**

27. In the photo, how many pairs of congruent angles are formed? What are these angle pairs called? **[Lesson 2.2]**

28. What is the meaning of the Transitive Property of Congruence? **[Lesson 2.5]**

29. Explain the meaning of the Parallel Postulate. **[Lesson 3.5]**

30. The measure of an exterior angle of a triangle is equal to the _?_ of the remote interior angles. **[Lesson 3.6]**

Look Beyond

31. Find the sum of the measures of the vertex angles of a 5-pointed star without using a protractor. Use the following hints.

- Recall the following theorem, which you proved in Lesson 3.5:

    The measure of an exterior angle of a triangle is equal to the sum of the measures of its remote interior angles.

- Using small triangles within the larger drawing, explain why two of the angles are labeled as 1 + 4 and 3 + 5.

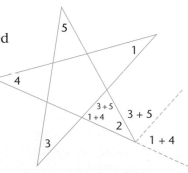

Exploring Midsegments of Triangles and Trapezoids

 Why *Can you think of a way to determine the width of the pyramid halfway up the stairs without actually measuring it at that point? The midsegment conjectures you make in this lesson will suggest a way.*

The Temple of the Giant Jaguar in Guatemala has four sloping trapezoidal sides. Along one side of the temple there is a stairway that rises 150 feet.

Exploration 1 Midsegments of Triangles

Geometry Graphics

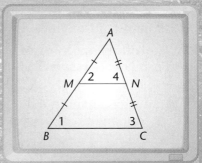

You will need
Geometry technology or
Ruler and protractor

1 Draw △*ABC*, which may be any shape and size. Find the midpoints of sides $\overline{AB}$ and $\overline{AC}$. Label them *M* and *N* as shown. Then draw the midsegment $\overline{MN}$.

2 Measure segments $\overline{MN}$ and $\overline{BC}$. What is the relationship between their lengths? Compare your results with those of your classmates.

3 Measure ∠1 and ∠2. What do you notice? Also measure ∠3 and ∠4. What do these measurements suggest about the relationship of $\overline{BC}$ and $\overline{MN}$? What postulate or theorem from an earlier lesson allows you to draw your conclusion?

4 Use your results from Steps 2 and 3 to write a conjecture about midsegments of triangles. ❖

José is on his school swim team. During the summer he enjoys training on a small lake near his house. To evaluate his progress he needs to know the distance across the lake as shown. How can he use the conjecture from Exploration 1 to find this distance?

José can select a point A where he can measure segments $\overline{AX}$ and $\overline{AY}$. Then he can find the midpoints of $\overline{AX}$ and $\overline{AY}$ and measure the distance between them. Since this distance is one-half XY, José can double the number to find his answer. ❖

•Exploration 2 Midsegments of Trapezoids

Geometry
Graphics

You will need
Geometry technology or
Ruler and protractor

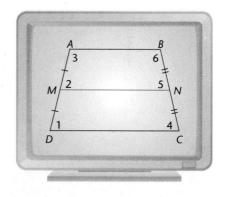

1 Draw trapezoid *ABCD*, which may be of any shape and size. Find the midpoints *M* and *N* of the nonparallel sides, and connect them. Draw the midsegment $\overline{MN}$.

2 Measure the lengths of $\overline{AB}$, $\overline{DC}$, and $\overline{MN}$.

3 Find the average length of the bases. Compare your results with those of your classmates. What do you notice about the length of the midsegment?

4 Measure $\angle 1$ and $\angle 2$. What do you notice? Also measure $\angle 4$ and $\angle 5$. What do these measurements suggest about the relationship of $\overline{CD}$ and $\overline{MN}$?

5 Measure $\angle 2$ and $\angle 3$. What do you notice? Also measure $\angle 5$ and $\angle 6$. What do these measurements suggest about the relationship of $\overline{AB}$ and $\overline{MN}$?

6 Use your results from Steps 2–5 to write a conjecture about midsegments of trapezoids. ❖

Archaeology The base of the Temple of the Giant Jaguar pyramid is a square 150 feet on a side. The top is a square 40 feet on a side. What is the width of the pyramid at a point midway between the base and the top?

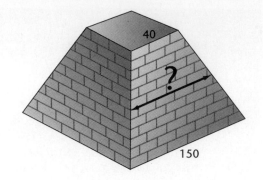

Solution ➤

Use the trapezoid midsegment conjecture:

$$\text{Length of midsegment} = \tfrac{1}{2} (\text{base 1} + \text{base 2})$$
$$= \tfrac{1}{2} (40 + 150)$$
$$= \tfrac{1}{2} (190)$$
$$= 95 \text{ ft} \; ❖$$

Exploration 3 *Making the Connection*

You will need

Geometry technology or
Graphics calculator
Ruler

Geometry Graphics

1 Draw trapezoid *ABCD*, which may be of any size and shape, with midsegment $\overline{MN}$. Then fill in a table like the one below by gradually reducing the length of one of the bases. (Use your own measurements for *AB* and *DC*.)

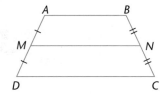

DC	AB	MN
6	5	?
6	4	?
6	3	?
6	2	?
6	1	?
6	.5	?
6	.1	?

You can use the table feature of a graphics calculator to find the values of MN as AB approaches zero. Enter the function as $y = (6 + x) \div 2$.

2 As *AB* shrinks to 0, what type of figure does the trapezoid become?

3 Write a formula for the length of the midsegment of a "trapezoid" with the measure of one base equal to 0. How does your formula relate to the Triangle Midsegment Conjecture?

EXERCISES & PROBLEMS

Communicate

1. How could you use the Trapezoid Midsegment Conjecture to find the length of the midsegment of a triangle?

2. Why is a triangle a "limiting case" of a trapezoid?

3. In the application on page 155, the triangle midsegment conjecture was used to find a distance that could not be measured directly. Describe a situation where the Trapezoid Midsegment Conjecture would be needed instead?

4. Do you think the midsegment conjectures would be useful in measuring a vertical distance, such as the height of a tree or a cliff? Explain your answer.

5. Suppose a student gave the following rule for finding the length of the midsegment of a trapezoid: "First you subtract the shorter base from the longer base. Then you take half the difference and add it to the shorter base." Would this method work? Discuss.

Practice & Apply

Find the indicated measures.

6. $AB = ?$

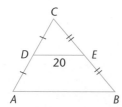

7. $DE = ?$

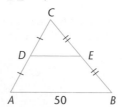

8. $EF = ?$

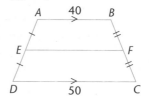

9. $AB = ?$

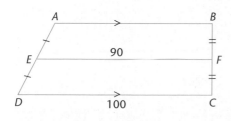

10. $HI = ?$
$FG = ?$
$DE = ?$

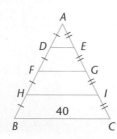

11. $EF = ?$
$GH = ?$
$IJ = ?$

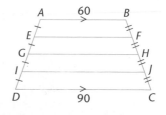

12. What pattern do you observe in the lengths of the segments in Exercise 10? Describe it.

13. What pattern do you observe in the lengths of the segments in Exercise 11? Describe it.

14. How would you find the lengths of the parallel segments of a triangle if its sides were divided into 2 congruent parts? 4 congruent parts? 16 congruent parts? *n* congruent parts? Is the rule the same for all?

15. Repeat Exercise 14 for trapezoids, instead of triangles.

16. The sides of △*AFG* are each divided into three congruent parts, with the points connected as shown. Write a conjecture about the connecting segments.

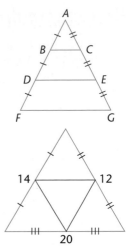

17. The figure at the right is formed by connecting the midpoints of all three sides. Find the perimeters of the inner and outer triangles. What is their relationship?

In Exercises 18–21 write an informal argument that explains why each special quadrilateral is formed.

18. Draw a scalene triangle. Connect the midpoints of two of the sides to the midpoint of the third side. What special quadrilateral is formed inside the triangle? Explain why.

19. Draw a right triangle that is not isosceles. Connect the midpoints of the two legs to the midpoint of the hypotenuse. What special quadrilateral is formed inside the triangle? Explain why.

20. Draw an isosceles triangle. Connect the midpoints of the two equal sides to the midpoint of the base. What special quadrilateral is formed inside the triangle? Explain why.

21. Draw an isosceles right triangle. Connect the midpoints of the legs to the midpoints of the hypotenuse. What special quadrilateral is formed inside the triangle? Explain why.

22. Structural Engineering Is $\overline{FC}$ the midsegment of trapezoid *ABDE*? Explain your answer. The overall trapezoidal shape of the tower gives it a broad, stable base. How are the trapezoids strengthened?

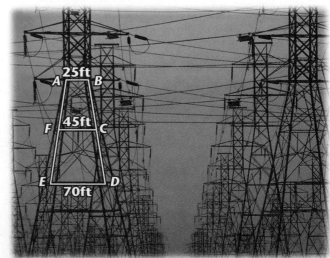

Flood Control Exercises 23–26 refer to the cross section of a dam shown that has the shape of a trapezoid. Determine the following:

23. The length of the cross section halfway up

24. The length of the cross section three-quarters of the way up

25. If the floodgates should be opened when the distance of the waterline across the dam is 400 meters or more, should they be opened before the water reaches the three-quarter level?

26. Is it possible for the median of a trapezoid to be congruent to its bases? Explain your answer.

The cross section is 456 m long across the top and 304 m long across the bottom.

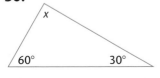 **Look Back**

Answer true or false for Exercises 27–29.   [Lesson 3.2]

27. All rhombuses are squares.

28. All rectangles are parallelograms.

29. All squares are rectangles.

In Exercises 30–32, find the missing angle measures. [Lesson 3.5]

30.

31.

32.

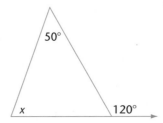

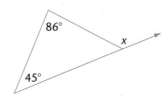

Look Beyond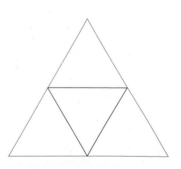

33. Draw a triangle and connect the midpoints of the sides. Cut out each of the four triangles that are formed, and formulate a conjecture about them.

34. Use the Triangle Midsegment Conjecture as a basis for an informal argument that each triangle has sides of the same length. (Hint: It may be helpful to write the measures of the inner and outer triangle sides directly onto the figure you have drawn.)

Analyzing Polygons Using Coordinates

Mathematics also has a method of indicating steepness. This method, though not as colorful as the names of roller coasters, is precise.

Amusement parks often have descriptive names for their roller coasters. Names such as "Shocker" or "Wild Thing" give riders an idea about the steepness of the falls they will encounter.

The **slope** of a line or surface tells you how it rises or falls. For example, a roller coaster hill with a slope of 4 is twice as steep as one with a slope of 2. If a slope of 1 describes a 45-degree incline, do you think a hill with a slope of 4 would be very steep?

Slope: A Measure of Steepness

Coordinate geometry is ideal for studying slope. You can define the slope of any line segment on a coordinate plane. The slope is the ratio of the vertical **rise** of the segment to the horizontal **run**.

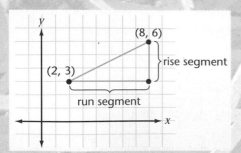

EXAMPLE 1

Find the slope of the segment with endpoints at (2, 3) and (8, 6).

Solution ➤

Draw a right triangle as shown. By counting squares, you can see that the rise is 3 and the run is 6. Thus, the slope is $\frac{3}{6} = \frac{1}{2}$, or 0.5. ❖

How could you determine the slope of the figure by using just the coordinates—that is, without drawing a picture? The answer is found in the following definition.

SLOPE FORMULA

The slope of a segment $\overline{PQ}$ with endpoints $P(x_1, y_1)$ and $Q(x_2, y_2)$ is the following ratio:

$$\frac{y_2 - y_1}{x_2 - x_1}$$

3.8.1

In the case of a vertical segment, the slope is undefined. Explain why.

EXAMPLE 2

Find the slope of segment $\overline{AB}$ with endpoints $A(-2, 3)$ and $B(-5, -3)$.

Solution ➤

$$\text{Slope} = \frac{y_2 - y_1}{x_2 - x_1} = \frac{-3 - 3}{-5 - (-2)} = \frac{-6}{-3} = 2 \; ❖$$

Parallel and Perpendicular Lines

ALGEBRA
Connection

Recall from algebra that an equation in the form $y = mx + b$ has m as its slope. Create a system of equations consisting of two lines with the same slope and try to find the solution to the system. What happens? Such equations are called **inconsistent**.

In working with a quadrilateral in a coordinate plane, it is often useful to characterize its sides as parallel or perpendicular. The following four theorems will help you characterize the sides of a quadrilateral in a coordinate plane. The proofs involve using algebraic systems of equations and are not given here.

EQUAL SLOPE THEOREM

If two nonvertical lines are parallel, then they have the same slope.

3.8.2

THE CONVERSE OF THE EQUAL SLOPE THEOREM

If two nonvertical lines have the same slope, then they are parallel. **3.8.3**

The slopes of perpendicular lines also have a special relationship to each other. This relationship is stated in the following two theorems.

SLOPES OF PERPENDICULAR LINES
If two nonvertical lines are perpendicular, then the product of their slopes is −1. **3.8.4**

THE CONVERSE OF SLOPES OF PERPENDICULAR LINES
If the product of the slopes of two nonvertical lines is −1, then the lines are perpendicular. **3.8.5**

The fact that the product of the slopes of perpendicular lines equals −1 reveals another relationship between the slopes. For the product of two numbers to equal −1, one number must be the negative reciprocal of the other. That is, if the slope of a line is $\frac{a}{b}$, then the slope of any line perpendicular to that line has slope $\frac{-b}{a}$. You can use this relationship to verify perpendicularity without multiplying slopes.

EXAMPLE 3

Plot quadrilateral *QUAD* with *Q*(1, 4), *U*(7, 8), *A*(9, 5), and *D*(3, 1). What type of quadrilateral is *QUAD*?

Solution ➤

Based on the figure, it appears that *QUAD* is a parallelogram, and perhaps a rectangle. You can test these conjectures by finding the slopes of each segment.

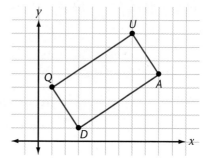

$$\text{Slope of } \overline{QU} = \frac{8 - 4}{7 - 1} = \frac{4}{6} = \frac{2}{3} \qquad \text{Slope of } \overline{DA} = \frac{5 - 1}{9 - 3} = \frac{4}{6} = \frac{2}{3}$$

$$\text{Slope of } \overline{DQ} = \frac{4 - 1}{1 - 3} = -\frac{3}{2} \qquad \text{Slope of } \overline{UA} = \frac{5 - 8}{9 - 7} = -\frac{3}{2} \; ❖$$

Since the opposite sides of the quadrilateral have the same slope, the figure is a parallelogram.

Since the slopes of adjacent sides of the quadrilateral are negative reciprocals, the figure is a rectangle.

EXERCISES & PROBLEMS

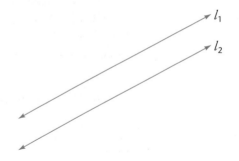

Communicate

Lines l_1 and l_2 with $l_1 \parallel l_2$ as shown.

1. Suppose l_1 has a slope of 2. What is the slope of l_2? Explain why.

Given: Points $A(2, 5)$ and $B(8, 15)$

2. Draw $\overline{AB}$ on graph paper and show by a diagram how to find the slope by counting squares.

3. Explain how the definition of slope allows you to compute the slope of $\overline{AB}$ without using a diagram.

Given: Lines p_1 and p_2 with $p_1 \perp p_2$ as shown.

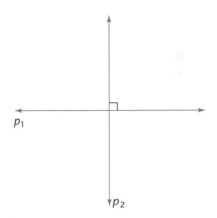

4. Suppose p_1 has slope m. What is the slope of p_2? Explain why.

5. Explain how you can use the slope to show that a quadrilateral is a rectangle.

Practice & Apply

In Exercises 6–8 use the points shown.

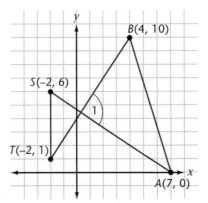

6. Find the slope of $\overline{SA}$.

7. Find the slope of $\overline{BT}$.

8. Is $\angle 1$ a right angle? Explain your answer.

For Exercises 9–12 plot the points on graph paper and connect them to form a polygon. Label the sides of the figure with their slopes. Then identify the type of polygon on the basis of the special properties it has and explain your reasoning. Write and then carry out a plan that proves that your identification is correct.

9. $C(0, -2)$, $R(5, -2)$, $A(5, 6)$, $B(0, 6)$

10. $S(0, 0)$, $A(2, 3)$, $I(5, 3)$, $N(7, 0)$

11. $R(0, 0)$, $A(-3, 4)$, $I(1, 7)$, $N(4, 3)$

12. $T(0, 6)$, $R(3, 9)$ $A(9, 3)$, $P(6, 0)$

Lesson 3.8 Analyzing Polygons Using Coordinates **163**

13. Draw a square on graph paper with the coordinates $A(-2, 1)$, $B(1, 5)$, $C(5, 2)$, and $D(2, -2)$. Use the coordinates to show that the diagonals of the square are perpendicular to each other. Explain your reasoning.

For Exercises 14–16 draw the indicated figure, using the given segment as one of its sides. Show the coordinates of the vertices of the figure and label the sides of the figure with their slopes.

14. Trapezoid *ABCD*. Endpoints of $\overline{AB}$ = $A(3, 5)$, $B(8, 5)$

15. Parallelogram *PLOG*. Endpoints of $\overline{PL}$ = $P(3, 2)$, $L(-1, 5)$

16. Rectangle *RCTG*. Endpoints of $\overline{RC}$ = $R(1, 6)$, $C(3, 2)$

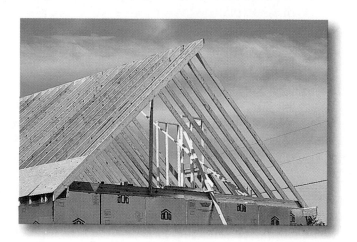

17. Construction   A house is 25.0 feet wide. The roof forms an isosceles triangle with the side walls. The height of the rectangular side wall is 10.5 feet, and the height of the peak of the roof is 23.0 feet. City building codes require that the pitch (slope) of a roof may be no more than 0.4. Use coordinates to determine whether the house is violating the building codes.

18. 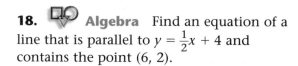 **Algebra**   Find an equation of a line that is parallel to $y = \frac{1}{2}x + 4$ and contains the point $(6, 2)$.

19. **Algebra**   Find an equation of a line that is perpendicular to $y = 3x - 1$ and contains the point $(6, 5)$.

Look Back

Given $l_1 \parallel l_2$ and $\triangle ABC$ as shown, find the indicated angle measures.   [Lesson 3.3]

20. $m\angle 1$ **21.** $m\angle 2$

22. $m\angle EBA$ **23.** $m\angle FBC$

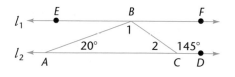

24. Explain the difference between a postulate and a theorem.   **[Lesson 2.2]**

25. Upon what postulate does the proof of the Triangle Sum Theorem depend?   **[Lesson 3.5]**

Given trapezoid *TRAP* shown, find the indicated angle measures.   [Lesson 3.5]

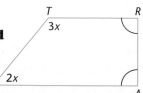

26. $m\angle R$ **27.** $m\angle T$

28. $m\angle P$ **29.** $m\angle A$

164   CHAPTER 3
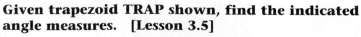

Look Beyond

30. 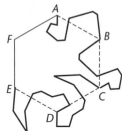 **Portfolio Activity**
Another tessellation pattern used by
Escher is known as a **rotation
tessellation**. Study the "reptile"
tessellation and see how many different
centers of rotation you can find.

You can make your own rotation
tessellation by following the steps below.
As in Lesson 3.2, use tracing paper, graph
paper, or computer software—and keep
trying until you get a figure you like.

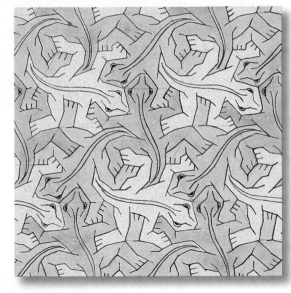

a. Start with a regular hexagon.
Replace one side of the hexagon, such
as $\overline{AB}$, by a broken line or a curve.
Rotate the curve about B so that A
moves to point C. Draw the broken line
or curve between B and C.

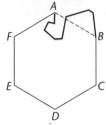
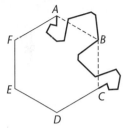

b. Draw a new curve at side $\overline{CD}$ and rotate
it around point D to side $\overline{DE}$.

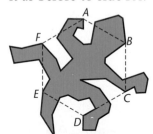

c. Draw a new curve at side $\overline{EF}$ and rotate
it as before to side $\overline{FA}$.

d. Your figure will now fit together with
itself on all sides. Make repeated
tracings of your figure in interlocking
positions. Add pictorial details to
make your figure more interesting.

String Figures

The following sequence and the resulting net have appeared in many parts of the world under different names: "Osage Diamonds" among the Osage Indians of North America, "The Calabash Net" in Africa, and "The Quebec Bridge" in Canada. In the United States it is commonly known as "Jacob's Ladder."

1. Start with a piece of string 4 to 5 feet in length. Tie the ends together. Loop the string around your thumb and little fingers as shown.

2. Use your right index finger to pick up the left palm string from below. In a similar way, use your left index finger to pick up the right palm string.

3. Let your thumbs drop their loop. Turn your hands so that the fingers face out. With your thumbs, reach under all the strings and pull the farthest string back toward you.

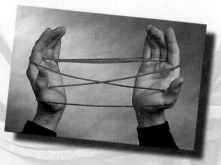

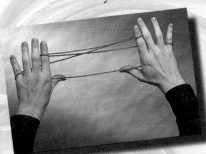

4. With your thumbs, go over the near index-finger string, reach under and get the far index-finger string, and return.

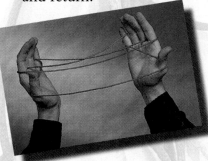

5. Drop the loops from your little fingers. Pass your little fingers over the index-finger string and get the thumb string closest to the little finger from below.

6. Drop the thumb loops. Pass your thumbs over the index-finger strings, get the near little-finger strings from below, and return.

7. Loosen the left index-finger loop with your right hand and place the loop over your thumb. Do the same with the right index-finger loop.

8. Each thumb now has two loops. Using your right hand, lift the lower loop of the left thumb up and over the thumb. Do the same with the lower loop of the right thumb.

9. Bend your index fingers and insert the tips into the triangles that are near the thumbs.

Gently take your little fingers out of their loops. Turn your hands so that your palms face away. The index-finger loops will slip off your knuckles. Straighten your index fingers. The finished net will appear.

Chapter 3 Review

Vocabulary

alternate exterior angles	133	corresponding angles	133	rhombus	124
alternate interior angles	133	exterior angles	147	rise and run	160
axis of symmetry	117	parallelogram	124	rotational symmetry	118
central angle	119	polygon	116	slope	160
center point	119	rectangle	124	slope formula	161
concave polygon	149	reflectional symmetry	117	square	124
consecutive interior angles	133	regular polygon	119	transversal	132
convex polygon	149	remote interior angles	147	trapezoid	124

Key Skills and Exercises

Lesson 3.1

➤ **Key Skills**

Identify reflection and rotation of planar figures.

The dotted lines show the axes of symmetry in this triangle.

Note that the triangle has rotational symmetry of 120°; that is, rotating the image 120° about its center point makes it coincide with the original image.

Sketch rotation or reflection of a figure.

Sketch a reflection of the pre-image shown over a horizontal axis of symmetry. Use the pre-image to sketch a figure with 90° rotational symmetry.

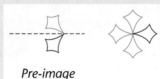

Pre-image

➤ **Exercises**

1. How many axes of symmetry does the figure below have? Copy the figure and draw all axes of symmetry.

In Exercises 2–3 use the pre-image shown.

2. Sketch a reflection of the pre-image over a vertical axis of symmetry.

3. Use the pre-image to sketch a new figure that has rotational symmetry.

Lesson 3.2

➤ **Key Skill**

Conjecture properties of quadrilaterals.

If a problem gives a "quadrilateral," all you know is that it is a four-sided polygon. If a problem gives a "square," you have much more information to solve that problem. You know that the polygon has four sides of equal length, four vertex angles of 90°, and perpendicular diagonals of equal length.

➤ Exercises

In Exercises 4–6 refer to figure *ABCD*.

   4. Given: *ABCD* is a rhombus. Find the measure of ∠*AEB*.

   5. Given: *ABCD* is a rectangle. Find the measure of ∠*BCD*.

   6. Given: *ABCD* is a rectangle. Name any sets of congruent line segments.

Lesson 3.3

➤ Key Skill

Identify angle relationships in parallel lines and transversals.

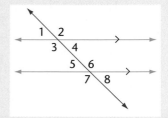

If you know one angle measure in the figure, you can
determine all the others.

If m∠4 = 40°, then m∠1, m∠5, and m∠8 = 40°; m∠2, m∠3, m∠6,
and m∠7 = 140°.

➤ Exercises

**In Exercises 7–8 refer to the parallel lines and transversal above. If
m∠3 = x and m∠4 = x − 60°, find the angle measures.**

   7. m∠5   **8.** m∠7

Lesson 3.4

➤ Key Skill

Use converse theorems to show that lines are parallel.

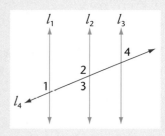

In the figure, if m∠1 = m∠2, then lines l_1 and l_2 are parallel by
the converse of the Corresponding Angles Theorem. Similarly,
if m∠3 = m∠4, then lines l_2 and l_3 are parallel.

➤ Exercise

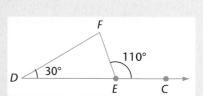

   9. ∠*UVX* and ∠*WXV* are right angles. Is this enough
information to show that $\overleftrightarrow{UV}$ and $\overleftrightarrow{XW}$ are parallel? Justify
your answer.

Lesson 3.5

➤ Key Skills

Use triangle sum theorems to find angle measures.
What is the measure of ∠*F*?

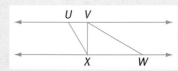

By the Exterior Angle Theorem m∠*D* + m∠*F* = m∠*CEF*.

30° + m∠*F* = 110°

       m∠*F* = 110° − 30°

       m∠*F* = 80°

➤ Exercises

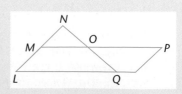

**In Exercises 10–12 refer to the figure on the right. $\overline{MP}$ is
parallel to $\overleftrightarrow{LQ}$; m∠*NOP* = 105°; m∠*MNO* = 50°. Find the
angle measure.**

   10. m∠*NMO*   **11.** m∠*MLQ*   **12.** m∠*LQO*

Lesson 3.6

➤ Key Skills

Find interior angle measures and exterior angle measures using the angle sums for polygons.

From the formula you discovered in this lesson, the sum of the interior angles of a regular octagon is 1080°. Thus the missing interior angle measure is 1080 ÷ 8 = 135.

You discovered in this lesson that the sum of the exterior angles of a polygon is always 360°. Because the given polygon is regular, simply divide 360° by the number of angles: 360° ÷ 8 = 45°.

Regular Octagon

➤ Exercises

In Exercises 13–14 find the missing angle measure.

13.

14.

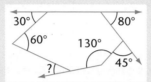

Lesson 3.7

➤ Key Skills

Use triangle and trapezoid midsegments to solve problems.

Find the measure of $\overline{FI}$.

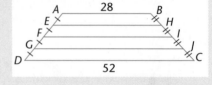

$$FI = \frac{(\text{base}_1 + \text{base}_2)}{2} = \frac{(AB + DC)}{2} = \frac{(28 + 52)}{2} = 40$$

➤ Exercises

In Exercises 15–17 find the measures. Use △ABC.

15. *EH* **16.** *DG* **17.** *FI*

Lesson 3.8

➤ Key Skills

Find the slope of a line.

You need two points on a line to determine its slope. A line runs through the origin and point (4, −4). Use the formula $\frac{y_2 - y_1}{x_2 - x_1}$ to find the slope.

$\frac{0 - (-4)}{0 - 4} = \frac{4}{-4} = -1$. Thus, the line has a slope of −1.

➤ Exercises

In Exercises 18–19 refer to the graph.

18. Find the slope of $\overleftrightarrow{DE}$. **19.** Find the slope of $\overleftrightarrow{FG}$.

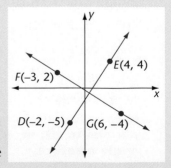

Applications

20. Paper Folding Use the properties of a square to fold and cut a square from an 8.5-by-11-inch sheet of paper. Use rotational symmetry to create a regular octagon by folding and cutting the square. What do the folds represent?

Chapter 3 Assessment

In Exercises 1 and 2, refer to the figure at right.

1. How many axes of symmetry does the figure have?

2. If the figure has rotational symmetry, what is the smallest positive angle of rotation?

In Exercises 3 and 4, use the pre-image below.

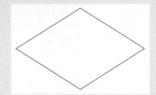

3. Sketch a reflection of the pre-image over a vertical line.

4. Sketch a figure that has rotational symmetry, using the pre-image.

5. In the trapezoid, what is the measure of ∠BCD?

   a. 45°   **b.** 135°

   c. 60°   **d.** 120°

6. △HIK and △JKI are equilateral triangles. Is this enough information to show that $\overleftrightarrow{HI}$ and $\overleftrightarrow{KJ}$ are parallel? Justify your answer.

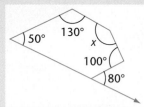

In Exercises 7–9, find the missing angle measures.

7.

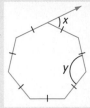

8.

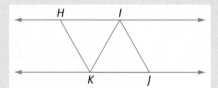

9.

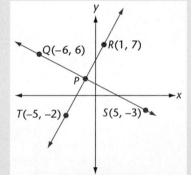

In Exercises 10 and 11, refer to the graph at right.

10. Is ∠SPR a right angle?

11. Give the equation for a line that is parallel to $\overleftrightarrow{SP}$.

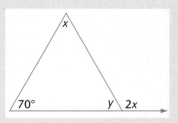

CHAPTER 4

Congruent Triangles

LESSONS

4.1 Polygon Congruence

4.2 *Exploring* Triangle Congruence

4.3 Analyzing Triangle Congruence

4.4 Using Triangle Congruence

4.5 Proving Quadrilateral Properties

4.6 *Exploring* Conditions for Special Quadrilaterals

4.7 Compass and Straightedge Constructions

4.8 *Exploring* Congruence in the Coordinate Plane

4.9 *Exploring* the Construction of Transformations

Chapter Project Flexagons

All around you– in nature, art, and human technology–you find things that are the same shape and size. Such things are said to be congruent. In geometry,too, where you can construct models of things in the real world, you find the same idea.

When manufactured items are mass produced, like cars coming off the assembly line, it is essential that certain standard parts all have an exact shape and size, to a high degree of precision. If they do not, they may not fit together properly with other parts, or there may be mechanical failures.

PORTFOLIO ACTIVITY

Use geometry technology to draw a triangle. Then, copy the triangle and use it to create your own geometric designs.

1. Make designs by overlapping the triangles. Do the overlapping triangles form other polygonal shapes?

2. Add color to your drawings. How does the color change the effect of your design?

Triangle congruence is basic to the study of more complex shapes.

173

Polygon Congruence

Andy Warhol,
One Hundred Cans, 1962

Artists often use congruent shapes as an element of composition. Do you think this painting would be as interesting without the repeating pattern?

 In earlier lessons you learned about congruent segments and angles. In this lesson you will develop a definition of congruent polygons.

Quadrilaterals 1 and 2 are congruent. If you slide one on top of the other, you will see that you can make them match exactly. This suggests a definition of congruent polygons.
Can you say what it is?

In the exploration that follows, you will begin to develop a mathematical definition of congruent polygons.

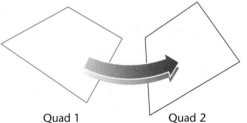

Quad 1   Quad 2

•Exploration *Polygon Congruence*

Geometry Graphics

You will need
Geometry technology or
Ruler and protractor
Scissors

1 Construct quadrilateral *ABCD* using the following set of measures for the sides and angles.

m∠*DAB* = 104°	*AB* = 5.4 cm
m∠*ABC* = 70°	*BC* = 7.0 cm
m∠*BCD* = 72°	*CD* = 4.9 cm
m∠*CDA* = 114°	*DA* = 3.7 cm

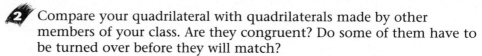

2 Compare your quadrilateral with quadrilaterals made by other members of your class. Are they congruent? Do some of them have to be turned over before they will match?

3 If all the sides and angles of two polygons are congruent, will the polygons match? Would they match if one or more of the sides and angles failed to be congruent?

Complete the following preliminary definition of congruent polygons: "Two polygons are congruent if and only if . . ." ❖

Naming Polygons

When naming polygons, the rule is to go around the figure, either clockwise or counterclockwise, and list the vertices in order. It does not matter which vertex you pick to start with. What are some possible names for the hexagon on the right? How many possibilities are there?

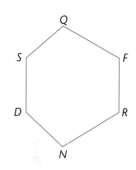

Corresponding Sides and Angles

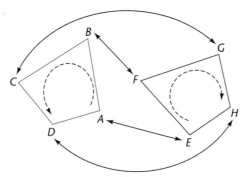

If two polygons have the same number of sides, it is possible to set up a correspondence between them by pairing off their parts. In quadrilaterals *ABCD* and *EFGH*, for example, you can pair off angles *A* and *E*, *B* and *F*, *C* and *G*, and *D* and *H*. Notice that you must go in order around each of the polygons.

The correspondence of the sides follows from the correspondence of the angles. In the present example, side $\overline{AB}$ corresponds to side $\overline{EF}$, and so on.

CRITICAL
Thinking

How many different ways are there of setting up a correspondence between the two quadrilaterals?

The polygons on the right are congruent. To show this fact mathematically, write a name for one of the polygons, followed by the congruence symbol. Then imagine the other polygon moved on top of the first one so that they match exactly. Finally, write the name of the second polygon to the right of the congruence symbol, with the angles listed in the order that they match:

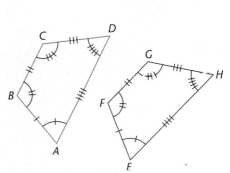

$$ABCD \cong EFGH$$

CRITICAL
Thinking

There is more than one way to write a congruence statement for polygons *ABCD* and *EFGH*. Complete the congruence statements below, and then write all the other possibilities.

$BCDA \cong ?$ $\qquad\qquad$ $CBAD \cong ?$

CONGRUENT POLYGONS

Two polygons are congruent if and only if there is a way of setting up a correspondence between their sides and angles, in order, so that

1. all pairs of corresponding angles are congruent, and
2. all pairs of corresponding sides are congruent. $\qquad$ **4.1.1**

Proving that polygons, particularly triangles, are congruent is one of the most important things you will do in geometry.

EXAMPLE

Prove that $\triangle REX \cong \triangle FEX$ in the figure as marked.

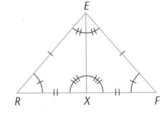

Solution ➤

List all the sides and angles that are given to be congruent.

$\angle R \cong \angle F$ $\qquad$ $\overline{RE} \cong \overline{FE}$
$\angle REX \cong \angle FEX$ $\qquad$ $\overline{RX} \cong \overline{FX}$
$\angle EXR \cong \angle EXF$

In all, there are six congruences that are required for triangles to be congruent—three pairs of angles and three pairs of sides. So one more pair of congruent sides is needed.

Notice that $\overline{EX}$ is shared by the two triangles. Use the Reflexive Property of Congruence to justify the statement that $\overline{EX} \cong \overline{EX}$. This gives the sixth congruence, so you can conclude that $\triangle REX \cong \triangle FEX$. ❖

The Penzoil Building in Houston, Texas, features trapezoidal faces. Are the trapezoids congruent? How many congruent sides and angles do you see?

EXERCISES & PROBLEMS

Communicate

Explain whether the following pairs of figures appear to be congruent. Use Definition 4.1.1 to justify your answer.

1.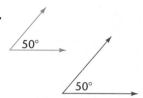

2.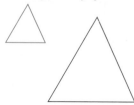

3.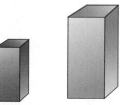

4.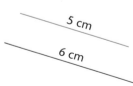

5.

6.

Practice & Apply

In Exercises 7–9, pentagon *UVWXY* ≅ pentagon *KMNTE*.

7. Draw a diagram and determine which angle is congruent to ∠N.

8. Describe how you would specify which angle is congruent to ∠V without drawing a diagram.

9. Find a segment that is congruent to each of the segments below.

   a. $\overline{WX} \cong$?   **b.** $\overline{ET} \cong$?   **c.** $\overline{UY} \cong$?

10. Given: △REV ≅ △FOT. Complete the following congruences.

   a. ∠FOT ≅ ?   **b.** ∠EVR ≅ ?   **c.** ∠TFO ≅ ?

Complete the congruences in Exercises 11–13.

11.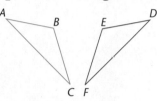

△ABC ≅ △DEF

$\overline{AB} \cong \overline{DE}$ ∠C ≅ ∠F
$\overline{AC} \cong$?   ? ≅ ∠D
$\overline{BC} \cong$?   ? ≅ ∠E

12.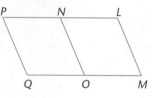

Quad LMON ≅ Quad NOQP

$\overline{NO} \cong \overline{LM}$ ∠PNO ≅ ∠NLM
? ≅ $\overline{MO}$ ∠NOQ ≅ ?
? ≅ $\overline{ON}$ ∠OQP ≅ ?
? ≅ $\overline{NL}$ ∠QPN ≅ ?

13. Pentagon *LMNOP* ≅ Pentagon *QRSTU*

$\overline{LM}$ ≅ $\overline{QR}$ ∠*PLM* ≅ ∠*UQR*

 ? ≅ $\overline{RS}$ ∠*LMN* ≅ ?

$\overline{NO}$ ≅ ?   ? ≅ ∠*RST*

 ? ≅ $\overline{TU}$ ∠*NOP* ≅ ?

$\overline{PL}$ ≅ ?   ? ≅ ∠*TUQ*

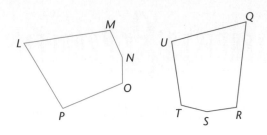

Given: △***XYZ*** ≅ △***FGH*** **and** △***FGH*** ≅ △***UMP***

14. What must be true about △*XYZ* and △*UMP*?

15. Name a property that justifies your conclusion.

16. Carpentry A small shop makes house-shaped mailboxes. The faces opposite each other are rectangles, pentagons, or triangles of the same size and shape, as are the two slanted portions of the roof.

    a. How many pairs of congruent polygons can you identify in this mailbox?

    b. List at least 10 pairs of congruent angles.

    c. List at least 10 pairs of congruent sides.

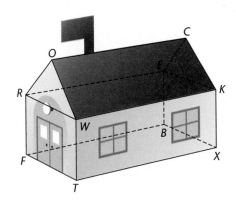

For Exercises 17–19, decide whether the polygons are congruent and write an explanation justifying your conclusion.

17.

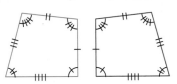

18.

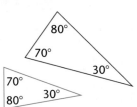

19.

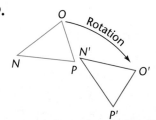

20. Quilting In the photograph of the quilt, how many different congruent shapes can you identify? How many different patterns of polygons did the quilter need to use to assemble the quilt? List them.

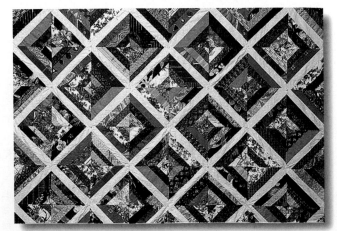

Given: $\angle L \cong \angle P$, $\angle M \cong \angle O$, $\angle MRL \cong \angle OQP$, $\overline{LM} \cong \overline{PO}$, $\overline{MR} \cong \overline{OQ}$, $LQ = 5$ cm, $QR = 3$ cm, $RP = 5$ cm

21. What is the length of $\overline{LR}$?

22. What is the length of $\overline{QP}$?

23. Is $\triangle LMR \cong \triangle POQ$? Why or why not?

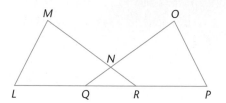

Look Back

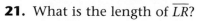

Given: *PARL* is a parallelogram. [Lessons 3.2, 3.3, 3.4]

24. How do you know that $\overline{AR} \parallel \overline{PL}$ and $\overline{AP} \parallel \overline{RL}$?

25. How do you know that $\angle RAL \cong \angle PLA$ and $\angle RLA \cong \angle PAL$?

26. How do you know that $\overline{AL} \cong \overline{AL}$?

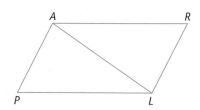

Determine whether each of the following is true or false.
[Lessons 3.2, 3.3, 3.4]

27. A square is a rhombus.   **28.** A rectangle is a parallelogram.

29. The diagonals of a parallelogram bisect each other.

30. Opposite angles of a rhombus are congruent.

Look Beyond

31. **Coordinate Geometry** Graph the following points on a coordinate grid and connect the points in the order given.

Point $A(1, 2)$ Point $B(2, 4)$ Point $C(4, 4)$ Point $D(3, 1)$

Next, graph and connect the points resulting from the following translation on each point.

$(x, y) \rightarrow (x + 4, y - 2)$

Example: point $A(1, 2) \rightarrow (5, 0)$

Is the new figure congruent to the first figure? Why or why not?

Graph and connect the points resulting from the following transformation of the first figure.

$(x, y) \rightarrow (3x, -2y)$

Example: Point $A(1, 2) \rightarrow (3, -4)$

Is this new figure congruent to the first? Why or why not?

LESSON 4.2

Exploring

Triangle Congruence

why *The structure of the Eiffel Tower includes thousands of triangles. Architects and engineers use triangular braces because triangles add rigidity. The rigidity of triangles has important mathematical consequences.*

Exploration 1 *Triangle Rigidity*

Geometry Graphics

You will need

Geometry technology or
Modeling materials, such as geostrips,
    straws, or string
Ruler
Scissors

1 Using modeling materials or geometry technology, construct each triangle.

    $\triangle ABC$: $AB = 8$ units, $BC = 9$ units, $CA = 10$ units

    $\triangle XYZ$: $XY = 6$ units, $YZ = 8$ units, $ZX = 10$ units

2 Using the same measurements, try to draw more than one $\triangle ABC$ and $\triangle XYZ$ that look different from the first ones you drew. Compare the triangles you constructed in Step 1 with the new triangles. Are all triangles ABC congruent to each other? are triangles XYZ?

Do you need to know the angle measures of a triangle to make a copy of it? Or is it enough to know the measures of the sides?

3 Given the lengths of the sides of a triangle, is there more than one shape that the triangle can have?

4 If the lengths of the sides of a triangle are fixed, there is just one shape the triangle can have. This property is what makes triangles rigid.

How do the triangles you drew show that triangles are rigid?

5 Write a rule, or **postulate**, that summarizes your results by completing the following statement.

If the __?__ of one triangle are congruent to the __?__ of another triangle, then the two triangles are congruent.

Give your postulate a descriptive name and record it in your notebook. ❖

CRITICAL *Thinking* Are any polygons other than triangles rigid? If you repeated the exploration using quadrilaterals instead of triangles, do you think quadrilaterals with the same names would all be congruent?

Useful Geometry Tools

If you use the definition of congruent polygons on page 176 to show that two triangles are congruent, you must show that three pairs of sides and three pairs of angles are congruent. The postulate above provides a shortcut for showing that two triangles are congruent by showing that three pairs of sides are congruent.

There are still other shortcuts for proving triangle congruence. In the next exploration you will discover two more of them.

•Exploration 2 *Two More Congruence Postulates*

Geometry Graphics

You will need
Geometry technology or
Ruler and protractor
Scissors

Part I

1 Draw and label each of the triangles.

△ABC: AB = 5 units, m∠B = 45°, BC = 5 units

△XYZ: XY = 5 units, m∠Y = 30°, YZ = 7 units

△MNO: MN = 8 units, m∠N = 120°, NO = 6 units

2 Cut out your triangles and compare them with triangles constructed by the other teams. If you are using a computer, draw additional triangles ABC, XYZ, or MNO using the above dimensions. Are all the triangles with the same names congruent?

Notice that in each of the triangles, the angles and the sides that you measured are arranged in a certain way. Describe that arrangement. (Hint: Ask yourself, what comes between what?)

Do you think that the above arrangements of measured sides and angles will determine the shape of a triangle? If you are not sure, try to find a set of similarly arranged measures that will not work. If you can draw two noncongruent triangles that have the same set of measures, you will have found a counterexample.

 Write a postulate that summarizes your results by completing the following statement:

If _?_ of one triangle are congruent to _?_ of another triangle, then the two triangles are congruent.

Give your postulate a descriptive name and record it in your notebook.

Part II

1 Construct and label each triangle.

$\triangle ABC$: m$\angle A$ = 60°, AB = 6 units, m$\angle B$ = 60°

$\triangle XYZ$: m$\angle X$ = 60°, XY = 6 units, m$\angle Y$ = 45°

$\triangle MNO$: m$\angle M$ = 120°, MN = 6 units, m$\angle N$ = 30°

2 Answer the same three questions (Steps 2–4) for these measurements that you answered in Exploration 1. ❖

Using Your New Postulates

The triangle congruence postulates you discovered in the explorations can save steps in proofs. But much more important, they allow you to determine congruence from limited information.

EXTENSION

In each of the pairs below, the triangles are congruent. Tell which triangle congruence postulates allow you to conclude that they are congruent, based on the markings on the figures. ❖

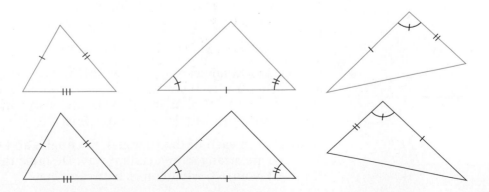

EXERCISES & PROBLEMS

Communicate

1. Explain triangle rigidity. What properties do triangles have that make them rigid?

2. Explain the congruence postulate you discovered in Exploration 1. Draw a diagram of two triangles that illustrates this postulate.

3. Explain the congruence postulate you discovered in Part I of Exploration 2. Draw a diagram of two triangles that illustrates this postulate.

4. Explain the congruence postulate you discovered in Part II of Exploration 2. Draw a diagram of two triangles that illustrates this postulate.

5. Use a counterexample to show that if corresponding angles of one triangle are congruent to corresponding angles of a second triangle, then the triangles are not necessarily congruent.

Practice & Apply

Decide whether each pair of triangles can be proven congruent using one of the three congruence postulates you discovered in the lesson. If so, write an appropriate congruence statement and name the postulate that it supports. If not, explain why.

6.
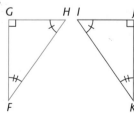

7.
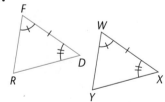

8.
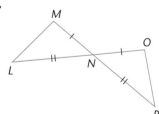

9.
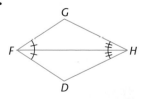

10.
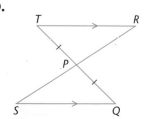

11.
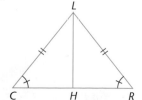

12. Construction In the house shown in the photo, the triangles in the roof structure must be congruent. How can you be sure that the triangles are congruent without measuring any angles? Which triangle congruence postulate would you be using?

13. Carpentry A carpenter is building a rectangular bookshelf and finds that it wobbles from side to side. To stabilize the bookshelf, he nails a board that attaches the top left-hand corner to the bottom right-hand corner. Why will this diagonal board stabilize the bookshelf?

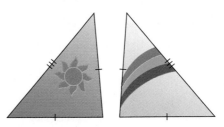

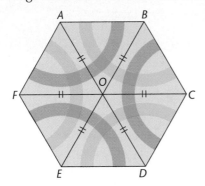

Quilting A quilter is making a quilt out of regular hexagons made from triangles as shown. Questions 14–16 refer to the quilt pattern.

14. Is △FOA ≅ △COD? Why or why not?

15. Is △BOA ≅ △COD? Why or why not?

16. Explain two ways to prove △FOE ≅ △COB.

Recreation In Exercises 17–20, pairs of boat sails in the shape of right triangles are shown. Based on the congruences marked, are the sails necessarily congruent?

17.

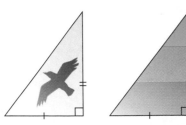

18.

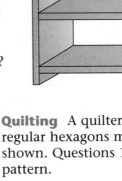

19.

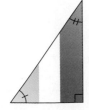

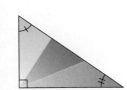

20.

21. Use the diagram to help you answer the questions.

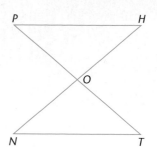

   a. If O is the midpoint of $\overline{PT}$, then what two segments are congruent?

   b. Because $\overline{PT}$ and $\overline{NH}$ intersect at point O, what angles must be congruent and why?

   c. If $\overline{PH} \parallel \overline{NT}$, then $\angle T$ is congruent to what other angle and why?

   d. If all of the above are true, what conclusion can you draw about these triangles?

Look Back

22. Points that lie on the same plane are called __?__. **[Lesson 1.1]**

23. If point B is between R and T, then what do you know about $RB + BT$ and why? **[Lesson 1.4]**

24. In the diagram, $\overrightarrow{AT}$ bisects $\angle RAM$, and m$\angle CAR = 122°$. Find the measures of $\angle RAM$, $\angle RAT$, and $\angle MAT$. **[Lesson 1.5]**

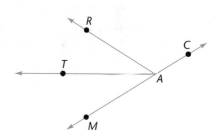

25. State two properties of the diagonals of a rhombus. **[Lesson 3.2]**

26. If the same-side interior angles of a transversal are congruent, then are the lines definitely parallel, possibly parallel, or never parallel? **[Lesson 3.4]**

Given: quadrilateral *ZMPA* ≅ quadrilateral *RIKL*. **[Lesson 4.1]**

27. Identify an angle that is congruent to $\angle M$.

28. Name a segment that is congruent to $\overline{MP}$.

29. Is it impossible, possible, or definite that $\angle R \cong \angle P$?

Look Beyond

Study the triangles at the right.

30. Are the two triangles congruent? Why or why not?

31. What additional information would help you prove that the triangles are congruent?

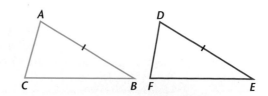

Analyzing Triangle Congruence

Why *In the last lesson you discovered three different three-part combinations of side and angle measures that determine the shape and size of a triangle. In this lesson you will learn two more.*

*By using directional measures from two different towers, forest rangers are able to pinpoint the location of a wisp of smoke. The method is known as **triangulation** because the given conditions determine the shape of a triangle.*

Three Congruence Postulates

The combinations of side and angle measures you learned in the previous lessons can be used to determine whether two triangles are congruent. In the explorations, you used these combinations to write your own congruence postulates. The traditional versions of the postulates are as follows:

SSS (SIDE-SIDE-SIDE) POSTULATE

If three sides in one triangle are congruent to three sides in another triangle, then the triangles are congruent.　　　　　　　　　　　　　　**4.3.1**

SAS (SIDE-ANGLE-SIDE) POSTULATE

If two sides and the angle between them in one triangle are congruent to two sides and the angle between them in another triangle, then the triangles are congruent.　　　　　　　　　　　　　　**4.3.2**

ASA (ANGLE-SIDE-ANGLE) POSTULATE

If two angles and the side between them in one triangle are congruent to two angles and the side between them in another triangle, then the triangles are congruent.　　　　　　　　　　　　　　**4.3.3**

Navigation

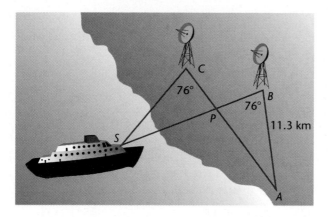

A ship at sea loses power and is stranded at *S*. The ship is located as shown, and point *P* is equidistant from stations *B* and *C*. The measurements shown are calculated from data obtained by radio and radar. How far is the ship from station *C*?

$\angle C \cong \angle B$ and $\overline{CP} \cong \overline{BP}$. $\angle CPS$ and $\angle BPA$ are congruent, because they are vertical angles. Therefore, by *ASA*, $\triangle PCS \cong \triangle PBA$. Since $\overline{BA}$ corresponds to $\overline{CS}$, the distance to the ship from the radar station is 11.3 km. ❖

Three Other Possibilities

You may have realized that the three-part combinations of sides and angles you studied in the last lesson are not the only combinations to consider. In fact, there are three others. Which of the following combinations do you think can establish triangle congruence?

1. AAA combination   Three pairs of angles

2. SSA combination   Two sides and an angle that is not between them (the angle is *opposite* one of the two sides)

3. AAS combination   Two angles and a side that is not between them

You can quickly deal with the first two of the combinations by finding counterexamples. A simple proof will show that the third combination works.

EXAMPLE 1

Show that the AAA combination is not a valid test for triangle congruence.

Solution ➤

In order to get a better idea of what you need to disprove, state the combination in the form of a conjecture, as follows:

> *Conjecture: If three angles of one triangle are congruent to three angles in another triangle, then the triangles are congruent.*

Counterexamples to this statement are easy to find. In the figures at right, there are three pairs of congruent angles (from one triangle to the next), but the two triangles are not congruent. Therefore, *the conjecture is false.* ❖

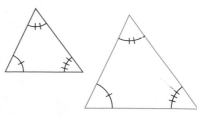

AAA Counterexample

EXAMPLE 2

Show that the AAS combination is a valid test for triangle congruence.

Solution ➤

The AAS combination can be converted to the ASA combination as illustrated below.

Notice that the given sides are not between the given angles.

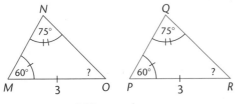

AAS example . . .

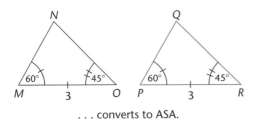

. . . converts to ASA.

The triangles on the right are an AAS combination. To convert this to an ASA combination, find the measures of the missing angles in the two triangles. Since the sum of the measures of the angles of a triangle is 180°:

$m\angle O = 180 - (60 + 75) = 45°$
$m\angle R = 180 - (60 + 75) = 45°$

The measures of the two missing angles are the same, so $\angle O \cong \angle R$. The given side of 3 units is between angles of 45° and 60° in each triangle. Therefore, the two triangles are congruent by the ASA postulate. ❖

CRITICAL Thinking

The two triangles below represent a "version" of AAS, but the triangles are not congruent. There's an important difference between the two arrangements. Can you say what it is?

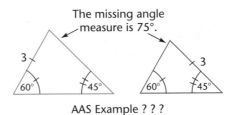

The missing angle measure is 75°.

AAS Example ? ? ?

For AAS to be a valid test, the congruent parts must correspond.

AAS (ANGLE-ANGLE-SIDE) THEOREM
If two angles and a side that is not between them in one triangle are congruent to the corresponding two angles and the side not between them in another triangle, then the triangles are congruent.   **4.3.4**

The first three combinations you studied were called postulates, but the AAS combination is called a theorem. Why?

Try This Which pairs of triangles can be proven congruent by the AAS Theorem?

a.   **b.**   **c.** 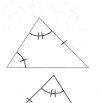 **d.**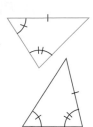

EXAMPLE 3

Show that the SSA combination is not a valid test for triangle congruence.

Solution ➤

It is a bit more difficult to find a counterexample for this conjecture, but the figures on the right do the trick. The sides and angles are congruent as required, but the triangles are obviously not congruent. Therefore, the conjecture is false.

SSA Counterexample

When you draw an SSA combination, there is often a way for the side opposite the given angle to pivot like a swinging door between two possible positions. ❖

"Swinging Door"

A Special Case of SSA

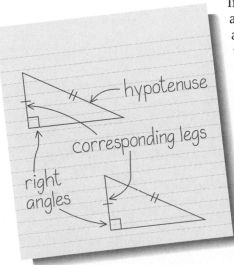

If the given angle in SSA is a right angle, the side opposite the angle cannot pivot in the "swinging-door" manner described above. So if the given angle is a right angle, SSA is a valid test for triangle congruence. In this case the test is called the Hypotenuse-Leg Theorem. A proof of this theorem will not be given here, but you will be able to prove it by adding an additional right triangle and using a property of isosceles triangles (see Lesson 4.4)

HL (HYPOTENUSE-LEG) THEOREM
If the hypotenuse and a leg of a right triangle are congruent to the hypotenuse and the corresponding leg in another right triangle, then the two triangles are congruent.   **4.3.5**

EXERCISES & PROBLEMS

Communicate

1. Explain the SSS Postulate.
2. Explain the SAS Postulate.
3. Explain the ASA Postulate.
4. Explain why SSA does not show congruence.
5. **Triangulation** Towers A and B each sight a smoke plume in the directions shown in the diagram. What combination of sides and angles (AAS, SAS, etc.) are known in the triangle?

   If a triangle has the given measures, is there more than one shape the triangle can have? Explain how the information determines the location of the smoke.

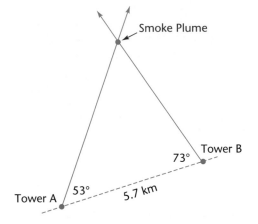

6. Which pairs of triangles are congruent and why?

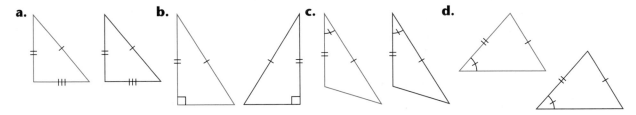

   a. **b.** **c.** **d.**

Practice & Apply

For Exercises 7–9, decide whether each pair of triangles are congruent. If so, write an appropriate congruence statement and name the postulate or theorem that supports it.

7.

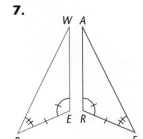

8.

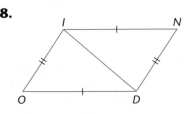

9.

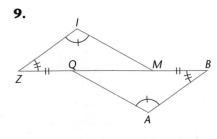

Copy the figures and label ∠U and ∠E as right angles for Exercises 10–12.

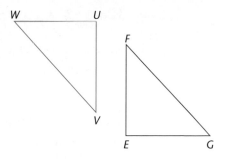

10. Identify the hypotenuse in each triangle.

11. Identify the legs in each triangle.

12. If $\overline{UV} \cong \overline{EF}$ and $\overline{UW} \cong \overline{EG}$, are the triangles congruent? Explain your answer.

Copy the figures and label ∠C and ∠T as right angles for Exercises 13–15.

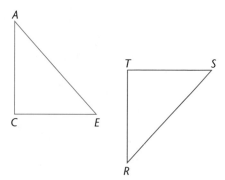

13. Identify the hypotenuse in each triangle.

14. Identify the legs in each triangle.

15. If $\angle A \cong \angle R$, are the triangles congruent? Explain your answer.

Triangle *TAL* is isosceles, with $\overline{AT} \cong \overline{AL}$. Copy the figure and label the congruent sides for Exercises 16–20.

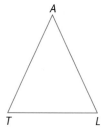

16. Draw $\overline{AK}$ with *K* on $\overline{TL}$ so that $\overline{AK} \perp \overline{TL}$. Label the right angles on your drawing.

17. You have now created two right triangles. Name the two right angles and identify the hypotenuse for each.

18. What must be true about the two hypotenuses? Explain your reasoning.

19. Why is $\overline{AK} \cong \overline{AK}$?

20. What do you know about $\triangle TAK$ and $\triangle LAK$? Why?

Use the diagram below for Exercises 21 and 22.

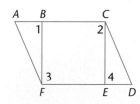

21. Given: $\angle A \cong \angle D$, $\overline{AF} \cong \overline{DC}$, $\overline{AC} \cong \overline{DF}$, $\angle BFA \cong \angle ECD$

  Prove: $\triangle AFB \cong \triangle DCE$

22. Given: $\angle 1 \cong \angle 4$, $\overline{AF} \cong \overline{CD}$, $\angle A \cong \angle D$

  Prove: $\triangle AFB \cong \triangle DCE$

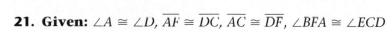

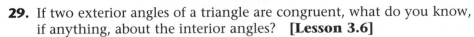

Look Back

23. Draw three lines intersecting in a point. **[Lesson 1.1]**

24. Draw three lines that do not intersect each other. **[Lesson 1.1]**

25. Draw two lines that intersect a plane without intersecting each other. **[Lesson 1.1]**

26. To the right is a drawing of a transversal intersecting parallel lines. Label the lines as parallel and identify the congruent angles. **[Lesson 3.3]**

27. Identify two angles in the diagram that are supplementary. **[Lesson 3.3]**

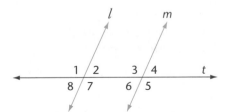

28. 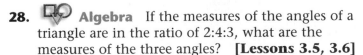 **Algebra** If the measures of the angles of a triangle are in the ratio of 2:4:3, what are the measures of the three angles? **[Lessons 3.5, 3.6]**

29. If two exterior angles of a triangle are congruent, what do you know, if anything, about the interior angles? **[Lesson 3.6]**

Look Beyond

30. Construction Triangles are often used in construction. Find a building, bridge, or house on which you can see the triangular supports, and draw an illustration of the supports. Identify the congruent triangles and explain why the triangles provide support.

The Forth Rail Bridge (1889) in Scotland is 5330 feet long and weighs 38,000 tons. Its strength rests in each of three cantilevered units that become a single truss. Can you see the triangles in the design?

4.4 Using Triangle Congruence

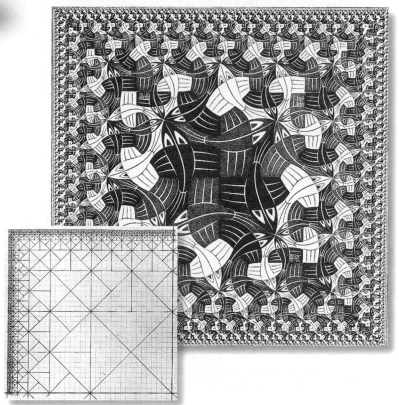

why *The design elements of this famous woodcut by Dutch artist M.C. Escher are mathematically related. The size of one part determines the size of another. By using chains of mathematical reasoning, you can deduce things about one part of a figure from information about another part.*

In this course you will use chains of mathematical reasoning to discover or prove things about geometric figures. Your triangle congruence postulates and theorem will be some of your most powerful tools for doing this.

"CPCTC"

According to the definition of congruent triangles, *if two triangles are congruent, then their corresponding parts are congruent.* Therefore, if $\triangle ABC \cong \triangle DEF$, you can conclude that $\overline{AB} \cong \overline{DE}$ based on knowing that $\triangle ABC \cong \triangle DEF$. What other pairs of congruent sides and angles must be congruent?

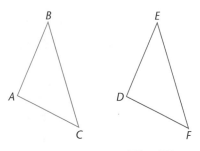

If $\triangle ABC \cong \triangle DEF$, then $\overline{AB} \cong \overline{DE}$, . . .

This idea is often stated in the following form: *Corresponding parts of congruent triangles are congruent,* abbreviated CPCTC. This statement is frequently used as a reason in proofs.

In each of the examples that follow, you will use a triangle congruence postulate or theorem to establish that two triangles are congruent. Then you will use CPCTC.

EXAMPLE 1

Given: $\overline{AB} \cong \overline{DE}$, $\overline{BC} \cong \overline{EF}$, $\overline{AC} \cong \overline{DF}$

Prove: $\angle A \cong \angle D$

Plan: Use a triangle congruence postulate to show that the two triangles are congruent. Then use CPCTC to establish $\angle A \cong \angle D$.

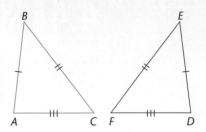

Solution ➤

Proof:

STATEMENTS	REASONS
1. $\overline{AB} \cong \overline{DE}$	Given
2. $\overline{BC} \cong \overline{EF}$	Given
3. $\overline{AC} \cong \overline{DF}$	Given
4. $\triangle ABC \cong \triangle DEF$	SSS
5. $\angle A \cong \angle D$	CPCTC ❖

In the next example, a simple congruence proof is a part of a larger proof.

EXAMPLE 2

Given: $\overline{AC} \cong \overline{BD}$, $\overline{CX} \cong \overline{DX}$, $\angle C \cong \angle D$

Prove: X is the midpoint of $\overline{AB}$.

Plan: Use congruent triangles to show that $\overline{AX} \cong \overline{BX}$ (and so, $AX = BX$). Then use the definition of a midpoint.

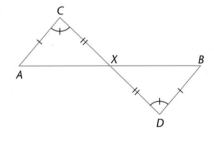

Solution ➤

Proof:

STATEMENTS	REASONS
1. $\overline{AC} \cong \overline{BD}$	Given
2. $\overline{CX} \cong \overline{DX}$	Given
3. $\angle C \cong \angle D$	Given
4. $\triangle ACX \cong \triangle BDX$	SAS
5. $\overline{AX} \cong \overline{BX}$ ($AX = BX$)	CPCTC
6. X is the midpoint of $\overline{AB}$	Def. midpoint ❖

CRITICAL *Thinking*

Why can't vertical angles $\angle AXC$ and $\angle BXD$ be used in this proof instead of $\angle C$ and $\angle D$?

Overlapping Triangles

How many different triangles can you find in the figures?

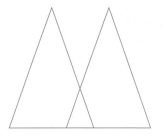

Figure 1

Figure 2

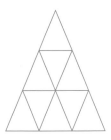

Figure 3

You will work with many diagrams such as Figure 2 in proving triangles congruent.

In the proof that follows, you will use the Overlapping Segment Theorem. Recall that for a figure like the one on the right you can conclude that $\overline{AC} \cong \overline{BD}$.

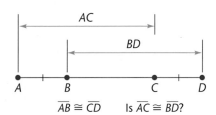

$\overline{AB} \cong \overline{CD}$ Is $\overline{AC} \cong \overline{BD}$?

EXAMPLE 3

Given: $\overline{AB} \cong \overline{CD}$ ($AB = CD$), $\overline{AE} \cong \overline{FD}$
($AE = FD$), $\angle A \cong \angle D$

Prove: $\overline{EC} \cong \overline{FB}$

Plan: Use the Overlapping Segments Theorem to show that $\overline{AC} \cong \overline{BD}$, and so $\triangle ACE \cong \triangle DBF$ by SAS. Then use CPCTC to show that $\overline{EC} \cong \overline{FB}$.

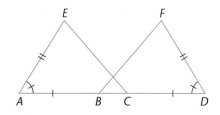

Solution ➤

Proof:

STATEMENTS	REASONS
1. $AB = CD$	Given
2. $AC = BD$ ($\overline{AC} \cong \overline{BD}$)	Overlapping Segment Thm.
3. $\overline{AE} \cong \overline{FD}$	Given
4. $\angle A \cong \angle D$	Given
5. $\triangle ACE \cong \triangle DBF$	SAS
6. $\overline{EC} \cong \overline{FB}$	CPCTC ❖

The Isosceles Triangle Theorem

An **isosceles triangle** is a triangle with at least two congruent sides. The two congruent sides are known as the **legs** of the triangle, and the remaining side is known as the **base**. The angles at the base are known as the **base angles**, and the angle opposite the base is known as the **vertex angle**.

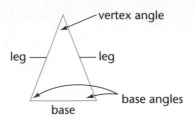

An **equilateral triangle** is a special type of triangle in which all three sides of the triangle are congruent. Is an equilateral triangle isosceles?

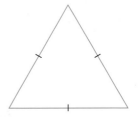

The following theorem, which is one of the great geometry classics, is one you may have already conjectured.

> **ISOSCELES TRIANGLE THEOREM**
> If two sides of a triangle are congruent, then the angles opposite those sides are congruent. **4.4.1**

Try This Prove the Isosceles Triangle Theorem using the plan provided below.

Given: $\overline{AC} \cong \overline{BC}$

Prove: $\angle A \cong \angle B$

Plan: Draw an angle bisector of the vertex angle and extend it to the base. Show that the two triangles that result are congruent by SAS. Then use CPCTC.

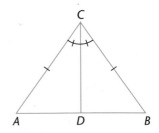

> **ISOSCELES TRIANGLE THEOREM: CONVERSE**
> If two angles of a triangle are congruent, then the sides opposite those angles are congruent. **4.4.2**

Construction   A resort owner wishes to place a gondola ride across a small canyon on her property. She needs to know the distance across the canyon in order to be well informed for construction bids. Study the diagram below. What is the distance across the canyon?

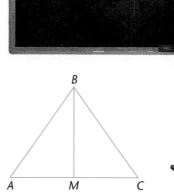

Since the 80° angle is an exterior angle, then $80° = 40° + m\angle x$, and so $m\angle x = 40°$ by the Exterior Angle Theorem. Since the base angles of the triangle are both 40°, then the sides opposite are congruent by the Converse of the Isosceles Triangle Theorem. Therefore, the distance across the canyon is 350 feet.

There are a number of corollaries to the Isosceles Triangle Theorem which you will also be able to prove.

COROLLARY	
An equilateral triangle has three angles that measure 60°.	**4.4.3**

COROLLARY	
The bisector of the vertex angle of an isosceles triangle is the perpendicular bisector of the base.	**4.4.4**

EXERCISES & PROBLEMS

Communicate

1–3. Give the reason for each of the statements.

STATEMENTS	REASONS
$\overline{BA} \cong \overline{BC}$	Given
M is the midpoint of $\overline{AC}$	Given
$\overline{AM} \cong \overline{MC}$	**(1)**
$\overline{BM} \cong \overline{BM}$	**(2)**
$\triangle ABM \cong \triangle CBM$	**(3)**

4–5. Complete the statement of the theorem below, which the above proof establishes.

THEOREM	
The median from the vertex to the base of an isosceles triangle **(4)** the triangle into **(5)**.	**4.4.5**

Practice & Apply

6–7. Give the reasons for each of the statements.

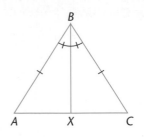

STATEMENTS	REASONS
$\overline{AB} \cong \overline{BC}$	Given
$\angle ABX \cong \angle CBX$	Given
$\overline{BX} \cong \overline{BX}$	____(6)____
$\triangle ABX \cong \triangle CBX$	____(7)____
$\overline{AX} \cong \overline{CX}$	____(8)____

9–10. Complete the statement of the theorem below, which the above proof establishes.

> **THEOREM**
> The __(9)__ of the vertex angle of an isosceles triangle __(10)__ .   **4.4.6**

11. Draw the altitude to the base of an isosceles triangle and prove a new theorem about the relationship between the two triangles formed.

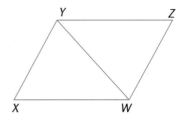

Use the diagram at the right for Exercises 12–17.

12. Given: $\overline{XY} \parallel \overline{ZW},\ \overline{YZ} \parallel \overline{XW}$
  Prove: $\overline{XY} \cong \overline{ZW},\ \overline{YZ} \cong \overline{XW}$

13–14. Complete the statement of the theorem below, which your proof in Exercise 12 establishes.

> **THEOREM**
> The __(13)__ of a parallelogram are __(14)__ .   **4.4.7**

15. Given: $\overline{XY} \parallel \overline{ZW},\ \overline{YZ} \parallel \overline{XW}$

  Prove: $\angle XYZ \cong \angle XWZ,\ \angle YXW \cong \angle YZW$

16–17. Complete the statement of the theorem below, which your proof in Exercise 15 establishes.

> **THEOREM**
> The __(16)__ of a parallelogram are __(17)__ .   **4.4.8**

18. Prove Corollary 4.4.3 (page 197) using paragraph form.

19. Prove Corollary 4.4.4 (page 197) using paragraph form.

Use the diagram at the right for Exercises 20–22.

20. Given: $\overline{AB} \parallel \overline{DE}$, C is the midpoint of $\overline{BE}$

    Prove: $\overline{AB} \cong \overline{DE}$

21. Given: $\triangle ABC \cong \triangle DEC$

    Prove: $\overline{AB} \parallel \overline{DE}$

22. Given: $\angle A \cong \angle D$, $\overline{AB} \cong \overline{DE}$

    Prove: C is the midpoint of $\overline{BE}$

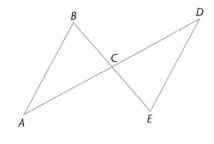

23. Given: $\angle A \cong \angle D$, $\overline{AB} \cong \overline{DE}$, $\overline{AF} \cong \overline{DC}$

    Prove: $\angle B \cong \angle E$

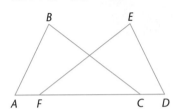

24. Given: $\overline{AB} \cong \overline{CB}$, $\overline{AD} \cong \overline{CE}$

    Prove: $\triangle BDE$ is isosceles

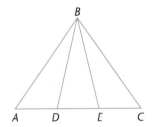

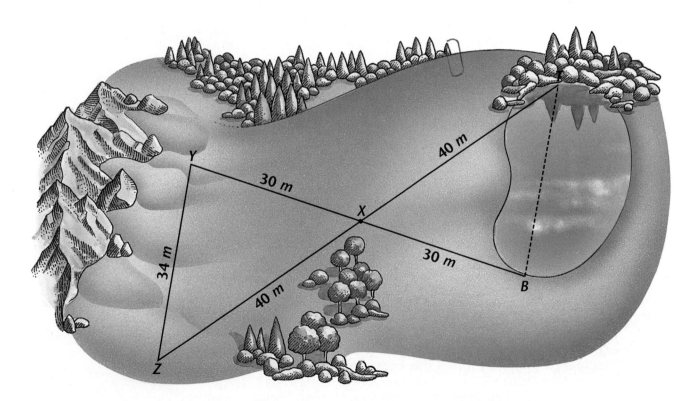

25. Surveying A surveyor needs to measure the distance across a pond from point T to point B. Describe how the measurements shown enable him to determine the distance.

26. What is the distance from T to B?

27. Traffic Signs A yield sign is an equilateral triangle. What else do you know about it because it is equilateral?

28. Traffic Signs A stop sign is a regular octagon; it is an eight-sided equilateral and equiangular polygon. Is it possible to have an equilateral octagon that is not equiangular? Is it possible to have an equiangular octagon that is not equilateral? Demonstrate and discuss your answers.

Look Back

Use the diagram for Exercises 29 and 30.   [Lesson 2.5]

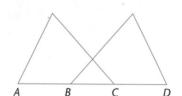

29. Given $\overline{AB} \cong \overline{DC}$, name another pair of congruent segments. What theorem guarantees this?

30. Given $\overline{AC} \cong \overline{DB}$, name another pair of congruent segments. What theorem guarantees this?

Use the diagram for Exercises 31 and 32.   [Lesson 2.5]

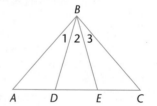

31. Given $\angle 1 \cong \angle 3$, name another pair of congruent angles. What postulate guarantees this?

32. Given $\angle ABE \cong \angle CBD$, name another pair of congruent angles. What postulate guarantees this?

Find the measure of each base angle of an isosceles triangle using the given vertex angle measures.   [Lesson 3.5]

33. 24°   **34.** 52°   **35.** $x°$

Look Beyond

36. Another proof of the Isosceles Triangle Theorem was "discovered" by a computer program. The proof was known to the Greek geometer Pappus of Alexandria (320 C.E.) but hadn't been noticed by mathematicians for centuries. It is much simpler than Euclid's proof (which is quite complicated) and even simpler than the one given in this chapter. See if you can rediscover it for yourself.

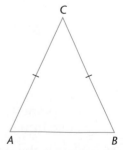

Given: $\overline{AC} \cong \overline{BC}$

Prove: $\angle A \cong \angle B$

Plan: Show that $\triangle ABC \cong \triangle BAC$

LESSON 4.5 Proving Quadrilateral Properties

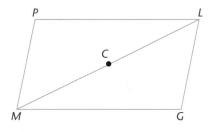

A property of rhombuses, which are special parallelograms, explains why this lamp assembly stays perpendicular to the wall as it moves.

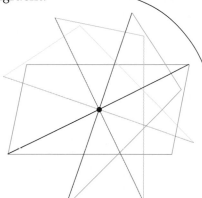

Why *In Chapter 3 you conjectured some properties of quadrilaterals. Now, with the right triangle congruence postulates and theorem, you are in a position to prove those conjectures—if they are true.*

In Lesson 3.2 you may have conjectured that a parallelogram has 180° rotational symmetry. If this conjecture is true, you should be able to rotate the parallelogram through its center of rotation so that △MPL will swing around to match the original position of △LGM. Therefore, the two triangles would seem to be congruent.

The above result can be stated as the conjecture shown below. It will be proved in Example 1. (Once proven, it becomes a theorem.)

> *A diagonal of a parallelogram divides the parallelogram into two congruent triangles.* **(Theorem 4.5.1)**

EXAMPLE 1

Given: Parallelogram *PLGM* with diagonal $\overline{LM}$

Prove: $\triangle LGM \cong \triangle MPL$

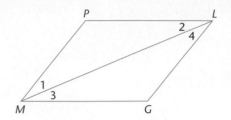

Solution ➤

Proof A (Paragraph form):

$\overline{PL}$ and $\overline{GM}$ are parallel, by definition of a parallelogram, so $\angle 3$ and $\angle 2$ are congruent because they are alternate interior angles. Similarly, $\angle 1$ and $\angle 4$ are congruent, because $\overline{PM}$ and $\overline{GL}$ are also parallel. Finally, diagonal $\overline{LM}$ is congruent to itself. Thus we have shown that two angles and the side between them are congruent in $\triangle LGM$ and $\triangle MPL$. We can conclude that the two triangles are congruent by ASA.

Proof B (Two-column form):

STATEMENTS	REASONS
1. *PLGM* is a parallelogram	Given
2. $\overline{PL} \parallel \overline{GM}$	Def. parallelogram
3. $\angle 3 \cong \angle 2$	$\parallel$s $\Rightarrow$ $\cong$ alt. int. $\angle$s
4. $\overline{PM} \parallel \overline{GL}$	Def. parallelogram
5. $\angle 1 \cong \angle 4$	$\parallel$s $\Rightarrow$ $\cong$ alt. int. $\angle$s
6. $\overline{LM} \cong \overline{LM}$	Reflexive prop. $\cong$
7. $\triangle LGM \cong \triangle MPL$	ASA ❖

CRITICAL *Thinking*

How could you use this result to show that $\angle P$ and $\angle G$ are congruent?

Another theorem, which you proved in Lesson 4.4, follows immediately from the theorem above. Explain.

The opposite sides of a parallelogram are congruent.

EXAMPLE 2

Construction John is building a rail for a stairway. Assume that the rail is parallel to the line of the steps and that the vertical bars are parallel to each other. How do you know that the vertical bars are congruent to each other?

Given: Parallelogram *PLGM* with diagonal $\overline{LM}$

Prove: $\overline{PL} \cong \overline{GM}$, $\overline{PM} \cong \overline{GL}$

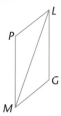

Solution ➤

Proof (Paragraph form):

Since a diagonal divides a parallelogram into congruent triangles, $\triangle LGM \cong \triangle MPL$. Therefore, $\overline{PL} \cong \overline{GM}$ and $\overline{PM} \cong \overline{GL}$ because they are corresponding parts of congruent triangles. ❖

Try This Write the Example 2 proof in two-column form.

Summary

> A diagonal of a parallelogram divides the parallelogram into two congruent triangles. Opposite angles of a parallelogram are congruent. Opposite sides of a parallelogram are congruent.

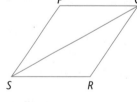

EXERCISES & PROBLEMS

Communicate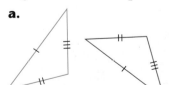

1. Given parallelogram *PQRS*, explain how you know $\triangle SPQ \cong \triangle QRS$.

2. Explain why $\angle QRS \cong \angle SPQ$.

3. Explain whether each pair of triangles could fit together to form a parallelogram and why.

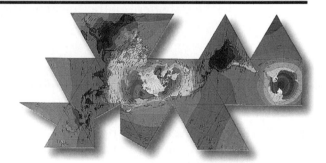

a.   **b.**   **c.**

  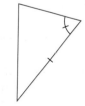 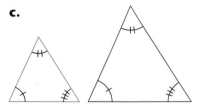

4. How many different pairs of congruent triangles are formed by the diagonals of a parallelogram?

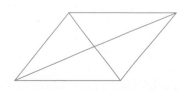

Practice & Apply

5–10. Draw a picture to show the given conditions, then complete the proof.

Given: Parallelogram *ABCD*, with diagonals $\overline{AC}$ and $\overline{BD}$ intersecting at point *E*

Prove: Diagonal $\overline{AC}$ bisects diagonal $\overline{BD}$ at point *E*.

Proof:

$\overline{AB}$ and $\overline{DC}$ are parallel by __(5)__ , so ∠BDC and ∠ABD are congruent __(6)__ angles. Also, ∠ACD and ∠CAB are congruent alternate interior angles. $\overline{AB} \cong \overline{DC}$, because __(7)__ sides of a parallelogram are __(8)__ . △ABE ≅ △CDE by __(9)__ . $\overline{BE} \cong \overline{DE}$, because __(10)__ . Therefore, point *E* is the midpoint of $\overline{BD}$ by the definition of a midpoint, and so $\overline{AC}$ bisects $\overline{BD}$ at point *E*.

11. Write a paragraph proof or a two-column proof for the following: Given parallelogram *ABCD*, with diagonals $\overline{AC}$ and $\overline{BD}$ intersecting at point *E*, prove $\overline{BD}$ bisects $\overline{AC}$ at point *E*.

12. Complete the statement of the theorem below, which your two proofs in Exercises 5–11 establish.

THEOREM	
The diagonals of a parallelogram __(12)__ each other.	**4.5.2**

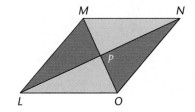

13–19. Fine Art An artist is preparing stained glass for a window using a repeating pattern of rhombuses with diagonals. The artist cuts several congruent triangles out of colored glass. Use the theorem you wrote in Exercise 12 for the following two proofs.

Given: Rhombus *LMNO* with diagonals $\overline{MO}$ and $\overline{LN}$ intersecting at point *P*

Prove: **A.** △MNP ≅ △ONP, **B.** △MNP ≅ △OLP

Part A:

STATEMENTS	REASONS
$\overline{MP} \cong \overline{PO}$	__(13)__
$\overline{PN} \cong \overline{PN}$	__(14)__
$\overline{MN} \cong \overline{ON}$	__(15)__
△MNP ≅ △ONP	__(16)__

Part B:

STATEMENTS	REASONS
$\overline{MP} \cong \overline{OP}$; $\overline{NP} \cong \overline{LP}$	__(17)__
$\overline{MN} \cong \overline{LO}$	__(18)__
△MNP ≅ △OLP	__(19)__

20. What do you need to add to the proof in Exercises 13–19 to prove the theorem below?

THEOREM
The diagonals and sides of a rhombus form four congruent triangles.   **4.5.3**

21–27. Use the theorem from Exercise 20 to help you prove the following.

Given: Rhombus *DEFG*, with diagonals $\overline{DF}$ and $\overline{GE}$ intersecting at *H*

Prove: The angles formed by the intersecting diagonals are right angles.

Proof:

$\triangle EHF \cong \triangle GHF$, because **(21)**. $\angle FHG \cong \angle FHE$, because **(22)**. Since $\angle FHG$ and $\angle FHE$ are a linear pair, the sum of the measures of the angles is **(23)**. Since $\angle FHG \cong \angle FHE$, their measures are equal and are therefore **(24)** degree angles or **(25)** angles. A similar relationship exists between $\triangle DHE$ and $\triangle DHG$. So $\angle DHE$ and $\angle DHG$ measure **(26)** degrees also. Therefore, $\angle DHE$, $\angle EHF$, $\angle FHG$, and $\angle GHD$ are **(27)** angles.

28. Complete the statement of the theorem below, which your proof in Exercises 21–27 establishes.

THEOREM
The diagonals of a rhombus are **(28)** .   **4.5.4**

29. Draw a picture to show the given conditions, then complete the proof. Hint: Since a rectangle is a parallelogram, you can use the theorems you proved about parallelograms in the Lesson 4.4 exercises.

Given: Rectangle *RSTU*, with diagonals $\overline{RT}$ and $\overline{US}$

Prove: $\overline{RT} \cong \overline{US}$.

30. Complete the statement of the theorem below, which your proof in Exercise 29 establishes.

THEOREM
The diagonals of a rectangle are **(30)** .   **4.5.5**

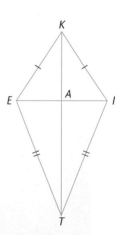

Kites A special quadrilateral called a kite has two pairs of congruent adjacent sides, and opposite sides are not congruent.

31. Using the given conditions and the diagram, write the required proof in paragraph or two-column form.

Given: Kite *KITE* with diagonals $\overline{KT}$ and $\overline{EI}$ intersecting at point *A*

Prove: $\overline{KT} \perp \overline{EI}$

32. Complete the statement of the theorem below, which your proof in Exercise 31 establishes.

THEOREM	
The diagonals of a kite are __(32)__ .	**4.5.6**

Look Back

33. Explain why if three angles of a triangle are congruent to three angles in another triangle, the triangles are not necessarily congruent. **[Lessons 4.2, 4.3]**

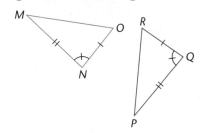

34. $\triangle ABC \cong \triangle DEF$ by which postulate? **[Lessons 4.2, 4.3]**

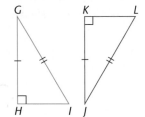

35. $\triangle GHI \cong \triangle JKL$ by which postulate? **[Lessons 4.2, 4.3]**

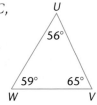

36. $\triangle MNO \cong \triangle PQR$ by which postulate? **[Lessons 4.2, 4.3]**

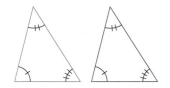

37. **Algebra** Given that $\triangle UVW \cong \triangle ABC$, find the value of x and y. **[Lesson 4.4]**

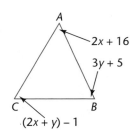

Look Beyond

38. Baseball A batter hits a ground ball at a 27° angle relative to the third-base line. The shortstop scoops up the ball and throws it to first base. The ball arrives at first base at an 18° angle relative to the first-to-second-base line. What angle did the shortstop turn to make the throw to first base?

Hint: A baseball diamond is a square.

206

Exploring

Conditions for Special Quadrilaterals

A rectangular foundation is essential to the construction of most modern buildings. Here workers are applying an important geometric principle to ensure that the foundation is a rectangle. In this lesson, you will discover what it is.

why *How do you know if a figure is a parallelogram, rectangle, or rhombus? Checking the definition works only if the required special properties are given. In earlier lessons you were told what kind of special quadrilateral a given figure was at the beginning of a problem. In this lesson, the procedure is reversed. You will learn to answer the question, "What does it take to make . . . ?"*

•Exploration *Parallelograms, Rectangles, and Rhombuses*

You will need

Toothpicks, popsicle sticks, or
    other materials to serve as
    sides for quadrilaterals or
Ruler and protractor

In the three parts that follow, decide whether each of the conjectures listed is true or false. If you believe a conjecture is false, prove that it is false by showing a counterexample. Make sketches of your results. If you believe a conjecture is true, consider how you would prove it.

Part I: What Does It Take to Make a Parallelogram?

For the following conjectures about parallelograms, answer True or False.

1 If one pair of opposite sides of a quadrilateral are congruent, then the quadrilateral is a parallelogram.

2 If two pairs of opposite sides of a quadrilateral are congruent, then the quadrilateral is a parallelogram.

3 If one pair of opposite sides of a quadrilateral are parallel and congruent, then the quadrilateral is a parallelogram.

4 If two pairs of sides of a quadrilateral are congruent, then the quadrilateral is a parallelogram.

5 If the diagonals of a quadrilateral bisect each other, then the quadrilateral is a parallelogram.

Part II: What Does It Take to Make a Rectangle?

For the following conjectures about rectangles, answer True or False.

1 If one angle of a quadrilateral is a right angle, then the quadrilateral is a rectangle.

2 If one angle of a parallelogram is a right angle, then the parallelogram is a rectangle.

3 If the diagonals of a quadrilateral are congruent, then the quadrilateral is a rectangle.

4 If the diagonals of a parallelogram are congruent, then the parallelogram is a rectangle.

5 If the diagonals of a parallelogram bisect the angles of the parallelogram, then the parallelogram is a rectangle.

6 If the diagonals of a parallelogram are perpendicular, then the parallelogram is a rectangle.

Part III: What Does It Take to Make a Rhombus?

For the following conjectures about rhombuses, answer True or False.

1 If one pair of consecutive sides of a quadrilateral are congruent, then the quadrilateral is a rhombus.

2 If one pair of adjacent sides of a parallelogram are congruent, then the parallelogram is a rhombus.

3 If the diagonals of a parallelogram are congruent, then the parallelogram is a rhombus.

4 If the diagonals of a parallelogram bisect the angles of the parallelogram, then the parallelogram is a rhombus.

5 If the diagonals of a parallelogram are perpendicular, then the parallelogram is a rhombus. ❖

In Lessons 4.4 and 4.5 you studied a number of theorems about the properties of parallelograms and other special quadrilaterals. For example,

The opposite sides of a parallelogram are congruent. **(Theorem 4.4.7)**

In each theorem, the special quadrilateral is the "given," and the conclusion follows. An "if-then" version of the theorem may make this relationship clearer:

If a quadrilateral is a parallelogram, then its opposite sides are congruent.

As you may have noticed, one of the conjectures you tested in the exploration is the converse of this theorem. Which one is it? ❖

Try This   Make a list of the exploration conjectures that you believe to be true. Write the converses of those conjectures. Which of these converses are theorems that you have already studied? State those theorems.

Cultural Connection: Africa   Mozambican workers traditionally use long poles to establish the sides of their house foundations. They make a rectangular foundation in the following two steps:

1. The workers use two pairs of poles of the desired lengths for congruent opposite sides. This guarantees that the base of the house is a parallelogram—but not necessarily a rectangle.

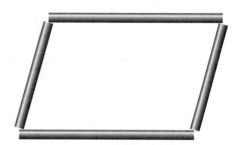

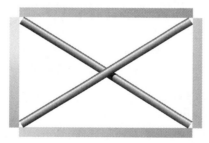

2. The workers measure the diagonals. If the diagonals are congruent, then the base of the house is a rectangle. If not, the workers reposition the poles as necessary to make the diagonals congruent. ❖

CRITICAL
Thinking

The lamp in the picture remains parallel to the wall as it moves up and down. How does the first part of the Mozambican procedure explain the motion?

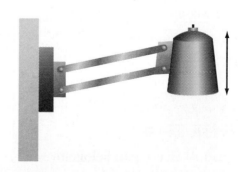

EXERCISES & PROBLEMS

Communicate

1. Summarize your investigations in this lesson by giving two or three answers to each question.

   a. What does it take to make a parallelogram?

   b. What does it take to make a rectangle?

   c. What does it take to make a rhombus?

2. All of the figures below are parallelograms. Decide whether they must also be rectangles or rhombuses based on the given information (figures are not drawn to scale).

a.
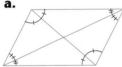

b.

c.

Practice & Apply

3–8. Complete the paragraph proof of the following theorem.

THEOREM

If two pairs of opposite sides of a quadrilateral are congruent, then the quadrilateral is a parallelogram.   **4.6.1**

Given: $\overline{BA} \cong \overline{SE}; \overline{BE} \cong \overline{SA}$

Prove: *BASE* is a parallelogram

Plan: Use congruent sides to get $\triangle BAS \cong \triangle SEB$. This gives congruent corresponding angles, which can be used to get parallel lines, satisfying the definition of a parallelogram.

Proof:

STATEMENTS	REASONS
$\overline{BA} \cong \overline{SE}; \overline{BE} \cong \overline{SA}$	_____ **(3)** _____
$\overline{BS} \cong \overline{SB}$	_____ **(4)** _____
$\triangle BAS \cong \triangle SEB$	_____ **(5)** _____
$\angle 1 \cong \angle 2; \angle 3 \cong \angle 4$	_____ **(6)** _____
$\overline{AS} \parallel \overline{EB}; \overline{BA} \parallel \overline{SE}$	_____ **(7)** _____
Quad *BASE* is a parallelogram	_____ **(8)** _____

9–17. Complete the proof of the following theorem.

THE "HOUSEBUILDER" RECTANGLE THEOREM

If the diagonals of a parallelogram are congruent, then the parallelogram is a rectangle. **4.6.2**

Given: *BASE* is a parallelogram; $\overline{BS} \cong \overline{EA}$

Prove: *BASE* is a rectangle.

Proof:

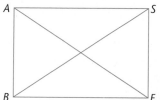

STATEMENTS	REASONS
BASE is a parallelogram; $\overline{BS} \cong \overline{EA}$	_____ **(9)** _____
$\overline{BA} \cong \overline{ES}$	_____ **(10)** _____
$\overline{BE} \cong \overline{EB}$	_____ **(11)** _____
$\triangle ABE \cong \triangle SEB$	_____ **(12)** _____
$\angle ABE \cong \angle SEB$	_____ **(13)** _____
$\angle ABE$ and $\angle SEB$ are supplementary	_____ **(14)** _____
$\angle ABE$ and $\angle SEB$ are right angles	_____ **(15)** _____
$\angle ASE$ and $\angle SAB$ are right angles	_____ **(16)** _____
BASE is a rectangle	_____ **(17)** _____

In Exercises 18–21 prove each of the following theorems, using either a paragraph or two-column form. (Set up your own "Given" and "Prove" statements).

18.

THEOREM

If one pair of opposite sides of a quadrilateral are parallel and congruent, then the quadrilateral is a parallelogram. **4.6.3**

19.

THEOREM

If the diagonals of a quadrilateral bisect each other, then the quadrilateral is a parallelogram. **4.6.4**

20.

THEOREM

If one angle of a parallelogram is a right angle, then the parallelogram is a rectangle. **4.6.5**

21.

THEOREM

If one pair of adjacent sides of a parallelogram are congruent, then the quadrilateral is a rhombus. **4.6.6**

22–30. Complete the proof of the following theorem.

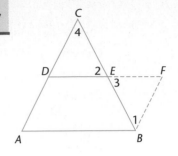

THE TRIANGLE MIDSEGMENT THEOREM
If a segment joins the midpoints of two sides of a triangle, then it is parallel to the third side and its length is one-half the length of the third side. **4.6.7**

Given: $\triangle ABC$, $\overline{AD} \cong \overline{CD}$, $\overline{BE} \cong \overline{CE}$

Prove: $\overline{DE} \parallel \overline{AB}$ and $DE = \frac{1}{2} AB$

Add two segments to $\triangle ABC$ as shown, with $\overline{EF} \cong \overline{DE}$.

STATEMENTS	REASONS
$\triangle ABC$, $\overline{AD} \cong \overline{CD}$, $\overline{BE} \cong \overline{CE}$, $\overline{EF} \cong \overline{DE}$ E is the midpoint of $\overline{DF}$	Given
$\angle 2 \cong \angle 3$	**(22)**
$\triangle CED \cong \triangle BEF$	**(23)**
$\angle 1 \cong \angle 4$	**(24)**
$\overline{AC} \parallel \overline{FB}$	**(25)**
$\overline{CD} \cong \overline{BF}$	**(26)**
$\overline{AD} \cong \overline{BF}$	Transitive Property of Congruence
$ABFD$ is a parallelogram	**(27)**
$\overline{DF} \cong \overline{AB}$ $(DF = AB)$	**(28)**
$DE = \frac{1}{2} DF$	Segment Addition, Substitution
$DE = \frac{1}{2} AB$	**(29)**
$\overline{DE} \parallel \overline{AB}$	**(30)**

Look Back

31. $\angle 1$ and $\angle 2$ form a linear pair. What special relationship exists between these two angles?
[Lesson 1.5]

32. Given: $\overline{BA} \cong \overline{BC}$; $\overline{BD} \cong \overline{BE}$
Prove: $\triangle ADC \cong \triangle CEA$
Hint: $\triangle ABC$ is isosceles.
[Lesson 4.4]

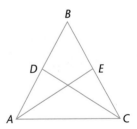

Look Beyond

33. Plot the 4 points in a coordinate plane and draw quadrilateral $QUAD$ with diagonal $\overline{UD}$.

$Q(2, 4)$ $U(2, 9)$ $A(10, 9)$ $D(10, 4)$

34. What type of quadrilateral is $QUAD$?

35. What is the perimeter of $QUAD$?

36. What is the area of $QUAD$?

37. What is the relationship between $\triangle QUD$ and $\triangle ADU$?

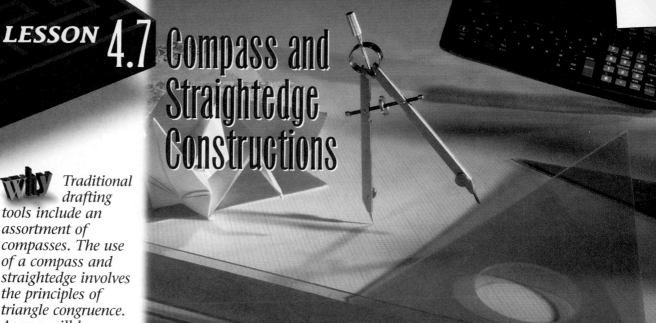

LESSON 4.7 Compass and Straightedge Constructions

Why *Traditional drafting tools include an assortment of compasses. The use of a compass and straightedge involves the principles of triangle congruence. As you will learn, compasses are useful for much more than drawing circles.*

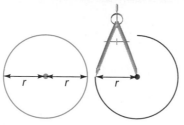

Using the same compass setting

Whenever you draw circles (or parts of circles) using the same compass settings, all the circles will have congruent radii. This is one of the keys to the classical constructions.

CONGRUENT RADII THEOREM
In the same circle, or in congruent circles, all radii are congruent. **4.7.1**

CRITICAL *Thinking*

Explain why Theorem 4.7.1 is true. (Use the fact that every point on a circle is the same distance from the center point.)

EXAMPLE 1

Given: $\overline{AB}$

Construct: A segment congruent to $\overline{AB}$

A •———————————• B

1. Using your straightedge, draw line l.
2. Select a point on l and label it O.
3. Set your compass equal to the distance AB. Use the point of your compass on O and draw an arc that intersects l. Label the intersection of the arc and the segment as point P.

Conclusion: $\overline{OP} \cong \overline{AB}$ ❖

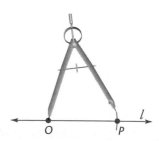

How can you use the Congruent Radii Theorem to justify the conclusion of this construction? Where have you used this construction method before?

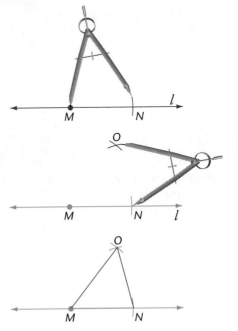

EXAMPLE 2

Given: △ABC

Construct: A triangle congruent to △ABC

1. Using your straightedge, draw a line and label it *l*. Select a point on *l* and label it *M*.

   Set your compass equal to the distance *AB*. Place the point of your compass on *M* and draw an arc that intersects *l*. Label the intersection of the arc and the segment *N*.

2. Set your compass equal to the distance *AC*. Place the point of your compass on *M* and draw an arc in the area above.

   Set your compass equal to the distance *BC*. Place the point of your compass on *N* and draw an arc that intersects the arc you drew in Step 2. Label the intersection of the two arcs *O*.

3. Using your straightedge, draw a line segment that connects points *M* and *O*. Then draw a segment that connects points *N* and *O*.

   Conclusion: △MNO ≅ △ABC ❖

What postulate justifies the final conclusion of the construction?

EXAMPLE 3

Given: ∠BAC

Construct: An angle bisector of ∠BAC using geometry software.

1. Construct ∠BAC.

2. Construct a circle with its center at *A*. Label the points of intersection between ∠BAC and the circle as points *D* and *E*.

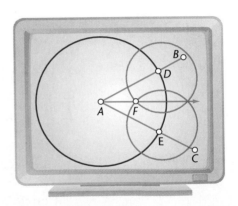

3. Using *D* and *E* as centers, construct two congruent circles. Use the measure menu to be sure that they are congruent. Make them large enough to intersect with each other. Label a point of intersection between them *F*.

4. Construct $\overline{AF}$. This is the angle bisector.

5. To check, measure ∠*BAF* and ∠*CAF*. ❖

EXAMPLE 4

Given: The diagram on the right, which illustrates the compass and straightedge method of construction of the bisector of ∠*BAC*

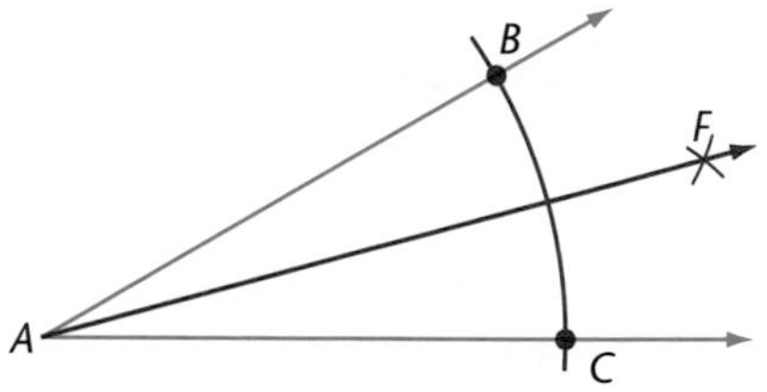

Prove: ∠*BAF* ≅ ∠*CAF*

Plan: Draw $\overline{BF}$ and $\overline{CF}$. Show $\overline{BF} \cong \overline{CF}$ since they are congruent radii, and △*BAF* ≅ △*CAF* by SSS. Then use CPCTC.

Proof:

STATEMENTS	REASONS
1. $\overline{AB} \cong \overline{AC}$	In the same or ≅ ⊙s all radii are ≅ (Definition of a circle).
2. $\overline{CF} \cong \overline{BF}$	In the same or ≅ ⊙s all radii are ≅ (Definition of a circle).
3. $\overline{AF} \cong \overline{AF}$	Reflexive Property of Congruence
4. △*BAF* ≅ △*CAF*	SSS
5. ∠*BAF* ≅ ∠*CAF*	CPCTC ❖

EXERCISES & PROBLEMS

Communicate

1. Given $\overline{AB}$ as shown, tell how you would construct $\overline{XY}$ congruent to $\overline{AB}$.

2. State the theorem that justifies your construction in Exercise 1.

3. Explain how the Congruent Radii Theorem makes constructions with a straightedge and compass possible.

4. Explain how you would construct an angle bisector with a straightedge and compass.

5. Using part of the technique from Example 2 in the lesson, explain how you would copy an angle using only a straightedge and compass.

Practice & Apply

Trace △GHJ onto your own paper and use it for Exercises 6–7.

6. Construct △DEF ≅ △GHJ.

7. Construct the angle bisector of ∠D in your constructed triangle.

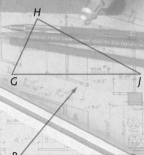

8. Trace ∠R onto your own paper and follow Steps a–f to construct ∠B ≅ ∠R.

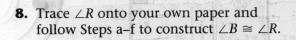

a. Using a straightedge, draw a ray with endpoint B.

b. Place your compass point on R in your original angle and draw an arc as shown. Label the intersection points Q and S.

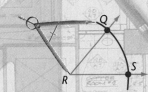

c. Use the same compass setting and place the point on B. Draw an arc crossing the ray and label the point of intersection C.

d. Set your compass equal to the distance QS in ∠R.

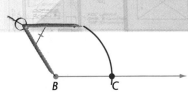

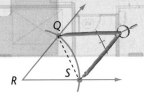

e. Keep the same compass setting and place the point on C. Draw an arc crossing the first arc you made on ray B⃗C. Label the point of intersection A.

f. Draw ray B⃗A, forming ∠B.

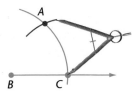

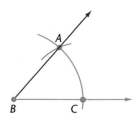

9. Refer to the construction in Exercise 8. Write a paragraph proving ∠ABC ≅ ∠QRS. (Hint: Construct segments $\overline{QS} ≅ \overline{AC}$.)

10. Trace $\overline{AC}$ onto your own paper and follow the steps to construct the perpendicular bisector of $\overline{AC}$.

a. Set your compass equal to a distance greater than half the length of $\overline{AC}$.

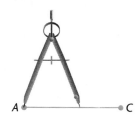

b. Place your compass point on A and draw an arc as shown.

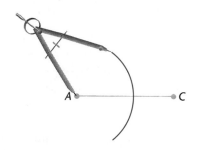

c. Using the same compass setting, place the compass point on C and draw a new arc. Label intersection points B and D.

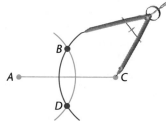

d. Use a straightedge to draw a line through points B and D. Label intersection E.

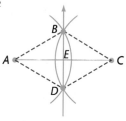

11–15. Using the construction in Exercise 10, prove that $\overline{BD}$ is the perpendicular bisector of $\overline{AC}$. Complete the paragraph proof.

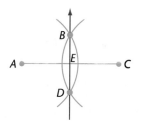

$\overline{AB} \cong \overline{BC} \cong \overline{CD} \cong \overline{DA}$, because **(11)**. Therefore, quadrilateral $ABCD$ is a rhombus, by **(12)**. **(13)**, because the diagonals of a rhombus are perpendicular to each other. A rhombus is also a parallelogram, by **(14)**. Therefore, $\overline{BD}$ and $\overline{AC}$ bisect each other, because **(15)**.

16. Draw a line l. Place point A above line l. Follow Steps a–d to construct a line perpendicular to l through A.

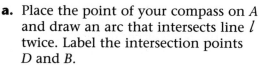

a. Place the point of your compass on A and draw an arc that intersects line l twice. Label the intersection points D and B.

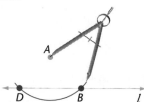

b. Place your compass point on D and draw an arc below line l. It is not necessary to keep the compass set at the same distance as Step a.

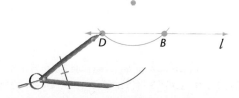

c. Using the same compass setting as Step b, place the compass point at *B* and draw an arc which intersects the arc you drew in Step b. Label the point of intersection *C*.

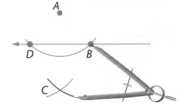

d. Using a straightedge, draw a line which passes through *A* and *C*. $\overline{AC}$ is perpendicular to *l*.

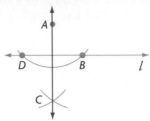

17–21. Kites You can create a kite by connecting the endpoints *A*, *B*, *C*, and *D* from the construction in Exercise 16. (Assume that a different compass setting was used to construct the arcs through *C* than was used to construct the arcs through *D* and *B*.) Prove that $\overline{AC} \perp \overline{DB}$.

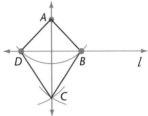

Proof:

STATEMENTS	REASONS
$\overline{AB} \cong$ __(17)__ and __(18)__ $\cong \overline{DC}$	**Congruent Radii Theorem**
$AB \neq DC$ and $AD \neq$ __(19)__	Different compass setting used to construct each segment
Quadrilateral *ABCD* is a kite	__(20)__
$\overline{AC} \perp \overline{DB}$	__(21)__

22. Draw a line and place point *M* anywhere above *l*. Follow the steps to construct a line parallel to *l* through point *M*.

a. Using a straightedge construct a transversal line to *l* through *M*. Label the point of intersection *P*.

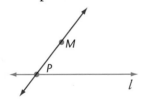

b. Place your compass point on *P* and draw an arc crossing both rays of ∠*P*. Label the intersection points *R* and *T*.

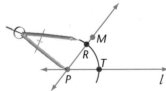

c. Using the same compass setting, place the compass point on *M* and draw an arc that crosses the transversal. Label the point of intersection *N*.

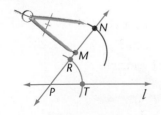

d. Set your compass to distance *RT*. Place the compass point on *N* and draw an arc that intersects the arc you drew in Step c. Label the point of intersection *O*.

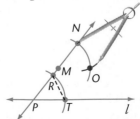

e. Using a straightedge draw the line which passes through points M and O. Label the line u. Line u is parallel to line l.

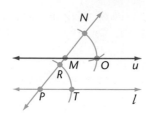

23. Write a two-column proof for the construction in Exercise 22. That is, prove that $u \parallel l$.

Look Back

Categorize each figure according to this list: parallelogram, quadrilateral, rectangle, square, rhombus, kite, or trapezoid. Use all appropriate terms with each. [Lessons 3.2, 3.3]

24. $\overline{AB} \parallel \overline{CD}$

25.

26.

27. $\overline{AB} \parallel \overline{CD}; \overline{AC} \parallel \overline{BD}$

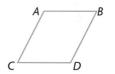

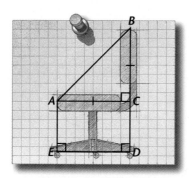

28. Interior Design Interior designers often use geometric figures to create interesting furniture designs. Identify at least 10 pairs of congruencies in the design of this chair. **[Lessons 4.1, 4.2, 4.3]**

Look Beyond

History The French Revolutionary general Napoleon was an amateur mathematician. Napoleon found the following construction especially intriguing.

29. a. Draw a circle using your compass.

b. Without changing your compass setting start at the top of the circle and draw an arc that crosses the circle. Move your compass point to the point of intersection and draw another arc. Continue until your are back at the top of the circle.

c. Connect the points of intersection. What have you created? How do the sides of this polygon relate to the radius of the circle?

30. Using the results from Exercise 29, construct an equilateral triangle inscribed in a circle.

31. Explain how you could construct a square inscribed in a circle.

Exploring

Congruence in the Coordinate Plane

Why *If you set up a coordinate plane with the same horizontal and vertical units, you can use techniques you learned in your study of algebra to analyze congruence.*

EXAMPLE

Given: △ABC with A(2, 1), B(5, 3), and C(5, 1); △XYZ with X(7, 2), Y(10, 4), and Z(10, 2)

Prove: $\overline{AB} \cong \overline{XY}$

Solution ➤

The triangles each have one horizontal segment and one vertical segment. Therefore, the angles between these segments are right angles, because of the way standard coordinate axes are set up.

Find the measures of the horizontal segments by finding the difference in the *x*-coordinates. Then find the measures of the vertical segments by finding the absolute value of the difference in the *y*-coordinates.

Horizontal distances: $AB = |5 - 2| = 3$, $XY = |10 - 7| = 3$

Vertical distances: $BC = |1 - 3| = 2$, $YZ = |2 - 4| = 2$

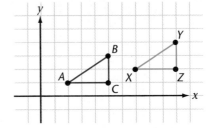

Notice which sides and angles are congruent in the given triangles. Therefore, △ABC ≅ △XYZ by SAS and $\overline{AB} \cong \overline{XY}$ by CPCTC. ❖

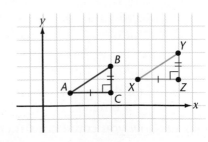

•Exploration 1 *Midpoints of Horizontal Segments*

You will need
Coordinate paper
Compass and straightedge or
Ruler

1 Plot each segment in a coordinate plane. Determine the coordinates of the midpoint of each segment. (You may use a compass and straightedge, a ruler, or paper folding.)

$\overline{AB}$, with $A(2, 3)$ and $B(10, 3)$ $\overline{XY}$, with $X(4, -2)$ and $Y(8, -2)$
$\overline{LN}$, with $L(-2, 5)$ and $N(10, 5)$ $\overline{CD}$, with $C(-10, 6)$ and $D(-2, 6)$

2 What rule can be applied to the *x*-coordinates of the endpoints of a horizontal segment to obtain the *x*-coordinates of the midpoint of the segment?

3 Apply your rule to find the coordinates of the midpoint of $\overline{KL}$, with $K(17, 432)$ and $L(95, 432)$. Find the average of the *x*-coordinates of *K* and *L* as follows: $(17 + 95) \div 2 = ?$ What do you notice about the result? ❖

CRITICAL *Thinking*

Write a formula for finding the midpoint of a horizontal segment with endpoints $A(x_1, y_1)$ and $B(x_2, y_2)$. Use the average of the *x*-coordinates. How could you find the formula for the midpoint of a vertical segment?

•Exploration 2 *Midpoints of General Segments*

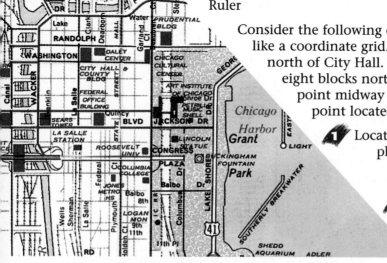

You will need
Coordinate paper
Ruler

Consider the following question. In some cities, streets are laid out like a coordinate grid. Ron lives three blocks east and two blocks north of City Hall. Ron's friend Jaime lives 11 blocks east and eight blocks north of City Hall. They want to meet at the point midway between their locations. Where is this point located.

1 Locate City Hall at the origin of the coordinate plane and let the direction of the positive *x*-axis represent east and that of the positive *y*-axis represent north. Plot $R(3, 2)$ and $J(11, 8)$.

2 Draw a segment with endpoints that pass through points *R* and *J*. Then draw a horizontal line through *R* and a vertical line through *J*. What kind of geometric figure is formed?

The "origin" of the Chicago grid is at the intersections of Washington and State streets.

3 Label the point where the horizontal and vertical lines intersect at Q. Find the coordinates of Q and label your drawing with them.

4 Use your knowledge of the midpoints of horizontal and vertical segments to find the coordinates of V (the midpoint of $\overline{JQ}$) and the coordinates of H (the midpoint of $\overline{RQ}$). Add these to your drawing.

5 Draw a vertical segment from H to intersect $\overline{RJ}$ and a horizontal segment from V to intersect $\overline{RJ}$. Label their intersection as point M. Does M lie on $\overline{RJ}$?

6 Measure the distances RM and MJ. What can you conclude about point M?

7 Find the coordinates of M and add them to your drawing. Compare the coordinates of M to the coordinates of H and V. What do you observe?

8 In your own words, write a rule for finding the midpoint of a segment that is not horizontal or vertical.

9 Express the rule you stated in words in Step 8 as an algebraic formula for finding the midpoint of a segment $\overline{AB}$ with endpoints $A(x_1, y_1)$ and $B(x_2, y_2)$. ❖

CRITICAL *Thinking*  How can you use congruent triangles in your exploration drawing to prove that your midpoint rule is correct? (*MVQH* is a rectangle.)

Exercises & Problems

Communicate

1. Explain how to find the midpoint of a horizontal segment.

2. Explain how to find the midpoint of a vertical segment.

3. Explain the formula for finding the coordinates of the midpoint of a segment in general.

4. $\overline{AB}$ and $\overline{CB}$ are drawn on a coordinate grid with the following coordinates: $A(2, 3)$, $B(7, 3)$, $C(7, 8)$. What can you conjecture about $\overline{AB}$ and $\overline{CB}$?

5. Given: $\triangle CAT$ with $C(2, 3)$, $A(2, 7)$, and $T(4, 3)$; $\triangle DOG$ with $D(2, -3)$, $O(2, -7)$, and $G(4, -3)$. Explain how to prove $\triangle CAT \cong \triangle DOG$.

Practice & Apply

Use a coordinate grid and determine whether the following pairs of segments are congruent.

6. $\overline{AB}$ and $\overline{CD}$, with $A(2, 6)$, $B(2, 10)$, $C(5, 1)$, and $D(5, 5)$

7. $\overline{PQ}$ and $\overline{RS}$, with $P(4, 3)$, $Q(4, 0)$, $R(-2, 5)$, and $S(-2, 12)$

8. $\overline{VE}$ and $\overline{HO}$, with $V(3, -1)$, $E(3, 5)$, $H(5, 4)$, and $O(11, 4)$

9. $\overline{TO}$ and $\overline{OP}$, with $T(-3, 0)$, $O(0, 5)$, and $P(3, 0)$

Use the midpoint formula from Exploration 2 to find the x- and y-coordinates of the midpoint for each segment.

10. $\overline{AB}$, with $A(1, 2)$ and $B(9, 2)$

11. $\overline{ST}$, with $S(4, 5)$ and $T(4, -3)$

12. $\overline{XY}$, with $X(4, 3)$ and $Y(9, 11)$

13. $\overline{PQ}$, with $P(-2, 5)$ and $Q(-8, -7)$

14. Plot the points and draw the figures on the coordinate plane:

$\triangle LCO$ and $\triangle RCO$ with $L(-5, 0)$, $C(0, 3)$, $O(0, 0)$, and $R(5, 0)$

15. Given the figures on the coordinate plane from Exercise 14, prove that $\triangle LCR$ is isosceles.

16. Spreadsheets Create a spreadsheet that will allow the user to specify the coordinates of the endpoints of a segment and that will compute the coordinates of the midpoint of the segment. Using the following column headings, insert the formulas for the x-coordinate and the y-coordinate of the midpoint in the appropriate cells of the spreadsheet.

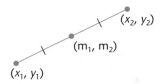

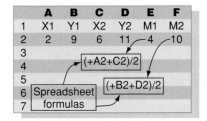

For Exercises 17–23, plot $\triangle PQR$, with $P(1, 2)$, $Q(3, 8)$, and $R(11, 4)$.

17. Let X be the midpoint of $\overline{PQ}$. Find its coordinates and plot it.

18. Let Y be the midpoint of $\overline{QR}$. Find its coordinates and plot it.

19. Draw midsegment $\overline{XY}$. What type of figure is $PXYR$? Justify your answer.

20. Let Z be the midpoint of $\overline{PR}$. Find its coordinates and plot it.

21. Draw midsegment $\overline{YZ}$. What type of figure is $PXYZ$? Justify your answer.

22. Find two segments in the figure that are congruent to $\overline{PZ}$.

23. What is the relationship between the length of $\overline{PR}$ and the length of $\overline{XY}$ based on information obtained above?

Technology   Use a graphics calculator to create the following figures. Write the equations of the lines that you plot.

24. A parallelogram that is not necessarilly a rhombus, with one pair of sides parallel to the x-axis

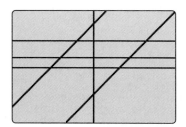

25. A parallelogram that is not necessarilly a rhombus, with no sides parallel to the coordinate axes

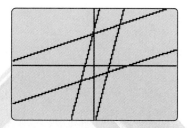

26. A rhombus that is not necessarilly a square

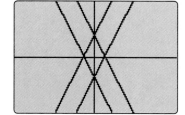

27. A square with no sides parallel to the coordinate axes

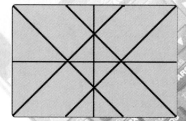

For Exercises 28–33, plot quadrilateral $QUAD$ in a coordinate plane. Determine what type of quadrilateral $QUAD$ is, and justify your answer.

28. $Q(-2, 1)$, $U(0, 4)$, $A(5, 4)$, $D(3, 1)$

29. $Q(1, 5)$, $U(3, 7)$, $A(3, 3)$, $D(1, 1)$

30. $Q(1, 3)$, $U(3, 6)$, $A(7, 7)$, $D(5, 4)$

31. $Q(-5, 1)$, $U(-2, 5)$, $A(2, 6)$, $D(3, 3)$

32. $Q(1, 6)$, $U(3, 10)$, $A(5, 6)$, $D(3, 2)$

33. $Q(-2, 1)$, $U(2, 5)$, $A(5, 2)$, $D(1, -2)$

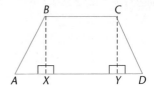

34. Write a two-column proof for the following:

Given: *ABCD* is an isosceles trapezoid with $\overline{BC} \parallel \overline{AD}$ and $\overline{AB} \cong \overline{DC}$; $\overline{BX} \perp \overline{AD}$; $\overline{CY} \perp \overline{AD}$

Prove: $\angle A \cong \angle D$

Hint: What must be true about $\overline{BX}$ and $\overline{CY}$?
[Lessons 4.5, 4.6]

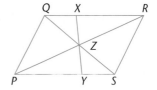

35. Write a two-column proof for the following:

Given: *PQRS* is a parallelogram; $\overline{QS}, \overline{PR}$, and $\overline{XY}$ intersect at *Z*

Prove: $\overline{XZ} \cong \overline{YZ}$
[Lessons 4.5, 4.6]

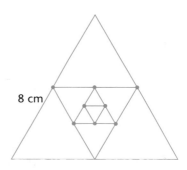

Look Beyond

Use the figure drawn in Exercise 36 to answer Exercises 37–40.

36. Draw an equilateral triangle 8 cm on a side. Then connect the midpoints of the sides of the triangle to form a second triangle inside the first. Next, draw a third triangle by connecting the midpoints of the sides of the second triangle. Continue with as many triangles as it is possible to draw.

37. What type of triangles are the second and third triangles?

38. What is the ratio of the perimeters of the first, second, and third triangles?

39. Find the sum of the perimeters of the second, third, fourth, and fifth triangles. You may wish to continue with additional triangles—sixth, seventh, eighth, etc.

40. How does the perimeter of the first triangle compare with the sum of the perimeters of the additional triangles? What happens to this answer as the perimeters of additional triangles are added to this sum?

Exploring

The Construction of Transformations

Why The transformations you have studied have all been rigid. That is, the size and shape of the objects that are transformed do not change. You can now prove that the transformations, as you have defined them, preserve congruence.

The ability to move in strict formation, without changing the size and shape of the pattern, is one of the requirements of a good synchronized swimming team.

In Lesson 1.6 you created your own definitions of three different transformations. A **translation**, for instance, is a transformation that moves every point of an object the same distance in the same direction. In the following exploration you will learn how to translate a segment by relocating and connecting the endpoints.

•Exploration 1 *Translating a Segment*

You will need
Compass and straightedge
Ruler

Make your own drawing like the one on the right. The slide arrow shows the direction and distance of the translation you are to construct. (The distance is the overall length of the arrow from its "head" to its "tail.")

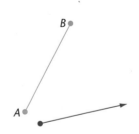

1 Construct a line l_1 through point A that is parallel to the slide arrow. Construct a line l_2 through point B that is parallel to the slide arrow. Are lines l_1 and l_2 parallel? What theorem justifies your answer?

2 Set your compass to the length s of the slide arrow. On the right hand side of $\overline{AB}$, construct points A' and B' that are a distance s from points A and B on lines l_1 and l_2, respectively.

3 Connect points *A'* and *B'*. Measure $\overline{A'B'}$. Is $\overline{A'B'}$ congruent to $\overline{AB}$?

4 Explain why the two segments must be congruent. Write a proof that they are. ❖

Exploration 1 shows you that the size and shape of a translated segment are exactly the same as in the pre-image. In Exploration 2, you will translate a polygon one line at a time, as in the previous method. Do you think each of the translated figures will have the same size and shape as each of the original figures?

•Exploration 2 *Congruence in Polygon Translations*

You will need
No special tools

1 If you translate each of the three segments that form triangle △*ABC* using the method of Exploration 1, would your new triangle be congruent to the original one? State the theorem or postulate that justifies your answer.

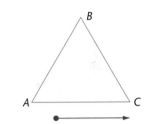

2 If you translate each of the four segments that form quadrilateral *ABCD* using the method of Exploration 1, would your new quadrilateral be congruent to the original one? State the theorem or postulate that justifies your answer. (Hint: Break the figure down into triangles by drawing a diagonal.)

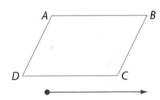

3 Show how you can break down any polygon into a number of triangles by connecting the vertices with segments. Include both convex and concave polygons in your discussion. What can you conclude about the translation of any polygon using the method of Exploration 1?

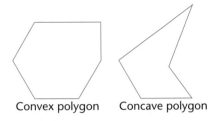

Convex polygon   Concave polygon

4 What can you conclude about the translation of an "open" figure such as the one on the right? ❖

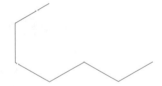

CRITICAL *Thinking*

Do you think your method of translating a figure can be applied to a curve? Hint: How can you use a number of segments to approximate a curve?

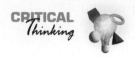

CRITICAL *Thinking* It has been said that the rigid transformations, which are known as isometries, preserve Angles, Betweenness, Collinearity, and Distance ("ABCD"). How do Explorations 1 and 2 show that translations preserve angles and distances?

If point X is between points A and B in a pre-image of a figure, can you be sure that the image point X' will be between image points A' and B' in a translated image of the figure? Put another way, can you be sure that translations preserve betweenness?

To answer the question, you will need a mathematical way to determine whether one point is between two other points. The following postulate, which is the converse of a postulate you have already studied (1.4.2), gives a method.

CONVERSE OF THE SEGMENT ADDITION POSTULATE ("BETWEENNESS")
Given three points P, Q, and R, if $PQ + QR = PR$, then Q is between P and R. **4.9.1**

Try This For the points R, S, and T, $RS = 8.7750$, $TR = 9.6540$, and $TS = 18.4290$. Assuming the distances are exact, are the points collinear? If so, which point is between the other two?

Exploration 3 *Betweenness in Translations*

You will need
No special tools

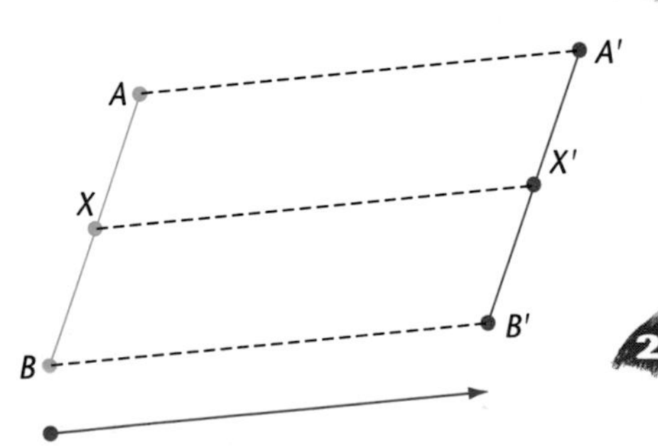

1. You are given a segment $\overline{AB}$ with a point X chosen anywhere between A and B. Translate each of the three marked points as indicated by the slide arrow. Then connect the points as in Exploration 1.

   Does X' lie on $\overline{A'B'}$? Is X' between A' and B'? You will answer these questions mathematically in the steps that follow.

2. What do you know about the distances AB and $A'B'$? About the distances AX and $A'X'$? about the distances BX and $B'X'$? Explain your reasoning.

3. From the Segment Addition Postulate you know that $AX + XB = AB$. What can you conclude about $A'X' + X'B'$? Explain your reasoning.

4. Does X' lie on $\overline{A'B'}$? (Is it collinear with A' and B'?) Is it between A' and B'? Explain your reasoning. ❖

You will consider the question of collinearity of points in translations in the exercise set.

EXERCISES & PROBLEMS

Communicate

1. In your own words, state a theorem describing the congruence relationship in Exploration 1.

2. In your own words, state a theorem describing the congruence relationship in Exploration 2.

3. Explain the difference between copying a segment or triangle by construction and translating a segment or triangle by construction.

4. Explain some construction techniques that could be used to construct a rotation and a reflection.

Practice & Apply

5. During a marching-band show, a group of band members moves in a triangle shape across the field as shown. Players X, Y, and Z are the section leaders. What must the section leaders do to ensure that $\triangle XYZ \cong \triangle X'Y'Z'$?

6. In Exercise 5, what must the other marchers who are not section leaders do to ensure that $\triangle XYZ \cong \triangle X'Y'Z'$?

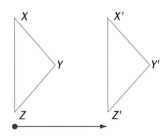

In $\triangle ABC$, B is *not* between A and C.

7. What seems to be true about $AB + BC$ compared with AC?

8. Suppose point B is relocated so that $AB + BC = AC$. What happens to $\triangle ABC$?

9. Use the results of Exercises 7 and 8 to complete the following statement, which is known as the Triangle Inequality Postulate.

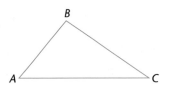

TRIANGLE INEQUALITY POSTULATE

The sum of the lengths of any two sides of a triangle is __?__.   **4.9.2**

---

10. If one point is between two others, can you conclude that the three points are collinear? (Consider what happens to the triangle in Exercise 8.) If a transformation preserves betweenness, must it also preserve collinearity? Explain.

In a known rotation, *every point of a figure is moved the same angle about a fixed point known as the center of rotation.* **In Exercises 11–14 you will use this idea to rotate a segment 40° about a point. (You will need a compass and protractor.)**

11. You are given a segment $\overline{AB}$ and point P not on the segment. Draw or trace $\overline{AB}$ and P on your own paper.

12. Draw $\overline{PA}$. Set your compass at distance PA and draw an arc through A. Use a protractor to draw $\angle A'PA$ with a measure of 40°.

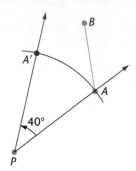

13. Draw $\overline{PB}$. Set your compass at distance PB and draw an arc through B. Use a protractor to draw $\angle BPB'$ with measure 40°.

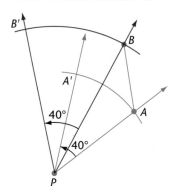

14. Draw $\overline{A'B'}$.

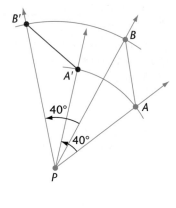

Answer the following questions (Exercises 15–17) about the above construction (Exercises 11–14).

15. What can you conclude about $\angle BPA$ and $\angle B'PA'$? (You will need to use the Overlapping Angle Theorem.) Explain.

16. What can you conclude about $\overline{AB}$ and $\overline{A'B'}$? Explain your reasoning.

17. How could rotations greater than 180° be drawn? (Hint: 360° − 290° = 70°)

In a reflection, *every segment connecting a point and its pre-image is bisected at right angles by a line known as the "mirror" of the transformation.* **In Exercises 18–27 you will use this idea in a proof.**

18. Draw or trace segment $\overline{AB}$, line l, and reflection $\overline{A'B'}$. Draw segments $\overline{AA'}$ and $\overline{BB'}$. Label the intersections with l as X and Y.

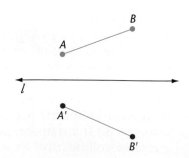

19. Label the right angles in your figure. Explain your reasoning.

20. Label the congruent segments in your figure. Explain your reasoning.

21. Label the parallel segments in your figure. Explain your reasoning.

22. Draw line *m* through *A* and parallel to *l*. Draw line *n* through *A'* and parallel to *l*. Label the points of intersection *C* and *C'* as shown.

23. What kind of quadrilaterals are *ACYX* and *A'C'YX*? Explain your reasoning.

24. What kind of quadrilateral is *ACC'A'*? Explain your reasoning.

25. What can you conclude about $\overline{BC}$ and $\overline{B'C'}$? Explain your reasoning.

26. What can you conclude about △*ABC* and △*A'B'C'*? Explain your reasoning.

27. What must be true about $\overline{AB}$ and $\overline{A'B'}$? Explain your reasoning.

28. Rotate a triangle about a point by 35°.

29. Use only a compass and a straightedge to rotate a triangle about a point. You may choose the amount of rotation. Remember to use the angle copying method from Lesson 4.7.

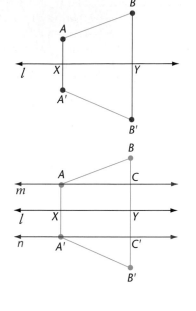

Look Back

30. Draw or trace the segment and construct the perpendicular bisector. **[Lesson 4.7]**

31. Draw or trace the angle and construct a copy of the angle. **[Lesson 4.7]**

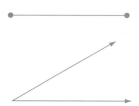

32. Draw or trace the triangle and construct a copy of the triangle. **[Lesson 4.7]**

Look Beyond

Use the drawing of two rooms of a house for exercises 33 and 34.

33. Find the number of square yards of carpet you would need to carpet the two rooms.

34. If carpet is $25.50 per square yard installed, how much will it cost to carpet the two rooms.

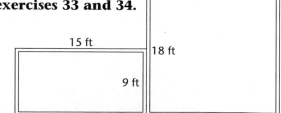

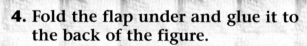

FLEXAGONS

In 1939, a university mathematics student was playing with a strip of paper trimmed from a notebook. He discovered something interesting—flexagons.

Flexagons are polygons made from folded paper that show different faces when "flexed." The instructions below for two flexagons will give you a chance to play with these unique figures yourself.

ACTIVITY 1: THE HEXAFLEXAGON

1. First, cut and mark a strip of two-colored paper with equilateral triangles as shown.

2. Fold the strip forward along $\overline{AB}$.

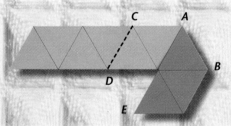

3. Fold the strip backward along $\overline{CD}$ and over point E.

4. Fold the flap under and glue it to the back of the figure.

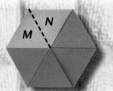

5. Now you are ready to flex your hexaflexagon. Squeeze triangles M and N together along the dotted line so that the center of the figure is lifted up. Close up the figure. Re-open the figure a new way. How are the colored faces now arranged? Repeat.

ACTIVITY 2: THE HEXAHEXAFLEXAGON

Here is a flexagon that will show twice as many faces as the first. The hexahexaflexagon is a bit more complicated than the hexaflexagon, so follow the instructions carefully.

6. Start with a strip of two-colored paper that is cut and marked with 19 equilateral triangles on each side. Number each triangle on the top side 1, 2, and 3 as shown, skipping the last triangle.

7. Without swapping the positions of the ends, turn the paper over. Number the triangles 4, 4, 5, 5, 6, 6 as shown, skipping the first triangle.

8. Fold the strip so that the 4s, 5s, 6s face each other as shown.

9. Now fold forward on $\overline{AB}$. Fold backward along $\overline{CD}$ and over point E.

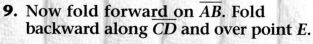

10. Glue the flap so that blank triangle meets blank triangle. The hexahexaflexagon flexes the same way as shown before.

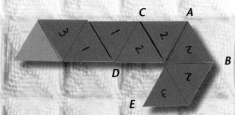

EXTENSION

11. How many faces will the hexahexaflexagon show?

12. What are the front-back face combinations?

13. Are there any combinations that are not possible?

14. Is there a pattern to the way the faces are revealed?

Chapter 4 Review

Vocabulary

base	196	CPCTC	193	rhombus	208
base angles	196	equilateral triangle	196	translation	226
betweenness	228	isosceles triangle	196	triangle rigidity	180
congruence	174	legs	196	triangulation	186
congruent polygons	176	parallelogram	208	vertex angle	196
corresponding sides, angles	175	rectangle	208		

Key Skills and Exercises

Lesson 4.1

➤ Key Skills

Name corresponding parts of congruent polygons.

Quadrilaterals $ABCD$ and $QRST$ are congruent. List all the corresponding sides and angles.

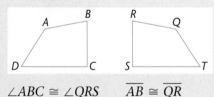

$\angle ABC \cong \angle QRS \qquad \overline{AB} \cong \overline{QR}$

$\angle BCD \cong \angle RST \qquad \overline{BC} \cong \overline{RS}$

$\angle CDA \cong \angle STQ \qquad \overline{CD} \cong \overline{ST}$

$\angle DAB \cong \angle TQR \qquad \overline{DA} \cong \overline{TQ}$

Show that polygons are congruent.

Prove that $\triangle EFG \cong \triangle GHE$.

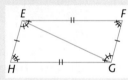

Five of the six necessary congruences are given:

$\angle HEG \cong \angle FGE \qquad \overline{EF} \cong \overline{GH}$

$\angle EGH \cong \angle GEF \qquad \overline{FG} \cong \overline{HE}$

$\angle GHE \cong \angle EFG$

By the Reflexive Property of Equality, $EG = EG$; therefore $\overline{EG} \cong \overline{EG}$. All six measures are congruent, so $\triangle EFG \cong \triangle GHE$.

➤ Exercises

1. $ABCDE \cong QRSTU$. Complete the congruences.

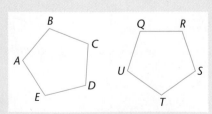

$\angle ABC \cong \angle QRS \qquad\qquad \overline{AB} \cong \overline{QR}$

$\angle BCD \cong \underline{\quad?\quad} \qquad\qquad \underline{\quad?\quad} \cong \overline{RS}$

$\angle CDE \cong \underline{\quad?\quad} \qquad\qquad \underline{\quad?\quad} \cong \overline{ST}$

$\angle DEA \cong \underline{\quad?\quad} \qquad\qquad \underline{\quad?\quad} \cong \overline{TU}$

$\angle EAB \cong \underline{\quad?\quad} \qquad\qquad \underline{\quad?\quad} \cong \overline{UQ}$

In Exercises 2–3 decide whether the triangles are congruent and justify your conclusion.

2.

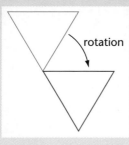

rotation

3.

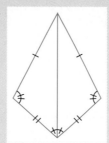

Lesson 4.2

➤ Key Skills

Apply SSS, SAS, and ASA congruence postulates.

With the information given you can prove three ways that $\triangle OPR \cong \triangle BNU$.

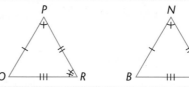

SSS: $\overline{BN} \cong \overline{OP}$, $\overline{NU} \cong \overline{PR}$, $\overline{UB} \cong \overline{RO}$
SAS: $\overline{OP} \cong \overline{BN}$, $\overline{PR} \cong \overline{NU}$, $\angle P \cong \angle N$ or
$\quad\quad \overline{OR} \cong \overline{BU}$, $\overline{RP} \cong \overline{UN}$, $\angle R \cong \angle U$,
ASA: $\angle P \cong \angle N$, $\angle R \cong \angle U$, $\overline{PR} \cong \overline{NU}$

➤ Exercises

In Exercises 4–6 decide whether the triangles are congruent and justify your answer.

4.

5.

6.

Lesson 4.3

➤ Key Skills

Apply AAS and HL congruence theorems.

With the information given you can prove two ways that $\triangle WXZ \cong \triangle YZX$.

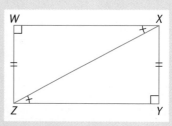

AAS: $\angle W \cong \angle Y$, $\angle WXZ \cong \angle YZX$, $ZX \cong XZ$
HL: $WZ \cong YX$, $ZX \cong XZ$

➤ Exercises

In Exercises 7–9 decide whether the triangles are congruent and justify your answer.

7.

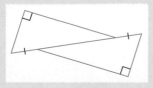

8.

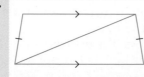

9.

Lesson 4.4

> ## Key Skill

Prove congruences in triangles.
In the figure, $\overline{AB} \cong \overline{DC}$ and $\overline{BD} \cong \overline{CA}$.
Prove $\angle ABC \cong \angle DCB$.

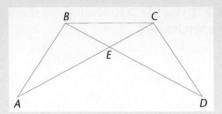

STATEMENTS	REASONS
1. $\overline{AB} \cong \overline{DC}$ and $\overline{BD} \cong \overline{CA}$	Given
2. $\overline{BC} \cong \overline{CB}$	Reflexive Property of Equality
3. $\triangle ABC \cong \triangle DCB$	SSS
4. $\angle ABC \cong \angle DCB$	CPCTC

> ## Exercises

10. Prove $\triangle ABD$ is isosceles.

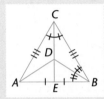

Lesson 4.5

> ## Key Skill

Prove properties of quadrilaterals.
FGHE is a rhombus. Prove that $\overline{ED} \cong \overline{GD}$.

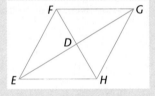

Since *FGHE* is a rhombus, $\overline{FG}$ is parallel and congruent to $\overline{HE}$. $\overline{FH}$, a transversal of $\overline{FG}$ and $\overline{HE}$, forms alternate interior angles that are congruent, so $\angle GFD \cong \angle EHD$. $\angle GDF \cong \angle EDH$ because they are vertical angles. By AAS $\triangle FGD \cong \triangle HED$. $\overline{ED} \cong \overline{GD}$ by CPCTC.

> ## Exercises

11. Prove that $\angle QPT \cong \angle SPT$.

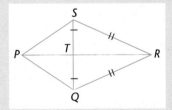

Lesson 4.6

> ## Key Skill

Determine whether a quadrilateral with given conditions is a special quadrilateral.

If two opposite sides of a quadrilateral are parallel and the other two sides are congruent, must the figure be a parallelogram? As the illustration shows, the answer is no, because the figure has the given conditions but is not a parallelogram.

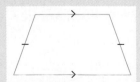

Based only on the given conditions, tell what special quadrilateral each figure must be. (A figure may be of more than one type.)

12.

13.

14.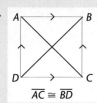

$\overline{AC} \cong \overline{BD}$

Lesson 4.7

➤ *Key Skills*

Construct copies of segments, angles, and triangles; construct a bisector of an angle.

Construct an angle twice the measure of $\angle A$.

Using $\overrightarrow{BC}$ as a base, construct $\angle CBX \cong \angle A$. Using $\overrightarrow{BX}$ as a base, construct $\angle XBY \cong \angle A$.

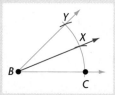

By the Addition Property of Equality $m\angle CBY = 2 \times m\angle A$.

➤ *Exercises*

15. Construct a line parallel to $\overleftrightarrow{AB}$ through a point C not on $\overleftrightarrow{AB}$.

Lesson 4.8

➤ *Key Skill*

Find the midpoints of segments on the coordinate plane.

EFGH is a square inscribed in square *ABCD*. Find the coordinates of each vertex of *EFGH*.

Find the midpoint of each side of *ABCD*.

midpoint of $\overline{AB} = E = (\frac{1}{2}[(-2) + 2], \frac{1}{2}[3 + 1])$

midpoint of $\overline{BC} = F = (\frac{1}{2}[2 + 4], \frac{1}{2}[3 + (-1)])$

midpoint of $\overline{CD} = G = (\frac{1}{2}[4 + 0], \frac{1}{2}[(-1) + (-3)])$

midpoint of $\overline{DA} = H = (\frac{1}{2}[0 + (-2)], \frac{1}{2}[(-3) + 1])$

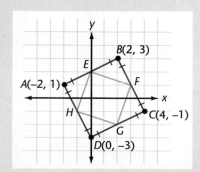

The coordinates of vertex E are (0, 2), of F are (3, 1), of G are (2, –2), and of H are (–1, –1).

16. In the figure, use the midpoint formula to show △*UTS*, △*VWS*, △*ZTS*, and △*SWZ* are congruent.

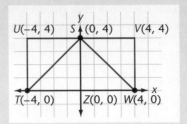

Lesson 4.9

➤ *Key Skills*

Construct the rigid transformations.

Construct a rotation of △*PQR* through point *P*.

Construct arcs centered at P through points Q and R. Locate points *Q'* and *R'* so that ∠*QPQ'* ≅ ∠*RPR'*.

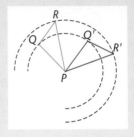

➤ *Exercises*

17. By construction, reflect the segment through a vertical line.

Applications

18. Surveying   For the figure, what measurement(s) would you need to make to determine *WX*, the distance across the lake?

19. Architecture   List five examples of congruent polygons in the building. Make your own sketch. Assume that the back and front are similar.

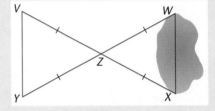

Arche de la Défense
Paris, France

20. Construct a congruent extension of this pattern.

Chapter 4 Assessment

1. $NMRT \cong SLKB$. Complete the congruences.

$\angle RTN \cong \angle KBS$ $\overline{RT} \cong \overline{KB}$

$\angle TNM \cong$ __?__   __?__ $\cong \overline{BS}$

$\angle NMR \cong$ __?__   __?__ $\cong \overline{SL}$

$\angle MRT \cong$ __?__   __?__ $\cong \overline{LK}$

In Exercises 2–3 decide whether the polygons are congruent and justify your conclusion.

2.

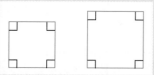

3.

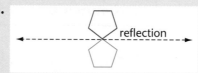

reflection

In Exercises 4–6 decide whether the triangles are congruent and justify your answer.

4.

5.

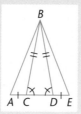

6.

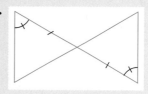

7. $\overline{AF}$, $\overline{BD}$, and $\overline{CE}$ are angle bisectors. Prove $\triangle ABC$ is equilateral.

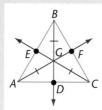

8. Which of the parallelogram(s) can be proven to be rectangles? rhombuses? squares?

a.   **b.**   **c.**   **d.**

$\overline{AC} \cong \overline{BD}$

9. Are segments $\overline{FG}$ and $\overline{GH}$ congruent? Justify your answer.

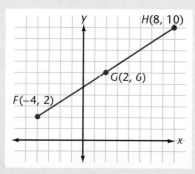

10. Construct a congruent triangle that shares point P.

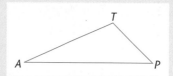

Chapters 1-4 Cumulative Assessment

College Entrance Exam Practice

Quantitative Comparison. Exercises 1–3 consist of two quantities, one in Column A and one in Column B, which you are to compare as follows.

A. The quantity in Column A is greater.
B. The quantity in Column B is greater.
C. The two quantities are equal.
D. The relationship cannot be determined from the information given.

	Column A	Column B	Answers
1.	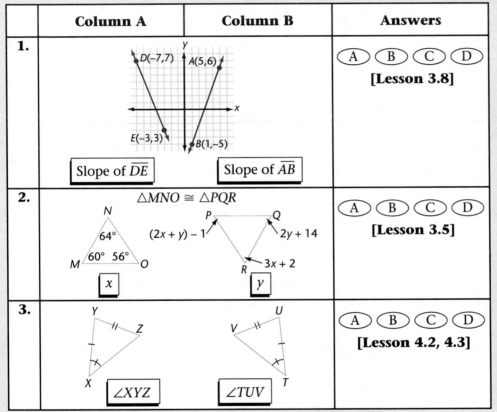 Slope of $\overline{DE}$	Slope of $\overline{AB}$	Ⓐ Ⓑ Ⓒ Ⓓ **[Lesson 3.8]**
2.	$\triangle MNO \cong \triangle PQR$ x	y	Ⓐ Ⓑ Ⓒ Ⓓ **[Lesson 3.5]**
3.	$\angle XYZ$	$\angle TUV$	Ⓐ Ⓑ Ⓒ Ⓓ **[Lesson 4.2, 4.3]**

4. Which set of points defines a line perpendicular to $\overline{MN}$?
  a. (0, 7), (8, −4)   **b.** (4, −7), (−4, 4)   **[Lesson 3.8]**
  c. (−7, 0), (4, 8)   **d.** (7, −4), (−4, 4)

5. What is the measure of $\angle ACB$?   **[Lesson 4.4]**
  a. 42°
  b. 126°
  c. 63°
  d. cannot be determined from the information given

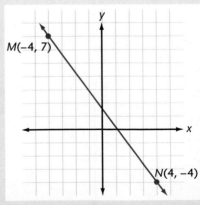

6. The missing angle measure is **[Lesson 3.6]**
  a. 120°   **b.** 180°
  c. 130°   **d.** 100°

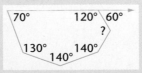

7. Supply the missing reason in the two-column proof.
[Lesson 4.2, 4.3]

Given: ∠NMP and ∠OPM are right angles,
and $\overline{NM} \cong \overline{OP}$. Prove that △MNP ≅ △POM.

STATEMENTS	REASONS
$\overline{NM} \cong \overline{OP}$	Given
m∠NMP = m∠OPM = 90°	Given
∠NMP ≅ ∠OPM	If ∠s have = measure, they are ≅.
$\overline{MP} \cong \overline{PM}$	Reflexive Property of Congruence
△MNP ≅ △POM	___(7)___

a. SSA   **b.** SSS   **c.** HL   **d.** SAS

8. Which is not a feature of every rhombus? [Lesson 4.6]
a. parallel opposite sides   **b.** equal diagonals
c. four equal sides   **d.** 90° intersection of diagonals

9. Write the converse of this statement: If trees bear cones, then the trees are conifers. **[Lesson 2.2]**

10. Using the rule $(x, y) \rightarrow (x + 2, y - 2)$, transform the pre-image at the right. What type of transformation results? **[Lesson 1.7]**

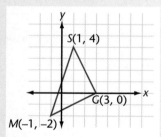

11. Lines l and m are parallel. Find m∠1. **[Lesson 3.3]**

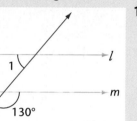

12. Are these triangles congruent? Justify your answer.
[Lesson 4.2, 4.3]

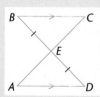

13. Find the midpoint of each side of △MSG.
[Lesson 4.8]

14. Construct a segment, a line, and a reflection of the segment over the line. **[Lesson 4.9]**

Free-Response Grid Questions 15–17 may be answered using a free-response grid commonly used by standardized test services.

15. QUAD is a rectangle. Find x.
[Lesson 4.5]

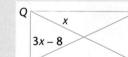

16. Find the interior angle measure of this regular polygon. **[Lesson 3.6]**

17. Find the measure of base $\overline{BC}$ of the trapezoid. **[Lesson 3.7]**

CHAPTER 5

LESSONS

5.1 *Exploring* Perimeter, Circumference, and Area

5.2 *Exploring* Areas of Triangles, Parallelograms, and Trapezoids

5.3 *Exploring* Circumferences and Areas of Circles

5.4 The "Pythagorean" Right-Triangle Theorem

5.5 Special Triangles, Areas of Regular Polygons

5.6 The Distance Formula, Quadrature of a Circle

5.7 Geometric Probability

Chapter Project
Area of a Polygon

Perimeter and Area

The map of Houston reminds us of the many uses of measurement. How far is it from Hobby Airport to the center of the city? How long will it take to get from Jacinto City to Memorial Park? How many miles does one inch on the map represent?

Think of other questions about measurement that the map brings to mind, even if you do not know the answers. We could not get very far or achieve very much—in travel, construction, sports, or science—if it were not for measurement.

In this chapter you will use perimeter, such as the distance along Route 610 on the map; and area, such as the number of square miles inside Route 610. You will also learn about the postulates, theorems, and formulas that make it possible to take and use measurements.

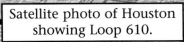
Memorial Park

Satellite photo of Houston showing Loop 610.

PORTFOLIO ACTIVITY

Lunes of Hippocrates

The figure shows a right isosceles triangle. Semicircles are constructed with centers at the midpoint of the hypotenuse and at the midpoint of each leg. The diameter of each semicircle has the length of a side of the triangle. The moonshaped crescents at the top of the figure are called **lunes**.

In this chapter you will learn how to find the area of the lunes—and the result may surprise you.

Houston Skyline

Jacinto City

Loop 610

Hobby Airport

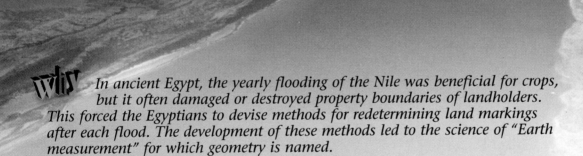

LESSON 5.1

Exploring Perimeter, Circumference, and Area

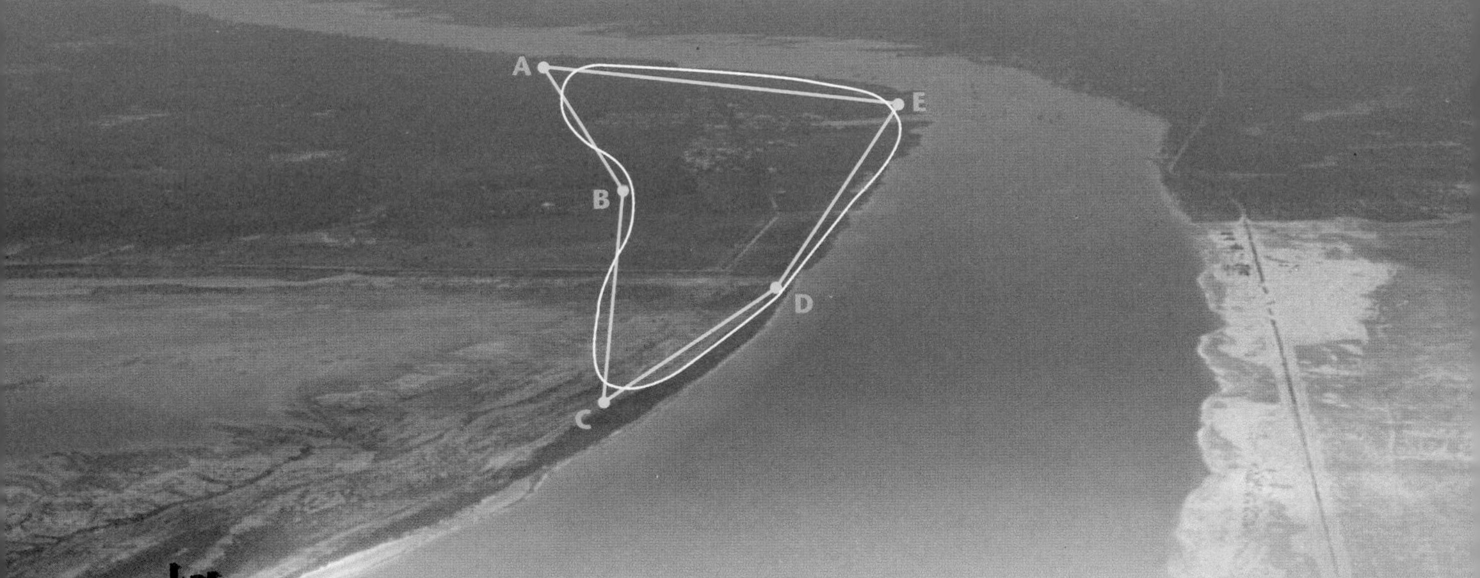

 In ancient Egypt, the yearly flooding of the Nile was beneficial for crops, but it often damaged or destroyed property boundaries of landholders. This forced the Egyptians to devise methods for redetermining land markings after each flood. The development of these methods led to the science of "Earth measurement" for which geometry is named.

Perimeter

Perimeter is the distance around a geometric figure that is contained in a plane. For a polygon, the perimeter is the sum of the lengths of its sides. The perimeter of a polygon can be used to approximate the perimeter of any closed figure on a plane. For example, the perimeter of the irregular curve is approximately equal to the perimeter of polygon *ABCDE*.

CRITICAL Thinking

How can you redraw the polygon in the figure above to get a better estimate of the perimeter of the curve?

Area

The **area** of a plane figure is the number of nonoverlapping unit squares that will cover the interior of the figure. Using squares as a unit of measure makes it easy to find the area of rectangle *ABCD*.

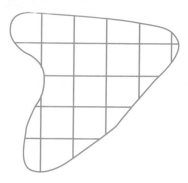

There are three unit squares in each of two rows. Thus, the area of rectangle *ABCD* is equal to 2×3, or 6 square units.

This leads us to the following Area Postulates.

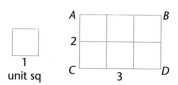

1
unit sq

THE AREA OF A RECTANGLE
The area of a rectangle with base *b* and height *h* is $A = bh$.   **5.1.1**

SUM OF AREAS
If a figure is composed of nonoverlapping regions *A* and *B*, then the area of the figure is the sum of the areas of regions *A* and *B*. (Regions are considered nonoverlapping if they share no common area, though they may share a common boundary.)   **5.1.2**

EXAMPLE

Find the area of the following figures. Explain your method in each case.

A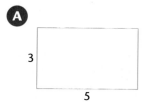

B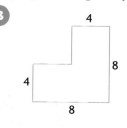

C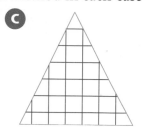

Solution ➤

A Multiply the base times the height. The area is 15 square units.

B Divide the figure into separate squares and/or rectangles and add the individual areas. (Three different ways of doing this are shown.) The area is 48 square units.

C One way of estimating the result is to take half the number of squares that are partially inside the figure and add this to the number of squares that are entirely inside the figure. The area is approximately 21 square units. ❖

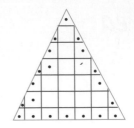

Place a dot in each partial square.

How are the two area postulates on page 245 used in the example?

Exploration 1 *Fixed Perimeter, Maximum Area*

Spreadsheet

You will need
Graph paper and
Graphics calculator or
Spreadsheet software

Gardening A gardener has material to make 24 ft of fencing for a miniature garden. What is the shape of a rectangle that will enclose the greatest area?

The formula for the perimeter of a rectangle is P = 2b + 2h.

**MAXIMUM
MINIMUM**
Connection

1 Trace three different rectangles that each have a perimeter of 24 units. What is the area of each one?

2 In the formula for the perimeter of a rectangle, substitute 24 for *P* and solve the equation for *h*.

3 Fill in a table like the one below. (Use a graphics calculator or a spreadsheet program if available.) What do you observe about the area values?

b	h = 12 − b	A = bh
1	11	11
2	?	?
3	?	?
•	•	•

 Plot your area values on a graph with the horizontal axis representing the length of the base and the vertical axis representing the area.

 What value of the base gives the maximum area? What is the value of the height?

 What is the shape of the rectangle that gives the maximum area? ❖

CRITICAL
Thinking

Do you think your result from Exploration 1 would hold true for a rectangle of any given perimeter? Explain.

•Exploration 2 *Fixed Area, Minimum Perimeter*

You will need
Graph paper
Spreadsheet software or
graphics calculator (optional)

Spreadsheet

A farmer wants to form a rectangular area of 3600 square ft with the minimum amount of fencing. What should the dimensions of the rectangle be?

 In the formula for the area of a rectangle ($A = bh$), substitute 3600 for A and solve for h.

ALGEBRA
Connection

 Fill in a table like the one below. Values for b should range from 10 to 100. (Use a graphics calculator or spreadsheet software if available.)

b	h = 3600 ÷ b	P = 2b + 2h
10	360	380
20	?	?
30	?	?
•	•	•

 Plot the perimeter values on a graph with the horizontal axis representing the length of the base and the vertical axis the perimeter.

 After finding the value to the nearest ten, find the value to the nearest one. What value of the base gives the minimum perimeter? What is the value of the height?

 What is the shape of the rectangle that gives the minimum perimeter for the given area? ❖

CRITICAL
Thinking

Do you think your result from Exploration 2 would hold true for a rectangle of any given area? Explain.

Lesson 5.1   Exploring Perimeter, Circumference, and Area   **247**

EXERCISES & PROBLEMS

Communicate

1. Explain how to estimate the perimeter of the swimming pool.

2. Explain how to estimate the top surface area of the airplane wing.

3. Explain why the definition of perimeter excludes overlapping pieces.

4. Many mathematics applications concern the area underneath a curve. Explain a way to estimate the area of the shaded part in the graph at right.

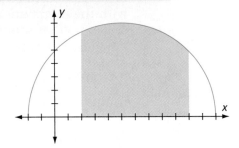

Practice & Apply

AD = 15 in.   AC = 13 in.
BD = 10 in.   DL = 11 in.
EI = 3 in.   CH = 4 in.

Use the figure and measurements above to answer Exercises 5–15.

5. The perimeter of rectangle *ADLI* = ___?___ .

6. The area of rectangle *ADLI* = ___?___ .

7. The perimeter of rectangle *CDLK* = ___?___ .

8. The area of rectangle *CDLK* = ___?___ .

9. The perimeter of hexagon *GHCDLJ* = ___?___ .

10. The area of hexagon *GHCDLJ* = ___?___ .

11. The perimeter of rectangle *BCHG* = ___?___ .

12. The area of rectangle *BCHG* = ___?___ .

13. The area of △*BHG* = ___?___ (use Exercise 12).

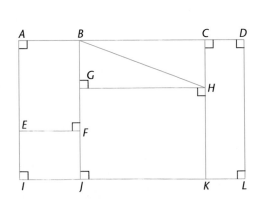

14. If points *I* and *D* were connected by a segment, what would the area of the resulting △*ADI* be?

15. What theorems or postulates from previous chapters helped you to determine the perimeters and areas in Exercises 5–14?

16. 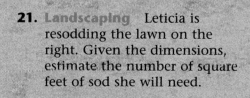 **Algebra**   Suppose the perimeter of a rectangle is 72 cm. The base measures three times the height. What are the lengths of the sides? What is the area?

17. **Algebra**   Suppose the perimeter of a rectangle is 80*x*. The base measures seven times the height. What are the lengths of the sides? What is the area? (Answers will be in terms of *x*.)

18. **Algebra**   Find the dimensions of a rectangle that has an area equal to its perimeter.

19. Construction   A house has a roof with dimensions as shown. The roof will be built with plywood covered by shingles. Plywood comes in pieces 8 ft by 4 ft. How many pieces of plywood will be needed to cover the roof?

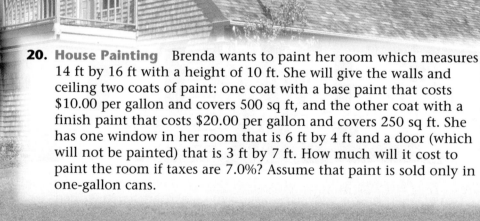

25 ft

42 ft

20. House Painting   Brenda wants to paint her room which measures 14 ft by 16 ft with a height of 10 ft. She will give the walls and ceiling two coats of paint: one coat with a base paint that costs $10.00 per gallon and covers 500 sq ft, and the other coat with a finish paint that costs $20.00 per gallon and covers 250 sq ft. She has one window in her room that is 6 ft by 4 ft and a door (which will not be painted) that is 3 ft by 7 ft. How much will it cost to paint the room if taxes are 7.0%? Assume that paint is sold only in one-gallon cans.

21. Landscaping   Leticia is resodding the lawn on the right. Given the dimensions, estimate the number of square feet of sod she will need.

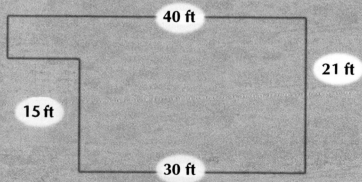

40 ft

21 ft

15 ft

30 ft

22. Solar Energy Mr. Venn is converting some of his house's electric power to solar energy. He wishes to provide an additional 15,000 BTUs of heat for the house through 2 solar panels. He knows that he needs 6 square feet of solar panel per 1000 BTUs. Mr. Venn also wants the 2 panels to be equal in size, with each 10 feet in length. What are the dimensions of the panels he needs?

23. **Maximum/Minimum** You have 200 feet of fencing material to make a pen for your livestock. If you make a rectangular pen, what is the maximum area you can fence in? Extend the table below to determine the answer.

base	height	perimeter	area
1	99	200	99
2	98	200	196
5	95	200	475
20	80	200	1600
•	•	200	•

24. Agriculture To care properly for your livestock, you must provide a certain amount of area per animal. Suppose you need an area of 3600 sq ft. What is the minimum amount of fencing you need for a rectangular pen? Extend the table below to determine an answer.

base	height	perimeter	area
1	3600	7202	3600
2	1800	3604	3600
5	720	1410	3600
20	360	740	3600
•	•	•	•

25. **Algebra** $2b + 2h = 100$ is the equation for a constant perimeter of 100. Solve for b or h and graph the equation. What type of function represents this relationship? What solution values for the function do not make sense for the perimeter example?

Look Back

26. Construct a Venn diagram to illustrate the relationships among parallelograms, rectangles, rhombuses, and squares. **[Lesson 3.2]**

27. Find the measure of an exterior angle of an equiangular triangle. **[Lesson 3.6]**

28. If the sum of the measures of three angles of a quadrilateral equals 300°, find the measure of the fourth angle. **[Lesson 3.6]**

29. Find the sum of the measures of the angles of a polygon with n sides. **[Lesson 3.6]**

30. Find the measure of an interior angle of a regular hexagon. **[Lesson 3.6]**

31. Find the slope of the segment connecting points (2, 3) and (4, −1). **[Lesson 3.8]**

32. Find the midpoint of the segment connecting the points (−4, −6) and (6, 4). **[Lesson 4.8]**

Look Beyond

33. Which has a greater area, a square with a side of 4 in. or a circle with a diameter of 4 in.? Explain your answer.

34. The square at right has 1-inch sides. Explain how you could estimate the area of the circle inside the square.

35. Use the formula for the area of a circle ($A = \pi r^2$) to confirm your estimate.

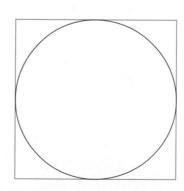

An Ancient Wonder

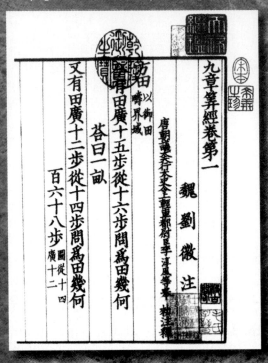

First page of the Nine Chapters

A hundred years before Euclid compiled his famous work, the *Elements*, exciting work in geometry and algebra was taking place half a world away in China. The mathematics of that time was later written, revised, and annotated to form *The Nine Chapters on the Mathematical Art*. Only recently has this ancient text received attention in the West.

Although a complete English translation of *The Nine Chapters on the Mathematical Art* may not yet be available, you can preview this mathematical treasure. On the following page, you will explore how to simplify complex problems by using a "patchwork" method—cutting up figures and rearranging the pieces.

Cooperative Learning

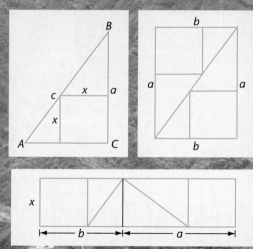

1. You can begin your exploration of *The Nine Chapters of the Mathematical Art* with this problem: Find the length of a side of a square inscribed in a right triangle.

   a. Draw and cut out two congruent copies of the triangle with the inscribed square. Use them to form a rectangle.

   b. Cut each triangle into three pieces as marked. Reassemble them as a long rectangle as shown.

   c. Compare the area of the rectangle you formed in Step a with the one you formed in Step b. How do they compare?

   d. In terms of *a* and *b*, what is the area of the rectangle in Step a? In terms of *a*, *b*, and *x*, what is the area of the rectangle in Step b?

   e. Why is the equation $x(a + b) = ab$ true for these figures? Use the equation to solve the original problem.

2. Now try a harder problem from *The Nine Chapters of the Mathematical Art*. (You can try this one by just imagining the cutting.) Find the radius of a circle inscribed in a right triangle.

   a. Why does $EH = r$?

   b. In rectangles *ABCD* and *EFGH*, why are the lengths marked *a* the same length? Why is *b* the same in both? Why is *c* the same length as diagonal from *B* to *D*?

   c. In terms of *a*, *b*, and *c*, what is the area of rectangle *ACBD*? In terms of *a*, *b*, *c*, and *r*, what is the area of rectangle *EFGH*?

   d. Why is the equation $r(a + b + c) = ab$ true for these figures? Use the equation to solve the original problem.

3. Use the patchwork method to find the formula for the area of an isosceles trapezoid.

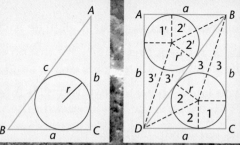

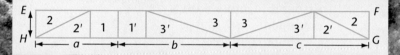

Exploring

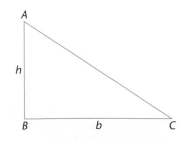

Areas of Triangles, Parallelograms, and Trapezoids

Designs drawn on grid paper, such as this knitting pattern, suggest methods of estimating the areas of geometric figures. But it is often more convenient to use exact formulas.

•Exploration 1 *Areas of Triangles*

You will need
Graph paper

Part I

1 Draw a rectangle on graph paper. Calculate its area.

2 Draw a diagonal of your rectangle. What kind of triangles are formed? Identify the heights or altitudes of each triangle.

3 What do you know about the two triangles from your study of special quadrilaterals? What is the area of each of the triangles?

4 If you are given a right triangle, can you form a rectangle by fitting it together with a congruent copy of itself? Illustrate your answer with examples.

ALGEBRA
Connection

5 Write a formula for the area of right triangle *ABC* in terms of its base, *b*, and its height (or altitude), *h*.

Part II

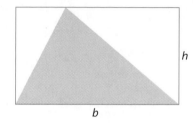

1 Make a copy of the drawing on graph paper.

2 Draw an altitude of the triangle from the top vertex of the triangle to the base of the rectangle. Is the altitude parallel to the sides of the rectangle? What theorem justifies your answer?

3 The altitude divides the rectangle into two smaller rectangles. Each of these rectangles is divided into two congruent triangles. What theorem justifies this fact?

4 What is the relationship between the area of the shaded part of the large rectangle occupied by the original triangle and the area of the unshaded parts? Explain your answer.

ALGEBRA
Connection

5 Write a formula for the area, A, of a triangle in terms of its base, b, and its height (or altitude), h. ❖

APPLICATION

Theatre Arts You are building a triangular flat as a set for a school play. The flat needs to be covered with cloth. What is the area you need to cover?

$h = 22$ in.

$b = 30$ in.

$A = \dfrac{1}{2} bh \qquad A = \dfrac{1}{2}(30)(22) \qquad A = 330$ sq in. ❖

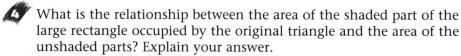

CRITICAL
Thinking

Triangle $\triangle ABC$ is obtuse. How can you use the method of Exploration 1, Part 2, to show that the formula for the area of a triangle holds for an obtuse triangle?

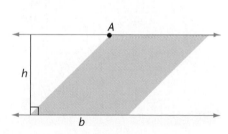

Exploration 2 *Areas of Parallelograms*

You will need
Graph paper
Scissors

1 Make a copy of the drawing on graph paper; b is the base of the parallelogram; h is the height of the parallelogram.

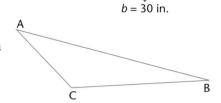

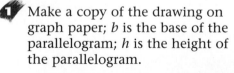

2 Draw an altitude in the parallelogram from point *A* to its base. What figure is formed?

3 Cut out the parallelogram. Cut off the right triangle, move it to the other side, and fit it with the parallelogram. What figure is formed? What is the area, *A*, of the parallelogram in terms of *b* and *h* of the original figure?

ALGEBRA
Connection

4 Write a formula for the area of a parallelogram in terms of its base, *b*, and its height, *h*. Explain your answer in terms of the figure you formed in Step 3.

5 How do you know that the triangle will always fit, as in Step 3? To answer this question, first prove that $\triangle AEB \cong \triangle DFC$. Then prove that quadrilateral *ADFE* is a rectangle. Write out your argument and save it in your portfolio. ❖

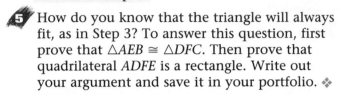

CRITICAL
Thinking

Use the fact that the area of a parallelogram is *bh* to find the formula for the area of a triangle.

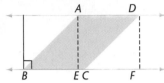

•Exploration 3 *Areas of Trapezoids*

You will need
Graph paper

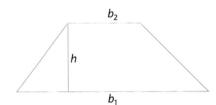

1 Make two copies of the trapezoid on graph paper: b_1 and b_2 are the bases of the trapezoid; *h* is the height of the trapezoid.

2 Find a way to fit the two trapezoids into a parallelogram.

3 Find several ways to explain why the area of a trapezoid is $A = \dfrac{(b_1 + b_2)h}{2}$. ❖

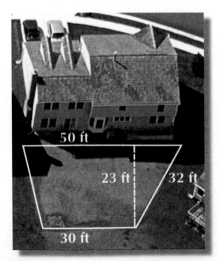

APPLICATION

In order to apply the correct amount of fertilizer, you need to know the area of Mrs. Zapata's lawn. The dimensions of the trapezoidal lawn are as shown. What is the area of the lawn?

$$A = \frac{(b_1 + b_2)h}{2}$$

$$A = \frac{(30 + 50)(23)}{2}$$

$$A = \frac{80}{2}(23)$$

$$A = 40(23)$$

$$A = 920 \text{ sq ft} ❖$$

EXERCISES & PROBLEMS

Communicate

1. The area of a parallelogram is 14 sq in. What are the areas of the two triangles formed by the diagonal? Explain your answer.

2. Is it possible for two parallelograms to have the same area and not be congruent? Explain your answer. Is it possible for two triangles to have the same area and not be congruent? Explain your answer.

3. Do you need to know the lengths of the sides of a trapezoid to find its area? Explain your answer.

Practice & Apply

$\overline{AC} \parallel \overline{IL}$ $\overline{DH} \parallel \overline{IL}$

$\overline{AI} \parallel \overline{JB}$ $\overline{KC} \parallel \overline{JB}$

$BK = 15$ units $KL = 14$ units

$IL = 32$ units $DH = 25$ units

$FK = 7$ units $EH = 17$ units

$JI = 8$ units $DG = 18$ units

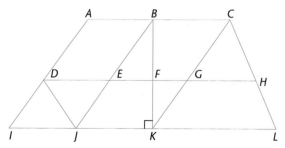

Use the diagram above to answer Exercises 4–15.

4. The area of $\triangle KCL$ = __?__.

5. The area of $\triangle BJK$ = __?__.

6. The area of $\triangle BCK$ = __?__.

7. The area of $\triangle DIJ$ = __?__.

8. The area of parallelogram $BCKJ$ = __?__.

9. The area of parallelogram $EGKJ$ = __?__.

10. The area of parallelogram $ABED$ = __?__.

11. The area of parallelogram $ACKI$ = __?__.

12. The area of trapezoid $EHLJ$ = __?__.

13. The area of trapezoid $BCHE$ = __?__.

14. The area of trapezoid $BCLJ$ − __?__.

15. The area of trapezoid $ACLI$ = __?__.

16. Which postulates and theorems from previous lessons helped you determine the areas in Exercises 4–15?

17. **Algebra** Find the base measure of a triangle with height 10 cm and area 100 cm².

18. **Algebra** Find the height of a parallelogram with a base 15 cm and area 123 cm².

19. **Maximum/Minimum** Each of the following triangles has the same perimeter. Which has the larger area? Explain your answer.

a.

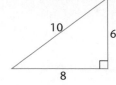

b.

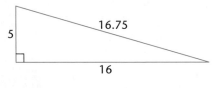

20. 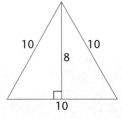 **Maximum/Minimum** Each of the following triangles has the same area. Which has the larger perimeter? Explain your answer.

a.

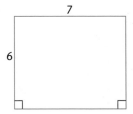

b.

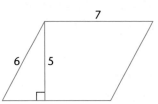

21.  **Maximum/Minimum** Each of the following parallelograms has the same perimeter. Which has the larger area? Explain your answer.

a.

b.

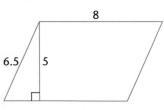

22. **Maximum/Minimum** Each of the following parallelograms has the same area. Which has the larger perimeter? Explain your answer.

a.

b.

23. Find the area of △*RST*.

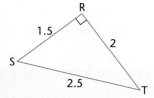

24. Decorating Joe and his friends have to decorate a float for the homecoming parade. The float is 6 ft high and the length and width are 12 ft by 8 ft. Joe needs to wrap the sides of the float with chicken wire to form a frame for the decorating materials. How many square feet of chicken wire does he need? If chicken wire costs $2.00 per square yard, what is the cost of the chicken wire?

25. **Algebra** A **kite** is a quadrilateral in which two pairs of adjacent sides are congruent. The diagonals of a kite are perpendicular. Use the formula for the area of a triangle to prove that the area of a kite is one-half the product of the length of the diagonals.

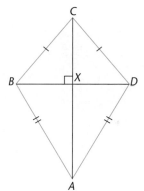

26. **Algebra** A line segment connecting the midpoints of the two nonparallel segments of a trapezoid is called the midsegment of a trapezoid. Develop a formula for the area of a trapezoid using the midsegment and the height of a trapezoid. (Hints: The midsegment is parallel to the bases. Use the triangles as shown in the diagram.)

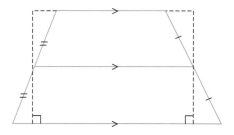

Look Back

27. Construct a Venn diagram to illustrate the relationship between scalene, isosceles, and equilateral triangles. **[Lesson 2.3]**

28. Given the congruences indicated in the diagram, prove $\triangle ABC \cong \triangle DEF$. **[Lesson 4.4]**

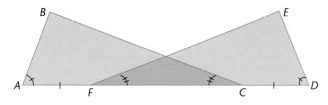

29. **Algebra** Find the area of a square whose side measures $x + y$. **[Lesson 5.1]**

30. Farming In order to fertilize a field, a farmer needs to estimate its area. Estimate the area of the field indicated by the diagram. If an acre has 43,560 square feet, how many acres need fertilizer? **[Lesson 5.1]**

Look Beyond

Cultural Connection: Africa There is a method for finding the area of a triangle using the lengths of the three sides. It is called Heron's Formula, named for a mathematician who lived in Alexandria in the first century. The formula is

$$A = \sqrt{s(s - a)(s - b)(s - c)}$$

where s is the semiperimeter—that is, half the perimeter—and a, b, and c are the lengths of the sides.

31. Find the semiperimeter of a triangle with sides 7 cm, 8 cm, and 9 cm.

32. Find the area of the triangle described in Exercise 31.

33. Technology Set up a spreadsheet to find the area of a triangle when the lengths of the sides are given using Heron's Formula.

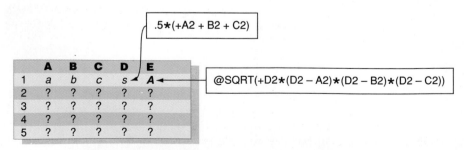

Exploring

Circumferences and Areas of Circles

why *How do the areas of the larger craters of the moon compare with the area of your home town or state? A detailed photograph can be used to answer these questions. What else would you need to know?*

When you draw a circle with a compass, the distance from the point of the compass to the pencil or pen does not change. Therefore, every part of the circle you draw is the same distance from the **center of the circle**, the point where the compass is fixed. This leads us to the following definition.

CIRCLE
A **circle** is the figure that consists of all the points on a plane that are the same distance r from a given point known as the **center** of the circle. The distance r is the **radius** of the circle. The distance $d = 2r$ is known as the **diameter** of the circle.   **5.3.1**

•Exploration 1 The Circumference of a Circle

You will need
String and ruler or
Tape measure
Various circular objects such as food cans, etc.

 The distance around a circle is called its **circumference**. Measure the circumferences and diameters of several circular objects. Record the results in a table.

object	C	d	ratio: $\frac{C}{d}$
1. can	31.4	10	?
2. ?	?	?	?
3. ?	?	?	?

 The ratio $\frac{C}{d}$ is known as π, pronounced "pie." Complete the table and find the average value of π according to your data.

 Compare your result with the results of your classmates. How close is the result to 3.14, the approximate value of π? Use the π key on your calculator. What value does the calculator show?

ALGEBRA
Connection

 Write a formula for the circumference, C, of a circle. Begin by expressing π as a ratio. Then solve for C. Now write the formula in terms of the radius, r. ❖

Try This   Use a calculator to find the circumferences of each of the following circles. Round your answers to two decimal places.

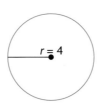

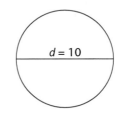

•Exploration 2 The Area of a Circle

You will need
Ruler and compass
Protractor (optional)
Scissors

Draw a circle. Label its radius *r*. Using any method you like, such as paper folding, divide the circle into eight congruent pie-shaped parts, or **sectors**.

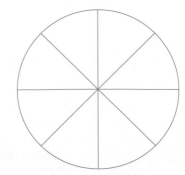

2 Cut out the sectors and reassemble them into a single figure as shown. If the curved parts of your figure were segments instead of curves, what kind of figure would you have?

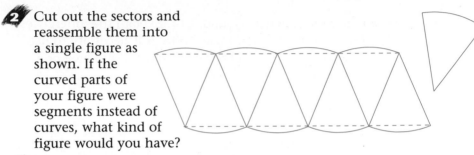

3 If you divided the circle into a larger number of congruent sectors, say 16, 32, or even more, would the curved parts seem more straight?

4 What geometric figures do your sectors in Step 3 resemble? The height of your assembled figure is approximately equal to *r*, the radius of the circle. What will happen to this approximation if the number of sectors increases infinitely?

5 The base of your assembled figure is equal to half the circumference of your original circle. Write an expression for the base of the figure in terms of π and the radius, *r*, of the original circle.

6 Write an expression for the area of the figure in terms of π and *r*. Should this be the formula for the area of a circle? Explain. ❖

CRITICAL *Thinking*   Why does the method you used become more realistic if you increase the number of sectors you cut?

Try This   Use your calculator to find the areas of each of the following circles. Round your answers to two decimal places.

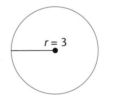 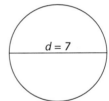

r = 3   *d* = 7

EXERCISES & PROBLEMS

Communicate

1. Suppose that you have 100 ft of fence to make a play area for your dog. Does a square yard or a circular yard provide the most area for your dog? Explain your answer.

2. When the cassette in the picture is rewinding, which point is moving faster? Explain your answer.

3. Sometimes π is approximated as $\frac{22}{7}$ and 3.14. Explain the differences between using one estimate or the other. Why is it necessary to use an estimate of π when calculating the circumference and area of a circle? What number does your calculator use for π?

4. Each of the figures below has an area of 9 sq units. Which figure has the smallest perimeter or circumference? Which figure has the largest perimeter or circumference? Explain your answers.

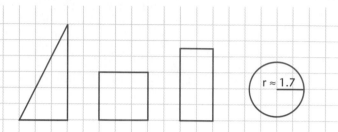

Practice & Apply

Algebra Find the circumference and area. Measurements are in inches. Use the π key on your calculator or use 3.14 for π. Round answers to the nearest tenth.

5. $r = 6$　　　　　　　**6.** $d = 4$　　　　　　　**7.** $d = 7$

8. If the area of a circle is 100π cm², find its radius and circumference.

9. If the circumference of a circle is 50π m, find its area.

10. Find the radius of a circle with area 221.7 m².

11. If a 12 in. pizza is enough to feed three people, will an 18 in. pizza be enough to feed six people? Explain why or why not.

12. Irrigation In some parts of the world, farmers irrigate the land using a circle pattern. The picture shows sections of land irrigated in this way. If the area inside the square is one square mile, what is the area, in square feet, of the cultivated circle?

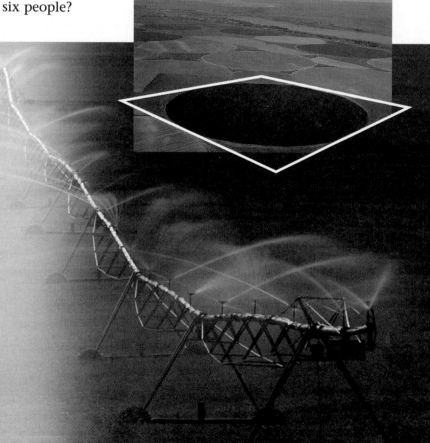

Automobile Engineering Tires are tested for how well they stick to pavement by driving them around a circular track. Use the dimensions in the diagram below for Exercises 13–15.

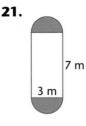

13. What is the circumference of the circle formed by the inside tire tracks?

14. What is the circumference of the circle formed by the outside tire tracks?

15. What can you conjecture about the speed of the inside tires compared with the outside tires based on the circumferences from Exercises 13 and 14?

$5\frac{1}{2}$ ft

$12\frac{1}{2}$ ft

Automobile Engineering Since the outside wheels of a car need to turn faster on a curve than the inside wheels, a car has a device called a "differential." The differential sends more power to the outside wheel, making it turn faster.

16. Suppose the inside wheel of a car has a radius of 15 in. and is turning in a circle with a circumference of 63 ft. How many revolutions will the tire make in one trip around the circle?

17. Suppose the outside wheel of the car in Exercise 16 has a radius of 15 in. and is turning in a circle with a circumference of 82 ft. How many revolutions will the tire make in one trip around the circle?

For Exercises 18–21, find the area of the shaded region.

18.

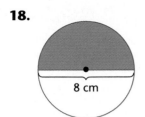

8 cm

19.

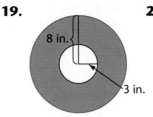

8 in.

3 in.

20.

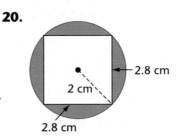

2.8 cm

2 cm

2.8 cm

21.

7 m

3 m

22. What happens to the area of a circle when the radius is doubled?

23. **Maximum/Minimum** You have 50 m of fence to enclose a play area. Which would be a larger play area, a circular area or a square area? Explain your reasoning.

24. **Maximum/Minimum** Suppose a circle and a square have an area of 300 cm². Which has the greatest perimeter or circumference? Explain your reasoning.

25. Sports A basketball rim is 18 inches in diameter. What is the area encompassed by the basketball rim?

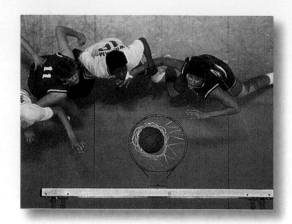

Look Back

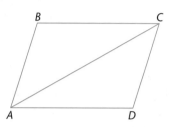

26. Given parallelogram *ABCD* with diagonal $\overline{AC}$, prove that $\triangle ABC \cong \triangle CDA$. **[Lesson 4.5]**

27. Find the area of a trapezoid with height 3 cm, and bases 6 cm and 5 cm. **[Lesson 5.2]**

Find the area of each figure below. **[Lesson 5.2]**

28.

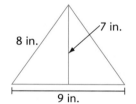

29.

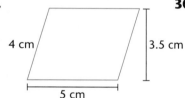

30.

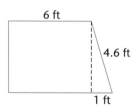

Look Beyond

 Algebra Recall from algebra that a perfect square that is a factor of a number under a radical sign may be removed by taking its square root. For example,

$$\text{Simplify } 2\sqrt{75} \quad 2\sqrt{75} = 2\sqrt{3 \times 25} = 2 \times 5\sqrt{3} = 10\sqrt{3}$$

Simplify the following.

31. $3\sqrt{75}$ **32.** $16\sqrt{32}$ **33.** $3\sqrt{500}$

 Algebra Find the positive solution for *x*.

34. $x^2 + 16 = 25$

35. $x^2 + 144 = 169$

36. $x^2 + 12.25 = 13.69$

37. Cultural Connection: Asia A yin-yang symbol consists of semicircles inside a circle, as shown. Which path from *A* to *B* is longer, the one through point *C* or the one through point *O*? Explain.

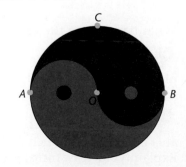

LESSON 5.4 The "Pythagorean" Right-Triangle Theorem

why *A 4000-year-old clay tablet from Babylon—what is now Iraq—has revolutionized our knowledge of ancient mathematics.*

When the Plimpton tablet was first found, no one understood the significance of the strange columns of numbers—until a mathematician who looked at it made an exciting discovery.

Exploration *Solving the Puzzle*

Scientific Calculator

You will need
A calculator

Cultural Connection: Asia The Babylonian tablet is a piece of a larger tablet. Part of it, including one column of numbers, has been broken off and lost. On the part that remains there are columns of numbers, including the two shown on the next page.

1 Work in pairs. Each person picks two numbers from 50 to 5000. Use a calculator to square each number. Subtract the smaller square from the larger square. Take the square root of the difference. Is the result a whole number?

2 Repeat Step 1 ten times. How often do you obtain a whole number?

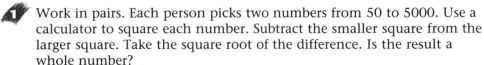

Lesson 5.4 The "Pythagorean" Right-Triangle Theorem **267**

119	169
3367	4825
4601	6649
12709	18541
65	97
319	481
2291	3541
799	1249
481	769
4961	8161
45	75
1679	2929
161	289
1771	3229
56	106

Columns II and III of Plimpton 322

3 Square the numbers in each row of the tablet. Subtract the smaller square from the larger square and take the square root of the difference. How often is the result a whole number?

4 Do you think the Babylonians knew which whole numbers had this property? Explain. ❖

From your exploration you should have discovered that the original complete table was a list of what we now call "Pythagorean" triples—that is, sets of whole numbers a, b, and c such that $a^2 + b^2 = c^2$. But surprisingly, the Babylonian table was produced over a thousand years before Pythagoras, the person to whom the relationship has been attributed.

From these numbers, and from other evidence on the tablet, it is clear that the Babylonians knew that the sides of a right triangle had the property that we now know as the "Pythagorean" relationship.

Proving the Relationship

The area of the large tilted square is equal to the area of the 4 right triangles plus the area of the small square.

At present there is no positive evidence that the Babylonians could prove the right-triangle relationship.

Cultural Connection: Europe The Pythagoreans were members of a secret society in ancient Greece. They brought to mathematics a sense of reverence and mystery. In spite of their widespread reputation, there is very little factual information about the society and its leader, Pythagoras (sixth century B.C.E.). We do not know how Pythagoras actually proved the theorem that now bears his name, or whether his particular proof—and many are possible—was original with him.

Cultural Connection: Asia The earliest proof that we now know of is found in an early Chinese source, the *Chiu Chang*. No proof is actually given; however, there is a diagram that suggests that the proof was known at least a hundred years before Pythagoras.

In the *Chiu Chang* diagram, four congruent right triangles have been assembled to form a large square with a smaller square in the center. The area of the larger square can be found by squaring the hypotenuse c of the right triangle. It can also be found by adding up the areas of the individual pieces of the figure. By setting these two equal to each other and simplifying, you obtain the famous result:

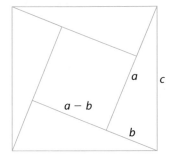

ALGEBRA
Connection

$$c^2 = 4(\tfrac{1}{2})ab + (a - b)^2$$
$$c^2 = 2ab + (a^2 - 2ab + b^2)$$
$$c^2 = 2ab + a^2 - 2ab + b^2$$
$$c^2 = a^2 + b^2$$

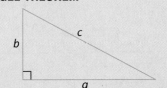

"PYTHAGOREAN" RIGHT-TRIANGLE THEOREM

For any right triangle, the square of the length of the hypotenuse is equal to the sum of the squares of the lengths of the legs.

*The **hypotenuse**, c, is the side opposite the right angle.*

5.4.1

CRITICAL
Thinking

Does the shape of the right triangle matter in the proof above? How do you know that the large figure is in fact a square? What happens to the smaller square in the center if the right triangles are isosceles?

EXAMPLE 1

A plowed field forms a right triangle, with its hypotenuse along one road and one of its legs along another. If the legs have the lengths shown in the figure, find the length of the boundary along the roads.

Solution ➤
$c^2 = a^2 + b^2$, so

$$c = \sqrt{a^2 + b^2} \; 5 \; \sqrt{1.7^2 + 3.4^2} \approx 3.8 \text{ mi}$$
$$a + c \approx 1.7 + 3.8 = 5.5 \text{ miles} \; ❖$$

Using the Converse of the Theorem

The converse of the "Pythagorean" Right-Triangle Theorem is also true. It is useful in proving that two segments or lines are perpendicular.

> ### THE "PYTHAGOREAN" RIGHT-TRIANGLE THEOREM: CONVERSE
> If the square of the length of one side of a triangle equals the sum of the squares of the lengths of the other two sides, then the triangle is a right triangle.
>
> **5.4.2**

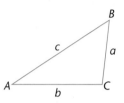

The figure at right shows the "Pythagorean" relationship. Two helpful inequalities can be derived from the relationship:

ALGEBRA
Connection

In any triangle, with c as its longest side,

If $c^2 = a^2 + b^2$, then $\triangle ABC$ is a right triangle.

If $c^2 > a^2 + b^2$, then $\triangle ABC$ is an obtuse triangle.

If $c^2 < a^2 + b^2$, then $\triangle ABC$ is an acute triangle.

EXAMPLE 2

A triangle has sides 7 inches, 8 inches, and 12 inches. Is the triangle right, obtuse, or acute?

Solution ➤

$12^2 \underline{\ ?\ } 7^2 + 8^2$

$144 > 113$. Therefore, the triangle is obtuse. ❖

EXAMPLE 3

If the bottom of a 15-foot ladder is 4 feet from the wall, how far up the wall does the ladder reach?

Solution ➤

$c^2 = a^2 + b^2$

$15^2 = 4^2 + b^2$

$225 = 16 + b^2$

$b^2 = 225 - 16$

$b = \sqrt{209} \approx 14.46$ ft ❖

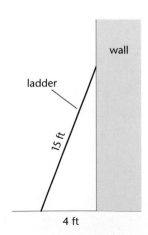

EXERCISES & PROBLEMS

Communicate

1. State the "Pythagorean" Right-Triangle Theorem in your own words.

2. Explain some of the practical uses of the "Pythagorean" Right-Triangle Theorem.

3. Explain how something called the "3-4-5 rule" would help carpenters square up corners.

4. What is the measure of the hypotenuse in the triangle at right?

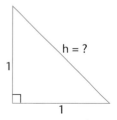

Practice & Apply

Algebra   For Exercises 5–9, two lengths of sides of a right triangle are given. Find the missing length; a = leg 1, b = leg 2, and c = hypotenuse. Leave answers in radical form.

5. $a = 3$, $b = 4$, $c = $ ___?___

6. $a = 10$, $b = 15$, $c = $ ___?___

7. $a = 46$, $b = 73$, $c = $ ___?___

8. $a = $ ___?___ , $b = 6$, $c = 8$

9. $a = 27$, $b = $ ___?___ , $c = 53$

10. What is the length of a diagonal of a square whose sides measure 5 cm?

11. The diagonal of a square measures 16 cm. Find its area.

12. **Sports**   A baseball diamond is a square with 90-foot sides. What is the approximate distance the catcher must throw from home to second base?

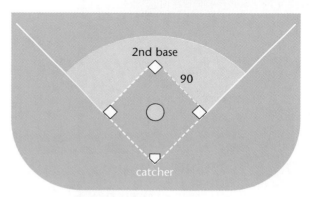

Algebra   Each of the following triples represents the sides of a triangle. Determine whether each triangle is right, obtuse, acute, or not a triangle. Give a reason for each answer.

13. 5, 9, 12

14. 13, 15, 17

15. 7, 24, 25

16. 7, 24, 26

17. 3, 4, 5

18. 25, 25, 30

Algebra Mathematicians have been fascinated with trying to generate "Pythagorean" triples. Two methods that generate sets of "Pythagorean" triples are shown below. Use each method to generate five sets of triples. Then use algebra to prove that the method will always work.

19. Method of Pythagoreans (*m* is an odd number greater than 1)

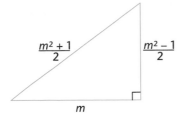

$\frac{m^2 + 1}{2}$ $\frac{m^2 - 1}{2}$ *m*

20. Method of Plato (*m* is any whole number greater than 1)

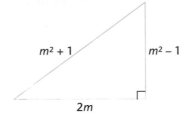

$m^2 + 1$ $m^2 - 1$ $2m$

21. Cultural Connection: Asia The *Sulbastutras* books of India, written around 800 B.C.E., demonstrate a method for constructing a square whose area is equal to the sum of the areas of two given squares. If the two given squares are *ABCD* and *EFGH*, and if $\overline{AR}$ is constructed so that *AR = EH*, explain why *BRST* is the desired square.

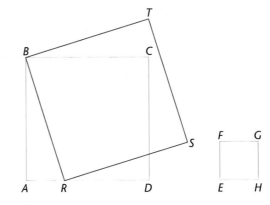

22. Use a method similar to the one in Exercise 21 to construct a square with double the area of a given square.

23. How high on the wall will the ladder reach? Assume that the base of the ladder is 8 feet off the ground.

Mathematicians have provided many possible proofs of the "Pythagorean" Right-Triangle Theorem. Three are presented below.

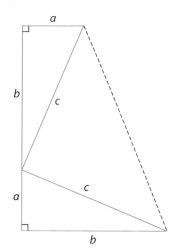

24. History   President Garfield, the twentieth president of the United States, devised a proof of the "Pythagorean" Right-Triangle Theorem using a trapezoid. Using the figure, find the area of the trapezoid in two different ways. Set your two expressions for the area equal to each other, and you will discover his proof.

25. 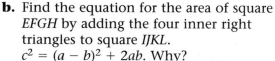 **Portfolio Activity**   This is a proof of the "Pythagorean" Right-Triangle Theorem represented pictorially. Explain how the figures prove the theorem. You may wish to sketch the squares on a separate piece of paper and cut them into pieces to help explain the proof.

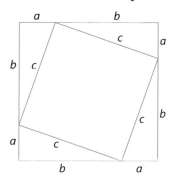 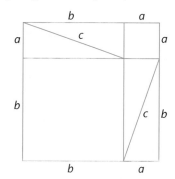

26. 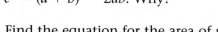 **Algebra**   The diagram from the *Chiu Chang* suggest at least two different proofs of the "Pythagorean" Right-Triangle Theorem.

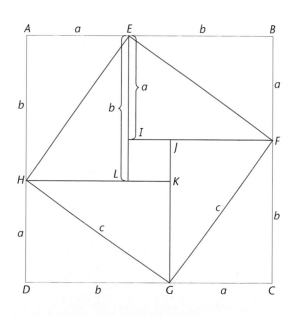

a. Find the equation for the area of square *EFGH* by subtracting the outer triangles of square *ABCD*:
$c^2 = (a + b)^2 - 2ab$. Why?

b. Find the equation for the area of square *EFGH* by adding the four inner right triangles to square *IJKL*.
$c^2 = (a - b)^2 + 2ab$. Why?

Simplify the equation in each part to obtain the desired results.

Why do you think the gridwork in the *Chiu Chang* was chosen with the small square in the center equal to one square unit?

**Find the area of the shaded region for each problem below.
Round answers to the nearest tenth.**

27.

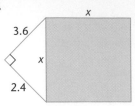

28.

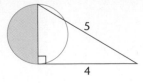

29.

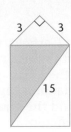

Look Back

**Answer Exercises 30–33 as true or false. If true, explain why;
if false, give a counterexample.   [Lesson 3.2]**

30. Every rhombus is a rectangle.

31. Every rhombus is a parallelogram.

32. If a diagonal divides a quadrilateral into two congruent triangles, then
the quadrilateral is a parallelogram.

33. The sum of the measures of the angles of a quadrilateral is 360°.

Algebra Recall from algebra that a fraction with a radical in
the denominator can be simplified by multiplying by a fraction equal
to 1. For example,

$$\text{Simplify } \frac{\sqrt{20}}{\sqrt{3}} \qquad \frac{\sqrt{20}}{\sqrt{3}} = \frac{\sqrt{20} \times \sqrt{3}}{\sqrt{3} \times \sqrt{3}} = \frac{\sqrt{60}}{3} = \frac{2\sqrt{15}}{3}$$

Simplify the following.

34. $\dfrac{\sqrt{30}}{\sqrt{5}}$

35. $\dfrac{\sqrt{72}}{\sqrt{6}}$

36. $\dfrac{\sqrt{6}}{\sqrt{6}}$

Look Beyond

37. Use the "Pythagorean" Right-Triangle
Theorem to find the distance between
points B and C.

38. Based on the result from Exercise 37, devise a
general formula for a distance between two
points on a coordinate plane. Consider the
distance between points K and I in Figure 2.

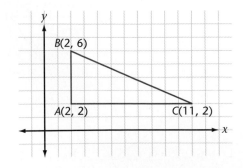

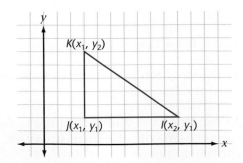

LESSON 5.5 Special Triangles, Areas of Regular Polygons

Why *The traditional tools of mechanical drawing include a T square or parallel bar and two special triangles. One triangle has angles measuring 30°, 60°, and 90°. The other has angles measuring 45°, 45°, and 90°. The properties of these triangles make them especially useful in geometry as well as in drawing.*

45-45-90 Right Triangles

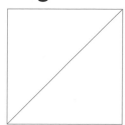

If you draw a diagonal of a square, two congruent isosceles triangles are formed. Since the diagonal is the hypotenuse of a right triangle, its length can be found by using the "Pythagorean" Right-Triangle Theorem.

EXAMPLE 1

ALGEBRA *Connection*

Use the "Pythagorean" Right-Triangle Theorem to find the length of the hypotenuse of triangle $\triangle ABC$. What is the ratio of the hypotenuse to the side?

Solution ➤

$h^2 = 10^2 + 10^2$

$h = \sqrt{200} = \sqrt{100} \times \sqrt{2} = 10\sqrt{2}$

The ratio of the hypotenuse to the side is $\frac{10\sqrt{2}}{10}$, or $\sqrt{2}$. ❖

CRITICAL *Thinking*

What is the length of the diagonal of a square with side s? What is the ratio of the diagonal to the side?

Notice that the diagonal of the square forms a right triangle with two 45° base angles. This triangle is known as a 45-45-90 right triangle. Thus, the hypotenuse can be found by applying the "Pythagorean" Right-Triangle Theorem.

45-45-90 RIGHT-TRIANGLE THEOREM
In any 45-45-90 right triangle, the length of the hypotenuse is $\sqrt{2}$ times the length of a leg.   **5.5.1**

30-60-90 Right Triangles

If you draw the altitude of an equilateral triangle, two congruent right triangles are formed. The acute angles of each right triangle measure 30° and 60°. These triangles are known as 30-60-90 right triangles. The length of the hypotenuse of each right triangle is two times the length of the shorter leg. (Why?)

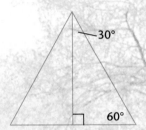

EXAMPLE 2

Find the missing lengths for the 30-60-90 right triangle to the right.

Solution ➤

ALGEBRA
Connection

In a 30-60-90 right triangle, the length of the hypotenuse is two times the length of the shorter leg. The length of the hypotenuse is 10.

Use the "Pythagorean" Right-Triangle Theorem to find the other leg.

$$5^2 + x^2 = 10^2$$

$$25 + x^2 = 100$$

$$x^2 = 100 - 25$$

$$x^2 = 75$$

$$x = \sqrt{75} = 5\sqrt{3} \approx 8.66 \; ❖$$

•Exploration 30-60-90 Right Triangles

You will need
No special tools

1 Use the "Pythagorean" Right-Triangle Theorem to fill in a table for 30-60-90 right triangles.

Shorter leg	Hypotenuse	Longer leg
1	2	?
2	?	?
3	?	?

2 Look for a pattern in the table regarding the longer leg and make a generalization.

3 If the length of the shorter leg is x:

a. What is the length of the hypotenuse?

b. What is the length of the longer leg?

4 Use the Exploration results to complete a drawing for a "general" 30-60-90 right triangle. ❖

30-60-90 RIGHT-TRIANGLE THEOREM

In any 30-60-90 right triangle, the length of the hypotenuse is 2 times the length of the shorter leg, and the longer leg is $\sqrt{3}$ times the length of the shorter leg.

5.5.2

APPLICATION

Measurement   Jake is measuring the height of a tall tree his grandfather planted as a boy. Jake uses a special instrument to find a spot where a 30° angle is formed by the ground and a line to the top of the tree. How tall is the tree if Jake is 80 feet from the base of the tree?

The line of sight is the hypotenuse of a 30-60-90 right triangle. The longer leg of the triangle is 80 feet. The height of the tree represents the shorter leg. If x is the height of the tree:

$$x\sqrt{3} = 80$$

$$x = \frac{80}{\sqrt{3}} \approx 46.18 \text{ feet.} ❖$$

Try This Find the missing measure *x*.

ALGEBRA
Connection

a.

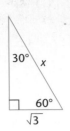

30°
x
2
60°
1

b.

30°
$2\sqrt{3}$
60°
x

c.

30°
x
60°
$\sqrt{3}$

Areas of Regular Polygons

To find the area of a regular hexagon, divide the hexagon into six congruent nonoverlapping equilateral triangles. Find the area of one triangle and multiply by 6 for the area of the hexagon. Note that the altitude of one equilateral triangle is the longer leg of a 30-60-90 right triangle. The altitude is equal to $\frac{1}{2}$ the length of a side multiplied by $\sqrt{3}$. (Why?)

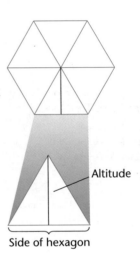

Altitude

Side of hexagon

EXAMPLE 3

Find the area of a regular hexagon with sides of length 20 cm.

Solution ➤

Divide the hexagon into 6 equilateral triangles. Since an altitude of one of the triangles forms the longer leg of a 30-60-90 right triangle, and half of the side of the hexagon forms the shorter leg, the length of the altitude is $10\sqrt{3}$.

Use the area formula to find the area of one of the triangles.

$$A = \tfrac{1}{2}(20)(10\sqrt{3})$$

$$A = 100\sqrt{3}$$

Since the hexagon is composed of 6 congruent triangles, the area of the hexagon is

$$6(100\sqrt{3}) = 600\sqrt{3}. \text{ ❖}$$

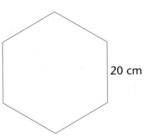

20 cm

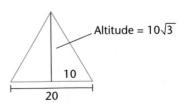

Altitude = $10\sqrt{3}$
10
20

The steps described in Example 3 can be applied to find the area of any regular polygon.

An *n*-sided polygon can be divided into *n* nonoverlapping congruent triangles. The altitude of each triangle is the segment from the center of the polygon to the midpoint of the side. In a regular polygon, this segment is called the apothem.

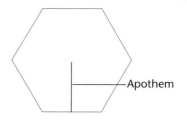
—Apothem

AREA OF A REGULAR POLYGON

The area of a regular polygon with apothem *a* and perimeter *P* is $A = \frac{1}{2} ap$, where *a* is the apothem and *p* is the perimeter.

5.5.3

Try This Find the area of a regular polygon with a perimeter of 40 inches and an apothem of 5 inches.

EXERCISES & PROBLEMS

Communicate

1. The leg of a 45-45-90 triangle is 6 centimeters long. Find the length of the hypotenuse and explain how you found it.

2. In a 30-60-90 triangle, the shorter leg has length 4 inches. Explain how to find the length of the longer leg and the length of the hypotenuse.

3. How do you find the length of the apothem of a regular polygon?

4. Explain how to find the area of a regular polygon?

5. What are the missing values in the chart for a 45-45-90 right triangle?

Leg 1	Leg 2	Hypotenuse
1	1	?
2	2	?
3	3	?
4	4	?
5	5	?

Practice & Apply

Find the area of each figure.

6.

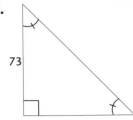

73

7.

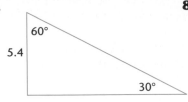

60°
5.4
30°

8.

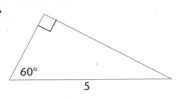

60°
5

For the given length, find each of the remaining two lengths.

9. $x = 6$

10. $y = 6$

11. $z = 14$

12. $y = 4\sqrt{3}$

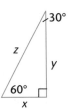

30°
z
y
60°
x

13. $p = 6$

14. $r = 6$

15. $q = 4\sqrt{2}$

16. $r = 10$

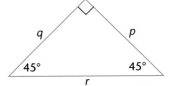

q
p
45°
45°
r

17. $k = 3.4$

18. $k = 6\sqrt{3}$

19. $g = 17$

20. $h = 2\sqrt{3}$

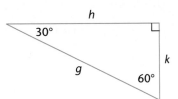

h
30°
g
k
60°

Civil Engineering An engineer is in charge of attaching guy wires to a tower.

21. One set of wires needs to extend at a 45° angle to the tower at a point *A* on the ground 80 feet from the tower. What is the length of each wire?

22. Another wire needs to make a 30° angle with the tower from a point on the ground, 80 feet from the tower. How high up the tower does the wire need to be?

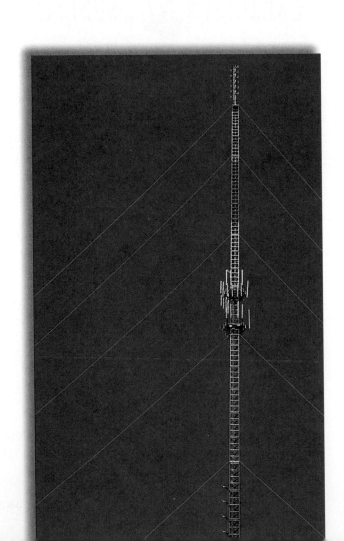

23. Find the area of a square if its perimeter is 16 cm and its apothem is 2 cm.

24. Use a protractor and ruler to construct a regular hexagon with sides 2 cm. Measure the apothem with your ruler and find the area of a hexagon.

25. Use a protractor and ruler to construct a hexagon with sides 6 cm. Measure the apothem with your ruler and find the area of the hexagon.

26. Find the area of a regular hexagon whose sides measure 10 cm.

27. Find the area and perimeter of an equilateral triangle whose sides measure 18 cm.

28. Find the area and perimeter of an equilateral triangle whose altitude measures 6 cm.

29. If the area of an equilateral triangle is 100 sq cm, find the length of a side.

Portfolio Activity   Refer to the figure and its description on page 242. Suppose the diameter of the large semicircle is 4.

30. Find the area of the large right triangle.

31. Find the area of the small semicircles.

32. Find the area of the large semicircle.

33. Combine the areas found to write an equation and find the area of the lunes.

34. What other area in the figure is the area of the lunes equal to?

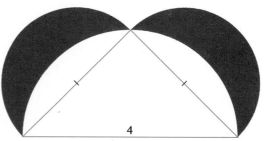

35. What other figures can you construct based on the idea of semicircles or circles constructed on segments of polygons? Discover what you can about these figures. Use them to create patterns and designs.

Look Back

Drafting   In drafting, a T square is used to help align drawings. The T square is held against the side of the drawing board and moved up and down. The angle of the straightedge stays perpendicular to the side of the drawing board.   **[Lesson 3.4]**

36. An artist makes a series of horizontal lines by using the straightedge of the T square and moving it up 1 inch for each new line. What is the relationship between the lines? Why?

37. With the 30-60-90 triangle positioned as shown, the artist uses the hypotenuse to draw an angle of 60° with one of the horizontal lines. Then she slides the triangle along the T square and draws another angle the same way. How are the two angles related? Why?

38. Find the smallest possible area for a rectangle with a perimeter of 100 m and sides that are whole numbers. Then find the greatest possible area. State the dimensions of each rectangle.   **[Lesson 5.1]**

39. The base of an isosceles triangle is 8 m and the legs are 6 m. Find the area and perimeter.   **[Lesson 5.2]**

40. Find the area of the triangles in the figure.   **[Lesson 5.2]**

41. Find the height of a trapezoid with area 103.5 sq ft, with bases 17.5 ft and 5.5 ft.   **[Lesson 5.2]**

42. Find the area of an isosceles right triangle with leg $\sqrt{2}$ cm.
[Lesson 5.4]

43. Find the area of a right triangle with legs a and b.   **[Lesson 5.4]**

44. In $\triangle PQR$, find PR.   **[Lesson 5.4]**

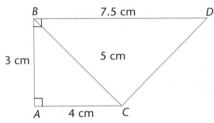

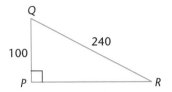

Look Beyond

45. The two figures below have the same volume. Do they have the same surface area? Give the surface area for each figure.

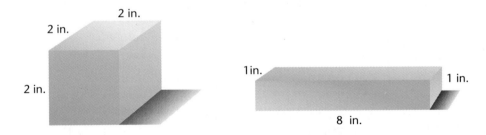

46. Sketch another figure with the same volume as the figures in Exercise 45. Does your new figure have a surface area larger or smaller than those of the two figures?

LESSON 5.6 The Distance Formula, Quadrature of a Circle

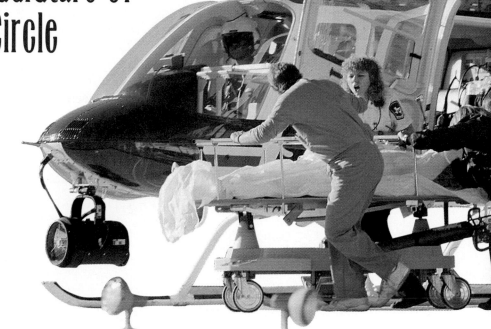

why *To reach its destination, a helicopter will travel the shortest distance between two points. The ability to compute the distance between two points is important to know, especially in emergency situations. In some cases, an estimate based on measurement is sufficient. However, in other cases, great precision is necessary.*

The Distance Formula

The distance between two points on the same vertical or horizontal line can be found by taking the absolute value of the difference between the x- and y-coordinates. The vertical distance between points A and B, or length AB, is $|7 - 3| = |3 - 7| = 4$.

The horizontal distance between points B and C, or length BC, is $|2 - 5| = |5 - 2| = 3$.

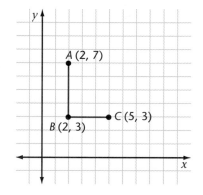

Since $\overline{AC}$ forms the hypotenuse of right triangle $\triangle ABC$, its length can be found using the "Pythagorean" Right-Triangle Theorem.

$$(AC)^2 = 4^2 + 3^2$$

$$(AC)^2 = 25$$

$$AC = \sqrt{25} = 5$$

The "Pythagorean" Right-Triangle Theorem can be used to find the distance between any two points on the coordinate plane.

DISTANCE FORMULA

The distance between two points (x_1, y_1) and (x_2, y_2) is

$$d = \sqrt{(x_2 - x_1)^2 + (y_2 - y_1)^2}$$

5.6.1

Proof:

Draw a right triangle with hypotenuse $\overline{AB}$, a right angle at C, and d equal to the distance between A and B.

The coordinates of C are x_2 and y_1. (why?)

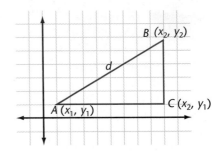

The length AC is $|x_2 - x_1|$.

The length BC is $|y_2 - y_1|$.

Using the "Pythagorean" Right-Triangle Theorem,

$$d^2 = |x_2 - x_1|^2 + |y_2 - y_1|^2$$

$$d^2 = (x_2 - x_1)^2 + (y_2 - y_1)^2$$

You can drop the absolute value symbols because the quantities are being squared.

Solving for d,

$$d = \sqrt{(x_2 - x_1)^2 + (y_2 - y_1)^2}$$

ALGEBRA
Connection

EXAMPLE

Navigation   A helicopter crew located 1 mile east and 3 miles north of Command Central must respond to an emergency located 7 miles east and 11 miles north of Command Central. How far must the helicopter travel to get to the emergency site?

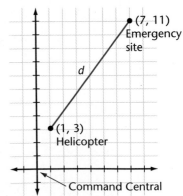

Solution ➤

Let $(0, 0)$ be the coordinates of Command Central. Then the helicopter crew coordinates are $(1, 3)$. The emergency site coordinates are $(7, 11)$. Using the distance formula,

$$d^2 = (7 - 1)^2 + (11 - 3)^2$$

$$d = \sqrt{6^2 + 8^2} = \sqrt{36 + 64} = \sqrt{100} = 10.$$

The helicopter must travel 10 miles to the emergency site. ❖

Try This Find the distance between the points $(-2, 4)$ and $(3, -9)$.

Midpoint Formula

An Exploration in Lesson 4.8 led to the midpoint formula. Recall that the midpoint of two points on the same line can be found by averaging the values of the x- and y-coordinates.

ALGEBRA
Connection

$x_1 = 2, x_2 = 8 \quad x_m = \frac{8 + 2}{2} \quad x_m = 5$

$y_1 = 6, y_2 = 2 \quad y_m = \frac{6 + 2}{2} \quad y_m = 4$

The midpoint coordinates are $(5, 4)$.

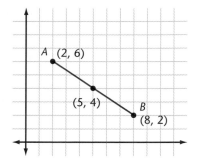

THE MIDPOINT FORMULA

For any two points (x_1, y_1) and (x_2, y_2) in the coordinate plane, the midpoint is given by

$$\left(\frac{x_1 + x_2}{2}, \frac{y_1 + y_2}{2} \right)$$

5.6.2

Try This Find the midpoint of $(8, 3)$ and $(2, 6)$. Show that the lengths of the two segments formed by the midpoint are equal using the distance formula.

Methods of Quadrature

The area of an enclosed region can be approximated by breaking it down into a number of rectangles. This technique, called quadrature, is particularly important for finding the area under a curve.

•Exploration •Estimating the Area of a Circle

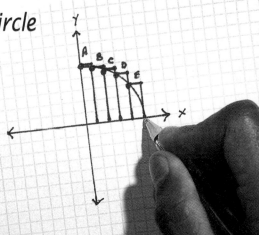

*Scientific
Calculator*

You will need
Graph paper
Scientific calculator
Spreadsheet software (optional)

Part I: Method A

1. Draw a quarter of a circle with a radius of five units and its center at the origin $(0, 0)$ of a coordinate plane. Add rectangles to completely cover the circle as shown.

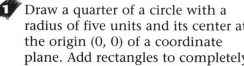

2 Find the coordinates of the indicated points using the "Pythagorean" Right-Triangle Theorem. The y-coordinate is the height of a rectangle. For point C, this is

$$y = \sqrt{5^2 - 2^2} = \sqrt{21}$$

3 Find the area of each rectangle. (This is simplified by the fact that the base of each rectangle is 1.)

4 Find the sum of the areas of the rectangles by completing the sequence below.

$$\sqrt{25} + \sqrt{24} + \sqrt{21} + \ldots = ?$$

5 Multiply your sum by 4. The result is an estimate of the area of a circle with radius 5. Does it overestimate or underestimate the area? Explain why.

6 Calculate the true value of the area to four decimal places using $A = \pi r^2$. Find the relative error of your estimate using the formula

$$E = \frac{|V_e - V_t|}{V_t} \times 100$$

where V_e = estimated value, V_t = true value, and E = percent of error.

Part II: Method B

Repeat Part 1 using an arrangement of rectangles like the one shown at left. Does the new method overestimate or underestimate the area of the circle? Explain why.

Part III: Combining Methods

Average your results from Parts 1 and 2. What is the relative error of your new estimate? ❖

CRITICAL *Thinking*
Do you think that the average of results from Methods A and B will always give more accurate estimates than either one by itself? Explain why or why not.

EXERCISES & PROBLEMS

Communicate

1. Explain why it is necessary to find the differences between *x*- and *y*-coordinates in the distance formula.

2. Explain how the distance formula is related to the "Pythagorean" Right-Triangle Theorem.

3. Explain two methods of estimating the shaded area under the curve.

4. Explain why the estimate for the area becomes more accurate as the rectangles are made smaller.

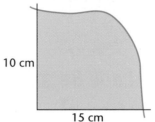

10 cm

15 cm

Practice & Apply

Find the distance between each pair of points.

5. (1, 4) and (3, 9)

6. (−3, −3) and (6, 12)

7. (−1, 4) and (−3, 9)

8. (−2, −3) and (−6, −12)

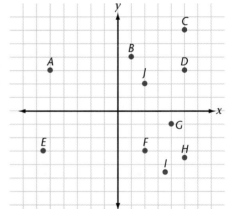

Algebra Use the diagram at right for Exercises 9–13.

9. Find the distance between *A* and *B*.

10. Find the distance between *B* and *I*.

11. Is the triangle formed by *B*, *C*, and *D* isosceles? Why or why not?

12. Is the quadrilateral formed by *F*, *G*, *H*, and *I* a square? Why or why not?

13. Find the lengths of the sides of the triangle formed by *E*, *J*, and *H*.

14. **Environmental Protection** In order to minimize the potential for environmental damage, an oil pipeline is being rerouted so that less of the pipe will fall inside a national forest. What is the distance between point *A* and point *B* in the map on the right?

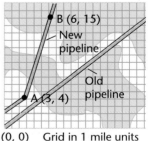

B (6, 15)

New pipeline

Old pipeline

A (3, 4)

(0, 0)   Grid in 1 mile units

15. Place a 5–12–13 right triangle on the coordinate system. Show coordinates for each point and the distances for the two legs and the hypotenuse.

16. Select three points on the coordinate plane that will be the vertices of an isosceles triangle. Prove that the triangle is isosceles. Use the distance formula.

17. **Algebra** Use the rectangular method to estimate the area under $y = x^2 + 2$ and the x-axis when $0 \leq x \leq 2$.

18. **Algebra** Use the rectangular method to estimate the area under $y = -x^2 + 2$ and the x-axis when $0 \leq x \leq 1\frac{1}{2}$.

Look Back

19. Prove that $WXYZ$ is a rectangle using the coordinates $W(-1, 1)$; $X(7, -5)$; $Y(10, -1)$; $Z(2, 5)$.   **[Lesson 3.8]**

20. Suppose you have 100 m of fence to enclose a play area. Which would be the larger play area, a square play area or a circular one? **[Lesson 5.1]**

21. Suppose a circle and a square both have an area of 225 sq cm. Which has the greater perimeter?   **[Lesson 5.1]**

22. The sides of an equilateral triangle measure 8 cm. Find the area and perimeter. Find the area if the sides are s units in length. **[Lesson 5.4]**

Find the missing side for each right triangle. Side c is the hypotenuse.   [Lesson 5.4]

23. $a = 4.5, b = 8, c = \underline{\ ?\ }$

24. $a = \underline{\ ?\ }, b = 4, c = 5$

Look Beyond

Technology   Use appropriate technology to solve Exercises 25–27.

25. Repeat the Exploration using a larger number of rectangles. If you wish, you can use rectangles with a base of 1 and increase the diameter of the circle. Does your accuracy increase with larger numbers of rectangles?

26. If you have access to a spreadsheet or other computer software, estimate using very large numbers of rectangles. How many rectangles do you need to get a 1 percent accuracy? 0.1 percent? .01 percent? Graph your accuracy for different numbers of rectangles.

27. Is it always necessary to have extremely accurate answers to real-life problems? Why might you sometimes prefer a rough estimate to a more precise one?

LESSON 5.7

Geometric Probability

According to the fossil record, the Earth experienced a sudden, drastic change about 65 million years ago—resulting in the extinction of great dinosaurs like the Tyrannosaurus rex. Many scientists believe that this upheaval was caused by an asteroid striking the Earth at the Yucatan Peninsula.

The surface of the Earth consists of about 30 percent land and 70 percent water. Using these figures, and assuming that a comet or asteroid would be equally likely to strike anywhere on the Earth, what is the probability that such an object would have struck land instead of water?

Why *Many scientists believe that an asteroid or comet caused the extinction of the dinosaurs. What is the likelihood that such an object would have struck land? You can use geometric probability to answer this question.*

The Basic Formula

Mathematical intuition should tell you that there is a 30 percent chance, or probability, that the object would have struck land. **Probability** is the ratio that compares the number of "successful" outcomes with the total number of possible outcomes. For example, if there are six marbles in a bag and two are blue, the probability of picking a blue marble at random is $\frac{2}{6}$, or $\frac{1}{3}$.

The same idea applies to areas. Since the surface area of the Earth is 100 percent and about 30 percent of the Earth's surface is covered by land, the probability that the asteroid would have struck land is 30 percent, $\frac{30}{100}$, or 0.3.

Geometric Probability

PROBABILITY
Connection

Given a figure like the one at right, the probability that a point randomly selected from the figure will be in Area *A* is

$$\frac{\text{Area } A}{\text{Area } B}.$$

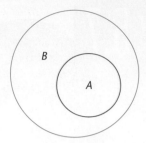

Area B is the total area of the figure.

EXAMPLE 1

If a dart lands in the shaded part of the board, a prize is given. What is the probability that a toss that hits the board at random will win?

Solution ➤

The area of the shaded part is 60.

The area of the entire region is 96.

The probability of landing on the shaded part is $\frac{60}{96}$, or $\frac{5}{8}$. ❖

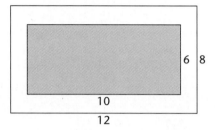

Try This Find the probability that a dart will land in the shaded area.

a.

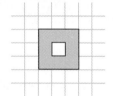

b.

c.

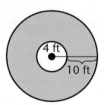

EXAMPLE 2

In a game, pennies are tossed onto a grid that has squares that are each the width of a penny. To win, a penny must not land on an intersection point of the lines that make up the grid. What is the probability that a random toss will lose?

Solution ➤

Imagine that circles the size of a penny are drawn around each intersection point on a grid. Any time the center of a penny falls within one of the circles, the penny will touch an intersection point.

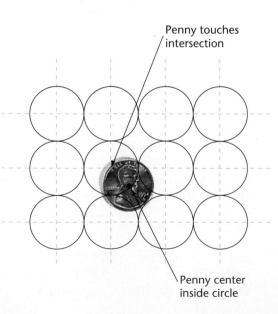

Penny touches intersection

Penny center inside circle

The center of a penny that falls on the grid will land somewhere within one of the squares. What is the probability that it will be within the shaded region of the square?

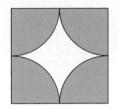

Let the radius of a penny equal 1 unit. Then the width of a square (and the diameter of a penny) will be 2. Thus,

Area of circle $\qquad A_c = \pi \times 1^2 = \pi$ sq units

Area of square $\qquad A_s = 2 \times 2 = 4$ sq units

The area of each quarter of a circle is $\frac{\pi}{4}$ sq units, so the area of the shaded region is π sq units. The probability P of the penny touching an intersection and of the player losing is

$$P = \frac{\text{Area of Shaded Region}}{\text{Area of Square}} = \frac{\pi}{4}. \ ❖$$

In the following exploration you will use the result of Example 2 to estimate the value of π.

•Exploration A "Monte Carlo" Method for Estimating π

You will need
Grid paper with squares equal in width to the diameter of a penny
Pennies

1 Toss a coin onto the grid paper. Notice whether the coin touches an intersection point of two lines or not. Repeat 20 times. Record the number of tosses that fall on intersection points and the number of tosses that do not.

2 Share your results with the rest of the class. Tally the totals for the entire class. Calculate the ratio of tosses that land on intersections in your experiment to the total number of tosses, and call it E.

3 For a large number of coin tosses, you may assume that the value of E will be close to the probability P that was calculated in Example 2. Using this assumption, calculate an estimate for π:

$E \approx \frac{\pi}{4}$ (assumed to be true for large numbers of tosses)

So $\pi \approx E \times 4$

4 Compare your estimate with the known value of π. Determine the relative error of your estimate using the formula

$$E = \frac{|V_t - V_e|}{V_t},$$

where E = relative error, V_t = true value, and V_e = estimated value. ❖

EXERCISES & PROBLEMS

Communicate

Use the diagram of $\overline{AB}$ for Exercises 1–4. *M* is the midpoint of $\overline{AB}$ and *C* is the midpoint of $\overline{AM}$. For each exercise, a point on $\overline{AB}$ is picked at random.

```
•——————•—————————•————————————————•
A       C         M                B
```

1. What is the probability that the point is on $\overline{AM}$? Explain.

2. What is the probability that the point is on $\overline{AC}$? Explain.

3. What is the probability that the point is on $\overline{AB}$? Explain.

4. What is the probability that the point is not on $\overline{AC}$? Explain.
(Hint: 1 minus the probability of the point being on $\overline{AC}$)

Practice & Apply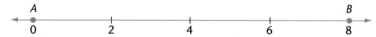

A point *Q* is selected at random from segment $\overline{AB}$ on the number line.

```
    A                                    B
←———•———+———————+———————+———•———→
    0       2       4       6   8
```

5. What is the probability that $0 \le Q \le 4$?

6. What is the probability that $1 \le Q \le 4$?

7. What is the probability that $5 \le Q \le 6.5$?

8. What is the probability that $0 \le Q \le 8$?

Probability Probabilities are usually expressed as decimals or fractions ranging from 0 to 1. Probabilities can also be expressed as percents. For example, in Exercise 3, the probability is 1, or 100%, that the point will be on $\overline{AB}$. In Exercise 1, the probability is .5, $\frac{1}{2}$, or 50%, that the point is on $\overline{AM}$. Convert the following probabilities to percents.

9. 0.75   **10.** $\frac{1}{4}$ **11.** $\frac{2}{3}$

Convert the following percentages to probabilities between 0 and 1.

12. 60%   **13.** 50%   **14.** $33\frac{1}{3}\%$

15. Meteorology The weather forecaster predicts an 80% chance of rain. Express this as a probability between 0 and 1.

Skydiving At the state fair, a sky diver jumps from an airplane and parachutes into a rectangular field as shown. Assume that he is equally likely to land anywhere in the field for Exercises 16–19.

16. What is the probability that he will land in the right triangle?

17. What is the probability that he will land in the square?

18. What is the probability that he will land in the circle?

19. The sky diver will be successful if he lands in one of the shaded figures and not successful if he misses the figures. What is the probability that the sky diver will miss all the shaded areas?

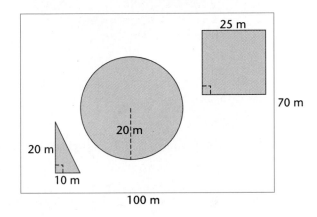

20. The probability of a success (P_s) and the probability of a failure or nonsuccess (P_f) is related by this formula: $P_f = 1 - P_s$. Find the probability that a penny will not land on a point of intersection in Example 2.

21. Suppose you randomly generate 20 ordered pairs with each coordinate being an integer from 0 to 6. For the graph on the right, what is the probability that an ordered pair lies inside the inner square?

22. Conduct an experiment for Exercise 21 generating the ordered pairs using dice, a computer, or a calculator. What is the experimental result, and how does it compare with the theoretical probability you found in Exercise 21?

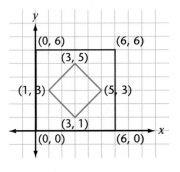

23. 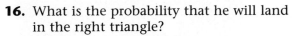 **Portfolio Activity** Design a dart-board in which the theoretical probability of a dart landing in a red circle is 50%.

24. 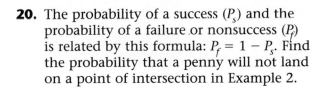 **Portfolio Activity** Design a dart-board in which the theoretical probability of a dart landing in a red triangle is $33\frac{1}{3}$%.

25. If the square measures 4 inches per side, and the circle has a diameter of 3 inches, what is the theoretical probability that a dart will land inside the circle? In throwing darts, why might experimental result be higher than theoretical probability?

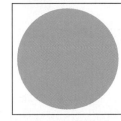

26. Transportation At a subway stop in New York City, a train arrives every 5 minutes, waits 1 minute, and then leaves. If you arrive at a random time, what is the probability that a train will be there in at most 2 minutes? The diagram will help you.

5 min.   5 min.   5 min.

1 min.   1 min.   1 min.

Look Back

Find the area for each figure indicated below.

27. Triangle: Base = 4 in, Height = 7.5 in.   **[Lesson 5.2]**

28. Parallelogram: Base = 4 cm, Height = 7.5 cm   **[Lesson 5.2]**

29. Regular Polygon: Apothem = 3 ft, Perimeter = 18 ft   **[Lesson 5.2]**

30. Trapezoid: Base$_1$ = 20 yd, Base$_2$ = 30 yd, Height = 12.6 yd
[Lesson 5.2]

31. Circle: Radius = 16 mm (use $\frac{22}{7}$ for π)   **[Lesson 5.3]**

Look Beyond

Fine Art   Perspective is important in painting and drawing. Use the following picture to answer the questions below.

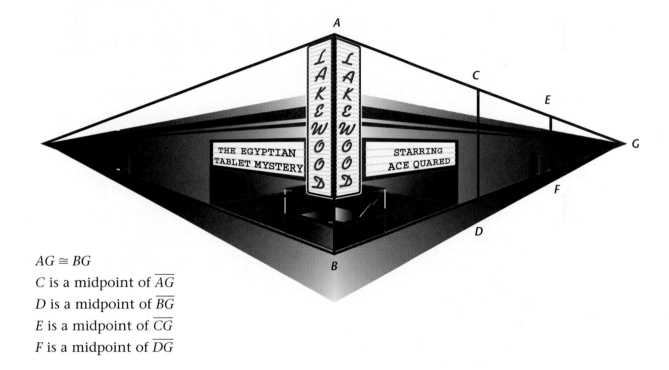

$AG \cong BG$

C is a midpoint of $\overline{AG}$

D is a midpoint of $\overline{BG}$

E is a midpoint of $\overline{CG}$

F is a midpoint of $\overline{DG}$

32. If AB = 8 in., CD = ___?___.

33. If AB = 8 in., EF = ___?___.

34. Explain your reasoning for Exercises 32 and 33.

AREA OF A POLYGON

This project challenges you to find a formula for the areas of polygons drawn on dot paper—surprisingly, one formula will work for all! Work with a team of 3 or 4 classmates. You will probably find it helpful to divide up the work among your team's members.

Compute the areas for figures a–f in each of the three parts below. Then, on a separate sheet of paper, complete a table like the ones given in each part. For each figure, N_b is the number of points on the boundary of the figure, and N_i is the number of points in its interior. A is the area of the figure.

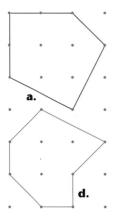

a.

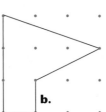

b.

c.

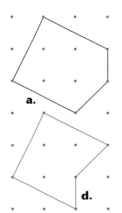

d.

e.

f.

Part A			
	N_b	N_i	A
a.	7	4	$6\frac{1}{2}$
b.	?	?	?
c.	?	?	?
d.	?	?	?
e.	?	?	?
f.	?	?	?

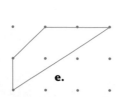

a.

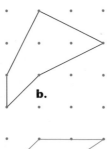

b.

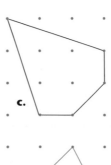

c.

d.

e.

f.

Part B			
	N_b	N_i	A
a.	?	?	?
b.	?	?	?
c.	?	?	?
d.	?	?	?
e.	?	?	?
f.	?	?	?

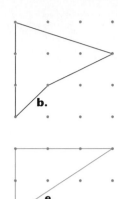

a.

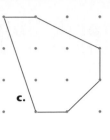

b.

c.

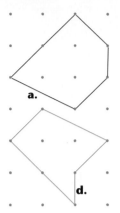

d.

e.

f.

Part C			
	N_b	N_i	A
a.	?	?	?
b.	?	?	?
c.	?	?	?
d.	?	?	?
e.	?	?	?
f.	?	?	?

EXTENSION

To help you determine the formula, each part focuses on a fixed number of boundary points.

1. What is the pattern to calculate the area?

2. Write the formula.

3. Test your formula by creating new figures on dot paper.

The formula you have discovered was originally discovered by G. Pick in 1899.

Chapter 5 Review

Vocabulary

area	245	hypotenuse	269	probability	289
circle	261	kite	259	radius	261
circumference	262	Monte Carlo method	291	sector	262
diameter	261	perimeter	244		

Key Skills and Exercises

Lesson 5.1

➤ *Key Skills*

Find the perimeter of a polygon.

In the figure, find the perimeter of the hexagon.

The perimeter equals the sum of all sides: $P = AB + BC + CD + DE + EF + FA$. We are given that $AB = DE = 8$ cm, $FA = EF = 5$ cm, and $CD = BC = 6$ cm. Thus, $P = 8 + 6 + 6 + 8 + 5 + 5 = 38$ cm.

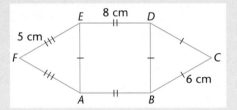

Find the area of a rectangle.

In the figure above, find the area of rectangle *ABDE*.

The formula for the area of a rectangle is $A = bh$. Here $\overline{BD}$ is the height; $BD = BC = 6$ cm. $\overline{ED}$ is the base (8 cm). Thus, $A = 8 \times 6 = 48$ cm^2.

➤ *Exercises*

Quadrilaterals ABFG and BCDE are rectangles.

1. Find the perimeter of *ACDEFG*.

2. Find the area of *ACDEFG*.

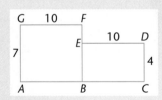

Lesson 5.2

➤ *Key Skills*

Solve problems using the formulas for the areas of triangles, parallelograms, and trapezoids.

This trapezoid has an area of 455 mm^2. What is its height?

The formula for the area of a trapezoid is $A = \dfrac{(b_1 + b_2)h}{2}$.
Substitute the given dimensions and solve the equation.

$$455 = \frac{(31 + 39)x}{2}$$
$$910 = 70x$$
$$x = 13$$

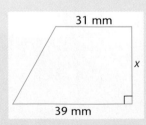

The trapezoid's height is 13 mm.

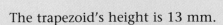

> **Exercises**

In Exercises 3–5, refer to the figure.

3. Find the area of parallelogram *PQSU*.

4. Find the area of trapezoid *PRSU*.

5. Find the area of △*QRS*.

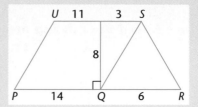

Lesson 5.3

> ### *Key Skills*

Solve problems using the formulas for the circumference and area of a circle.

A wheel makes 25 revolutions to go 4 meters. Find the radius of the wheel.

We know the circumference of the wheel is $\frac{4}{25} = .16$ meters, or 16 centimeters. We use the formula for circumference, 2π times the radius, to find the radius.

$16 = 2\pi r \approx 2(3.14)r$

$r \approx 2.5$ centimeters.

> ### *Exercises*

6. Find the area of a circle whose diameter is 21 inches.

7. If the circumference of a circle is 3.14 yards, find its radius.

Lesson 5.4

> ### *Key Skills*

Solve problems using the "Pythagorean" relationship.

The "Pythagorean" relationship holds for any right triangle. The relationship of the length of the sides of △*FGH* is $a^2 + b^2 = c^2$.

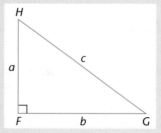

> ### *Exercises*

In Exercises 8–9, find the missing measures. Round to the nearest tenth.

8.

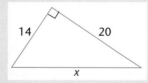

9.

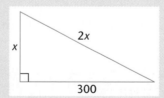

Lesson 5.5

> ### *Key Skills*

Use the properties of 45-45-90 and 30-60-90 right triangles to solve problems.

Find *HI* and *HJ*. Since two angles of △*HIJ* measure 30° and 60°, it is a 30-60-90 right triangle. In that type of triangle, the shorter leg is half the length of the hypotenuse, and the longer leg is $\sqrt{3}$ times the length of the shorter leg. In △*HIJ*, *HJ* = 7.5, and *HI* = 7.5$\sqrt{3}$.

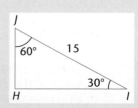

➤ Exercises

In Exercises 10–11, find the missing measures.

10.
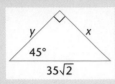

11.
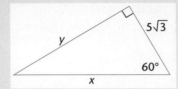

Lesson 5.6

➤ Key Skills

Determine the distance and the midpoint between two points on a coordinate plane.

Find the distance between points (6, 5) and (–6, 10).

You can use the distance formula, $d = \sqrt{(x_2 - x_1)^2 + (y_2 - y_1)^2}$.

$d = \sqrt{(-6 - 6)^2 + (10 - 5)^2} = \sqrt{(-12)^2 + (5)^2} = \sqrt{169} = 13$

The midpoint for two points is given by $\left(\frac{x_2 + x_1}{2}, \frac{y_2 + y_1}{2}\right)$. Thus,

$x = \frac{-6 + 6}{2} = 0$, and $y = \frac{10 + 5}{2} = 7.5$. The midpoint is (0, 7.5).

➤ Exercises

In Exercises 12–13, refer to the figure.

12. Find the perimeter of $\triangle ABC$.

13. $\triangle DEF$ has its vertices at the midpoints of the sides of $\triangle ABC$. Give the coordinates of $\triangle DEF$'s vertices.

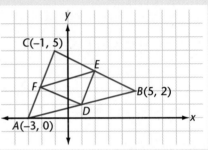

Lesson 5.7

➤ Key Skills

Find the probability of an event, using the ratio of two areas.

In the sketch, the circle represents a tree's crown, and the rectangle represents a reclining chair. If the tree drops an apple, what are the chances the chair will be hit? (Assume that the apple is equally likely to hit anywhere under the tree.)

Find the ratio of the area of the rectangle to the area of the circle.

$$\frac{12 \text{ ft}^2}{49\pi \text{ ft}^2} \approx \frac{12}{154} \approx .08$$

The chair has an 8% chance of being hit by an apple.

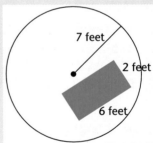

➤ Exercises

14. A random-number generator gives you ten pairs of integers, ranging from 0 to 8. What is the probability that $\triangle RNG$ includes one of the pairs?

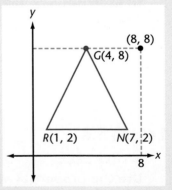

Applications

15. You have a 20-foot ladder to reach a roof that is 15 feet high. The top of the ladder must extend at least 3 feet above the roof's edge before you can safely step from the ladder onto the roof. To the nearest foot, what is the farthest from the house wall that you could set the bottom of the ladder?

300 CHAPTER 5

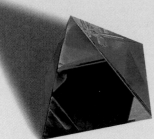

Chapter 5 Assessment

In Exercises 1–2, refer to the figure at the right.

1. Find the perimeter of *ABCDE*.
2. Find the area of *ABCDE*.
3. Find the area of trapezoid *MNOP*.

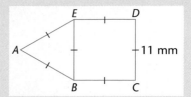

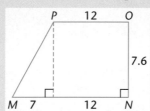

4. If the circumference of a circle is 30 millimeters, what is its area?

   a. 706.9 mm²   **b.** 289.4 mm²   **c.** 72.3 mm²   **d.** 94.2 mm²

In Exercises 5–6, find the missing measures.

5.

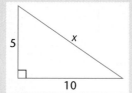

6.

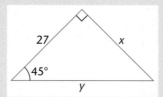

7. Find the area of a regular hexagon with a side of 6 inches and an apothem of $3\sqrt{3}$ inches.

In Exercises 8–9, refer to the figure at the right.

8. Find the perimeter of quadrilateral *ABCD*.

9. △*EFG* has its vertices at the midpoints of three sides of *ABCD*. Give the coordinates of △*EFG*'s vertices.

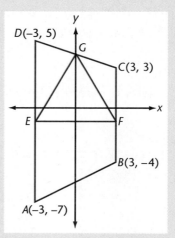

10. A point *P* is selected at random from the segment below. What is the probability that $2 \leq P \leq 8$?

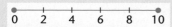

CHAPTER 6

Shapes in Space

LESSONS

6.1 *Exploring* Solid Shapes

6.2 *Exploring* Spatial Relationships

6.3 An Introduction to Prisms

6.4 Three-Dimensional Coordinates

6.5 Lines and Planes in Space

6.6 Perspective Drawing

Chapter Project
Building and Testing an A-Shaped Level

What gives this picture such a dramatic effect? Even though it is printed on a flat, two-dimensional plane, it has the illusion of depth. In this chapter you will find out how to give drawings this three-dimensional quality.

As you study geometric figures in space, you will discover that three-dimensional figures have properties that are similar to those of two-dimensional figures. In this chapter you will apply what you already know about points and lines in a plane to points, lines, and planes in space.

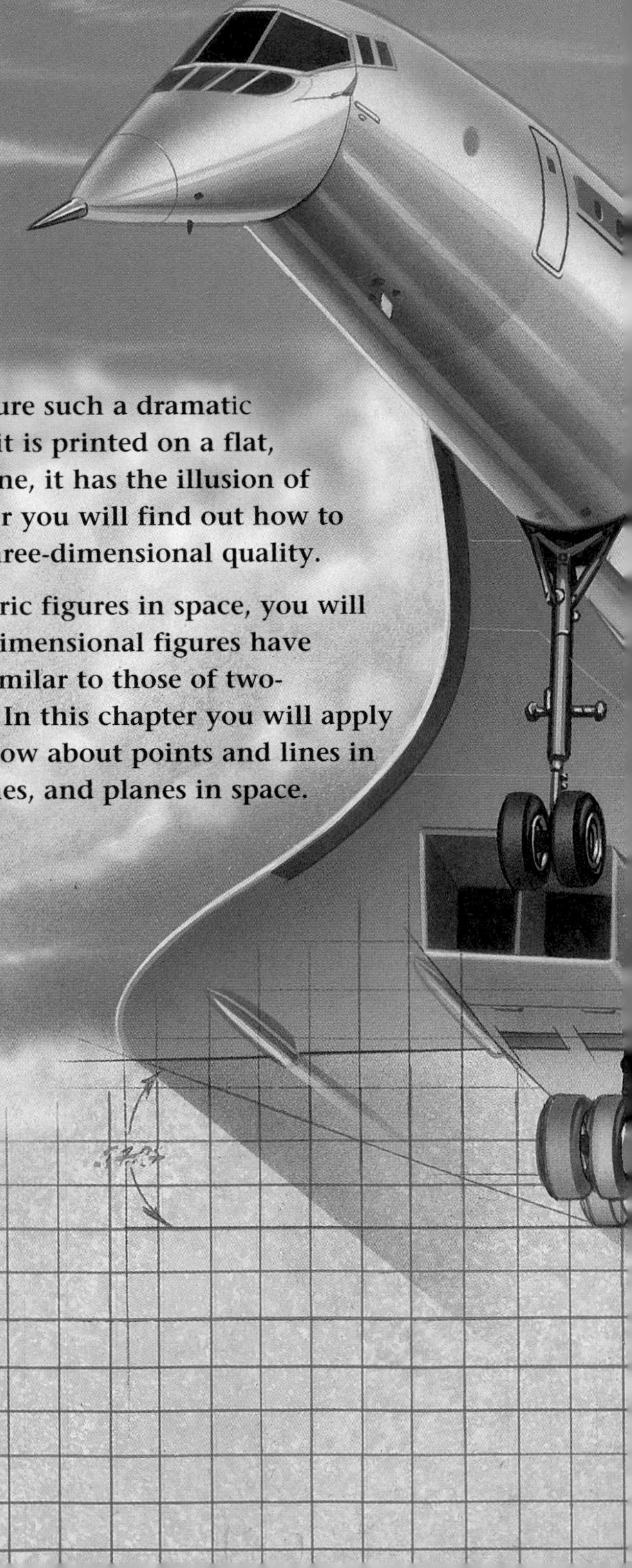

PORTFOLIO ACTIVITY

Create your own drawings that have the illusion of three dimensions. You may use pencil, paints, computer graphics, or any medium you wish. When you are finished, describe the techniques you used to give the appearance of depth. Include in your descriptions any special tools you may have used.

303

LESSON 6.1

Exploring

Solid Shapes

why *Architects must be skilled in rendering their conceptions of solid shapes. There are many other careers that require this skill. Perhaps you have already applied some of these skills in some of your school courses.*

Mechanical Drawing An **isometric drawing** is a type of three-dimensional drawing. In an isometric drawing of a cube, each of the nonvertical edges makes a 30° angle with a horizontal line on the plane of the picture.

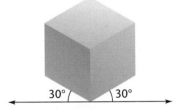

Drawing Cubes

Isometric grid paper can be used to draw a cube.

Ordinarily, not all six faces of a cube are visible. To make the drawing of the cube more realistic, the "hidden" segment can be erased. Three faces of the cube are visible: top, left, and right. The hidden parts of the cube can be represented with dashed lines, as though the faces in front were semitransparent.

A cube can be drawn so that it appears as if you are looking up at the bottom face.

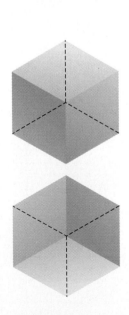

•Exploration 1 Using Isometric Grid Paper

You will need
Isometric grid paper

1 Draw each of the solid figures on isometric grid paper. Then redraw the figure with the red cube or cubes removed.

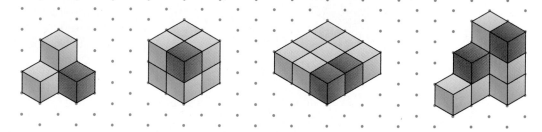

2 Draw each of the solid figures on isometric grid paper. Then add a cube to each red face and draw the new figure.

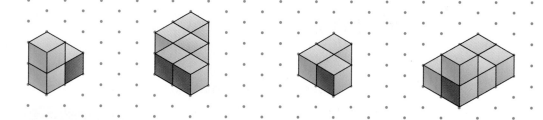

3 Were any of the drawings more difficult to make than others? What special problems did you encounter? Discuss. ❖

•Exploration 2 Using Unit Cubes

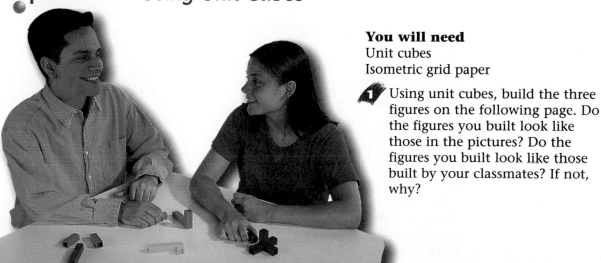

You will need
Unit cubes
Isometric grid paper

1 Using unit cubes, build the three figures on the following page. Do the figures you built look like those in the pictures? Do the figures you built look like those built by your classmates? If not, why?

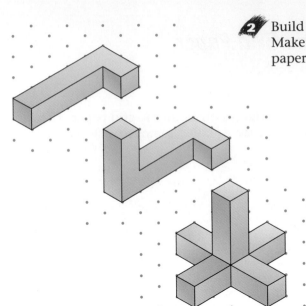

2 Build shapes of your own design from unit cubes. Make drawings of the shapes on isometric grid paper.

3 Do the representations in the illustrations leave any guesswork about the solid figures? Could there be parts of the figures that do not show in the drawings? Could there be any parts that do not show in the designs you created? Discuss. ❖

Orthographic Projections

Mechanical Drawing

An **orthographic projection** shows an object as it would appear if viewed from one of its sides. Such views are especially useful when an object's faces are at right angles to each other.

Typically, a figure may be drawn in six different orthographic projections or views: front, back, right, left, top, and bottom. For the figure at right, all of these views are shown. Dashed lines are used to show edges of the figure that cannot be seen in a particular view.

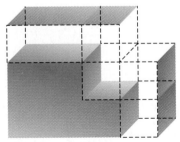

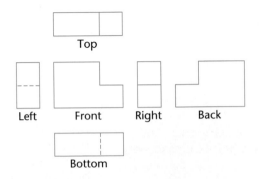

You may find it helpful to think of the different views of an object as an unfolded box with different views or projections on each face.

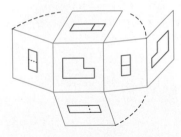

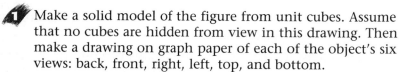

Exploration 3 Volume and Surface Area

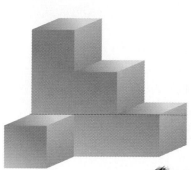

You will need
Graph paper
Unit cubes

1 Make a solid model of the figure from unit cubes. Assume that no cubes are hidden from view in this drawing. Then make a drawing on graph paper of each of the object's six views: back, front, right, left, top, and bottom.

2 The volume of a solid figure is the number of unit cubes that it takes to completely fill it. How can you determine the volume of the figure you built from centimeter cubes?

3 The total area of the exposed surfaces of a figure is called its surface area. How can you use the six views of the object you built to determine its surface area? ❖

In the six views you drew, each of the exposed cube faces of the object appeared in one—and only one—view. Do you think this would be true for any object you might build? Discuss, using drawings or models to illustrate your point.

EXERCISES & PROBLEMS

Communicate

1. Explain how a cube can be drawn so that three "hidden" faces are apparently visible.

Study the isometric drawing of a cube to answer Exercises 2–5.

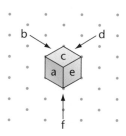

2. Which is the front-left face?

3. Which is the back-right face?

4. Which is the top face?

5. Which is the back-left face?

Use the isometric drawing of a rectangular solid to answer Exercises 6–9. Assume that the drawing shows all cubes.

6. How many small cubes are used?

7. How many faces does the solid have?

8. How many cube faces are exposed in the solid?

9. Draw six orthographic views of the solid.

The red face of the solid is made up of two cube faces.

Practice & Apply

10. Use isometric grid paper to draw four different rectangular solids using eight cubes each.

11. How many faces does each solid have? How many cube faces are exposed?

12. Draw four more figures by removing exactly two cubes from each figure you drew in Exercise 10.

13. Repeat Exercise 11 for the four new solids you drew in Exercise 12.

14. 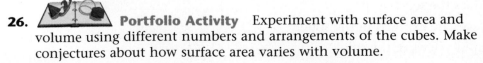 **Portfolio Activity** Using isometric grid paper, design a building using 12 cubes.

15. Using isometric grid paper, draw six orthographic views of the building you designed in Exercise 14.

Study the isometric drawing to answer Exercises 16–20. Assume that no cubes are hidden from view in the drawing.

16. How many cubes are used in the drawing?

17. How many faces does the solid have?

18. Sketch the shape of the right-front face.

19. Does any other face of the solid have this shape?

20. Are any two faces of the solid congruent? Explain your answer.

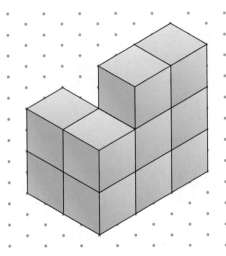

For Exercises 21–26, study the diagram and definitions.

The surface area of a solid built from unit cubes is the total area of all the exposed faces of the unit cubes.

The volume of a solid built from unit cubes is the total number of unit cubes.

21. What is the surface area of the rectangular solid in the figure if a side of a cube is one unit?

22. What is the volume of the rectangular solid in the figure?

23. Use the drawing as a base. On isometric grid paper, add two cubes to the drawing.

24. What is the surface area of the solid you drew in Exercise 23?

25. What is the volume of the solid you drew in Exericse 23?

26. 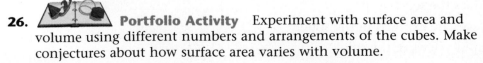 **Portfolio Activity** Experiment with surface area and volume using different numbers and arrangements of the cubes. Make conjectures about how surface area varies with volume.

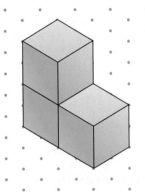

Recreation A wall tent is a good model of how planes intersect in space. Use the isometric drawing of a wall tent to answer Exercises 27 and 28.

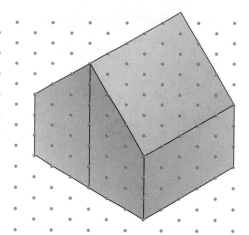

27. Sketch your own isometric drawing of a wall tent. Identify and label the various geometric shapes you find.

28. Think about the definitions, postulates, and theorems you have learned so far. State some conjectures about relationships that occur in the geometry of a wall tent.

29. When you look at the picture below you may see a cube that "pops out" toward you. But this would be spacially illogical with the rest of the picture. What does the picture "really" show?

Make orthographic projections of the building in the views indicated below.

30. Front view   **31.** Right side view

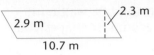

Look Back

Find the area for each figure shown. [Lessons 5.2 and 5.3]

32.

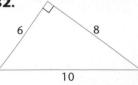

33.

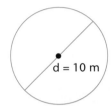

34.

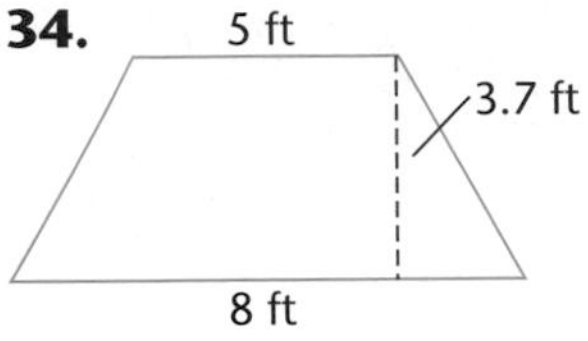

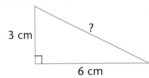

35.

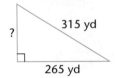

Find the length of the missing side of each right triangle below. [Lesson 5.4]

36.

3 cm   ?

6 cm

37.

?   315 yd

265 yd

Look Beyond

Use the orthographic projection at right for Exercises 38–41.

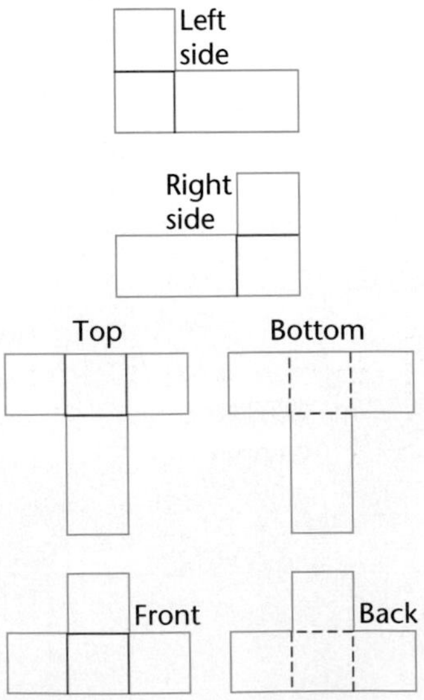

38. Use isometric grid paper to draw two different views of the rectangular solid.

39. Are any cubes hidden by your drawing?

40. How many faces does the solid have?

41. How many congruent faces does the solid have? Explain your answer.

Left side

Right side

Top   Bottom

Front   Back

LESSON 6.2
Exploring

Spatial Relationships

why *An interplay of lines and planes is apparent in real-world objects such as this quartz crystal cluster. The relationships among lines and planes in three dimensions is essential to understanding the structure of matter.*

In the explorations that follow, you will discover and develop ideas about how points, lines, and planes relate to each other in three-dimensional space.

•Exploration 1 *Parallel Lines and Planes in Space*

You will need
No special tools

Part I

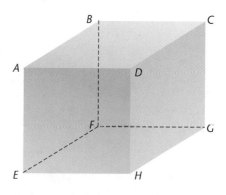

Salt crystals

1 Sketch a cube and label its vertices as shown. Identify the segments that form the vertical edges.

Do segments $\overline{AE}$ and $\overline{CG}$ seem to be coplanar? Do they seem to be parallel? (Do you think they would meet if they were extended infinitely?)

2 Which segments in the cube seem to be parallel? List them.

3 Are there segments in the cube that are not parallel and yet would never meet if they were extended infinitely in either direction? Explain how this can be. Such segments (or lines or rays) are said to be **skew**. Make a list of 4 pairs of skew segments in the cube.

Part II

1 How many faces does a cube have? Which faces seem to be parallel to each other?

2 Write your own definition of parallel planes by completing the sentence: "Two planes are parallel if and only if . . ." **(Def 6.2.1)**

3 How many sets of parallel planes are there in a cube? List the parallel planes in the figure you drew in Part I. ❖

•Exploration 2 *Segments and Planes*

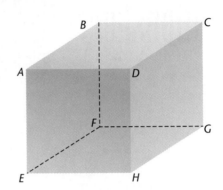

You will need
No special tools

Part I

1 Each of the segments in the cube is perpendicular to two different planes. Make a list of segments and the planes to which they are perpendicular.

2 What do you think it means for a segment or a line to be perpendicular to a plane?

Draw a line *l* on a piece of paper and label a point on it as *P*. Place your pencil on the point and hold it so that it is perpendicular to the paper. Is the pencil also perpendicular to line *l*?

3 Is it possible to tilt your pencil so that it is still perpendicular to line *l* but not to the paper? Make a sketch illustrating your answer.

4 Draw a new line *m* through point *P*. Place your pencil on *P* so that it is perpendicular to *l* and *m* at point *P*. What is the relation of the pencil to the plane of the paper?

5 Draw several other lines through point *P*. When your pencil is placed on *P* so that it is perpendicular to the paper, is it also perpendicular to these other lines?

6 Write your own definition of a line perpendicular to a plane by completing the following sentence: "A line is perpendicular to a plane if and only if. . ." **(Def 6.2.2)**

Part II

1 Each of the segments in the cube is parallel to two different planes. Make a list of segments and the planes to which they are parallel. What do you think it means for a segment or a line to be parallel to a plane?

2 Draw a line *l* on a piece of paper. Hold your pencil above *l* so that it is parallel to it. Does the pencil seem to be parallel to the plane of the paper?

3 Turn your pencil in a different direction so that it is still parallel to the paper but no longer parallel to the line. Do you think there is another line in the paper that is parallel to the pencil?

4 Write your own definition of a line (or segment or ray) parallel to a plane by completing the following sentence: "A line is parallel to a plane if and only if . . ." **(Def 6.2.3)** ❖

•Exploration 3 *Measuring Angles Formed by Planes*

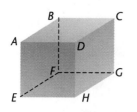

You will need
Stiff folding paper or an index card
Scissors

1 Some of the faces of a cube form right angles (are perpendicular to each other). Each face of the cube is perpendicular to how many other faces of the cube?

2 Draw a horizontal line *l* on a piece of paper. Mark and label points *A*, *B*, and *C*, with *B* between *A* and *C*. Make a crease through B so that line *l* folds onto itself. What is the relationship between *l* and the line of the crease?

3 Open the paper slightly. The angle formed by the sides of the paper is known as a **dihedral angle**. The measure of the dihedral angle is the measure of ∠*ABC*.

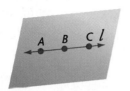

4 Write your own definition of the measure of a dihedral angle by completing the following sentence: "The measure of a dihedral angle is the measure of the angle formed by . . ." **(Def 6.2.4)**

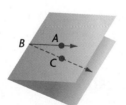

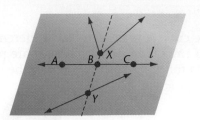

5 Open the paper and flatten it out. Add points X and Y to the line of the crease you made. Draw two rays from each point, with one ray on either side of the crease, as shown.

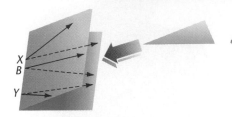

6 Fold the paper again on the same crease. Cut out pieces of paper that fit neatly into the angles formed by the different rays. Compare the shapes of the pieces of paper. Do the angles that are formed by the new rays have the same or different measures from $\angle ABC$? Discuss. ❖

Use pieces of paper to take the measure of the angles.

CRITICAL
Thinking

By measuring angles formed by different rays in Exploration 3, you can get a variety of results. What is the smallest angle you could measure? What is the largest?

EXERCISES & PROBLEMS

Communicate

Use the figure for Exercises 1–3.

1. Name three pairs of parallel segments, and explain why they are parallel.

2. Name two pairs of parallel planes, and explain why they are parallel.

3. Name three pairs of perpendicular planes, and explain why they are perpendicular.

4. If two lines are perpendicular to the same line, are the two lines parallel to each other? Why or why not?

5. If a line not on a plane is perpendicular to a line on a plane, is the line perpendicular to the plane also? Why or why not.

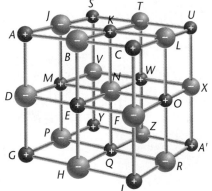

Sodium chloride (table salt) crystal lattice

Practice & Apply

Use the figure for Exercises 6–10.

6. Name all pairs of parallel planes.

7. Name two segments skew to $\overline{LM}$.

8. Name all planes perpendicular to plane *HIJ*.

9. Name two segments that are perpendicular to plane *JKO*.

10. Name two segments that are parallel to plane *HKO*.

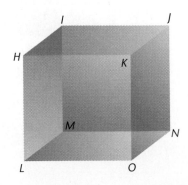

Explain why the following statements are true or false for a figure in three-dimensional space.

11. In three-dimensional space, if two lines are parallel to a third line, then the lines are parallel.

12. If two planes are parallel to a third plane, then the planes are parallel.

13. If two planes are perpendicular to a third plane, then the planes are parallel.

14. If two planes are perpendicular to the same line, then the planes are parallel.

15. If two lines are perpendicular to the same plane, then the lines are parallel.

Navigation Ships navigate on an ocean surface, which can be modeled with a plane. Airplanes navigate in three-dimensional space. Exercises 16–19 concern navigation of ships and airplanes.

16. Two airplanes flying at the same altitude are flying in the same ___?___ .

17. Airplane A is flying northwest at 10,000 ft, and airplane B is flying north at 5,000 ft. The lines formed by the flight paths are ___?___ .

18. Ship A is going south at 20 knots, and ship B is going southeast at 15 knots. The lines formed by the ships' paths will eventually ___?___ .

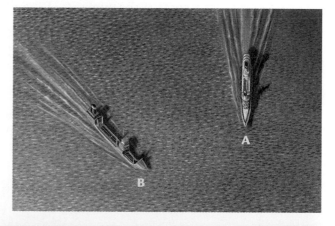

19. An airplane takes off from an airport runway. The airplane's nose and wingtips define the "plane of flight." Describe the dihedral angles formed by the airplane's plane of flight and the runway. Is that angle acute, right, or obtuse? Explain your answer.

Use Exercise 20 to answer Exercises 21–23.

20. Use a couple of note cards to construct the following. Fold a note card in half and mark it with two segments $\overline{AB}$ and $\overline{CD}$ such that $\overline{AB}$ is perpendicular to the folded edge, and $\overline{CD}$ is not perpendicular. Cut halfway into each segment from the folded edge. Use two other pieces of note card to model intersecting planes as shown.

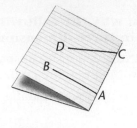

21. Which insert forms a plane which is not perpendicular to the edge?

22. Which insert can be used to measure the dihedral angle formed by the two upright planes?

23. Which of the two angles formed by the inserts is larger and why?

Look Back

Use a compass and straightedge to construct the following.
[Lesson 4.7]

24. Trace ∠ABC on your own paper and construct a copy of the angle.

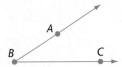

25. Trace segment $\overline{DE}$ on your own paper and construct the perpendicular bisector.

26. Trace segment $\overline{FG}$ and point H on your own paper and construct a line parallel to the segment through point H.

27. Trace triangle *JKL* on your own paper and construct a copy of the triangle.

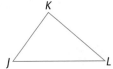

Look Beyond

28. Arrange the figure below into three triangles by moving only two toothpicks.

29. Arrange the figure below into three squares by moving only three toothpicks.

An Introduction to Prisms

*Many familiar objects have the shape of three-dimensional geometric figures known as **prisms**. As you study prisms you will apply your knowledge of congruent segments, angles, and polygons.*

A beam of laser light is bent as it passes through the glass, triangular prisms in this student's experiment. Since laser light consists of just one color, it is not broken up into a spectrum.

The figures below are prisms. A prism is named according to the shape of its base.

Triangular prism

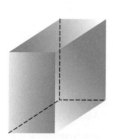

Rectangular prism

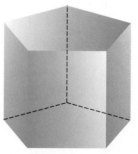

Pentagonal prism

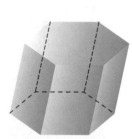

Hexagonal prism

A prism consists of a polygonal region, its translated image, and connecting line segments. The picture on the right suggests a way to think of constructing a prism.

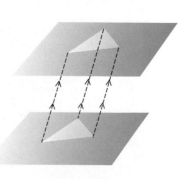

1. Start with a polygonal region in a plane.

2. Translate onto a parallel plane.

3. Connect the image and pre-image with segments.

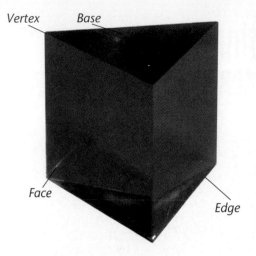

Vertex   Base

Face

Edge

Each flat surface of a prism is called a **face**. Each line segment formed by the intersection of two faces is called an **edge**. Each point of intersection of the edges is called a **vertex**. The polygonal region and its translated image are each called a **base**. Each face of the prism that is not a base is called a **lateral face**. The triangular prism shown has two bases, three lateral faces (five faces total), and nine edges.

Each lateral face of a prism is a quadrilateral. The edges of the quadrilaterals that are part of the bases of the prism are congruent and parallel. Therefore, each face is a parallelogram. Explain why, referring to the description of the construction of a prism on the previous page.

•Exploration· *The Lateral Faces of Prisms*

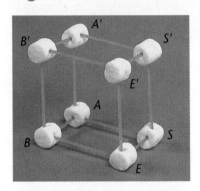

You will need
Thick, uncooked spaghetti
Miniature marshmallows

Part I

Use thick, uncooked spaghetti and miniature marshmallows to build a model of the rectangular prism shown.

1. What type of quadrilateral is each of the following? Explain your reasoning.

   a. *BASE*   b. *B′E′EB*   c. *E′S′SE*

2. List all pairs of congruent quadrilaterals.

   Rotate the lateral edges clockwise relative to *BASE* as shown.

3. What type of quadrilateral is each of the following? Explain your reasoning.

   a. *BASE*   b. *B′E′EB*   c. *E′S′SE*

4. Test your congruence statements from Step 2. Which ones are still true?

Part II

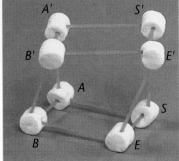1. Return your prism model to its original upright position. Rotate, or twist, the upper base to the right or left as shown. Is your new figure still a prism? Explain why or why not.

2. Are the segments that form the "faces" of your twisted figure coplanar?

3. Why, do you think, are prisms defined so that one base is a translation of the other?

Is it possible to manipulate your prism so that it remains a prism, but so that none of its lateral faces are rectangles? If your answer is "yes," explain why. ❖

Right prism

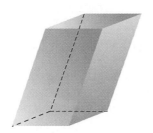

Oblique prism

In Exploration 1, you encountered two types of lateral faces: rectangles and nonrectangular parallelograms. Prisms that appear to lean in one direction have at least one lateral face that is not a rectangle. Prisms that appear to be perfectly upright have lateral faces that are all rectangles. This gives us an additional classification for prisms:

A **right prism** is a prism in which all lateral faces are rectangles.

An **oblique prism** is a prism that has at least one nonrectangular parallelogram as a lateral face.

Why is an upright prism called a "right" prism?

Try This Use spaghetti and marshmallows to construct each of the prisms illustrated at the beginning of the lesson. Classify each as oblique or right.

The Diagonal of a Right Rectangular Prism

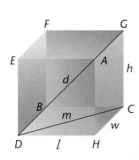

The length of the diagonal of a right rectangular prism with length l, width w, and height h is given by:

$$d = \sqrt{l^2 + w^2 + h^2}$$

Proof:

STATEMENTS	REASONS
$d^2 = m^2 + h^2$	"Pythagorean" Right-Triangle Theorem
$m^2 = l^2 + w^2$	"Pythagorean" Right-Triangle Theorem
$d^2 = l^2 + w^2 + h^2$	Substitution
$d = \sqrt{l^2 + w^2 + h^2}$	Take the square root of both sides

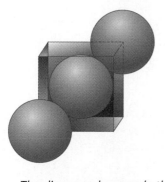

The diagram shows only the atoms along the diagonal. An iron crystal would have six additional atoms on the remaining corners of the cube.

APPLICATION

Iron forms a cubic crystal with a unit-cell cube measuring 291 picometers (1×10^{-12} meters) on each edge. The unit cell is the smallest repeating unit of a crystal structure. The diagonal of an iron-crystal unit contains the diameters of two atoms, as shown. What is the radius of an iron atom?

First, find the length of the diagonal. Each edge of the unit-cell cube is 291 pm.

$$d = \sqrt{l^2 + w^2 + h^2}$$

$$d = \sqrt{291^2 + 291^2 + 291^2}$$

$$d = \sqrt{3(84681)}$$

$$d = 504 \text{ pm}$$

Since the diagonal contains two diameters or four radii, divide by 4.

$$504 \div 4 \approx 126 \text{ pm}$$

The radius of an iron atom is approximately 126 picometers. ❖

EXERCISES & PROBLEMS

Communicate

Classify each prism and explain your classification.

1.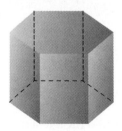

2.

3.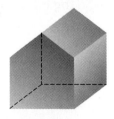

4.

5. Name each face of the oblique prism, identifying the bases. What kind of polygon is each face?

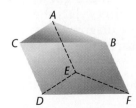

Practice & Apply

6. Draw a right triangular prism.

7. Draw an oblique pentagonal prism.

Use the right triangular prism for Exercises 8–13.

8. Name two faces that must be congruent.

9. Name all segments congruent to $\overline{BE}$.

10. What type of quadrilateral is *ACFD*?

11. Name all segments congruent to $\overline{EF}$.

12. List all pairs of congruent lateral faces.

13. List the angles whose measure must be 90°.

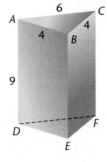

Use the oblique rectangular prism for questions 14–19.

14. Name two rectangles that must be congruent.

15. Name all segments congruent to $\overline{GK}$.

16. What type of quadrilateral is *GJNK*?

17. What type of quadrilateral is *JIMN*?

18. List the lateral faces that are congruent to each other.

19. List the angles whose measures are greater than 90°.

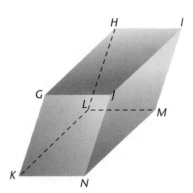

For Exercises 20–24, fill in the table below for each prism indicated.

Number of edges in the base	Number of faces	Number of vertices	Number of edges
20. 3	?	?	?
21. 4	?	?	?
22. 5	?	?	?
23. 6	?	?	?
24. *n*	?	?	?

25. The base of a prism is a 20-sided polygon. How many faces, vertices, and edges does the prism have?

Find the length of the diagonal of each right rectangular prism given below.

26. $l = 10$; $w = 5$; $h = 12$ **27.** $l = 4.5$; $w = 2.3$; $h = 10$

28.

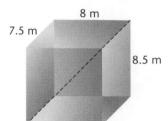

8 m
7.5 m
8.5 m

29.

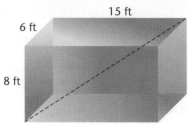

15 ft
6 ft
8 ft

Look Back

Explain why each of the statements in Exercises 30–33 is true or false.   [Lesson 4.6]

30. The diagonals of a rectangle that is not a square bisect each other.

31. The diagonals of a rectangle that is not a square bisect the angles.

32. The diagonals of a rectangle that is not a square are perpendicular to each other.

33. The diagonals of a rectangle that is not a square divide it into four congruent triangles.

34. A rectangular pizza is cut into four pieces along its diagonals. Which pieces are larger? Explain.   **[Lesson 4.6]**

Find the midpoint of a segment connecting each of the pairs of coordinates below.   [Lesson 4.8]

35. $(3, 4)$, $(-3, -4)$ **36.** $(4, -2)$, $(-2, 3)$ **37.** $(2, -2)$, $(5, 6)$

Look Beyond

38. Cultural Connection: Europe   The Swiss mathematician Leonard Euler (1707–1783) proved an interesting relationship between the parts of a prism: F, the number of faces; V, the number of vertices; and E, the number of edges. The relationship is called "Euler's Formula." Rediscover this formula by examining the relationship among the numbers in the last three columns of the table you created for Exercises 20–24. For a given prism, compare the number of faces and vertices against the number of edges. Write your formula in terms of F, V, and E.

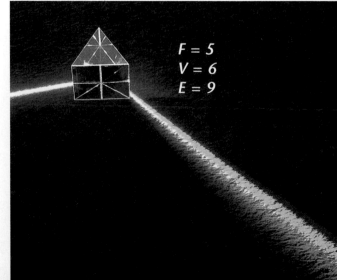

F = 5
V = 6
E = 9

Did the Camera Lie?

How would you feel if a mistake you made was shown in slow motion on national television? Nobody likes to be second-guessed, especially by 25 million people. It's no wonder that baseball umpires objected to an overhead camera that showed replays of close pitches during the 1993 World Series.

The umpires claimed that TV viewers did not get a true picture because the camera was not directly over home plate. Did the umpires have a valid point? To decide for yourself, first study the diagram and the information on the following pages. Then read the newspaper article about the controversy.

According to USA Today and CBS Sports, the overhead camera at the Sky Dome in Toronto was on a catwalk about 280 ft above the pitcher's mound. The mound is about 60 ft from home plate.

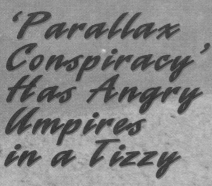

'Parallax Conspiracy' Has Angry Umpires in a Tizzy

From the *Albuquerque Journal*, October 21, 1993

"Parallax", the Major League Umpires Association charged Tuesday night in a sharply worded letter of outrage. The unsigned statement said this:

"The 'overhead' camera creates what experts have termed a parallax (the apparent change in the position of an object resulting from the change in the direction or position from which it is viewed.)

"This parallactic effect coupled with the camera's two-, rather than three-, dimensional capabilities creates the impression that pitches which are actually over the plate are outside.

"This distortion is being used selectively by self-appointed experts to create controversy, support their own views, and to undermine the credibility of the umpires."

A photo from the overhead camera's position

WHAT DO YOU THINK?

The photos at the right illustrate parallax. You can apply this to the baseball situation by imagining that the far finger represents the edge of home plate and the closer finger represents the ball. The two different eye views represent two different views of the pitch.

The umpires argue that the camera's position causes it to view the ball and the plate differently. Complete the activity and determine for yourself whether the umpires have a case.

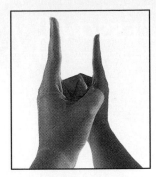

To see the parallax effect, line up your index fingers with your left eye. Then look at them with your right eye.

COOPERATIVE LEARNING

1. Make a three-dimensional model to see what the camera sees. Answer these questions to help you set up the model.

   a. Use your eye level when standing to represent the TV camera and the floor to represent the field. Based on your height, what is the scale of your model?

   b. About how wide should home plate be in your model? What size is your model baseball? (A standard baseball is 2.9 inches in diameter.)

   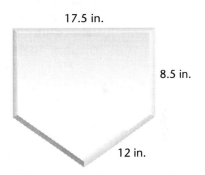

   17.5 in.

   8.5 in.

   12 in.

   c. How high should the ball be placed to represent a pitch that is 2 ft above the plate?

   d. To correctly model the position of the camera, your eyes should be directly above the pitcher's mound. In a real baseball field, the pitcher's mound is 60 ft. 6 in. from home plate. What is this distance in your model?

2. Comparing the view of the umpire with the view of the camera, is the camera in a better or worse position to judge strikes? Explain your answer.

3. The umpire has two eyes which bring a three-dimensional capability to the observation of strikes. How does this affect the umpire's ability to judge strikes, compared to the two-dimensional imaging of the camera?

4. Suppose the camera shows a pitch to be 8 inches outside the strike zone and the umpire calls the pitch a strike. Do you think the parallax effect could cause this much difference? Explain your reasoning.

Three-Dimensional Coordinates

why *By using a three-dimensional system, it is possible to locate a point anywhere in space. One of the systems astronomers use to locate objects in space involves rectangular coordinates.*

By using the *x*- and *y*-coordinate axes on a plane, you can give the location of a point anywhere on the plane. Since two numbers are required to do this, a plane is said to be **two-dimensional**. By adding a third coordinate axis at right angles, you can give the location of a point anywhere in space. Three numbers are now required, and so space is said to be **three-dimensional**.

The Arrangement of the Axes

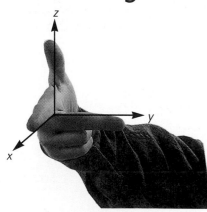

A number of arrangements of axes in space are possible. The one most commonly used by scientists and mathematicians is called the **right-handed system**. Let your right forefinger represent the positive direction of the *x*-axis.

Point with your forefinger and hold your thumb and middle finger so that they each make right angles with the line of your forefinger. Your middle finger will be pointing in the positive direction of the *y*-axis, and your thumb will be pointing in the positive direction of the *z*-axis.

CRITICAL *Thinking*

Normally, a right-handed system of axes is pictured with the *y*-axis to the right on the page. Do you think there are other ways of drawing a right-handed system?

Is this a right-handed system?

EXAMPLE 1

Locate the point $P(1, 2, 3)$ in a three-dimensional coordinate system.

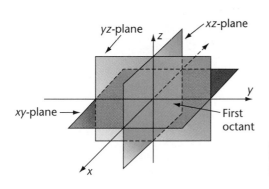

Solution ➤

1. Place your pencil at the origin. Count one unit in the x-direction. Make a mark or dot at the new position.

2. From the new position, count two units in the y-direction, drawing a light or dashed line to represent your path. Make a new mark or dot at the new position.

3. From the new position, count three units in the z-direction, drawing a light or dashed line to represent your path. Label your final position as point $P(1, 2, 3)$. ❖

The Arrangement of the Octants

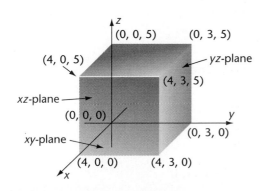

The x- and y-axes in the coordinate plane divide the plane into four quadrants. In a similar way, the x-, y-, and z-axes divide space into eight **octants**. The first octant is the one in which all three coordinates of the points in it are positive.

The octants can be referred to using the words *top, bottom, front, back, right,* and *left.* For example, the first octant is the top-front-right octant. Any point in this octant has three positive coordinates $(+, +, +)$. What are the signs of the coordinates for points in the other seven octants?

Notice that there are three separate planes determined by the axes of a three-dimensional coordinate system.

EXAMPLE 2

Draw a box in the first octant of a right-handed coordinate system. Place the box so that three of its faces are in the planes determined by the coordinate axes. Label the vertices with their coordinates. What is true of the points in the three axis planes?

Solution ➤

One possible solution is given in the diagram. In the xy-plane, all the z-coordinates are 0. In the xz-plane, all the y-coordinates are 0. In the yz-plane, all the x-coordinates are 0. ❖

The Distance Formula in Three Dimensions

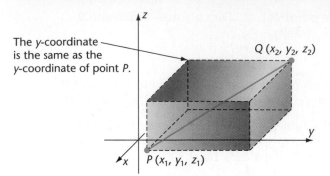

The y-coordinate is the same as the y-coordinate of point P.

$Q(x_2, y_2, z_2)$

$P(x_1, y_1, z_1)$

In a rectangular prism with length a, width b, and height c, the length of the diagonal can be found using the following formula.

$$l = \sqrt{a^2 + b^2 + c^2}$$

Any line in an x-y-z coordinate space can be viewed as a diagonal of a rectangular prism. The difference in the x-coordinates is the length of one side of the prism, and the differences in the y- and z-coordinates are the lengths of the other two sides.

Therefore, for points $P(x_1, y_1, z_1)$ and $Q(x_2, y_2, z_2)$, the distance PQ is

$$PQ = \sqrt{(x_2 - x_1)^2 + (y_2 - y_1)^2 + (z_2 - z_1)^2}.$$

EXAMPLE 3

ALGEBRA *Connection*

Find the length of the segment with endpoints $R(4, 6, -9)$ and $S(-3, 2, -6)$.

Solution ➤

$$RS = \sqrt{(-3 - 4)^2 + (2 - 6)^2 + (-6 - (-9))^2}$$

$$= \sqrt{(-7)^2 + (-4)^2 + (3)^2}$$

$$= \sqrt{49 + 16 + 9} = \sqrt{74} \approx 8.6 \ \diamond$$

Try This Find the length of the segment with endpoints $C(3, -4, -5)$ and $D(2, 0, -1)$.

EXERCISES & PROBLEMS

Communicate

In a three-dimensional coordinate system, on which axis does each point lie?

1. $(0, 0, 7)$ **2.** $(0, 3, 0)$ **3.** $(-5, 0, 0)$

In a three-dimensional coordinate system, on which coordinate plane does each point lie?

4. $(2, 3, 0)$ **5.** $(-3, 0, -2)$ **6.** $(-2, 4, 0)$

7. A rectangular prism has one vertex at the origin and an opposite vertex at $(2, -3, -5)$. Three of its faces are parallel to the planes determined by the coordinate axes. Explain how to find the volume of the prism.

8. In the illustration, a white sphere is illuminated by two different light sources, one red and one blue. Where the red and blue lights mix, magenta (a pink color) is produced. What can you tell about the location of the light sources? Express your answer in terms of the coordinate axes.

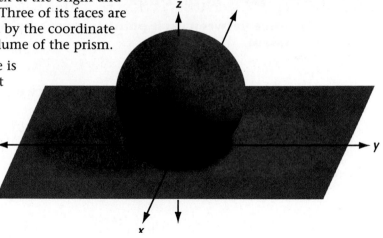

Practice & Apply

Locate each point on a three-dimensional coordinate axis.

9. $(1, 2, 3)$

10. $(1, -2, 3)$

11. $(1, -2, -3)$

12. $(-1, -2, -3)$

Determine the octant that would contain each of the following points if x, y, and z are positive and not zero.

13. $(-x, y, z)$

14. $(-x, -y, z)$

15. $(x, y, -z)$

16. $(-x, -y, -z)$

Sketch a graph of each of the following.

17. Point A; $(3, -2, 1)$

18. Point B; $(4, -6, 2)$

19. Segment CD; $C(1, 4, -2)$, $D(2, -3, 0)$

20. Plane EFG; $E(4, -2, 3)$, $F(0, 1, 0)$, $G(-3, -1, 5)$

Find the distance between each pair of points.

21. $(2, 1, 3)$, $(5, -2, 7)$

22. $(1, 1, 1)$, $(-1, -1, -1)$

23. $(7, -6, 5)$, $(6, 4, 3)$

24. $(2, 0, 1)$, $(-5, -6, -5)$

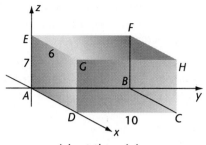

The grid units of the coordinate system are the same as the measurement units of the box. Determine the coordinates of each point.

25. Point F

26. Point H

27. Point C

28. Point E

A is at the origin.

In a three-dimensional coordinate system, the midpoint of a segment is defined for points (x_1, y_1, z_1) and (x_2, y_2, z_2) as $\left(\frac{x_1 + x_2}{2}, \frac{y_1 + y_2}{2}, \frac{z_1 + z_2}{2} \right)$. Find the midpoint of each segment below.

29. $A(3, 2, 1)$, $B(1, 2, 3)$

30. $A(5, -2, 3)$, $B(6, -7, 4)$

31. $A(2, -1, 0)$, $B(0, 0, 1)$

32. $A(-1, -4, -5)$, $B(6, 1, 0)$

33. $A(2, 0, 2)$, $B(-1, 5, -3)$

34. $A(1, 1, 1)$, $B(-1, -1, -1)$

Find the distance between each point in a rectangular coordinate system. [Lesson 4.8]

35. (4, 7), (−3, 2)   **36.** (21, 37), (−2, 15)

Find the midpoint of the segment joining the two points. [Lesson 4.8]

37. $A(3, -4)$, $B(1, -6)$ **38.** $A(-2, 15)$, $B(11, -7)$

39. Draw a right triangular prism. [Lesson 6.3]

40. Draw an oblique pentagonal prism. [Lesson 6.3]

Look Beyond

SUN-CENTERED COORDINATES OF MAJOR PLANETS FOR AUG. 17, 1990 IN ASTRONOMICAL UNITS		
	x	y	z
Earth/Moon	0.8189001	−0.5462572	−0.2368435
Mars	1.3894497	0.1451882	0.0290001
Jupiter	−2.238098	4.328104	1.909756
Saturn	3.934523	−8.443261	−3.656253

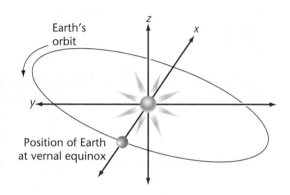

Astronomy The table gives three-dimensional, x-y-z coordinates for the planets using a coordinate system with the Sun at the origin. The x-axis is defined as lying in the direction of the Earth's position at the first moment of spring (the vernal equinox).

The xy-plane of the system is the equatorial plane, which corresponds to the plane of the equator of the Earth at the time of the equinox.

The unit of measure used in the table is the astronomical unit (au), which equals the mean distance of the Earth from the sun, about 92,900,000 mi, or 149,600,000 km.

The values given for the Earth/Moon are actually for a point called the center of mass of the two bodies, which is located between their centers.

41. Calculate the distance between the Earth/Moon center and Jupiter on August 17, 1990. Convert the astronomical units into miles.

42. Calculate the distance between Mars and Saturn on August 17, 1990. Convert the astronomical units to kilometers.

43. Calculate the distance between the Sun and Mars on August 17, 1990. Find a book that gives the mean distances of planets from the Sun. Does your calculation match? Why or why not?

LESSON 6.5 Lines and Planes in Space

why *Planes in space may form three-dimensional shapes, as in this drawing. Each plane in the drawing can be described mathematically.*

The Equation of a Plane

ALGEBRA *Connection*

Recall from algebra that the standard form of the equation of a line in a plane is

$$Ax + By = C.$$

The equation of a plane in space resembles the equation of a line, with an extra variable for the added dimension of space:

$$Ax + By + Cz = D.$$

For example,

$$2x + 5y - 3z = 9$$

is the equation of a plane where $A = 2$, $B = 5$, $C = -3$, and $D = 9$.

Using Intercepts in Graphing

In the coordinate plane, a line crosses the x- and y-axes at one or two points called **intercepts**. In coordinate space, a plane has one, two, or three intercepts at the x-, y-, and z-axes.

EXAMPLE 1

Sketch the graph of the plane with the equation $2x + y + 3z = 6$.

Solution ➤

ALGEBRA
Connection

Find the intercepts where the plane crosses each of the axes. The x-intercept, for example, is the point where the plane crosses the x-axis. At this point, the y- and z-coordinates are 0. To find this point, set $y = 0$ and $z = 0$ and solve the equation for x:

$$2x + 0 + 3(0) = 6$$

$$2x = 6$$

$$x = 3$$

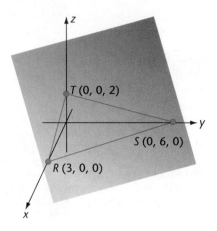

The coordinates of the x-intercept are (3, 0, 0). Similarly, the coordinates of the y-intercept are (0, 6, 0), and the coordinates of the z-intercept are (0, 0, 2). Plot these on the axes. These three noncollinear points determine the plane. To sketch the plane, connect the points with segments and add shading. ❖

What Happens to the Equation of a Line in Space?

ALGEBRA
Connection

$2x + 4y = 8$ is the equation of a line in the plane. But in space it is a special case of the equation of a plane in which the coefficient of z is equal to 0.

Notice that this equation of a plane is unaffected by the values of z. Thus, for any values of x and y that satisfy the equation, any value of z will also work. So every point above and below a point on the line is in the graph of the plane.

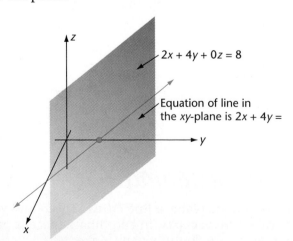

Lines in Space: A Step-by-Step Procedure

ALGEBRA
Connection

Imagine that you are able to plot one point of a graph every second according to a set of instructions. Your instructions give you one rule for the *x*-coordinates, another for the *y*-coordinates, and (if you are working in three dimensions) another for the *z*-coordinates.

Let $t = 1, 2, 3, \ldots$ represent the time (in seconds) at which you plot each point. Your rules for each coordinate will have the following form.

x = [an expression involving t]

y = [an expression involving t]

z = [an expression involving t]

(Once you get the idea, you can also think of 0 and negative values for t.)

EXAMPLE 2

Use the following rules to plot a line on a plane, where $t = \{1, 2, 3, \ldots\}$.

$x = 2t$

$y = 3t + 1$

Solution ➤

Fill in a table like the one below. Then plot the graph on an *xy*-coordinate plane. ❖

t	x	y
1	2	4
2	4	7

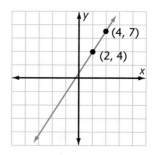

EXAMPLE 3

Use the following rules to plot a line in space.

$x = 2t + 1$

$y = 3t - 5$

$z = 4t$

Solution ➤

Fill in a table like the one below. Then plot the graph. ❖

t	x	y	z
1	3	−2	4
2	5	1	8

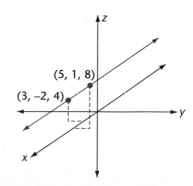

EXERCISES & PROBLEMS

Communicate

1. State the standard form for an equation of a plane.

2. Describe the characteristics of a plane with an equation in which the coefficient of z is 0.

3. Describe the characteristics of a plane with an equation in which the coefficient of x is 0.

4. Describe the characteristics of a plane with an equation in which the coefficients of y and z are 0.

Practice & Apply

Algebra   Use intercepts to sketch a graph of each of the planes represented by the equations below.

5. $3x + 2y + 7z = 4$ 6. $2x - 4y + z = -2$ 7. $x - 2y - 2z = -4$

8. $-3x + y = 4$ 9. $x - 2y = 2$ 10. $x = -4$

Algebra   Plot the line in space for each of the following rules.
Let $t = 1, 2, 3, \ldots$

11. $x = t$
$\quad y = 2t$
$\quad z = 3t$

12. $x = t + 1$
$\quad y = 2t + 1$

13. $x = \frac{2t}{3}$
$\quad y = t$
$\quad z = 1 - t$

14. $x = 3t$
$\quad z = 4t - 7$

Algebra   The **trace** of a plane is the line of intersection of the plane with the xy-plane. To find the equation of the trace, remember that a point lies on the xy-plane if and only if the z-coordinate is 0. Set the z-coordinate equal to 0 for a given equation of a plane, and the resulting equation of a line is the trace of the plane. For example:

$$2x + y + 3z = 6 \qquad \text{original equation}$$
$$2x + y + 3(0) = 6 \qquad \text{substitute 0 for } z$$
$$2x + y = 6 \qquad \text{equation of the trace}$$

Find the equation of the trace for each equation of a plane below.

15. $x + 3y - z = 7$ 16. $5x - 2y + z = 2$

Algebra Sketch each of the following planes and indicate the trace.

17. $2x + 7y + 3z = 2$ **18.** $-4x - 2y + 2z = 1$

19. Navigation Suppose an airplane's initial climb out of takeoff is determined by the following parameters. Plot the line for the following rules:

$$x = -t$$
$$y = t$$
$$z = 0.5t$$

If the positive x-axis is to the north, in which direction did the airplane take off?

 **Look Back**

20. Given: $\angle 1 \cong \angle 2$; $\overline{EH} \cong \overline{FG}$
    Prove: $\overline{EF} \cong \overline{HG}$ **[Lesson 3.3]**

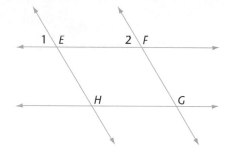

21. Given rectangle $ABCD$ with diagonals $\overline{AC}$ and $\overline{BD}$, prove $\triangle ADC \cong \triangle BCD$.   **[Lesson 4.6]**

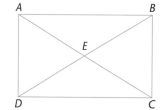

22. Given rectangle $ABCD$ with diagonals $\overline{AC}$ and $\overline{BD}$ intersecting at E, prove $\triangle AED \cong \triangle BEC$. **[Lesson 4.6]**

23. Given quadrilateral $DART$ with $\overline{DA} \cong \overline{TR}$ and $\overline{DT} \cong \overline{AR}$, prove $DART$ is a parallelogram. **[Lesson 4.6]**

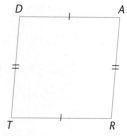

Look Beyond

24. Suppose a segment has endpoints with coordinates $(4, 0, 0)$ and $(4, 7, 0)$. What figure is formed by rotating the segment about the x-axis? Sketch the figure.

25. What figure is formed by rotating the segment from Exercise 24 about the z-axis? Sketch the figure.

Perspective Drawing

European Renaissance artists of the fourteenth through sixteenth centuries rediscovered, from classical Greek and Roman art, how to create the illusion of depth in drawings and paintings. Compare the earlier work above with the Renaissance painting below it.

Why Objects that are far away appear smaller than they would if they were close to you. The study of perspective in drawing will show you the rules for making things (and parts of things) appear to be in the proper relationship to each other.

Perspective Drawings: Windows to Reality

Modern perspective drawing methods, discovered by the Italian architect Brunelleschi (1377–1446), are based on the idea that a picture is like a window.

The artist creating the picture, or the person looking at the finished product, is thought of as looking through the picture to the reality it portrays. (The word perspective comes from the Latin words meaning "looking through.")

When someone looks at an object, there is a line of sight from every point

Picture-plane "window" containing projected image

Albrecht Dürer

on the object to the eye. Imagine a plane, such as an empty canvas, intersecting the lines of sight. The points of intersection on the plane make up the image of the object, and the image is said to be **projected** onto the plane.

Albrecht Dürer (1471–1528), who visited Italy to learn the techniques of perspective drawing, produced a number of works that showed artists employing these techniques.

Parallel Lines and Vanishing Points

Segments $\overline{AB}$ and $\overline{CD}$ on the sidewalk are actually the same length. But when they are projected onto the student's picture plane, the image of one segment is longer than the image of the other.

Have you ever noticed how the rails of a railroad track or the sides of a highway seem to meet as they recede into the distance? The point at which parallel lines seem to meet, often on the horizon, is known in perspective drawing as a **vanishing point**.

The idea of a picture as a window provides a way to understand why parallel lines seem to meet as they vanish into the distance.

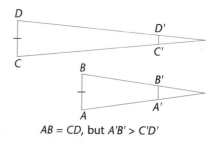

$AB = CD$, but $A'B' > C'D'$

Perspective Drawing Theorems

As Renaissance artists studied vanishing points in nature and in their perspective drawings, they developed a number of rules and procedures for making realistic drawings without using grids.

The following principles, which are stated as theorems, are basic to an understanding of how perspective drawings are made. They can be informally demonstrated using sketches and your earlier geometry theorems.

THEOREM: SETS OF PARALLEL LINES

In a perspective drawing, all lines that are parallel to each other, but not to the picture plane, will seem to meet at the same point.   **6.6.1**

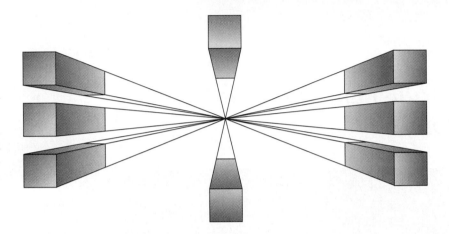

CRITICAL *Thinking*

Do you think the point at which a set of parallel lines seem to meet would have to be somewhere in the drawing? Explain, using illustrations.

In the following theorem, the "ground" is assumed to be flat.

THEOREM: LINES PARALLEL TO THE GROUND

In a perspective drawing, a line on the plane of the ground will meet the horizon of the drawing if it is not parallel to the picture plane. Any line parallel to this line will meet at the same point on the horizon.   **6.6.2**

CRITICAL *Thinking*

Parallel lines that are parallel to the picture plane of a perspective drawing are usually represented with no vanishing point. In most cases this procedure causes no problems. Can you think of situations in which it would result in unrealistic drawings?

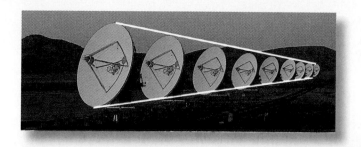

In perspective drawings, the principles of vanishing points apply even when there are no parallel lines actually shown in the drawings. In the row of telescope dishes, for example, there are imaginary lines through points at the tops and at the bottoms of the reflectors. These lines meet at the horizon. Explain why.

Much of the subject matter in early perspective drawings was architectural. Buildings and houses are ideally suited to the development of the theories of perspective drawing because they usually contain many lines that are parallel to each other and to the ground. Eventually, perspective drawing began to influence architecture, as the Church of San Lorenzo in Florence, Italy, illustrates.

In his design for this church, Brunelleschi used the principles of perspective drawing. The vanishing point of the structural elements appears to be on the altar.

CRITICAL *Thinking*

The two polar bears in the picture are actually the same size—measure them! And yet one appears much larger than the other. Explain why.

EXERCISES & PROBLEMS

Communicate

Complete the phrase in Exercises 1–3.

1. In a perspective drawing, all lines that are parallel to each other but not parallel to the picture plane . . .

2. In a perspective drawing, all lines parallel to the ground but not parallel to the picture plane . . .

3. In a perspective drawing, lines not parallel to the ground or the picture plane . . .

4. Architects and interior designers sometimes use perspective techniques to create illusions of depth or height. Explain how these illusions work.

5. In perspective drawing, explain what is meant by "vanishing point."

Practice & Apply

The exercises below take you through steps to produce various types of perspective drawing.

6. Follow the steps to produce a one-point perspective drawing of a cube.

    a. Draw a square and a line above the square as the horizon line. Mark a vanishing point on the horizon line centered above the box.

    b. From each corner of the box, lightly draw lines to the vanishing point you marked in Step a.

    c. Fill in the box by drawing lines parallel to the horizon line "behind" the original square.

    d. You may erase the perspective lines, if you wish, to complete your drawing.

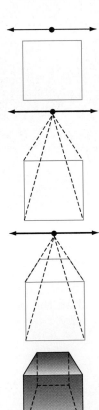

7. Follow the steps from Exercise 6, but draw the cube as if it were viewed from below. Set the horizon line and vanishing point below the square.

8. Follow the steps from Exercise 6, but set the vanishing point to the right or left instead of the center.

For Exercises 9 and 10, trace the figure on your own paper.

9. Locate the vanishing point.

10. Draw the horizon line.

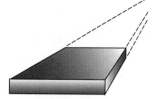

11. So far you have learned about one-point perspective. Perspective drawing can also be done with two-point perspective. Follow the steps below to complete a two-point perspective drawing.

a. Draw a vertical line. This is the front edge of your box. Draw a horizon line and two vanishing points on the line to the left and right of the box.

b. Lightly draw lines back to the vanishing points from the vertical line as shown.

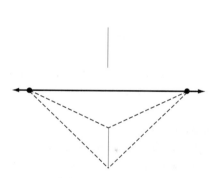

c. Draw vertical segments to complete the sides of the box.

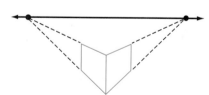

d. Complete the other edges of the box as shown.

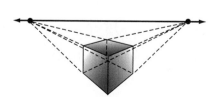

e. You may erase the perspective lines to complete the drawing.

For Exercises 12 and 13, trace the figure on your own paper.

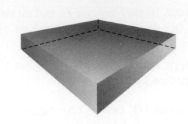

12. Locate the vanishing points.

13. Draw the horizon line.

14. 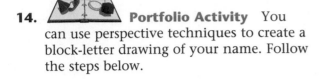 **Portfolio Activity** You can use perspective techniques to create a block-letter drawing of your name. Follow the steps below.

   a. Draw "flat" block letters and a horizon line with a vanishing point.

   b. Draw lines from all corners of the letters and appropriate curved edges as shown.

   c. Fill in the edges.

   d. You may erase the perspective lines, if you wish, to complete your drawing.

15. 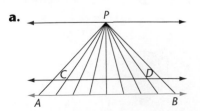 **Portfolio Activity** Use a one-point or two-point perspective to draw a city. Start with boxes for buildings and add details.

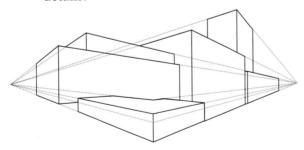

Fine Art Tiling was particularly intriguing to the artists who first completed perspective studies. The pictures (a–d) suggest a technique for creating a proper tile pattern. Study the pictures to answer Exercises 16–18.

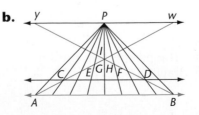

a.

b.

c.

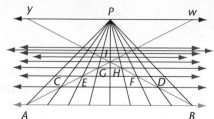

d.

16. A method for finding the lines parallel to $\overleftrightarrow{AB}$ is suggested by the diagonals. Explain how the diagonals can be used to determine the parallel lines.

17. Explain how the tile pattern could be viewed from a corner using two-point perspective. How would the intersecting lines be determined?

18. Create your own one- or two-point perspective drawing of a square tile pattern.

Look Back

Locate and sketch each point on a three-dimensional coordinate system. [Lesson 6.4]

19. $(5, -1, -2)$ **20.** $(13, 0, 0)$ **21.** $(-2, 0, 5)$

Find the distance between the following points in a three-dimensional coordinate system. [Lesson 6.4]

22. $(4, 3, 2), (-5, 2, -1)$ **23.** $(-1, 0, 1), (15, 6, -2)$

Find the midpoint of the segments joining the following points in a three-dimensional coordinate system. [Lesson 6.4]

24. $(5, 5, 5), (-3, -3, -3)$ **25.** $(0, 0, 0), (-1, 10, 9)$

Look Beyond

26. Explain how the projective process used by the artist in the photo is similar to perspective drawing techniques.

27. Explain how it is different from perspective drawing techniques.

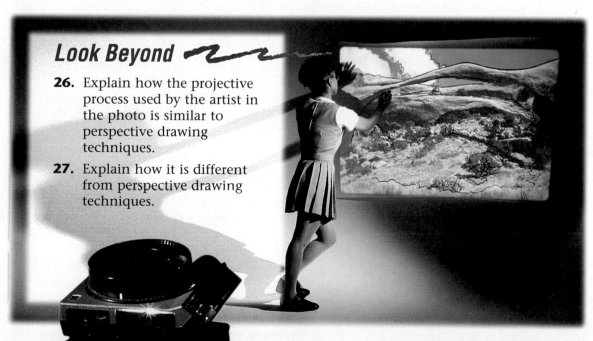

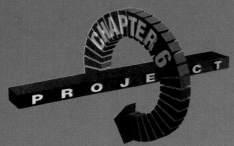

BUILDING AND TESTING AN A-SHAPED LEVEL

Many carpentry tools use basic properties of geometry. One example is the early Egyptian A-shaped level which has been used for thousands of years for construction and surveying. In this project you will build and test the various properties of an A-shaped level.

A-shaped Level from the tomb of Senedjem

Activity 1

Part of the A-shaped level is made with a string and a weight forming what is called a plumb line. Test the behavior of a plumb line by attaching a washer or similar weight to a string as shown. Suspend your plumb line above a surface that you believe is level, such as the floor of your classroom. Gradually lower the plumb line until the weight makes contact with the floor. Imagine that the weight is touching the floor at point *F*, the intersection of two segments as shown below. Then answer the questions that follow.

1. m∠*AFP* = ?

2. m∠*AFQ* = ?

3. Hold a pencil at arm's length and suspend the plumb line directly above the pencil. If you hold the pencil so that it is level (parallel to the floor), what is the measure of the angles formed between the plumb line and the pencil?

4. Formulate a conjecture:

   A segment is level if and only if it is __?__ to a plumb line.

Activity 2

An A-shaped level is easy to make from three popsicle sticks, glue, a thumbtack, string, and a small weight. Build and test an A-shaped level by following the steps below.

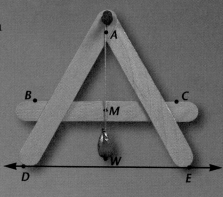

1. Glue two popsicle sticks together to form the legs of the level.

2. Glue the crossbar to the legs forming an isosceles triangle above the crossbar and two congruent segments below it.

3. Carefully mark the midpoint M on the crossbar.

4. Put a thumbtack at the vertex of your level and tie your plumb line to the thumbtack.

5. Place your A-shaped level along several different segments that you believe are level. Observe and record the behavior of your plumb line $\overline{AW}$. What is the relationship between $\overline{AW}$ and M?

6. Place your A-shaped level along several different segments that you believe are not level. Observe and record the behavior of $\overline{AW}$. What is the relationship between $\overline{AW}$ and M?

7. Use your results and the picture above formulate a conjecture:

   $\overline{DE}$ is level if and only if __?__ (a statement about $\overline{AW}$ and M).

Activity 3

1. What type of quadrilateral is $DBCE$?

2. If $\overline{AW} \perp \overline{BC}$, must $\overline{AW}$ also be perpendicular to $\overline{DE}$? Explain your reasoning.

3. If $\overline{AW} \perp \overline{DE}$, must $\overline{AW}$ also be perpendicular to $\overline{BC}$? Explain your reasoning.

Chapter 6 Review

Vocabulary

base	318	isometric drawing	304	right prism	319
dihedral angle	313	lateral face	318	right-handed system	326
edge	318	oblique prism	319	skew segments	312
face	318	octant	327	vanishing point	337
intercept	331	orthographic projections	306	vertex	318

Key Skills and Exercises

Lesson 6.1

➤ **Key Skills**

Create and interpret isometric drawings.
Assuming that no cubes are hidden in this drawing, what is the surface area of the solid?

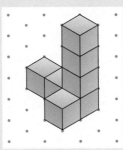

A cube standing alone would have six faces showing, each face with a surface area of 1 cm². The cubes in this drawing show different numbers of faces: three cubes show five faces, three cubes show four faces, and one cube shows three faces. Adding up the exposed faces at 1 cm² each gives a total of 30 cm².

Orthographic projections that show each face of the object exactly once demonstrate that 30 cube faces make up the total surface area.

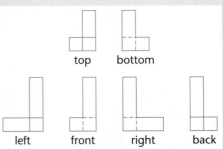

➤ **Exercises**

In Exercises 1–2, refer to the isometric drawing. Assume that no cubes are hidden in the drawing.

1. What is the volume (in cubes) of the solid?

2. Draw orthographic projections of the faces of the object.

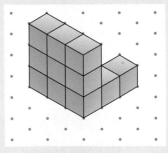

Lesson 6.2

➤ **Key Skills**

Identify spatial relationships among lines and planes in space.
In the sketch of a cube, name one set of parallel planes, perpendicular planes, skew segments, and segments perpendicular to a plane.

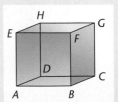

Three points name a plane, so you can arbitrarily choose three points from the four planes available. Planes *ADE* and *BCF* are parallel. Both *ADE* and *BCF* are perpendicular to *ABF*, *DCG*, *EFG*, and *ABC*. $\overline{EH}$ and $\overline{BF}$ are a set of skew segments. Perpendicular to *ADE* and *BCF* are $\overline{AB}$, $\overline{HG}$, $\overline{DC}$, and $\overline{EF}$.

➤ Exercises

In Exercises 3–5, refer to the figure.

3. Name all the pairs of parallel planes.

4. Name a pair of skew segments.

5. Name all the pairs of perpendicular planes.

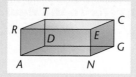

Lesson 6.3

➤ Key Skills

Recognize congruences in a prism.

For the oblique prism at the right, name the segments that are congruent to $\overline{AD}$ and the lateral faces that are congruent to each other.

You're given that $\overline{EH} \cong \overline{FG}$ and $\overline{AD} \cong \overline{BC}$. Because the bases of a prism are congruent, $\overline{EH} \cong \overline{AD}$; thus $\overline{EH} \cong \overline{FG} \cong \overline{AD} \cong \overline{BC}$. Lateral faces $ADHE$ and $BCGF$ are congruent because the base edges, lateral edges, and angles are congruent.

Find the length of a diagonal of a right rectangular prism.

A right rectangular prism has a length of 20, a height of 24, and a width of 10. What is the length of its diagonal?

The length of the diagonal of a right rectangular prism is $\sqrt{l^2 + w^2 + h^2}$. In this case, $d = \sqrt{20^2 + 10^2 + 24^2} = \sqrt{1076} \approx 32.8$.

➤ Exercises

In Exercises 6–7, refer to the figure.

6. Name the segments congruent to $\overline{RC}$.

7. Name the lateral faces that are congruent.

8. A right rectangular prism has a length of 12, a width of 2, and a height of 12. Find the length of the prism's diagonal.

Lesson 6.4

➤ Key Skills

Describe and sketch lines and planes on a three-dimensional coordinate system.

The figure below represents a cube, drawn in a three-dimensional coordinate system. If $\overline{OP}$ connects the midpoints of $\overline{RQ}$ and $\overline{TS}$, find its end points and its length.

The midpoint formula for a segment is

$$\left(\frac{x_1 + x_2}{2}, \frac{y_1 + y_2}{2}, \frac{z_1 + z_2}{2} \right).$$

Using this formula for $\overline{RQ}$ and $\overline{TS}$, we get $(-2.5, 0, 5)$ and $(-2.5, 5, 0)$ as the endpoints of segment $\overline{OP}$. Using the distance formula,

$$d = \sqrt{(x_2 - x_1)^2 + (y_2 - y_1)^2 + (z_2 - z_1)^2},$$

$$OP = \sqrt{(-2.5 - (-2.5))^2 + (0 - 5)^2 + (5 - 0)^2}$$

$$= \sqrt{50} = 5\sqrt{2}$$

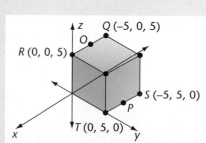

> ### Exercises
> **In Exercises 9–10, refer to the figure.**

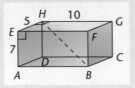

9. Sketch the box on a coordinate system. Give the coordinates for each vertex.

10. Find the length of $\overline{BH}$.

Lesson 6.5

> ### Key Skills

Plot planes on a three-dimensional coordinate system.

Sketch the graph of the plane with the equation $x + 2y - 3z = 18$.

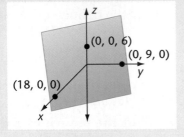

You need three noncollinear points to graph a plane. You can use the three points where the plane intercepts the three axes. When y and z are 0, $x + 2(0) + 3(0) = 18$. Thus $x = 18$. So the coordinates for the x-intercept are (18, 0, 0). By a similar process, you find the y-intercept is (0, 9, 0) and the z-intercept is (0, 0, 6). Now you can plot the points and sketch the graph of the plane.

> ### Exercises

11. Sketch the graph of the plane with the equation $2x - y + z = 7$.

Lesson 6.6

> ### Key Skills

Use vanishing points to make perspective drawings.

The drawings below illustrate two ways to make the letter T look three-dimensional. Each uses only one vanishing point: in the left-hand drawing the point is above and to the right of the T, in the right-hand drawing beside the T to the left.

> ### Exercises

12. Make a perspective drawing of a cube. Show the horizon, the vanishing point(s), and the perspective lines you use.

Applications

13. **Geology** Topaz is a mineral that forms orthorhombic crystals. This type of crystal has noncongruent edges that meet at right angles. Use isometric grid paper to sketch an orthorhombic crystal.

14. Here are two views of a table. Make a perspective drawing of it.

side   end

Chapter 6 Assessment

In Exercises 1–2, refer to the isometric drawing. Assume that no cubes are hidden in the drawing.

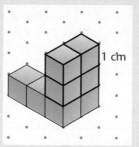

1 cm

1. What is the surface area of the solid?
2. Draw orthographic projections of the faces of the object.

In Exercises 3–5, refer to the figure below.

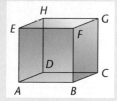

3. Name a pair of parallel planes.
4. Name a pair of skew segments.
5. Name a segment perpendicular to plane *ABC*.

In Exercises 6–8, refer to the right rectangular prism below.

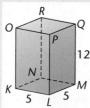

6. Name the segments congruent to $\overline{RQ}$.
7. Name the lateral faces that are congruent.
8. Find the length of the prism's diagonal.
9. Sketch the box below on a coordinate system. Give the coordinates for each vertex.

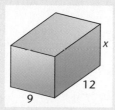

10. Sketch the graph of the plane with the equation $3x - y + 4z = 24$.

Chapters 1-6   Cumulative Assessment

College Entrance Exam Practice

Quantitative Comparison   Questions 1–4 consist of two quantities, one in Column A and one in Column B, which you are to compare as follows:

A. The quantity in Column A is greater.

B. The quantity in Column B is greater.

C. The two quantities are equal.

D. The relationship cannot be determined from the information given.

	Column A	Column B	Answers
1.	length of $\overline{AB}$	length of $\overline{XZ}$	(A) (B) (C) (D) **[Lesson 5.5]**
2.	distance between (9, 9, 9) and (0, 0, 0)	distance between (15, 0, 19) and (6, 9, 10)	(A) (B) (C) (D) **[Lesson 7.7]**
3.	area of *EFGH*	area of *JKLM*	(A) (B) (C) (D) **[Lesson 5.2]**
4.	area of the circle	area of the square	(A) (B) (C) (D) **[Lesson 5.3]**

5. What is the measure of an interior angle of a regular nonagon? **[Lesson 3.6]**

    a. 40°   **b.** 100°   **c.** 140°   **d.** 160°

6. The area of a circle is 154 square inches. What is its circumference? **[Lesson 5.3]**

    a. 22 inches   **b.** 44 inches

    c. 51 inches   **d.** 77 inches

7. What are the coordinates of the midpoint of a segment whose endpoints are (9, 12) and (−12, −9)? **[Lesson 5.7]**

    a. (−1.5, 1.5)   **b.** (4.5, 6)   **c.** (−6, −4.5)   **d.** (10.5, −10.5)

8. What are the coordinates of the midpoint of a segment whose endpoints are (2, 10, 0) and (11, 10, −6)? **[Lesson 6.4]**

    a. (6.5, 0, −3)   **b.** (4.5, 10, 3)

    c. (6.5, 10, −3)   **d.** (5.5, 5, 3)

9. In the oblique rectangular prism below, what is the measure of ∠x? **[Lesson 3.5]**

    a. 60°

    b. 80°

    c. 100°

    d. 120°

10. What is the surface area of the stack of cubes at the right? Assume that no cubes are hidden. **[Lesson 6.1]**

    a. 38 cm²   **b.** 45 cm²

    c. 54 cm²   **d.** 66 cm²

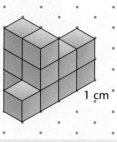

11. "Skew lines are not parallel and never intersect." Is this statement a definition? Explain. **[Lesson 2.3]**

12. A circle and a square both have an area of 441 square inches. What are their perimeter and circumference? **[Lesson 5.3]**

13. Use the slope formula to show that quadrilateral *ABCD* is a rectangle. **[Lesson 3.8]**

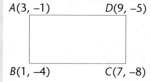

14. Find the area of the regular hexagon. **[Lesson 5.5]**

15. Draw a rhombus and construct a reflection of it over a horizontal line. **[Lesson 4.9]**

Free-Response Grid Questions 16–19 may be answered using a free-response grid commonly used by standardized test services.

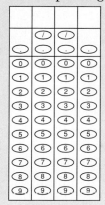

16. Find the area of triangle PEG. **[Lesson 5.2]**

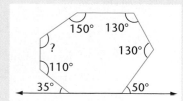

17. Find the total area of the three rectangles. **[Lesson 5.1]**

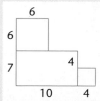

18. Find the missing angle measure. **[Lesson 3.6]**

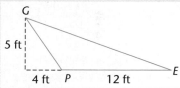

19. A point *P* is selected at random from the segment below. What is the probability that 2 ≤ *P* ≤ 2.5? **[Lesson 5.7]**

CHAPTER 7

Surface Area and Volume

LESSONS

7.1 *Exploring* Surface Area and Volume

7.2 Surface Area and Volume of Prisms

7.3 Surface Area and Volume of Pyramids

7.4 Surface Area and Volume of Cylinders

7.5 Surface Area and Volume of Cones

7.6 Surface Area and Volume of Spheres

7.7 *Exploring* Three-Dimensional Symmetry

Chapter Project
Polyhedra

For seven days, 120 workers and 90 rock climbers worked to cover the Reichstag building in Berlin with shiny silver fabric. A million square feet of the fabric and over 10 miles of blue plastic cord were required for the feat.

Why would anyone want to wrap a huge building? This is a question conceptual artist Christo, who conceived of the huge project 24 years before he was allowed to do it, has heard many times. What do you think? Is it art?

The amount of material needed to cover an object is known as its *surface area*. Christo certainly needed to determine the surface area of the Reichstag to prepare for the event.

PORTFOLIO ACTIVITY

The base of the Great Pyramid of Giza is about 756 feet on each side. The pyramid itself is about 481 feet. If you wanted to wrap it in cloth, how much cloth would it take? How does this compare with the amount of cloth Christo used to wrap the Reichstag?

The Pyramids of Gisa. The Great Pyramid is the one farthest away.

LESSON 7.1

Exploring Surface Area and Volume

why *The shapes and sizes of plants are partly determined by the necessary ratio of surface area to volume. Sometimes a large ratio is advantageous, while other times a small ratio is better.*

Desert plants must conserve water but have plentiful light. Tropical plants have plentiful water but must often compete for light. How does climate affect the shapes of desert and tropical plants?

Exploration 1 Ratio of Surface Area to Volume

You will need
Isometric grid paper or centimeter cubes
Spreadsheet software or graphics calculator (optional)

Part I

1 Draw or build rectangular prisms with the shape $n \times 1 \times 1$ and fill in a table like the one below. A spreadsheet may be used to complete the data table.

2 If $n = 100$, what is the ratio of surface area to volume? What happens to the ratio of surface area to volume as the length of n increases? Is there a number that the ratio approaches? ❖

ALGEBRA
Connection

Length n	Surface Area $2lw + 2lh + 2wh$	Volume lwh	$\dfrac{SA}{Vol}$
1	6	1	?
2	?	?	?
3	?	?	?
n	?	?	?

354 CHAPTER 7

CRITICAL *Thinking*

On sunny days a snake may lie in the sun to absorb heat. At night, a snake may coil up tightly to retain its body heat. Explain why this strategy works.

Part II

Draw or build rectangular prisms with the shape $n \times n \times n$ and fill in a table like the one below.

MAXIMUM MINIMUM *Connection*

Length n	Surface Area $2lw + 2lh + 2wh$	Volume lwh	$\dfrac{SA}{Vol}$
1	6	1	?
2	?	?	?
3	?	?	?
n	?	?	?

1 If $n = 100$, what is the ratio of surface area to volume? As the value for n gets larger, what happens to the ratio of surface area to volume?

2 What conclusions can you draw about the surface area–volume ratios of smaller cubes compared with the ratios of larger cubes? ❖

CRITICAL *Thinking*

In cold weather, why might a very large animal be able to maintain its body heat more easily than a smaller one? Why do you think small, warmblooded animals, such as mice and birds, need to have higher metabolisms than larger ones, such as cows or elephants?

Exploration 2 *Maximizing Volume*

You will need

Graphics calculator or graph paper

Graphics Calculator

1 Make an 11 × 14-inch piece of paper into an open box by cutting squares out of the corners and folding up the sides. How large should the squares be to maximize the volume of the box? Fill in a table like the one shown to help determine the answer.

Side of Square x	Length l	Width w	Height h	Volume lwh
1	12	9	1	108
2	?	?	?	?
3	?	?	?	?
x	?	?	?	?

2 Use a graphics calculator to graph the Side-of-Square data on the *x*-axis and Volume data on the *y*-axis. Hint: Note that if a square with side length *x* is removed, the length of the box is 14 − 2*x*.

MAXIMUM
MINIMUM
Connection

3 Trace along the graph and determine the following:

a. *x*-value that produces the largest volume;

b. largest volume. ❖

 CRITICAL *Thinking*

Why are food processing companies concerned with maximizing volume for a given surface area for product packaging?

APPLICATION

Product Packaging

A cereal company is choosing between two box designs with dimensions as shown

2.5 in.   8 in.

8 in.

Box A

2 in.   8 in.

10 in.

Box B

The company would like to keep the amount of cardboard used for the box to a minimum because it is more cost effective and better for the environment. The company must choose between two box designs. Which box provides less surface area for the same volume?

Both boxes provide a volume of 160 cubic inches. The surface area of box A is

$2(2.5)(8) + 2(2.5)(8) + 2(8)(8) = 208$ square inches.

The surface area of box B is

$2(2)(8) + 2(2)(10) + 2(8)(10) = 232$ square inches.

Box A has less surface area. ❖

EXERCISES & PROBLEMS

Communicate

For each situation, determine whether maximum volume or maximum surface area is appropriate. Explain your reasoning.

1. building a storage bin with a limited amount of lumber

2. designing soup cans to hold 15 oz of soup

3. building solar energy panels

4. designing the "heat" tiles on the space shuttle

5. designing tires for 4-wheel-drive vehicles

Practice & Apply

Use the charts from Explorations 1 and 2 for Exercises 6–10.

6. If a rectangular prism has the dimensions $3 \times 1 \times 1$, what is the ratio of surface area to volume?

7. If a rectangular prism has the dimensions $3 \times 3 \times 3$, what is the ratio of surface area to volume?

8. Find the dimensions of a cube with a volume of 1000 cubic centimeters. What is the ratio of surface area to volume?

9. Find the ratio of surface area to volume for a cube with a volume of 64 cubic inches.

10. What is the surface area of the cube in Exercise 9?

11. Biology Flatworms absorb oxygen through their skin. Explain why a flatworm's shape helps it to live.

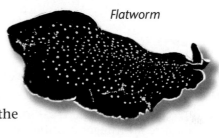

Flatworm

12. Physiology Human lungs are subdivided into thousands of little air sacs. How does this help the absorption of oxygen when we breathe?

13. Botany The short, squat form of cactus (barrel cactus) has a unique function. Explain how the shape helps a cactus thrive in the desert.

14. Botany Why do tall trees with large, broad leaves not usually grow in the desert?

15. ⬛ **Maximum/Minimum** Compare what happens to the ratio of surface area to volume of a rectangular prism of dimensions $n \times 1 \times 1$ to that with dimensions $n \times n \times n$ as n becomes large.

16. ⬛ **Maximum\Minimum** Using the result of Exercise 15, make a conjecture about what type of rectangular prism has the smallest surface area for a given volume.

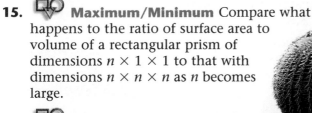

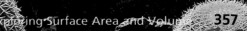

17. Cooking A roast is finished cooking when the inside of the meat reaches a certain temperature. Explain why a large roast takes longer to cook than a small roast.

18. Physiology Considering surface area and volume, why does thoroughly chewing your food help make digestion easier?

19. Aeronautical Engineering Thrust is the force caused by air escaping from the back of a jet engine. Thrust propels the jet forward. Currently, two different engine designs are used on commercial aircraft. One engine pushes a relatively small volume of air very quickly to produce thrust. Another engine pushes a much larger volume of air more slowly to produce thrust. Which engine do you think is more fuel efficient? Why? Which engine is quieter? Why?

Look Back

20. Show that AAA cannot be used to prove triangles congruent. **[Lesson 4.3]**

21. If the perimeter of a rectangle is 20 m, what are five possible areas if the sides measure an integer number of meters? **[Lesson 5.2]**

22. If two triangles have the same area, are they necessarily congruent? If not, use a counterexample to show why not. **[Lesson 5.2]**

23. The circumference of a circle is 10π. Find the area. **[Lesson 5.3]**

24. The legs of an isosceles right triangle each measure 6 cm. Find the measure of the hypotenuse. **[Lessons 5.4, 5.5]**

25. Find the area of a trapezoid in which one base measures 10 in., the other base measures 7 in., and the height measures 8 in. **[Lesson 5.2]**

Look Beyond

26. What information would you need in order to compute the surface area of a triangular pyramid?

LESSON 7.2 Surface Area and Volume of Prisms

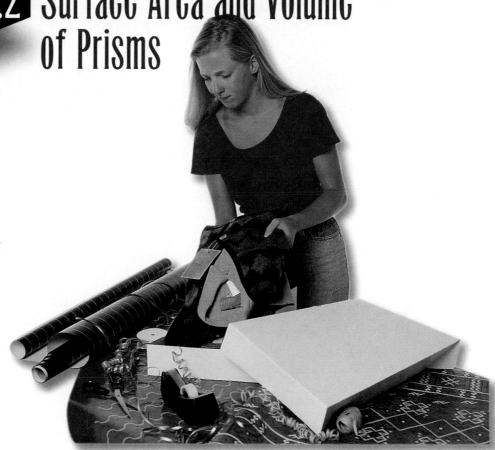

why *The surface area of an object is the area of the material it would take to cover it completely, without overlapping. You can use the surface area of a box to estimate the amount of wrapping paper you would need to wrap it.*

You may recall from Lesson 6.3 that a **prism** is a polyhedron with two bases that form polygons that are parallel and congruent. The lateral faces are parallelograms formed by the segments connecting the corresponding vertices of the bases. A **right prism** has lateral edges that are perpendicular to the planes of the bases.

An **altitude** of a prism is any segment perpendicular to the planes containing the bases with endpoints in these planes. The **length of an altitude** is the height of a prism.

The **lateral area** of a prism is found by adding all the areas of the lateral faces. The **surface area** of a prism is found by adding the lateral area to the areas of the two bases of a prism. Since the bases of a prism are congruent, the base area is twice the area of one base.

To analyze the surface area of certain solid figures, you can think of them as **nets**—flat figures that can be folded to enclose a particular solid figure. By calculating the area of the net, you can determine the surface area of a solid.

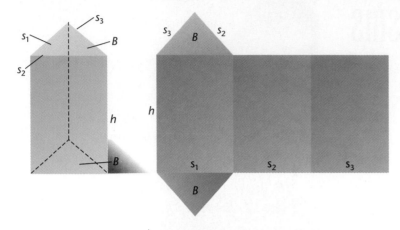

To find a formula for the surface area of a right triangular prism with height h and the area of a base B, think of cutting the prism and flattening it to form a net. To find the lateral area L, find the area of the rectangular faces.

$$L = h(s_1 + s_2 + s_3)$$

But $(s_1 + s_2 + s_3)$ is the perimeter p of a base, so

$$L = hp.$$

To find the total surface area S of the prism, add $2B$ to the lateral area.

$$S \text{ (surface area)} = L \text{ (lateral area)} + 2B \text{ (area of the bases)} = hp + 2B$$

LATERAL AREA OF A RIGHT PRISM

The lateral area L of a right prism with height h and perimeter of a base p is

$$L = hp \qquad\qquad \textbf{7.2.1}$$

SURFACE AREA OF A RIGHT PRISM

The surface area S of a right prism with lateral area L and the area of a base B is

$$S = L + 2B \qquad\qquad \textbf{7.2.2}$$

CRITICAL Thinking  Explain why some of the letters in the formulas above are lowercase and other letters are uppercase.

Volumes of Right Prisms

To find the volume V of a right rectangular prism, use the formula

$$V = lwh,$$

where l is the length, w is the width, and h is the height of the prism.

EXAMPLE 1

Marine Biology Find the volume of an aquarium that is 110 ft by 50 ft by 7 ft in size. How many gallons of water will the aquarium hold? (1 gal ≈ 1.6 cu ft.) If the aquarium is completely filled, how much will the water weigh? (Weight of 1 gal of water ≈ 8.33 lb)

Solution ➤

The volume of the aquarium is

$$V = lwh = 110 \times 50 \times 7 = 38,500 \text{ cu ft.}$$

To find the volume in gallons, divide by 1.6:

$$V \text{ (in gallons)} = \frac{38,500}{1.6} \approx 24,062.5 \text{ gal}$$

To find the weight, multiply by 8.33:

$$\text{Weight} = 24,062.5 \times 8.33 \approx 200,441 \text{ lb} \; ❖$$

In the formula for the volume of a right rectangular prism, $V = lwh$, notice that lw is the area of a base. Setting $B = lw$, the formula becomes $V = Bh$. This formula is true for all prisms.

CRITICAL *Thinking*

Write a volume formula for a right prism with bases that are regular polygons.

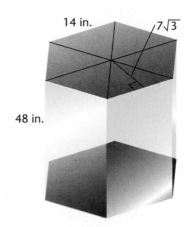

14 in.

$7\sqrt{3}$

48 in.

EXAMPLE 2

An aquarium is in the shape of a right regular hexagonal prism. Each side of its base is 14 in. If the aquarium is 48 in. tall, find the volume.

Solution ➤

Find the area of a base of the aquarium. Notice that each base can be divided into six equilateral triangles. The altitude of each triangle is $7\sqrt{3}$ in., which is also the measure of the apothem of the hexagon. The perimeter of the hexagon is 84 in.

The area of the base is

$$\tfrac{1}{2}ap = \tfrac{1}{2} \times 7\sqrt{3} \times 84 = 509.22 \text{ sq in.}$$

To find the volume of the aquarium, multiply this result by the height h:

$$V = Bh = 509.22 \times 48 = 24,442.56 \text{ cu in.} \; ❖$$

Lesson 7.2   Surface Area and Volume of Prisms   **361**

Oblique Prisms

In an oblique prism the lateral edges are not perpendicular to the planes of the bases. For such prisms there is no simple general formula for the surface area. However, the formula for the volume is the same as for a right prism. To understand why this is so, consider the following idea.

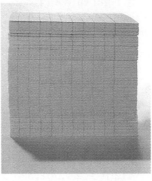

Stack a set of index cards in the shape of a right rectangular prism. If you push the stack over into an oblique prism, the volume of the figure does not change because the number of cards does not change.

In the picture, both stacks have the same number of cards, and so each prism is the same height. Also, since every card has the same size and shape, they all have the same area. Any card in either stack represents a cross section of each prism.

This idea illustrates an important geometry concept.

CAVALIERI'S PRINCIPLE

If two solids have equal heights, and if the cross sections formed by every plane parallel to the bases of both solids have equal areas, then the two solids have the same volume.   **7.2.3**

The solids pictured above will have equal volumes if all the cross-sectional areas are equal.

Since an oblique prism can be compared to a right prism that has been pushed over like the deck of cards, the base area, the height, and the volume do not change. Therefore, the formula for the volume of an oblique prism is the same as the formula for the volume of a right prism.

VOLUME OF A PRISM

The volume V of a prism with height h and the area of a base B is

$$V = Bh$$ **7.2.4**

EXERCISES & PROBLEMS

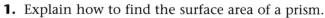

Communicate

1. Explain how to find the surface area of a prism.

2. Explain the formula for the volume of a right prism.

3. Explain Cavalieri's principle and how it can be used to find the volume of an oblique prism.

4. Can Cavalieri's principle be used for two prisms of the same height if the base of one is a triangle and the base of the other is a hexagon? Why or why not?

5. What does doubling the height of a prism do to the volume if the base is unchanged?

Practice & Apply

Draw the net for each of these figures. Assume them to be right (not oblique).

6. Cube (square prism)

7. Equilateral triangular prism

8. Regular hexagonal prism

9. Rectangular prism

10. Find the surface area of a rectangular prism measuring 18 cm by 15 cm by 14 cm.

11. Find the height of a rectangular prism with a surface area of 286 sq in. and base measuring 7 in. by 9 in.

12. Find the surface area of a right regular hexagonal prism whose altitude is 20 cm. The apothem of the hexagonal base is $4\sqrt{3}$.

13. **Recreation**   Suppose a tent in the shape of an isosceles triangular prism is resting on a flat surface. Find the surface area of the tent, including the floor.

14. Find the surface area of a right regular hexagonal prism. The sides of the regular hexagon are 18 cm and its altitude is 25 cm.

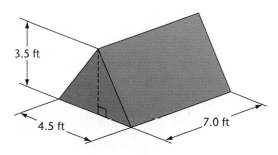

3.5 ft

4.5 ft

7.0 ft

Find the volume of each prism.

15. $B = 7$ cm^2, $h = 5$ cm

16. $B = 9$ m^2, $h = 6$ m

17. $B = 17$ cm^2, $h = 23$ cm

18. $B = 32$ ft^2, $h = 17$ ft

Find the volume of each prism.

19.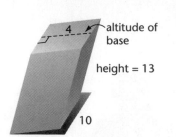

altitude of base

4

height = 13

10

20.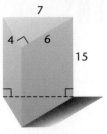

7

4 ⌐ 6

15

21. If a cube has a volume of 343 cu yds, what is its surface area?

22. Find the volume of a right trapezoidal prism. The parallel sides of the trapezoids measure 6 m and 8 m and the trapezoid's altitude is 7 m. The altitude of the prism is 18 m.

23. Find the volume of a right hexagonal prism. The length of the sides of the hexagonal bases are 8 cm and the altitude of the prism is 12 cm.

24. Find the volume of a right triangular prism whose base is an isosceles right triangle with hypotenuse 10 cm. The altitude of the prism is 23 cm.

25. Find the surface area and volume of a right triangular prism whose base is an isosceles triangle with sides 4 in., 4 in., and 6 in. The height of the prism is 13 in.

26. **Portfolio Activity**   Using 100 sq in. of cardboard, a pair of scissors, and masking tape, make a box without a top. What are the measures of its sides? What is the volume?

27. If the edges of a cube are doubled, what happens to the surface area? to the volume? If the edges are tripled, what happens to the surface area? to the volume?

28. When does a cube have the same numerical value for its surface area and volume?

29. **Product Packaging**   Find the surface area and volume of each box of breakfast food.

3 in.

5 in.

8 in.

8 in.

10 in.

2 in.

30. Architecture Estimate how much glass was used to cover the outside walls of this building. The distance from one floor to the next is 12 feet and the base of the building is a square which measures 48 feet on each side.

 **Look Back**

Simplify each radical expression. Leave answers in simplified radical form. [Lesson 5.4]

31. $\sqrt{20}$

32. $(\sqrt{18})(\sqrt{2})$

33. $(5\sqrt{7})^2$

34. $(2\sqrt{2})(3\sqrt{8})$

In a 30-60-90 triangle, for Exercises 35 and 36, the hypotenuse measures 10 in. [Lesson 5.5]

35. Find the measure of the shorter leg.

36. Find the measure of the longer leg.

37. Find the apothem of a regular hexagon with each side measuring 18 in.

Look Beyond

38. **Coordinate Geometry** Point (x, y) is in a plane while point (x, y, z) is in space. Find the volume of the prism shown.

39. **Coordinate Geometry** Find the length of the diagonal joining $(4, 0, 0)$ and $(0, 5, 6)$.

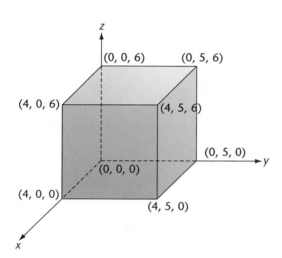

Surface Area and Volume of Pyramids

Across cultures, the pyramid has an enduring allure. From ancient Egyptian tombs to the Trans-America Tower in San Francisco, the pyramid shape has strength and simple beauty.

A **pyramid** consists of a **base**, which is a polygon, and a number of **lateral faces**, which are triangles. The lateral faces meet at a point called the **vertex** of the pyramid. The **altitude** of a pyramid is the segment from the vertex perpendicular to the plane of the base. The height of a pyramid is the length of its altitude.

A **regular** pyramid is a pyramid whose base is a regular polygon. Regular pyramids may be **right** or **oblique**. In a right pyramid the altitude intersects the base at its center, which is the point equidistant from the vertices. In an oblique pyramid the altitude intersects the plane of the base at some point other than the center.

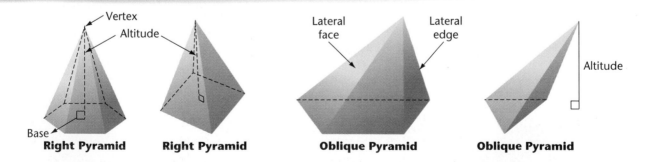

Vertex
Altitude
Right Pyramid
Base

Right Pyramid

Lateral face
Lateral edge
Oblique Pyramid

Altitude
Oblique Pyramid

In a right regular pyramid, all the faces are congruent triangles. In an oblique pyramid, the faces are not congruent.

CRITICAL Thinking

What difficulty is there in defining a right pyramid whose base is not a regular polygon?

The Surface Area of a Pyramid

A pyramid can be "unfolded" and analyzed as a net as shown. The slant height l of a pyramid is the length of the altitudes of the lateral faces (the triangles). The figure shows a right regular pyramid with a square base. To find the surface area of the pyramid, add the lateral area (the area of the faces) to the area of the base.

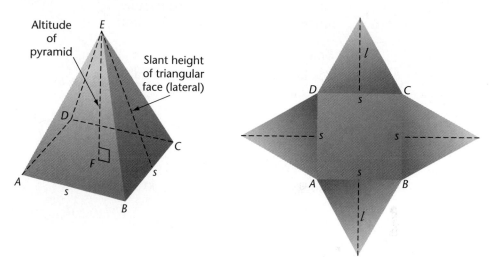

ALGEBRA
Connection

Surface Area = Lateral Area + Area of the Base

$$S = L + B$$

$$S = 4(\tfrac{1}{2}sl) + s^2 = \tfrac{1}{2}l(4s) + s^2$$

Since $4s$ is the perimeter p of the base,

$$S = \tfrac{1}{2}lp + s^2.$$

For any right regular pyramid, the lateral area is $\tfrac{1}{2}lp$. The total surface area is found by adding the base area.

LATERAL AREA OF A RIGHT REGULAR PYRAMID

The lateral area L of a right regular pyramid with slant height l and perimeter p of a base is

$$L = \tfrac{1}{2}lp$$

7.3.1

Note: For pyramids other than right regular pyramids, the areas of the triangular faces must be calculated individually and then added together.

SURFACE AREA OF A PYRAMID

The surface area S of a pyramid with lateral area L and area of a base B is

$$S = L + B$$

7.3.2

Construction The roof of a gazebo is a right square pyramid. The square base of the pyramid is 12 ft on each side and the slant height is 16 ft. Find the area of the roof. If roofing material costs $3.50 per sq ft, how much will it cost to cover the roof.

Solution ➤

The area of the roof is the lateral area L of the pyramid.

$L = \frac{1}{2} l p$

$\quad = \frac{1}{2} 16(4 \times 12)$

$\quad = 384$ sq ft

384 sq ft × $3.50 = $1,344.00 ❖

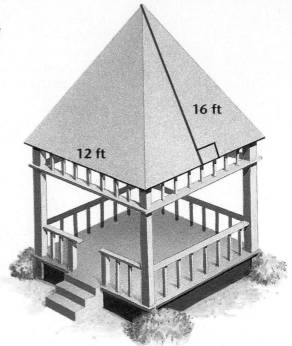

The Volume of a Pyramid

To find the volume of a pyramid, compare the pyramid with a prism that has the same base and height as the pyramid.

Exploration *Pyramids and Prisms*

You will need
Stiff construction paper
Scissors, tape, ruler
Dry cereal or packing material

1 Construct a right square prism and a right square pyramid with the same base and the same height. Draw each of the nets on construction paper and tape the tabs in place. Seal the necessary edges with tape.

2 Fill the pyramid with dry cereal or packing material. Pour the contents into the right prism. Repeat as many times as necessary to fill the prism completely. How many times did you have to fill the pyramid in order to fill the prism?

3 Make a conjecture about the relationship of the volume of a pyramid to the volume of a prism with the same base and height. Express your conjecture as a formula for the volume of a pyramid. ❖

The relationship can be studied mathematically by dividing a prism into three pyramids that have the same volume. Although the pyramids are not all congruent to one another, it can be shown that, in pairs, they have equal volumes.

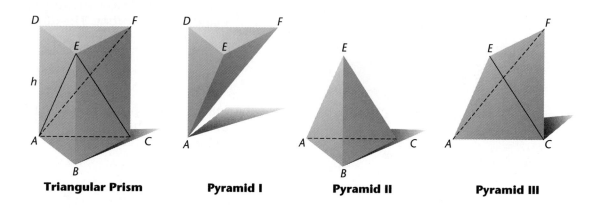

Triangular Prism   Pyramid I   Pyramid II   Pyramid III

(Each pyramid is named by its vertex, followed by the polygon that forms its base.)

Consider the two following ways of pairing the pyramids.

A. I   Pyramid *ADEF* has base *DEF*, altitude *h*.

  II Pyramid *EABC* has base *ABC*, altitude *h*.

  Since the bases are congruent and the altitudes are equal, these two pyramids have equal volumes.

B. I   Pyramid *EABC* has base *EBC*.

  II Pyramid *AEFC* has base *EFC*.

  These two pyramids have congruent bases, since diagonal $\overline{EC}$ divides rectangle *BEFC* into two congruent triangles. The vertices of the pyramids are the same, as are the altitudes. Therefore, the two pyramids also have equal volumes.

By transitivity, all three pyramids have the same volume, and each must be one-third of the original prism. This suggests the following formula:

VOLUME OF A PYRAMID

The volume *V* of a pyramid with area of its base *B* and altitude *h* is

$$V = \frac{1}{3}Bh$$

7.3.3

CRITICAL Thinking

Does the proof above work for all pyramids? If not, what is needed to make it a completely general proof?

Cultural Connection: Africa   What is the weight of the pyramid of Khufu?

The pyramid of Khufu is constructed of stones with weight of approximately 167 pounds per cubic foot. The dimensions of the pyramid are as follows:

Base side = 775.75 ft

Height = 481 ft

What is the volume of the pyramid of Khufu in cubic feet? What is the weight of the pyramid of Khufu in pounds? in tons? (1 ton = 2000 lb)

The volume of the pyramid is

$$V = \tfrac{1}{3}Bh = \tfrac{1}{3}(775.75)^2 (481) \approx 96{,}486{,}686 \text{ cubic feet.}$$

The weight in pounds is

$$96{,}486{,}686 \cdot 167 \approx 16{,}113{,}000{,}000 \text{ lb, which is } 8{,}056{,}638 \text{ tons.} \;\; \diamondsuit$$

EXERCISES & PROBLEMS

Communicate

1. Explain how to find the total surface area of a right pyramid.
2. Explain how to find the volume of a pyramid.
3. In a pyramid, which is longer in measure, its altitude (height) or its slant height? Why?
4. Explain the relationship between the volume of a rectangular pyramid and the volume of a rectangular prism.
5. Where does the altitude of a regular right pyramid intersect the base?

Practice & Apply

Draw the net for each pyramid below.

6. a square pyramid
7. an equilateral triangular pyramid
8. a regular hexagonal pyramid
9. a regular pentagonal pyramid

Find the total surface area of each right pyramid.

10.

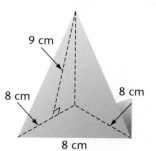

9 cm

8 cm   8 cm

8 cm

11.

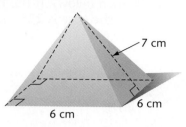

7 cm

6 cm   6 cm

12.

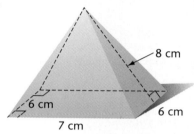

8 cm

6 cm   6 cm

7 cm

13.

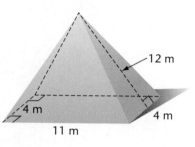

12 m

4 m   4 m

11 m

For Exercises 14–15, find the volume. Draw the figure and label the parts.

14. a right rectangular pyramid with base 5 m by 7 m and height 11 m

15. a right triangular pyramid whose base dimensions are 5, 12, and 13 and whose altitude is 10

Find the volume for each right pyramid.

16.

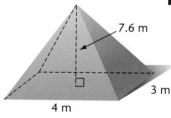

7.6 m

3 m

4 m

17.

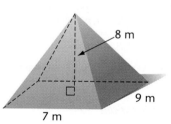

8 m

9 m

7 m

18.

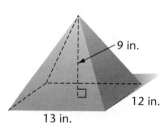

9 in.

12 in.

13 in.

19. Find the surface area and volume of a right square pyramid. The base measures 18 cm on each side. The slant height measures 15 cm.

20. Find the surface area and volume of a right square pyramid. The base measures 1 ft on each side. The slant height measures 10 in.

21. Construction An architect is designing a building to resemble the entrance of the Louvre in France (shown below). If the height of the structure is 15 meters, and the area of the base is 225 square meters, what will be the volume of the building?

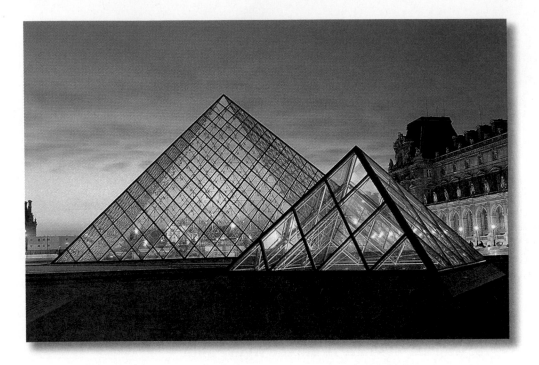

22. A right pyramid has an isosceles right triangle as its base. The legs are 20 cm and the volume of the pyramid is 3000 cubic cm. Find the altitude of the pyramid.

Look Back

23. Angles *Y* and *Z* are complementary. Angle *Y* measures 51°. What is the measure of angle *Z*? **[Lesson 1.5]**

A quadrilateral has coordinates (0, 0), (5, 0), (2, 6), and (7, 6).

24. Find the perimeter of the quadrilateral. **[Lesson 5.6]**

25. Find the area of the quadrilateral. **[Lesson 5.6]**

26. Give the coordinates of a quadrilateral that has the same area as the one above, but a different perimeter. **[Lesson 5.6]**

Look Beyond

27. Consider pyramids with bases of 3, 4, 5, 6, 7, 8, and *n* sides. Count the vertices, faces, and edges. Make a table. Does the generalization for *n* satisfy Euler's formula?

Surface Area and Volume of Cylinders

Why The gasoline you buy at a pump is stored in underground, cylindrical tanks. How could you use the dimensions of an underground gas tank to estimate the number of cars that could be filled from it?

A cylinder is a three-dimensional figure with two congruent circular **bases** in parallel planes, connected by a curved **lateral surface**.

The **altitude** of a cylinder is a segment that is perpendicular to the planes of the bases, from the plane of one base to the plane of the other. The length of the altitude is the height of the cylinder.

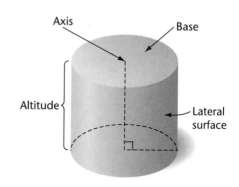

The **axis** of a cylinder is the segment joining the centers of the two bases.

If the axis of a cylinder is perpendicular to the bases, then the cylinder is a **right cylinder**. If not, it is an **oblique cylinder**.

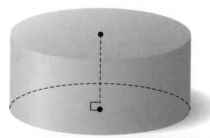
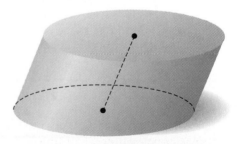

Cylinders and Prisms

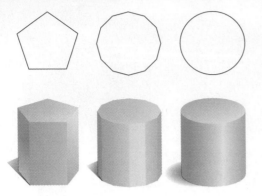

As the number of sides of a regular polygon increases, the figure becomes more and more like a circle.

Similarly, as the number of sides of a regular polygonal prism increases, the figure becomes more and more like a cylinder.

The illustrations suggest the formulas for the surface areas and volumes of prisms and cylinders are similar.

The Surface Area of a Cylinder

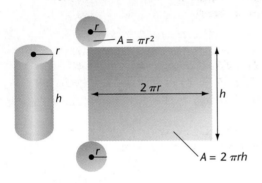

$A = \pi r^2$

$2\pi r$

$A = 2\pi rh$

The surface area of a right cylinder is found by a method similar to that used for a right prism. The net for a right cylinder includes the two circular bases and the flattened lateral surface, which becomes a rectangle. The length of this rectangle is the circumference of the base. The width of the rectangle is the height of the cylinder.

The lateral area of the cylinder equals $2\pi rh$, and the area of each base is πr^2. Thus,

ALGEBRA *Connection*

$$S \text{ (surface area)} = L \text{ (lateral area)} + 2B \text{ (area of bases)}$$
$$= 2\pi rh + 2\pi r^2$$

LATERAL AREA OF A RIGHT CYLINDER

The lateral area L of a right cylinder with radius r and height h is

$$L = 2\pi rh$$

7.4.1

SURFACE AREA OF A RIGHT CYLINDER

The surface area S of a right cylinder with radius r, height h, area of a base B, and lateral area L is

$$S = L + 2B \qquad \text{or} \qquad S = 2\pi rh + 2\pi r^2$$

7.4.2

Simplify the formula $S = 2\pi rh + 2\pi r^2$ by factoring. Is the new formula easier or harder to compute? Is it easier or harder to remember?

APPLICATION

Find the surface area of a right circular cylinder with a radius of 4 cm and a height of 7 cm.

Solution ➤

The surface area is

$S = 2\pi rh + 2\pi r^2$, or

$S = 2\pi(4)(7) + 2\pi(4)(4) \approx 276.46 \text{ cm}^3$ ❖

The Volume of a Cylinder

In the following exploration, recall the method you used to find the area of a circle.

•Exploration• *Analyzing the Volume of a Cylinder*

You will need
No special tools

The formula for the area of a circle was found by cutting a large number of sections and fitting them together to approach a rectangular shape. The same idea can be used with a cylinder. Use the figure to derive the formula for the volume of a cylinder.

1 What geometric solid does the cylinder approximate when the sections are organized as they are above?

2 Using the length, width, and height of the figure above, write a formula for the volume of a cylinder.

3 Summarize your exploration by writing a theorem for the volume of a right cylinder. ❖

Is the formula for the volume of an oblique cylinder the same as or different from the formula for a right cylinder? What principle justifies your answer? Illustrate your answer with sketches.

VOLUME OF A CYLINDER

The volume V of a cylinder with radius r, height h, and area of a base B, is

$$V = Bh \quad \text{or} \quad V = \pi r^2 h$$

7.4.3

EXAMPLE

Gasoline Storage   Find the volume of a cylindrical underground gasoline tank with a radius of 6 ft and an altitude of 25 ft. How many gallons of gasoline will the tank hold? (1 gal = 0.13368 cu ft)

Solution ➤

$$V = \pi r^2 h = \pi(6^2)(25) = 900\pi \approx 2827.4 \text{ cu ft}$$

To find the amount in gallons, divide by 0.13368.

$$V \approx \frac{2827.4}{0.13368} \approx 21,151 \text{ gal} \quad \diamondsuit$$

EXERCISES & PROBLEMS

Communicate

1. State the formula for the volume of a cylinder.
2. Explain how to find the surface area of a cylinder.
3. Explain how a cylinder is like a rectangular prism.
4. What does doubling the radius of a cylinder do to the volume?
5. What does doubling the height of a cylinder do to the surface area?

Practice & Apply

Draw a net for each of the cylinders shown. Label the known parts.

6.

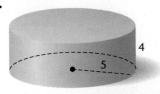
4
5

7.

15
4

8. Find the surface area for the figure in Exercise 6.

9. Find the surface area for the figure in Exercise 7.

10. Find the volume of the figure in Exercise 6.

11. Find the volume of the figure in Exercise 7.

12. The surface area of a cylinder is 200 sq cm. The diameter and altitude are equal. Find the radius.

13. Find the surface area of a right cylinder whose diameter is 10 cm and whose altitude is 20 cm.

14. Find the surface area of a right cylinder whose radius is 10 cm and whose height is 30 cm. If the dimensions of the cylinder are doubled, what happens to the surface area?

15. Find the volume of a cylinder whose diameter is 5 m and whose height is 8 m.

16. If the dimensions of the cylinder in Exercise 15 are doubled, what happens to the volume?

17. If the volume of a cylinder is 360π cubic mm and its altitude is 10 mm, find the circumference of the base.

18. Georgia wants to wrap a large Christmas present in a cylindrical package. She bought 1 roll of wrapping paper which contains 30 square feet of paper. If the diameter of the cylinder is 2 feet and the height is 4 feet, will she have enough wrapping paper?

19. Which cylinder has the larger lateral area? Which has the larger surface area? Explain your answers.

   cylinder *x*: radius = 3, altitude = 6   cylinder *y*: radius = 6, altitude = 3

20. A cylindrical glass is full of water and is to be poured into a rectangular pan. The base of the pan is 15 in. by 10 in. and the height of the pan is 3 in. The height of the cylinder is 15 in. What must the radius of the cylinder be so that the water fills the pan but does not spill?

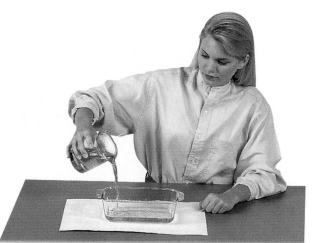

21. A straw is 25 cm long and 4 mm in diameter. How much liquid can the straw hold?

22. **Portfolio Activity** Find a cylindrical can of vegetables or fruit in your home. Take the measurements to find the surface area (sq in.) and volume.

23. **Consumer Awareness** Many grocery shoppers assume that a taller can or a wider can holds more. Why is this not always true?

24. Find the volume of a semicircular cylinder formed by cutting a cylinder in half on its diameter. The diameter is 8 ft and the height of the cylinder is 10 ft.

25.  **Maximum/Minimum** A city zoo needs a cylindrical aquarium that holds 100,000 cubic feet of water. In order to save money, the city wants to use the least amount of materials possible to build the aquarium. What is the smallest surface area the aquarium can have and still hold 100,000 cubic feet of water?

26. **Portfolio Activity** Take an 11 × 14-inch sheet of paper and roll it up two different ways. Which cylinder has the larger volume? Explain your reasoning.

Look Back

Find the surface area and volume for each figure.
[Lessons 7.2, 7.3]

27.

6
4
12

28.
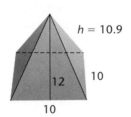
h = 10.9
12
10
10

29.
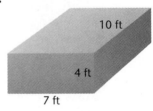
10 ft
4 ft
7 ft

30.
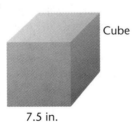
Cube
7.5 in.

31. State Cavalieri's principle and use a diagram to support it.
[Lesson 7.2]

Look Beyond

32. In a guessing contest at a school fair, students are asked to guess the number of marbles in a jar. The inner dimensions of the jar are 12.4 cm in diameter and 28.6 cm tall. The diameters of the marbles average 1.55 cm. List as many different ways as you can for estimating the number of marbles in the jar. State your own estimate of the number. Your teacher has the actual number.

Surface Area and Volume of Cones

why As a volcano erupts and deposits lava and ash over a period of time, it forms a cone. Volcanic cones may be of different shapes and sizes, depending on factors such as the rate at which the lava and ash are deposited, how fast the lava cools, etc. However, all of them can be modeled by geometric cones. By using the properties of geometric cones, geologists can study the physical structure of volcanoes.

A circular **cone** is an object that consists of a circular **base** and a curved **lateral** surface which extends from the base to a single point called the **vertex**.

The **altitude** of a cone is the segment from the vertex perpendicular to the plane of the base. The height of the cone is the length of the altitude.

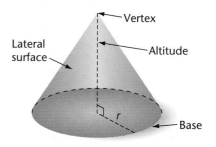

If the altitude of a cone intersects the base of the cone at its center point, the cone is a **right cone**. If not, it is an **oblique cone**.

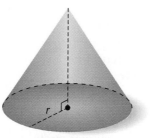

Right Cone

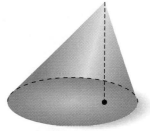

Oblique Cone

Just as a cylinder resembles a prism, a cone resembles a pyramid. As the number of sides of the base of a pyramid increases, the figure becomes more and more like a cone.

The illustrations suggest that the formulas for the surface areas and volumes of prisms and cylinders are similar.

•Exploration 1 *The Surface Area of a Right Cone*

You will need
No special tools

The surface area of a right cone is found by a method similar to the one used for a right pyramid. The net for a cone includes the circular base and the flattened lateral surface, which becomes a portion of a circle known as a sector. The radius l of the sector is the slant height of the cone.

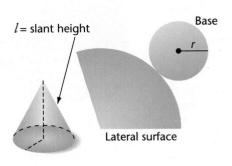

l = slant height

Base

r

Lateral surface

The surface area of a cone can be found by adding the area of the lateral surface and the area of the base.

1 The lateral surface of a cone is part of a larger circle. The arc labeled *c* matches the circumference of the base of a cone in a three-dimensional figure. So *c* is equal to the circumference of the base. Find *c* using *r*, the radius of the base.

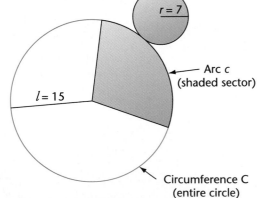

r = 7

Arc *c* (shaded sector)

l = 15

Circumference C (entire circle)

2 The radius of the larger circle is *l*. Find *C*, the circumference of the larger circle.

3 Divide *c* by *C*. This number tells you what fractional part the lateral surface occupies in the larger circle.

4 Find the area of the larger circle. Multiply this number by the fraction from Step 3. The result is the lateral surface of the cone. This is the lateral area *L*.

5 Find *B*, the area of the base of the cone. Add this to the area of the lateral surface *L* (Step 4). What does your answer represent? ❖

EXAMPLE 1

Find the surface area of a cone with the measures shown.

Solution ➤

Step 1 $c = 2\pi r = 14\pi$

Step 2 $C = 2\pi l = 30\pi$

Step 3 $\dfrac{c}{C} = \dfrac{14\pi}{30\pi} = \dfrac{7}{15}$

Step 4 area of larger circle

$$\pi l^2 = 225\pi$$

$$L = \frac{7}{15} \cdot 225\pi = 105\pi$$

Step 5 $B = \pi r^2 = 49\pi$

$$B + L = 49\pi + 105\pi = 154\pi \text{ m}^2 \approx 483.8 \text{ m}^2 \text{ ❖}$$

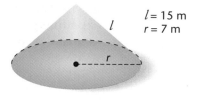

$l = 15$ m
$r = 7$ m

•Exploration 2 *The Surface Area Formula for a Cone*

You will need
No special tools

1 Follow the steps in the example above and on the previous page to find the surface area of a general cone. That is, use the variables r, h, and l instead of numerical measures. Your answer will be a formula for the surface area of a cone.

2 Use algebra to simplify your formula. Show your steps in an organized way and save your work in your notebook.

3 Compare your results with the formula given below. Are they the same? Explain. ❖

SURFACE AREA OF A RIGHT CONE

The surface area S of a right cone with radius of base r, height h, and slant height l is

S (surface area) = L (lateral area) + B (base area)

$$S = \pi r l + \pi r^2 \qquad\qquad \textbf{7.5.1}$$

Lesson 7.5 • Surface Area and Volume of Cones **381**

The Volume of a Cone

In Lesson 7.3 you found that a pyramid had one-third the area of a prism with the same base and height. Now imagine performing a similar experiment with a cone and a cylinder. The result of the experiment should be the same— you can try it—because cones and cylinders are like many-sided pyramids and prisms.

This leads us to the following formula for the volume of a cone.

VOLUME OF A CONE

The volume V of a cone with radius r and height h is

V (volume) $= \frac{1}{3} B$ (base area) $\times h$ (height): $V = \frac{1}{3} \pi r^2 h$ **7.5.2**

CRITICAL *Thinking*

Is the volume of an oblique cone the same as the formula for a right cone? What principle justifies your answer? Illustrate your answer with sketches.

EXAMPLE 2

Geology   A vulcanologist is studying a violent eruption of a cone-shaped volcano. The original volcano cone had a radius of 5 miles and a height of 2 miles. The eruption removed a cone from the top of the volcano. This cone had a radius of 1 mile and a height of $\frac{1}{2}$ mile. What percentage of the total volume of the original volcano was destroyed by the eruption?

Solution ➤

The volume of the original volcano is

$V = \frac{1}{3} \pi r^2 h = \frac{1}{3} \pi (5^2)(2) \approx 52.4$ cubic miles.

The volume of the destroyed cone is

$V = \frac{1}{3} \pi r^2 h = \frac{1}{3} \pi (1^2)(0.5) \approx 0.52$ cubic miles.

The percentage of the original volcano is

$\left(\frac{0.52}{52.4}\right)(100) \approx 1\%$. ❖

EXERCISES & PROBLEMS

Communicate

1. Explain how to find the surface area of a cone.

2. Explain how to find the volume of a cone.

3. In a cone, which is longer in measure, the altitude or the slant height? Explain your reasoning.

4. What happens to the lateral area of a cone if the radius is doubled? if the slant height is doubled? if both are doubled?

5. What happens to the volume of a cone if the radius is doubled? if the slant height is doubled? if both are doubled?

Practice & Apply

Find the surface area for each right cone.

6.
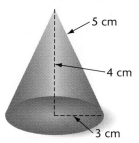
5 cm

4 cm

3 cm

7.
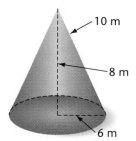
10 m

8 m

6 m

8.
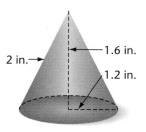
2 in.→

1.6 in.

1.2 in.

9. A right cone has a radius of 5 in. and a surface area of 180 sq in. What is the slant height of the cone?

Find the volume of each right cone.

10.
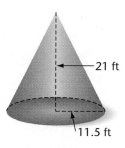
21 ft

11.5 ft

11.
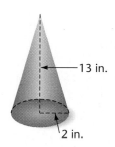
13 in.

2 in.

12.
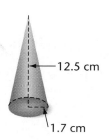
12.5 cm

1.7 cm

13. Find the surface area of a right cone whose base area is 25π sq cm and whose altitude is 13 cm.

14. An oblique cone has a volume of 1000 cm³ and a height of 10 cm. What is the radius of the cone?

15. 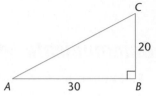 **Portfolio Activity**   Draw a circle with a radius of 5 cm and cut it out with scissors. At the center of the circle, draw a right angle and cut it out. Use tape and make a cone. What is the area of the sector that was cut out? What is the surface area of the cone?

16. A cone and a cylinder have bases with the same area and equal altitudes; why doesn't Cavalieri's principle hold?

17. If the triangular region *ABC* is rotated about *AB*, what solid figure is formed? Find its surface area.

18. A right triangle has sides 10, 24, and 26. Rotate the triangle about each leg and find the volume of the figures formed.

Manufacturing   Cone-shaped paper cups can be manufactured from patterns in the shape of sectors of circles. The figures below show two patterns. The first is cut with a straight angle. The second is cut with an angle measuring 120°. Use the information provided for Exercises 19–21.

19. If each sector has a radius of 6 cm, what will be the area of each sector? (Hint: What part of the circle is used by the sector?)

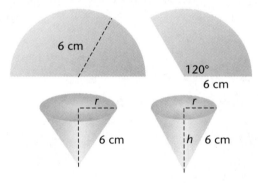

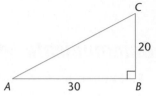 **Algebra**   To find the volume of a cone-shaped cup formed from the first sector, we must first consider that the length of the arc of the sector becomes the circumference of the base of the cone.

Arc length $= \frac{1}{2} \times 2\pi \times 6 = 6\pi =$ circumference of cone base.

Using the circumference formula again, we find the radius of the base.

If $C = 2\pi r = 6\pi$, then $r = 3$.

Using the Pythagorean Theorem, we can find the height of the cone:

$h = \sqrt{36 - 9} = \sqrt{27} = 3\sqrt{3}$.

We can now find the volume of the cone.

$$V = \frac{1}{3}\pi r^2 h = \frac{1}{3}\pi \times 3^2 \times 3\sqrt{3}$$

$$= \pi \times 9 \times \sqrt{3}$$

$$= 48.97 \text{ cm}^3$$

20. Find the volume of the cup formed by the sector measuring 120°.

21. If the radius of the larger sector increases by 50%, by how much does the volume of the cup formed from it increase?

22. Small Business The owner of an ice cream parlor is choosing bowls for her new store. Of the two bowls shown, which has the greater volume? Express the difference in the volumes in terms of π. How do the amounts of ice cream heaped above the tops of the cups compare?

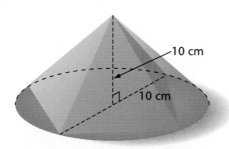

23. A right square pyramid is inscribed in a right cone of the same height. A radius of the base and the altitude of the cone are each 10 cm long. What is the ratio of the volume of the pyramid to the volume of the cone?

24. Find the surface area and volume of a cone with a radius of 6 cm and an altitude of 9 cm.

Maximum/Minimum A cone has a slant height of 10. What should the radius and height be to maximize its volume? Fill in the chart for Exercises 25–31.

Radius	Height	Volume
1	$\sqrt{99}$	$\frac{1}{3}\pi \times 1^2 \times \sqrt{99} \approx 10.4$
2	$\sqrt{96}$	**25.** _____ ? _____
3	**26.** _____ ? _____	**27.** _____ ? _____
4	**28.** _____ ? _____	**29.** _____ ? _____
x	**30.** _____ ? _____	**31.** _____ ? _____

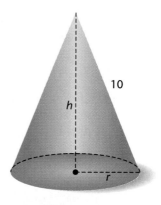

Technology Use a graphics calculator or software and the information from Exercises 25–31 to graph the radius on the *x*-axis and the volume on the *y*-axis. Set the viewing window so that Ymax is at least 410. Trace the graph and find the following.

32. the *x*-value that produces the largest volume

33. the largest volume

34. Recreation At an amusement park, an oblique conical tower is to be built. The vertex is to be directly above the edge of the circular base. The dimensions are as shown. What is the volume of the cone?

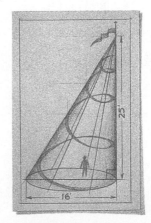

Look Back

Use the given coordinates to find the coordinates of point Q.
[Lesson 3.8]

35.
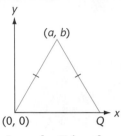
Isosceles Triangle

36.
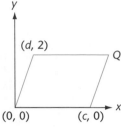
Parallelogram

37. Prove that a diagonal divides a parallelogram into two congruent triangles.   **[Lesson 4.6]**

38. Find the distance between (2, 4) and (–5, 8).   **[Lesson 5.6]**

39. 🖎 **Maximum/Minimum**   A manufacturing company makes cardboard boxes of varying sizes by cutting out square corners from rectangular sheets of cardboard that are 12 in. by 18 in. The cardboard is then folded to form a box without a lid. If a customer wants a box with the greatest possible volume, what dimensions should be used? Use your graphics calculator.   **[Lesson 7.1]**

12 in

18 in.

Look Beyond

40. Scale Models   A model train engine is $\frac{1}{36}$ the size of a real train engine. If the model's width is 4 inches and its length is 16 inches, what are the width and length of the real engine?

Surface Area and Volume of Spheres

Why *To make a hot-air balloon, individual pieces of cloth are sewn together. How much cloth do you think is needed? The "envelope" of an inflated balloon is not a perfect sphere, but you can use the information about spheres to get good approximate answers to this and other questions about spherical objects.*

The Volume of a Sphere

To find the formula for the volume of a sphere, we first show that a sphere has the same volume as a special cylinder with a double cone cut out of it. Then, by using the formulas for cones and cylinders, we derive the formula for the volume of each figure. We begin with a numerical calculation before generalizing.

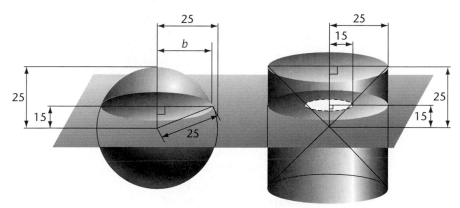

The sphere has a radius of 25 units. The cylinder has a radius of 25 units and a height of 50 units, and a double cone is cut out of it. A plane cuts through both figures 15 units above the center of each. We can show that the area of the shaded circular disk in the sphere is equal to the area of the shaded **annulus** (the ring-shaped figure) in the cylinder.

The large triangle formed in the cylinder by the side of the cone, the base radius, and half the height is isosceles with two legs 25 units in length. Since the plane is parallel to the bases, congruent corresponding angles are formed with the side of the cone, so the small triangle is similar to the large triangle, and is also isosceles with legs 15 units in length. Thus the radius of the inner circle of the annulus is 15 units.

ALGEBRA
Connection

Area of the circle in the sphere	**Area of the annulus**
	area of large circle – area of small circle
$b^2 + 15^2 = 25^2$	$= \pi 25^2 - \pi 15^2$
$b^2 + 225 = 625$	$= \pi(25^2 - 15^2)$
$b^2 = 400$	$= \pi(625 - 225)$
$b = 20$ units	$= \pi(400)$
Area $= \pi 20^2 = 400\pi$ sq units	Area $= 400\pi$ sq units

Now generalize the procedure to get the proof for a sphere and a cylinder of radius r.

Area of the circle	**Area of the annulus**
$b^2 + y^2 = r^2$	Area $= \pi r^2 - \pi y^2$
$b^2 = r^2 - y^2$	
$b = \sqrt{r^2 + y^2}$	
Area $= \pi\left(\sqrt{r^2 - y^2}\right)^2 = \pi(r^2 - y^2)$	Area $= \pi(r^2 - y^2)$

We now know that the result is true for all planes parallel to the bases of each figure. The result is true for all values of y.

The corresponding cross sections have equal areas.

Therefore, the conditions of Cavalieri's Principle are satisfied. And so the volume of the sphere equals the volume of the cylinder with the double cone removed. That is,

$$V \text{ (Sphere)} = V \text{ (Cylinder)} - V \text{ (Cones)}$$
$$V \text{ (Sphere)} = \pi r^2(2r) - 2(\tfrac{1}{3}\pi r^2)(r)$$
$$= 2\pi r^3 - \tfrac{2}{3}\pi r^3$$
$$= \tfrac{4}{3}\pi r^3$$

> **VOLUME OF A SPHERE**
> The volume V of a sphere with radius r is
> $$V = \frac{4}{3}\pi r^3$$
>
> **7.6.1**

CRITICAL *Thinking* When the cutting plane in the proof of the volume formula cuts through the center of the sphere, the distance y is 0. What happens to the annulus? How does this affect the calculations? Explain.

EXAMPLE 1

Hot Air Ballooning   If the envelope of a hot-air balloon has a radius of 27 ft when fully inflated, approximately how many cubic ft of gas can it hold?

Solution ➤

$V = \frac{4}{3}\pi r^3$

$\quad = \frac{4}{3}\pi 27^3$

$\quad = \frac{4}{3}(19,683)\pi$

$\quad = 26,244\pi$ cubic ft $\approx 82,448$ cubic ft ❖

The Surface Area of a Sphere

Cartography   In previous lessons you analyzed the surface areas of three-dimensional figures by unfolding them to form a net on a flat surface. But, as map makers well know, a sphere will not unfold smoothly onto a flat surface. The most common map of the world uses the Mercator projection of the Earth's surface onto a flat plane. On this map, the landmasses near the North and South Poles, such as Greenland, appear much larger than they should.

A formula for the surface area can nevertheless be derived using some clever techniques.

Imagine the surface of a sphere as a large number of congruent regular polygons, as in the geodesic dome in the photo. The smaller these polygons are, the more closely they will approximate the surface area of the sphere.

Doppler radar

Consider each polygon to be the base of a pyramid with its vertex at the center of the sphere. The volumes of these pyramids taken together will approximate the volume of the sphere. The height of each pyramid is the radius of the sphere. Therefore, the volume of each pyramid is $\frac{1}{3}Br$, where B is the area of the base of the pyramid and r is the radius of the sphere. So,

$$\text{Volume of sphere} \approx \frac{1}{3}B_1 r + \frac{1}{3}B_2 r + \ldots \frac{1}{3}B_n r$$
$$= \frac{1}{3}r(B_1 + B_2 + \ldots B_n).$$

If the total area of the bases of the pyramids is assumed to equal the surface area S of the sphere, then

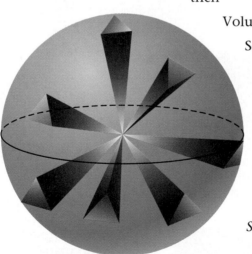

$$\text{Volume of sphere} = \frac{1}{3}r(S).$$

Solving for the surface area S,

$$S = \frac{3V}{r},$$

where V is the volume of the sphere.

You now have a formula for the surface area of a sphere in terms of its volume. Substituting the formula for V from above,

$$S = 3\frac{V}{r}$$
$$= 3\,\frac{\left(\frac{4}{3}\pi r^3\right)}{r} \text{ (Substitution)}$$
$$= 4\pi r^2$$

SURFACE AREA OF A SPHERE

The surface area S of a sphere with radius r is

$$S = 4\pi r^2$$

7.6.2

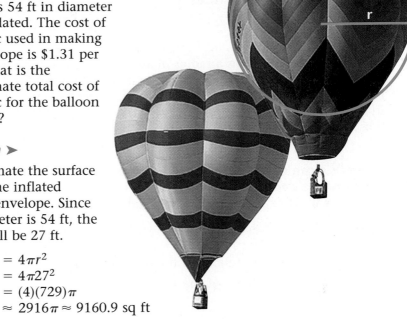

EXAMPLE 2

Hot Air Ballooning The envelope of a hot-air balloon is 54 ft in diameter when inflated. The cost of the fabric used in making the envelope is $1.31 per sq ft. What is the approximate total cost of the fabric for the balloon envelope?

Solution ➤

Approximate the surface area of the inflated balloon envelope. Since the diameter is 54 ft, the radius will be 27 ft.

$$S = 4\pi r^2$$
$$= 4\pi 27^2$$
$$= (4)(729)\pi$$
$$\approx 2916\pi \approx 9160.9 \text{ sq ft}$$

Next, multiply the surface area of the materials by the cost to find the approximate cost of the fabric.

9160.9 sq ft · $1.31 ≈ $12,000 ❖

EXERCISES & PROBLEMS

Communicate

1. Explain how to find the surface area of a sphere.

2. Explain how to find the volume of a sphere.

3. What happens to the surface area of a sphere when the radius is doubled?

4. What happens to the volume of a sphere when the radius is doubled?

5. Explain how the "Pythagorean" Right-Triangle Theorem is used to help derive the formula for the volume of a sphere.

Practice & Apply

Find the volume and surface area of each sphere. All measurements are in centimeters.

6. $r = 8$ **7.** $r = 7$ **8.** $d = 10$ **9.** $r = 9$

10. Sports   A basketball has a radius of approximately 4.75 in. when filled. How much material is needed to make one (round to the nearest tenth)? How much air will it hold? If the basketball is stored in a box that is a cube whose edges are 9.5 in., what percent of the box is not filled by the basketball?

11. Sports   A can of tennis balls has 3 balls in it. The approximate height of the stacked balls is 9 in. and the approximate radius of each ball is 1.5 in. How much space do the tennis balls occupy (to the nearest tenth)?

12. Sports   Find the surface area of the baseball.

13. Sports   Find the volume of the softball.

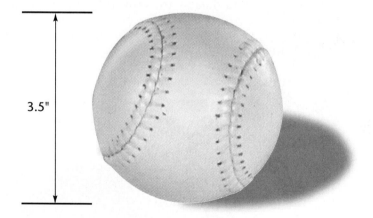

3.5"

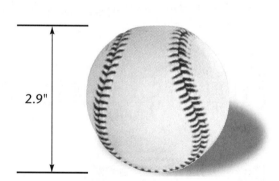

2.9"

Food   Hosea buys an ice cream cone. The ice cream is a sphere with a radius of 1.25 in. The cone has a height of 8 in. and a diameter of 2.5 in.

14. What is the volume of the ice cream?

15. What is the volume of the cone?

16. If the ice cream sits on top of the cone to form a hemisphere, what is the total surface area of the ice cream cone?

17. Suppose that a cube and a sphere both have a volume of 1000 cubic units. Compare the surface areas.

18. Suppose a cube and a sphere both have the same surface area. Compare their volumes. Let the surface area equal 864 sq units.

19. What is the volume of the largest ball that would fit into a cubical box with edges 12 in.?

Look Back

Find the surface area of each figure.   [Lessons 7.2, 7.4, 7.5]

20. right rectangular prism: l = 3 in., w = 10 in., h = 5 in.

21. right cone: r = 15 cm, slant height = 45 cm

22. right cylinder: r = 9 m, h = 10 m

Find the volume of each figure.   [Lessons 7.3, 7.4, 7.5]

23. oblique rectangular pyramid: l = 15 ft, w = 7 ft, h = 12 ft

24. right cone: r = 5 in., h = 10 in.

25. oblique cylinder: r = 7.5 m, h = 20 m

Look Beyond

A cube is inscribed inside a sphere with a diameter of 2 units.

26. What is the volume of the cube?

27. What is the ratio of the volume of the sphere to the volume of the cube?

28. What is the surface area of the cube?

29. What is the ratio of the surface area of the sphere to the surface area of the cube?

TREASURE
of King Tut's Tomb

Funerary mask of gold, lapis, cloisonné, and quartz

GEM-STUDDED RELICS AMAZE EXPLORERS

SPECIAL CABLE TO THE NEW YORK TIMES

LONDON, Nov. 30, 1922 — The Cairo correspondent of *The London Times*, in a dispatch to his paper, describes how Lord Carnarvon and Howard Carter unearthed below the tomb of Rameses VI, near Luxor, two rooms containing the funeral paraphernalia of King Tutankhamun, who reigned about 1350 B.C.

A sealed outer door was carefully opened, then a way was cleared down some sixteen steps and along a passage of about twenty-five feet, A door to the chambers was found to be sealed as the outer door had been and as the

Excerpt from Howard Carter's notes.

The excitement of unearthing the tomb was followed by three years of hard work. Unpacking a box of priceless relics took three painstaking weeks. Items stored for thousands of years are easily damaged forever if not handled with great care.

A crumpled robe in Box 21 presented a dilemma for Howard Carter. Should he preserve the garment as is and lose the chance to learn its full design? Or should he handle it and sacrifice the cloth in order to examine the complete robe?

...by sacrificing the cloth, picking it carefully away piece by piece, we could recover, as a rule, the whole scheme of decoration. Later, in the museum, it will be possible to make a new garment of the exact size, to which the original ornamentation—bead-work, gold sequins, or whatever it may be—can be applied. Restorations of this kind will be far more useful, and have much greater archaeological value, than a few irregularly shaped pieces of preserved cloth and a collection of loose beads and sequins.

Cooperative Learning

The bead pattern on the ceremonial robe, enlarged

The robe in Box 21, which had a cylindrical shape, was covered by a pattern of glass beads and gold sequins, with a lower band of pendant strings. Howard Carter recorded how he figured out the robe's size. For the following activity, some of the data have been removed from his notes.

1. Figure out the missing numbers in Howard Carter's notes and show how you found them.

2. Based on its size, do you think the robe was an adult's garment or a child's? Why?

We know the distance between the pendant strings of the lower band to have averaged 8 mm. There are 137 of these pendants. Therefore the circumference of the lower part of the garment must have been __(a)__ m, which would make one width about __(b)__ cm.

Now there are 3054 gold sequins. Three sequins in the pattern require 9 square centimeters. Therefore, total area of network = __(c)__ square meters. As circumference was __(d)__, this would make the height work out at about __(e)__ cm.

Thus size of garment would work out at about __(f)__ X __(g)__.

Incised cartouche of King Tutankhamun, gold, and gems

Gold collar of Vulture Goddess, Nekhbet

LESSON 7.7

Exploring

Three-Dimensional Symmetry

Why So far, the definitions of symmetry have been limited to a plane. But three-dimensional figures, like the tiger in the photo, may also have symmetry. You are ready to extend your ideas of symmetry to solid or three-dimensional figures.

Tiger, Tiger, burning bright,
In the forests of the night;
What immortal hand or eye,
Could frame thy fearful symmetry?
                              —*William Blake*

A three-dimensional figure may be reflected through a plane, just as a two-dimensional figure may be reflected through a line. What happens to each point in the pre-image of a figure as a result of a reflection through a plane? In Exploration 1 you will investigate this question.

Exploration 1 *Reflections in Coordinate Space*

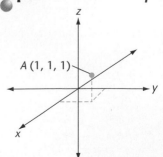

You will need
No special tools

Part I

1 Graph the point $A(1, 1, 1)$ in a three-dimensional coordinate space. Use dashed lines to make the location of the point in space evident.

2 Multiply the x-coordinate of point A by -1 and graph the resulting point. Label the new point A'.

3 The point A' is the reflection image of point A through a plane. Name this plane. If you connect points A and A' to form a segment $\overline{AA'}$, what is the relationship of this segment to the reflection plane?

4 Write your own definition for the reflection of a point through a plane in space.

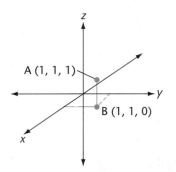

5 Experiment with other reflections of point A. What happens if you multiply the y-coordinates by -1? the z-coordinates?

6 Now study the reflection of an entire segment, such as $\overline{AB}$, through a plane. (For example, multiply the x-coordinates of points A and B by -1 and connect the endpoints of the resulting points.) Experiment with the reflections of other figures, such as cubes.

7 Write your own definition of the reflection of a figure in space through a plane.

Part II

1 Fill in the table below. Use the terms *front, back, left, right, top,* and *bottom* to describe the octant of a point.

		Octant of image	Coordinates of image
A(2, 3, 4)	Reflection image through the xy-plane	front-right-bottom	(2, 3, −4)
	Reflection image through the xz-plane	?	?
	Reflection image through the yz-plane	?	?
B(−4, 5, 6)	Reflection image through the xy-plane	?	?
	Reflection image through the xz-plane	?	?
	Reflection image through the yz-plane	?	?

2 Generalize your findings, given $P(x, y, z)$.

a. What are the coordinates of the reflection image of P through the xy-plane?

b. What are the coordinates of the reflection image of P through the xz-plane.

c. What are the coordinates of the reflection image of P through the yz-plane. ❖

Exploration 2 *Reflectional Symmetry in Space*

You will need
No special tools

1 The three-dimensional figures below have reflectional symmetry. Explain why. Where is the reflection "mirror" in each case?

2 Write your own definition of reflectional symmetry in space. Use your earlier definition of reflectional symmetry in the plane as a model. ❖

Exploration 3 *Rotations in Coordinate Space*

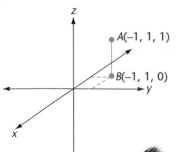

You will need
No special tools

1 Graph segment $\overline{AB}$ in a three-dimensional coordinate space. Use dashed lines to make the location of the segment evident.

2 Multiply the *x*- and *y*-coordinates of points *A* and *B* by -1 and graph the resulting points *A′* and *B′*. Connect the new points to form segment $\overline{A'B'}$.

3 Segment $\overline{A'B'}$ is a reflection image of segment $\overline{AB}$ about the *z*-axis. What is the reflection "mirror" in each case?

4 Through what point on the *z*-axis has endpoint *A* been rotated? endpoint *B*?

5 Imagine the segment rotating about the z-axis, as suggested by the picture. Does it seem to you that each of the rotation images of point *A* is in the same plane? of point *B*? What is the relationship between these planes and the z-axis?

6 Write your own definition of the rotation of a figure about an axis in space. ❖

•Exploration 4 *Rotational Symmetry in Space*

You will need
No special tools

1 These three-dimensional figures each have rotational symmetry. Explain why. Where is the axis of rotation in each case?

2 Write your own definition of rotational symmetry in space. Use your earlier definition of reflectional symmetry in the plane as a model. ❖

Revolutions in Coordinate Space

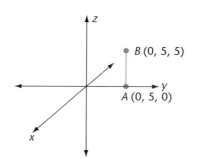

If you rotate a plane figure about an axis in space, the result is called a **solid of revolution**.

EXAMPLE

You are given the segment $\overline{AB}$ from *A*(0, 5, 0) to *B*(0, 5, 5). Sketch, describe, and give the dimensions of the figure that results if

A $\overline{AB}$ is rotated about the z-axis.

B $\overline{AB}$ is rotated about the y-axis.

Lesson 7.7 Exploring Three-Dimensional Symmetry **399**

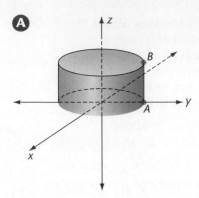

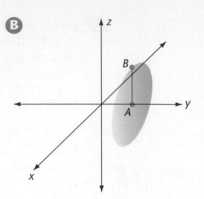

$\overline{AB}$ rotated about the z-axis forms a cylinder with radius 5 and height 5.

$\overline{AB}$ rotated about the y-axis forms a circle with radius 5, and its interior.

Try This Describe or sketch the solid of revolution that would be formed by rotating each of the plane figures about the dotted line.

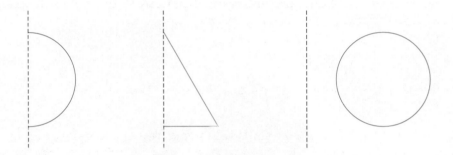

EXERCISES & PROBLEMS

Communicate

1. Describe the similarities and differences between three-dimensional symmetry and two-dimensional symmetry.

2. What geometric object is formed by rotating rectangle *ABCD* about $\overline{AD}$?

3. What geometric figure is formed by rotating $\triangle EFG$ about $\overline{EG}$?

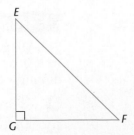

4. Describe the effect of multiplying each of the coordinates of the endpoints of a segment by −3.

5. Name some objects in your classroom that have three-dimensional rotational symmetry.

Practice & Apply

Draw three-dimensional coordinate systems and graph the segments with the given endpoints. Reflect each segment by multiplying each *y*-coordinate by −1.

6. (4, −2, 3); (−2, −3, 2) **7.** (−5, 2, 1); (1, 1, 1) **8.** (1, −2, −3); (−1, 5, 2)

What is the octant and coordinate of the reflection image if each of the points below is reflected over the *xy*-plane?

9. (6, 5, 8) **10.** (−2, 3, −1) **11.** (1, 1, 1)

What is the octant and coordinate of the reflection image if each point below is reflected over the *xz*-plane?

12. (6, −2, 8) **13.** (−4, −1, −1) **14.** (1, 0, 1)

15. Identify five physical objects that have rotational symmetry in space.

A potter is making pots according to certain patterns. For each half below, sketch what the complete pot will look like. (Rotate about the dotted line.)

16.

17.

18.

19.

For Exercises 20–23, segment $\overline{AB}$ has coordinates $A(5, 0, 10)$ and $B(5, 0, 0)$.

20. What figure is formed by rotating $\overline{AB}$ about the *x*-axis?

21. What is the area of the figure formed by rotating $\overline{AB}$ about the *x*-axis?

22. What figure is formed by rotating $\overline{AB}$ about the *z*-axis?

23. What is the volume of the figure formed by rotating $\overline{AB}$ about the *z*-axis?

For Exercises 24 and 25, $\overline{CD}$ has coordinates $C(4, 0, 0)$ and $D(0, 0, 4)$.

24. What figure is formed by rotating $\overline{CD}$ about the *z*-axis?

25. What is the volume of the figure formed by rotating $\overline{CD}$ about the *z*-axis?

26. 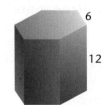 **Maximum/Minimum** The area of a right triangle for a given perimeter is maximized when the triangle is a 45-45-90 triangle. Suppose you rotate a 45-45-90 triangle about a leg to create a cone. Does the cone have the maximum volume for the given slant height? Why or why not? (See Exercises 25–33 in Lesson 7.5, p. 385.)

Look Back

Find the volume and surface area of each prism. **[Lesson 7.2]**

27.

7
11
15

Right-triangular prism

28.

6
12

Regular hexagonal prism

29. The truck's storage space measures 8 m × 5 m × 4 m. What is the volume of the storage space? **[Lesson 7.2]**

30. Find the surface area and volume of a right square pyramid with area of base $B = 36$ cm^2, height $h = 5$ cm, and slant height $l = 5.83$ cm. **[Lesson 7.3]**

31. Noticing the barbecue smokers next to the potting soil, Ms. Solis decides to make a planter of her old smoker at home. She notices that the center of a smoker comes to her waist, or half her height, which is 5'-0". Using this information to estimate the dimensions, determine how much soil it will take to fill the smoker to the centerline of the barrel from which it is made.   **[Lesson 7.4]**

Look Beyond

A spinning fan creates the illusion of being a solid. A strobe light, which emits a flash at certain time intervals, can "freeze" the action.

32. Suppose a fan is turning at 12 revolutions per second. How often will the strobe need to flash to make the fan appear to stop?

33. If a strobe flashes at 36 flashes per second, at what speeds can a fan be going and still appear to be stopped?

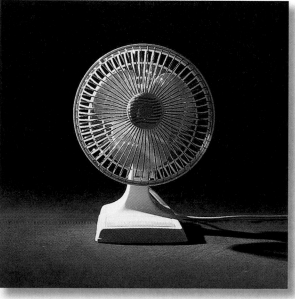

POLYHEDRA

A **polyhedron** is a geometric solid with polygons as faces. The plural of polyhedron is **polyhedra.**

Polyhedra are named for the number of faces they have. If all the faces of a polyhedron are congruent, the polyhedron is called a **regular polyhedron.** The five regular polyhedra shown below are also known as **Platonic solids** after Plato of ancient Greece.

| Tetrahedron (4 faces) | Hexahedron (6 faces) | Octahedron (8 faces) | Dodecahedron (12 faces) | Icosahedron (20 faces) |

Activity 1: Building Polyhedra

1. Copy each of the net patterns shown below to make models of the five Platonic solids. The tabs are for glue or tape.
2. Design your own pattern for a regular hexahedron and another Platonic solid of your choice. Verify your patterns by assembling the polyhedra.

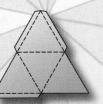

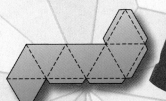

Tetrahedron   Hexahedron   Octahedron

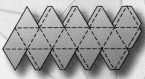

Dodecahedron   Icosahedron

Activity 2: "Pop-up" Dodecahedron

1. Make two copies of the pattern below out of cardboard. Fold and crease the cut-outs on the dotted lines.

2. Holding the two patterns together, place a rubber band around them as shown. Release your model to form a dodecahedron.

Chapter 7 Review

Vocabulary

altitude	359	net	360	right cone	379
annulus	387	oblique pyramid	366	right cylinder	373
axis	373	oblique cone	379	right prism	359
base	366	oblique cylinder	373	solid of revolution	399
cone	379	prism	359	sphere	387
lateral area	359	pyramid	366	surface area	359
lateral face	366	regular pyramid	366	vertex of a pyramid	366
lateral surface	373	right pyramid	366		

Key Skills and Exercises

Lesson 7.1

➤ **Key Skills**

Solve problems using the ratio of surface area to volume.

A cube has a volume of 27,000 cubic millimeters. What is the ratio of its surface area to its volume?

The length of the cube's side must be 30 mm ($30 \times 30 \times 30 = 27,000$). The area of a face of the cube is 900 mm². Multiply that by 6 to get the surface area, 5,400 mm². Thus the ratio is $5,400 \div 27,000$, or 0.2.

➤ **Exercise**

1. Find the ratio of surface area to volume for the rectangular prism and the cube. Which ratio is larger?

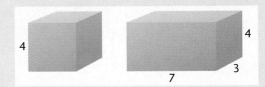

Lesson 7.2

➤ **Key Skills**

Find the surface area of a right prism. $S = L + 2B$ or $ph + 2B$

The surface area of the prism is the lateral area plus the area of the bases. Multiply the height, 20 cm, by the perimeter of the base, 12 cm, to get lateral area, 240 cm². The bases happen to be right triangles; the base area is $\frac{1}{2}$ (3×4), or 6 cm². Total surface area is $240 + 6 + 6 = 252$ cm².

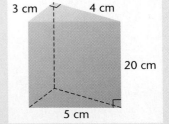

Find the volume of a prism. $V = Bh$

The volume of the prism equals the area of the base times the height of the prism. The triangular base has an area of 6 cm², and the height is 20 cm. Therefore, the volume of the prism is $6 \times 20 = 120$ cm³.

➤ **Exercises**

In Exercises 2–3, refer to the right rectangular prism.

2. Find the surface area of the prism. **3.** Find the volume of the prism.

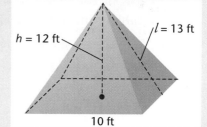

9 in.

23 in.

11 in.

Lesson 7.3

➤ Key Skills

Find the surface area of a right pyramid. $S = L + B$ or $\frac{1}{2}pl + B$

The surface area of the pyramid is the lateral area plus the area of the base. Multiply the slant height, 13 ft, by the perimeter of the base, 40 ft, and by $\frac{1}{2}$ to get lateral area, 260 ft². The base is a square; the base area is thus 100 ft². Total surface area is 260 + 100 = 360 ft².

$l = 13$ ft

$h = 12$ ft

10 ft

Find the volume of a pyramid. $V = \frac{1}{3}Bh$

The volume of the pyramid above equals the area of the base times the height times $\frac{1}{3}$, or $100 \times 12 \times \frac{1}{3} = 400$ ft³.

➤ Exercises

In Exercises 4–5, refer to the right regular pyramid.

4. Find the surface area of the pyramid.

5. Find the volume of the pyramid.

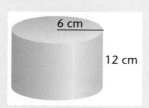

$h = 12$ ft $l = 14.8$ ft

20 ft

Lesson 7.4

➤ Key Skills

Find the surface area of a right cylinder. $S = L + 2B$, or $2\pi rh + 2\pi r^2$

The surface area of the cylinder above is the lateral area plus the area of the bases. Multiply the height, 35 m, by the circumference of the base, 44 m, to get lateral area, 1,540 m². The base area is πr^2, or approximately 154 m². Total surface area is 1,540 + 154 + 154 ≈ 1,848 m².

7 m

35 m

Find the volume of a cylinder. $V = Bh$, or $V = \pi r^2 h$

The volume of the cylinder above equals the area of the base times the height of the cylinder, or $154 \times 35 ≈ 5,390$ m³.

➤ Exercises

In Exercises 6–7, refer to the right cylinder.

6. Find the surface area of the cylinder.

7. Find the volume of the cylinder.

6 cm

12 cm

Lesson 7.5

➤ Key Skills

Find the surface area of a right cone.
$S = L + B$, or $S = \pi rl + \pi r^2$

The surface area of the cone is the lateral area plus the area of the base. Multiply the slant height, 5 in., by the radius of the base, 3 in., by π, ~3.14, to get lateral area, 47.1 in.². The base area is approximately 28.3 in.². Total surface area is 47.1 + 28.3 ≈ 75.4 in.².

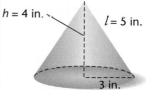

$h = 4$ in. $l = 5$ in.

3 in.

Find the volume of a cone. $V = \frac{1}{3}Bh$, or $V = \frac{1}{3}\pi r^2 h$

The volume of the cone above equals the area of the base times the altitude of the cone times $\frac{1}{3}$, or approximately 37.7 in.³.

➤ Exercises

In Exercises 8–9, refer to the right cone.

8. Find the surface area of the cone.

9. Find the volume of the cone.

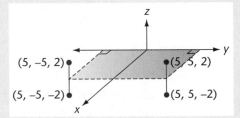

Lesson 7.6

➤ Key Skills

Find the surface area of a sphere. $S = 4\pi r^2$

For a sphere with radius 21 feet, the surface area is

$4 \times \frac{22}{7} \times (21)^2 = 5{,}544$ square feet.

Find the volume of a sphere. $V = \frac{4}{3}\pi r^3$

For a sphere with radius 21 feet, the volume is approximately

$\frac{4}{3} \times \frac{22}{7} \times (21)^3 = 38{,}808$ cubic feet.

➤ Exercises

10. Find the surface area of a sphere whose radius is 4 nanometers.

11. Find the volume of a sphere whose radius is 2.5 millimeters.

Lesson 7.7

➤ Key Skills

Reflect a line in a three-dimensional coordinate system.

In the coordinate system at the right, a segment with endpoints at (5, 5, –2) and (5, 5, 2) is reflected through the *xz*-plane.

Sketch a solid of revolution, given a line.

If you rotate the segment above about the *z*-axis, you produce the lateral surface of a cylinder of height 4 and radius 5 shown at the right.

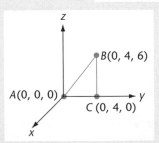

➤ Exercises

In Exercises 12–13, refer to the figure at the right.

12. Draw a reflection of segment $\overline{AB}$ through the *xy*-plane. Give the coordinates of the reflection's end points.

13. Find the volume of the figure created by rotating the figure around the *y*-axis.

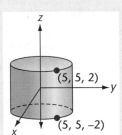

Applications

14. **Manufacturing** A processing plant needs storage tanks to hold at least half a million gallons of waste water. For a combination of visual and engineering reasons, the interior of the tanks must be no higher than 25 feet and no wider than 50 feet. How many cylindrical tanks must be built to hold the waste water? (1 cubic foot = 7.48 gallons)

15. Find the surface area of the truncated cone at the right.

Chapter 7 Assessment

1. Find the ratio of surface area to volume for a sphere with a radius of 120 feet.

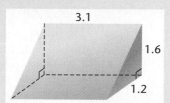

In Exercises 2–3, refer to the right rectangular prism.

2. Find the surface area of the prism.
3. Find the volume of the prism.

In Exercises 4–5, refer to the pyramid below.

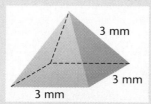

4. Find the surface area of the pyramid.
5. Find the volume of the pyramid.

In Exercises 6–7, refer to the right cylinder.

6. Find the surface area of the cylinder.
7. Find the volume of the cylinder.

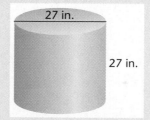

In Exercises 8–9, refer to the right cone below.

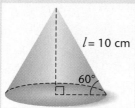

8. Find the surface area of the cone.
9. Find the volume of the cone.

10. Give the coordinates for the end points of the reflection through the yz-plane of the segment to the right.

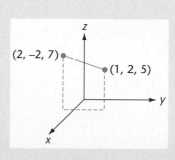

CHAPTER 8

Similar Shapes

What geometric idea is illustrated by the different sizes on these pages? The car and the model are not congruent because they do not have the same size. However, they do have the same shapes. Such figures are known as **similar figures.** You will find many opportunities to study similar figures both in nature and in art.

Artists and designers often enlarge or reduce figures without changing their shapes. For example, a muralist may cover an entire wall with a painting made from a small photograph. You will learn the mathematics of such procedures in this chapter.

LESSONS

8.1 Transformations and Scale Factors

8.2 *Exploring* Similar Polygons

8.3 *Exploring* Triangle Similarity Postulates

8.4 The Side-Splitting Theorem

8.5 Indirect Measurement, Additional Similarity Theorems

8.6 *Exploring* Area and Volume Ratios

Chapter Project
Indirect Measurement and Scale Models

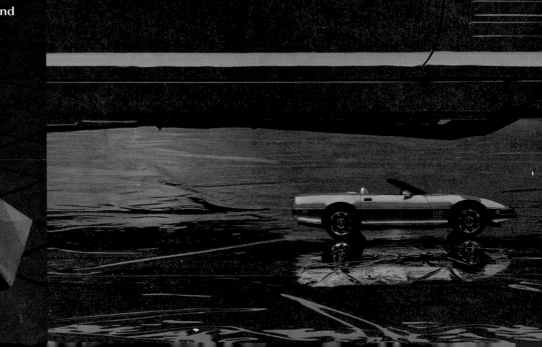

PORTFOLIO ACTIVITY

The use of grids for copying, enlarging, or reducing figures was known to the ancient Egyptians, as markings in tombs reveal. The method is still used today.

Using grids, draw an enlargement or reduction of a photo or design that you find interesting.

411

Exploring Transformations and Scale Factors

Why *The principle of the camera obscura was discovered by the Iraqi scientist Ibn al Haitham (965–1039 C.E.), probably from thinking about inverted images of objects projected through small holes in tents. Transformations can be used to explain why the images are inverted.*

The camera obscura, which was the forerunner of the modern camera, and the pinhole used by these students to observe an eclipse of the sun operate on the same principles.

Scale Factors

What happens to a point when you multiply both coordinates of a point by the same real number?

Pre-image   **Image**

$A(2, 3)$ $\rightarrow$ $A'(2 \times 4, 3 \times 4) = A'(8, 12)$

Multiply by 4

The result is a *transformation* of the original point. The number that multiplied the coordinates is called the **scale factor**.

•Exploration 1 *Transforming a Point Using a Scale Factor*

Geometry Graphics

You will need
Geometry technology or
Graph paper
Ruler and calculator (optional)

In this exploration, you will observe what happens when you transform a segment using different scale factors.

Part I

1 Draw segment $\overline{AB}$ on a coordinate plane, with endpoints $A(2, 4)$ and $B(6, 1)$. Multiply the coordinates of A and B by a scale factor of 2.

2 Plot your transformed points A' and B' on your graph. Connect them to form segment $\overline{A'B'}$.

3 Use scale factors of 0.5 and -1 to create additional new segments $\overline{A''B''}$ and $\overline{A'''B'''}$. Construct these segments on your graph and label each one with its scale factor.

Part II

1 Look at just the point A and its images A', A'', and A'''. What is the simplest geometric figure that contains all of these points and the *origin O as well*? Add this figure to your graph.

2 Use the Distance Formula or a ruler to find the distances OA and OA'. Find the ratio $\frac{OA'}{OA}$. How does this ratio compare with the scale factor that gives the point A'?

Repeat for all the transformations of point A. Then copy and complete the table.

A	Scale factor	A'	OA	OA', etc.	$\frac{OA'}{OA}$, etc.
$(2, 4)$	-1	?	?	?	?
$(2, 4)$	0.5	?	?	?	?
$(2, 4)$	2	?	?	?	?
$(2, 4)$	n	?	?	?	?

3 Write a conjecture about the effect of a scale factor transformation on a point. Include all your findings from Steps 1 and 2.

4 What will happen if you transform a point by a scale factor of n?

Pre-image		Image
$P(x, y)$	$\rightarrow$	$P'(x \times n, y \times n) = P'(xn, yn)$
	Multiply by n	

What is the distance of the original point from the origin? of the transformed point from the origin? What is the ratio of the second distance to the first distance?

Part III

 Consider each segment in the table below. Find the length of each of the segments. How do the lengths of the transformed segments compare to the length of the original segment $\overline{AB}$? Copy and complete the table.

A	B	Scale factor	A′, etc.	B′, etc.	Length AB	Length A′B′	$\frac{A'B'}{AB}$, etc.
(2, 4)	(6, 1)	−1	?	?	?	?	?
(2, 4)	(6, 1)	0.5	?	?	?	?	?
(2, 4)	(6, 1)	2	?	?	?	?	?
(2, 4)	(6, 1)	n	?	?	?	?	?

 Find the slope of each of the segments on your graph.

3 Write a conjecture about the effect of a scale factor transformation on a segment. Include all of your findings from Steps 1 and 2.

4 What is the effect of a scale factor transformation on a segment if the scale factor is n? Let $P(x_1, y_1)$ and $Q(x_2, y_2)$ be endpoints of the segment. What is the length of the original segment? of the transformed segment? What is the slope of each? How do these compare? ❖

Dilations

Muscles in the iris dilate the pupil to let in more light and contract to let in less light.

The transformations you have been exploring in this lesson are known as **dilations**. The name applies to all transformations of this kind. However, when transformed figures are reduced in size, they are often called **contractions**.

Every dilation has a point known as the **center of dilation**. In Exploration 1, that point was the origin.

If you want to use a center of dilation that is not at the origin, how can you adapt the procedure in Exploration 1 to your purposes? You will need to do some other transformations in the process.

•Exploration 2 *Constructing a Dilation*

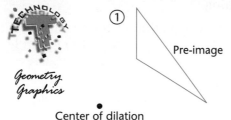

Geometry Graphics

①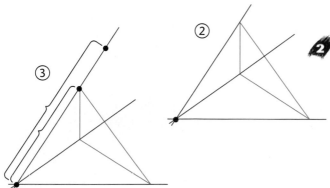

Pre-image

• Center of dilation

②

③

You will need
Geometry technology or
Straightedge and ruler

1 Draw a plane figure, such as a triangle. To construct a dilation of the figure, first decide on a center of dilation and a scale factor. Place the center of dilation somewhere on your drawing. Draw lines from the center of dilation through each vertex of the figure to be transformed.

2 Measure the distance from the center of dilation to a vertex. Multiply this distance by the scale factor.

Using the new distance you just determined, plot the image point on the line containing the center and the vertex. (Measure the distance from the center of dilation. If the number is negative, measure *away* from the pre-image point.)

3 Repeat Step 2 for each of the remaining pre-image points.

4 Connect the image points to form the transformed image of the original figure. ❖

CRITICAL *Thinking*

How could you construct an approximate dilation of a figure that was composed of curves rather than segments?

•Exploration 3 *The Position of Transformed Points*

You will need
No special tools

In Exploration 1, you may have noticed that the pre-images and images of a point you transformed using different scale factors were collinear, and that the line passed through the origin (0, 0).

ALGEBRA *Connection*

You can prove both of these results using the equation of a line,

$$y = mx + b,$$

where m is the slope of the line and b is the y-intercept.

Recall from algebra that when you are given points (x_1, y_1) and (x_2, y_2) on a line, the equation of the line can be written as

$$y = \left(\frac{y_2 - y_1}{x_2 - x_1}\right) x + b.$$

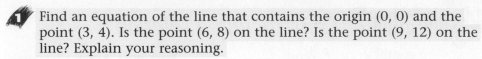

1 Find an equation of the line that contains the origin (0, 0) and the point (3, 4). Is the point (6, 8) on the line? Is the point (9, 12) on the line? Explain your reasoning.

2 Generalize as follows: Find the equation of the line that contains the points (0, 0) and (a, b). Is the point (2a, 2b) on the line? (3a, 3b)? (na, nb)? ❖

EXERCISES & PROBLEMS

Communicate

1. What is a scale factor? How does it affect the points in a transformation?

2. What is a dilation? How is a dilation different from other transformations you have studied?

3. What is the result of a dilation when the scale factor is one? when the scale factor is zero? when the scale factor is negative? when the scale factor is a fraction?

4. In the photo, what would the scale factor tell you about the relationship of the pre-image to the dilated images? Assume the center bowl is the pre-image.

Practice & Apply

In Exercises 5–7, the dashed-line pre-image has been transformed to form the solid-line image. What is the scale factor for each dilation?

5.

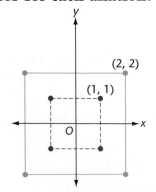

6.

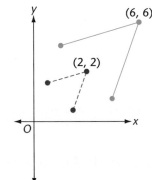

7.

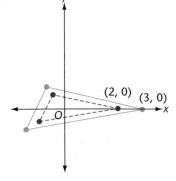

Transform the points on graph paper using the given scale factors.

8. (3, 5) scale factor 3

9. (−2, 6) scale factor $\frac{1}{3}$

10. (4, −12) scale factor $-\frac{1}{3}$

Trace the figures below onto your own paper. Draw a dilation of each figure. Choose a point for the center of dilation and use the scale factors indicated for each dilation.

11. scale factor 3   **12.** scale factor 4   **13.** scale factor −2   **14.** scale factor −3

 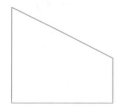 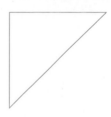 

Optics   In the drawing of a camera obscura, the projected image is an example of a dilation.

15. What part of the camera acts as the center of dilation?

16. Is the scale factor positive or negative? Explain your answer.

17. Explain why the projected image is inverted.

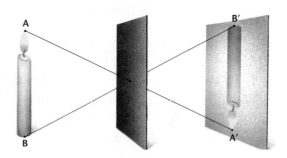

Draw the image of △ABC under the given mapping.

18. $(x, y) \rightarrow (2x, 2y)$

19. $(x, y) \rightarrow (3x, 3y)$

20. $(x, y) \rightarrow \left(1\frac{1}{2}x, 1\frac{1}{2}y\right)$

21. $(x, y) \rightarrow \left(\frac{3}{4}x, \frac{3}{4}y\right)$

22. What scale factor will determine the given dilation image of △ABC if the sides of the image are three times as long as the sides of the original figure?

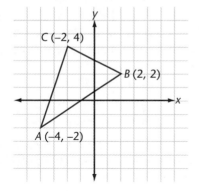

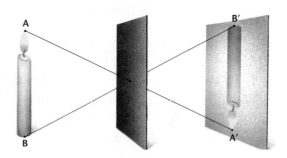 **Algebra**   Find an equation of the line that contains the following points.

23. (0, 0), (2, 5)   **24.** (3, 7), (1, 6)   **25.** (−8, 2), (4, 5)   **26.** (a, b), (4a, 4b)

The following points were transformed from point A using different scale factors. Show that these points are collinear.

27. A(5, 9)   A′(10, 18)   A″(25, 45)

28. B(6, 10)   B′(3, 5)   B″(24, 40)

29. $C\left(\frac{2}{3}, \frac{1}{5}\right)$ $C'\left(\frac{8}{3}, \frac{4}{5}\right)$ $C''\left(2, \frac{3}{5}\right)$

30. D(−3, 7)   D′(−6, 14) $D''\left(\frac{-3}{4}, \frac{7}{4}\right)$

Lesson 8.1   Exploring Transformations and Scale Factors   **417**

Draw a dilation of each figure using the given scale factor.

31. scale factor 3 **32.** scale factor $\frac{3}{4}$ **33.** scale factor 2 **34.** scale factor $\frac{1}{2}$

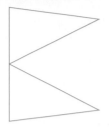 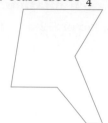 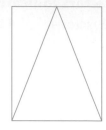 

Look Back

35. The base of an isosceles triangle is 6 meters and the legs are each 8 meters. Find the area and perimeter of the triangle. **[Lessons 5.1, 5.2]**

36. A leg of a 45-45-90 triangle is 7 cm long. What is the length of the hypotenuse? **[Lesson 5.4]**

37. Earth Science If the circumference of a great circle of the Earth is about 40,000 km, what is the surface area of the Earth? **[Lesson 7.6]**

38. Earth Science The atmosphere of the Earth has an altitude of about 550 km. Use Exercise 37 to find the volume of the Earth and its atmosphere. **[Lesson 7.6]**

39. A steel gas tank has the shape of a sphere. A radius of the inner surface of the tank is 2 feet long. The tank itself is made of $\frac{1}{4}$-inch-thick steel. Find the difference between the outside area and inside area of the tank. **[Lesson 7.6]**

The Earth's atmosphere, as photographed by a Russian cosmonaut.

40. If a gallon of paint will cover 400 square feet, how many gallons of paint will be needed to paint both the inside and the outside of the tank described in Exercise 39? **[Lesson 7.6]**

Look Beyond

41. Draw the dilation for the box using a scale factor of 2. Write the coordinates for the vertices.

42. How do the lengths of the sides of the pre-image compare to the lengths of the sides of the image?

43. Find the volumes of both boxes. How do they compare?

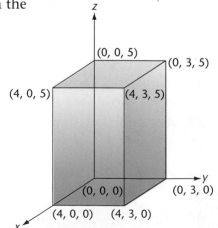

LESSON 8.2

Exploring Similar Polygons

why *Graphic designers use the geometric-concept of similarity to enlarge and reduce figures in flyers, posters, and newsletters.*

Figures that have the same shape, but not necessarily the same size, are called **similar figures**. In the exploration that follows, you will develop a definition for **similar polygons**.

•Exploration *Defining Similar Polygons*

You will need
Ruler and protractor

1. $\triangle ABC$ and $\triangle DEF$ seem to have the same shape. In mathematical terms, they appear to be similar. (Notation: $\triangle ABC \sim \triangle DEF$.) Measure the sides and angles in the two figures.

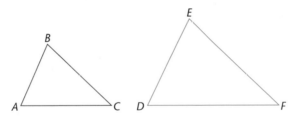

2. Are any of the angles congruent? If so, state which ones.

3 Find the following ratios of the lengths of the sides:

$$\frac{AB}{DE} \qquad \frac{BC}{EF} \qquad \frac{AC}{DF}$$

What do you notice about the ratios?

4 When the ratios of corresponding sides of two polygons are all the same, the sides are said to be **proportional**. Do the sides of the figures in Step 1 seem to be proportional? Explain.

5 Is proportionality of corresponding sides enough to guarantee that two polygons are similar? Is congruence of corresponding angles enough? Use the figures below to illustrate your answer.

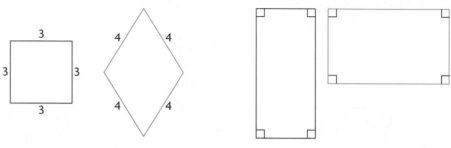

6 Write your own definition of *similar polygons*. Hint: Use the definition of *congruent polygons* as a pattern to make sure that the sides of the two figures are paired properly. ❖

CRITICAL *Thinking*

Can triangles be used in Step 4 of the exploration, instead of quadrilaterals, to illustrate the argument? Explain why or why not.

A formal definition of *similar polygons* is stated below. How does your definition from the exploration compare with it?

SIMILAR POLYGONS
Two polygons are similar if, and only if, there is a way of setting up a correspondence between their vertices so that

1. the corresponding angles are congruent, and
2. the corresponding sides are proportional.   **8.2.1**

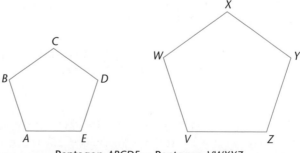

Pentagon *ABCDE* ~ Pentagon *VWXYZ*

Proportions and Scale Factors

Recall from Lesson 8.1 that transforming a figure by a scale factor (dilating the figure) causes the length of its sides to be multiplied by that scale factor. What scale factor transforms $\triangle ABC$ to $\triangle XYZ$? $\triangle XYZ$ to $\triangle ABC$?

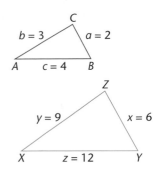

The sides of $\triangle ABC$ and $\triangle XYZ$ are in proportion. That is,

$$\frac{x}{a} = \frac{6}{2} = 3,$$

$$\frac{y}{b} = \frac{9}{3} = 3,$$

$$\frac{z}{c} = \frac{12}{4} = 3,$$

and so, $\dfrac{x}{a} = \dfrac{y}{b} = \dfrac{z}{c}$,

which is the required condition for proportionality of sides. Notice that each of the ratios (the fractions in the above example) reduces to 3, which is the scale factor of the transformations.

It is often useful to think of similar figures in terms of dilations. When you do, you may find it convenient to place the sides of the pre-image figure in the denominators of the fractions and the sides of the image figure in the numerators as in the example above. Then, the value of each fraction is the scale factor. (But, as you will see later, either way of arranging numerators and denominators results in a true statement.)

Similiar polygons appear in many structures.

Try This Find the scale factor that would transform $\triangle ABC$ into $\triangle DEF$.

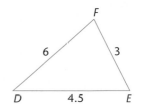

 CRITICAL *Thinking* If you are given that the sides of two figures are proportional, is it necessary to find all the ratios of the sides to determine the scale factor? Explain why or why not.

The Properties of Proportions

When working with similar figures it is often helpful to know the following properties of proportions.

ALGEBRA
Connection

PROPERTIES OF PROPORTION

Cross-Multiplication Property

If $\frac{a}{b} = \frac{c}{d}$ and $b, d \neq 0$, then $ad = bc$.

8.2.2

Reciprocal Property

If $\frac{a}{b} = \frac{c}{d}$ and $a, b, c, d \neq 0$, then $\frac{b}{a} = \frac{d}{c}$.

8.2.3

Exchange Property

If $\frac{a}{b} = \frac{c}{d}$ and $a, b, c, d, \neq 0$, then $\frac{a}{c} = \frac{b}{d}$.

8.2.4

"Add-One" Property

If $\frac{a}{b} = \frac{c}{d}$ and $b, d \neq 0$, then $\frac{a+b}{b} = \frac{c+d}{d}$.

8.2.5

APPLICATION

Some students are sizing photos for the yearbook. An original slide measures 1.25 inches wide by 0.75 inches long. In the yearbook, the photo needs to fill a space that is 5 inches wide. How would the students determine the length of the sized photo?

To find the length, they would set up a proportion and solve for the missing length. $\frac{x}{5} = \frac{0.75}{1.25}$

Method A

By observing the denominators, determine that the scale factor is 4. Multiply the numerator and denominator by this factor.

$$\frac{x}{5} = \frac{0.75 \times 4}{1.25 \times 4} = \frac{3}{5} \quad \text{So, } x = 3.$$

Method B

$$\frac{x}{5} = \frac{0.75}{1.25}$$

$$1.25x = 3.75$$

$$x = 3 \text{ in. long}$$

Using the Reciprocal Property

ALGEBRA
Connection

You are given the following proportionality statement for two triangles.

$$\frac{a}{x} = \frac{b}{y} = \frac{c}{z}$$

Is the following statement also true?

$$\frac{x}{a} = \frac{y}{b} = \frac{z}{c}$$

To answer the question, consider the first two members of the original statement.

$$\frac{a}{x} = \frac{b}{y}$$

By the Reciprocal Property, you get the following proportion.

$$\frac{x}{a} = \frac{y}{b}$$

Apply the Reciprocal Property to the second and third members of the original statement to obtain the following.

$$\frac{y}{b} = \frac{z}{c}$$

Then you can conclude that $\frac{x}{a} = \frac{y}{b} = \frac{z}{c}$.
(Why?)

Using the Exchange Property

You are given the following proportionality statement for two triangles.

$$\frac{a}{x} = \frac{b}{y} = \frac{c}{z}$$

By using the Exchange Property on two members of the statement at a time, you can show that the following statements are also true.

$$\frac{a}{b} = \frac{x}{y} \qquad\qquad \frac{b}{c} = \frac{y}{z} \qquad\qquad \frac{a}{c} = \frac{x}{z}$$

The three statements just proven show another way of thinking about similarity. That is, the parts of one figure are in the same ratios within the figure as the corresponding parts within the other figure.

Architecture Amber and Adrienne are making a scale model of a building with a rectangular foundation. If they want the longer sides of the model to be 24 inches, what should the length of the shorter sides be?

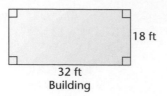

32 ft
Building

Since the model and the building must have the same shape, the figures are similar.

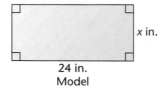

24 in.
Model

$$\frac{x}{18} = \frac{24}{32}$$

$$\frac{x}{18} = \frac{3}{4}$$

$$4x = 54$$

$$x = 13.5 \text{ in.} \quad \blacklozenge$$

Exercises & Problems

Communicate

1. If $\triangle RTW \sim \triangle IOU$, name all pairs of proportional sides.

2. If quadrilateral $KMPQ \sim$ quadrilateral $RTAW$, show the ratios between the sides.

3. In $\triangle FRG$ and $\triangle NBY$, $\frac{FR}{BY} = \frac{RG}{YN} = \frac{FG}{BN}$. State the similarity showing the correct correspondence.

4. In $\triangle ZXC$ and $\triangle VML$, $\frac{CX}{ML} = \frac{XZ}{LV} = \frac{ZC}{VM}$. State the similarity showing the correct correspondence.

5. If $\triangle WER \cong \triangle POI$, is $\triangle WER \sim \triangle POI$? Explain.

6. Use the diagram at right to determine whether quadrilateral $GHJK \sim$ quadrilateral $KLMN$. Explain your conclusion.

7. Explain why the "Add-One" Property has that name.

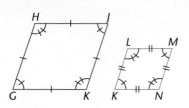

Practice & Apply

8. Write the ratio $\frac{3}{4} = \frac{15}{20}$ using the Cross-Multiplication Property.

9. Write the ratio $\frac{5}{2} = \frac{10}{4}$ using the Reciprocal Property.

10. Write the ratio $\frac{2}{5} = \frac{4}{10}$ using the Exchange Property.

11. Write the ratio $\frac{6}{9} = \frac{2}{3}$ using the "Add-One" Property.

12. If $\frac{x}{4} = \frac{y}{8}$, find $\frac{x}{y}$.

Algebra Solve for x.

13. $\frac{6x}{24} = \frac{27}{9}$

14. $\frac{4.8}{x} = \frac{6}{8.4}$

15. $\frac{\frac{2}{5}}{8} = \frac{\frac{7}{10}}{x}$

16. $\frac{6}{x} = \frac{x}{150}$

17. $\frac{3}{x-4} = \frac{7}{x+5}$

18. $\frac{5-2x}{12} = \frac{3x+1}{4}$

19. Rectangle $TGHF \sim$ rectangle $NBKJ$. Use proportions to find GH.

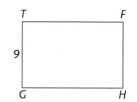

Algebra Tell whether the proportion is true for all values of the variables. If the proportion is false, give a numerical counterexample.

20. If $\frac{x}{y} = \frac{r}{s}$, then $\frac{x+c}{y} = \frac{r+c}{s}$.

21. If $\frac{x}{y} = \frac{r}{s}$, then $\frac{x+a}{y+a} = \frac{r+a}{s+a}$.

22. If $\frac{x}{y} = \frac{r}{s}$, then $\frac{x}{x+y} = \frac{r}{r+s}$.

23. If $\frac{x}{y} = \frac{r}{s} = \frac{m}{n}$, then $\frac{x}{y} = \frac{x+r+m}{y+s+n}$.

The Aransas National Wildlife Refuge, on the Texas coast, contains important nesting sites for endangered whooping cranes. The delicate salt-marsh environment must be carefully maintained in order to help this species, and other species of wildlife, survive.

Wildlife Management   Use the map for Exercises 24–26.

24. Use the scale of miles and the map above to estimate the number of square miles contained in the wildlife refuge. If there are 640 acres in 1 square mile, how many acres does the refuge contain? Describe your method for estimating the area of the refuge.

25. Animal overpopulation within the refuge must be monitored in order to prevent diseases and environmental damage. A deer census reveals that there are an average of 3 deer on every 20 acres of the refuge. Give an estimate for the number of deer that live in the wildlife refuge.

26. Most of the endangered whooping cranes spend their summers in Canada and their winters in or near the Aransas National Wildlife Refuge. If there are 123 whooping cranes alive in the wild and 98 of them are spotted inside the wildlife refuge, what percentage of the whooping crane population is inside the refuge?

27. Marine Biology   In a marine environment, 1 pound of sea water consists of 6.5% minerals and organic material. How many pounds of pure water are there in 57 pounds of sea water?

28. Landscaping Using the landscape design plans and the given scale, determine how many feet apart the trees must be planted. Express your answer in meters and feet (1m = 3.28 ft). Scale: 1 cm = 4.5 m.

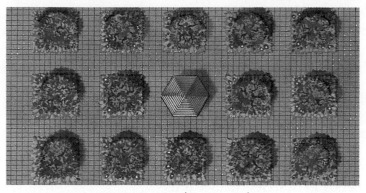

1.7 cm

29. Fine Art Brenda is attempting to paint an accurate reproduction of the *Mona Lisa* from a print that hangs in her art class. The dimensions of the print are 16 in. by 24 in., but Brenda's canvas is 15 in. by 20 in. What modifications must she make to her canvas to ensure that her painting is proportional to the print?

30. Fine Art On the classroom print of the *Mona Lisa*, the length of the woman's face from the hairline to the chin is 6 in. The width of the face is 4 in. What width and length must Brenda use for the dimensions of the woman's face in her reproduction?

31. Advertising An athletic shoe manufacturer wants to put the company logo on a billboard. An artist has an 8-inch-by-12-inch picture of the logo and must reproduce it onto the billboard using a scale factor of 25. What are the dimensions of the billboard logo?

32. Model-Building An artist builds a scale model of a clipper ship. The lifeboats on the model ship are 1 inch long, but on the real clipper ship they were about 24 feet long. If the length of the model is 10 inches, what was the length of the real ship?

33. What is the scale factor used in Exercise 32 for the model ship?

34. The perimeter of a rectangle is 60 cm. The ratio of the base to the altitude of the rectangle is 5 to 7. Find the lengths of the base and altitude.

35. 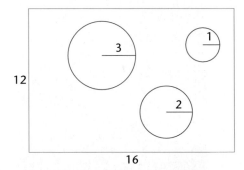 **Portfolio Activity** The ratio $\frac{1 + \sqrt{5}}{2}$ is called the Golden Ratio because rectangles with length and width in the ratio $\frac{l}{w} = \frac{1 + \sqrt{5}}{2}$ are thought to be pleasing to the eye. Make a collage of rectangles whose lengths and widths use this ratio.

Look Back

36. **Algebra** The angles of a triangle have degree measures $x + 5$, $5x + 12$, and $2x + 3$. Find the measures of the angles. **[Lesson 3.6]**

37. Find the measure of the base angles of an isosceles triangle if its vertex angle is 92°. **[Lesson 4.4]**

38. A right triangle has legs of length 5 cm and 7 cm. Find the length of its hypotenuse. **[Lesson 5.4]**

39. A right triangle has a hypotenuse of length 7 cm and a leg of length 5 cm. Find the length of the other leg. **[Lesson 5.4]**

40. **Probability** A blindfolded person throws a dart at the target at right. Assuming that it is equally likely that the dart will land anywhere on the rectangle, find the probability of the following: **[Lesson 5.7]**

  a. The person will hit a circle.

  b. The person will hit the biggest circle.

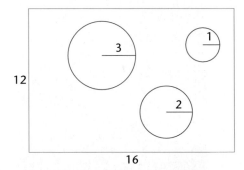

Look Beyond

Indirect Measurement Anthony wants to find the width of the river. First he pulls the visor of his cap down over his eye until a spot R, on the opposite shore, is in his line of vision. Without changing the position of his cap, he turns and sights along the visor to another spot, A, on his side of the river.

41. Explain how Anthony can now find the width of the river.

42. Which segments in the figure will have the same length as $\overline{NR}$?

43. Which postulate can be used to prove that $\triangle SNR \cong \triangle SNA$? Explain your answer.

44. What are some possible problems with using this method of indirect measurement? How accurate do you think this method is?

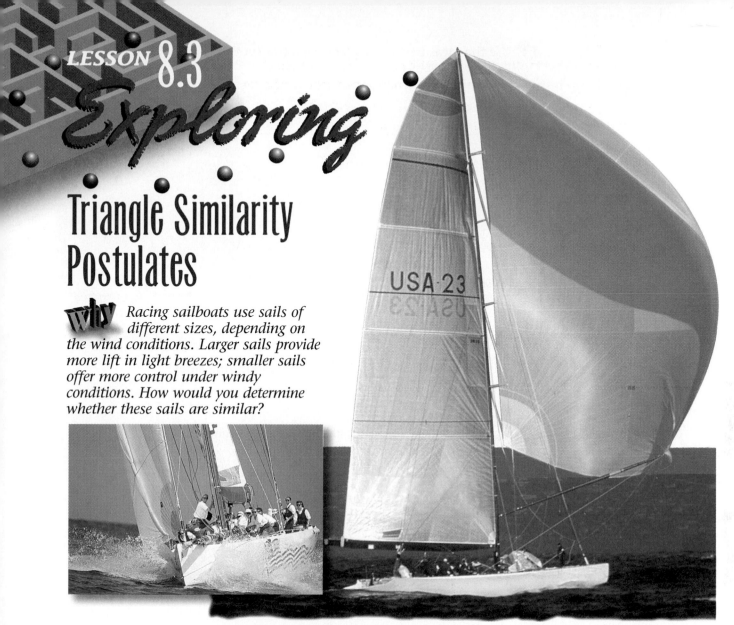

LESSON 8.3

Exploring

Triangle Similarity Postulates

Why *Racing sailboats use sails of different sizes, depending on the wind conditions. Larger sails provide more lift in light breezes; smaller sails offer more control under windy conditions. How would you determine whether these sails are similar?*

The explorations below suggest different shortcuts for determining whether triangles are similar. The shortcuts can be formalized into triangle similarity postulates. To provide a basis for the explorations, recall the definition of similar polygons from the previous lesson (Definition 8.2.1).

•Exploration 1 *Similar Triangles Postulate 1*

Geometry Graphics

You will need
Geometry technology or
Ruler, protractor, and compass

 1 Construct △*ABC* with m∠*A* = 45° and m∠*B* = 65°. Measure the sides and remaining angle.

2 Construct △*DEF* with m∠*D* = 45°, m∠*E* = 65°, and sides greater than the sides of △*ABC*. Measure the sides and remaining angle.

Lesson 8.3 Exploring Triangle Similarity Postulates **429**

3 Use the measures to calculate the ratios on the chart below.

	△DEF	△ABC	Ratio of the sides $\frac{\triangle DEF}{\triangle ABC}$
Measures of corresponding sides	DE = ___?___ EF = ___?___ DF = ___?___	AB = ___?___ BC = ___?___ AC = ___?___	$\frac{DE}{AB}$ = ___?___ $\frac{EF}{BC}$ = ___?___ $\frac{DF}{AC}$ = ___?___
Measures of corresponding angles	m∠D = 45° m∠E = 65° m∠F = ___?___	m∠A = 45° m∠B = 65° m∠C = ___?___	

4 What is the relationship between corresponding angles?

5 What is the relationship between corresponding sides?

6 Are the triangles similar according to the definition of *similar polygons*?

7 State a postulate for similarity between two triangles based on the result of the exploration. ❖

Try This   Tell whether each pair of triangles is similar and why.

1.

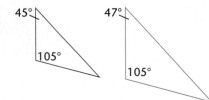

2.

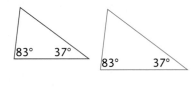

Exploration 2 *Similar Triangles Postulate 2*

You will need

Geometry technology or
Ruler, protractor, and compass

Geometry Graphics

1 Construct △ABC with
AB = 2 cm, BC = 3 cm,
and AC = 4 cm. Measure
the angles.

2 Construct △DEF with
DE = 6 cm, EF = 9 cm,
and DF = 12 cm. Measure
the angles.

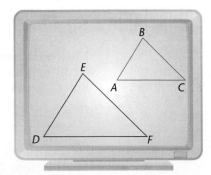

3 Use the measures to calculate the ratios on the chart below. For example, $\overline{DE}$ corresponds to $\overline{AB}$, so the ratio is $\frac{6}{2}$ or $\frac{3}{1}$.

	△*DEF*	△*ABC*	Ratio of the sides $\frac{\triangle DEF}{\triangle ABC}$
Measures of corresponding sides	$DE = 6$ $EF = 9$ $DF = 12$	$AB = 2$ $BC = 3$ $AC = 4$	$\frac{3}{1}$ $\frac{3}{1}$?
Measures of corresponding angles	m∠D = ? m∠E = ? m∠F = ?	m∠A = ? m∠B = ? m∠C = ?	

4 What is the relationship between corresponding angles?

5 What is the relationship between corresponding sides?

6 Are the triangles similar according to the definition of *similar polygons*?

7 State a postulate for similarity between two triangles based on the result of the exploration. ❖

Try This Tell whether each pair of triangles is similar and why.

1.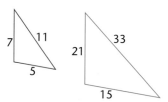

2.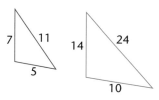

•Exploration 3   *Similar Triangles Postulate 3*

You will need

Geometry technology or Ruler and protractor

Geometry Graphics

1 Construct △*ABC* with $AB = 3$ cm, m∠$B = 90°$, and $BC = 4$ cm. Measure the sides and the angles.

2 Construct △*DEF* with $DE = 6$ cm, m∠$E = 90°$, and $EF = 8$ cm. Measure the sides and the angles.

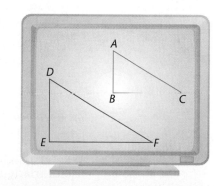

3 Use the measures to calculate the ratios on the chart below. For example, $\overline{DE}$ corresponds with $\overline{AB}$, so the ratio is $\frac{8}{4}$ or $\frac{2}{1}$.

	△DEF	△ABC	Ratio of the sides $\frac{\triangle DEF}{\triangle ABC}$
Measures of corresponding sides	$DE = 6$	$AB = 3$	$\frac{2}{1}$
	$EF = 8$	$BC = 4$	$\frac{2}{1}$
	$DF =$ __?__	$AC =$ __?__	__?__
Measures of corresponding angles	$m\angle D =$ __?__	$m\angle A =$ __?__	
	$m\angle E =$ __?__	$m\angle B =$ __?__	
	$m\angle F =$ __?__	$m\angle C =$ __?__	

4 What is the relationship between corresponding angles?

5 What is the relationship between corresponding sides?

6 Are the triangles similar according to the definition of *similar polygons*?

7 State a postulate for similarity between two triangles based on the result of the exploration. ❖

Try This Tell whether each pair of triangles is similar and why.

1.

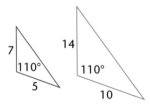

2.

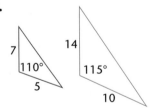

The three similarity postulates you explored above could be called the Angle-Angle, Side-Side-Side, and Side-Angle-Side postulates.

Summary

Two triangles are similar if
1. two pairs of corresponding angles are congruent, or
2. all three pairs of corresponding sides are in the same proportion, or
3. two pairs of corresponding sides are in the same proportion and the included angles are congruent.

CRITICAL
Thinking

Why are Angle-Side-Angle and Angle-Angle-Side not included in a list of useful triangle similarity postulates?

EXERCISES & PROBLEMS

Communicate

1. Explain the similarity postulate that you discovered in Exploration 1. What would you call this postulate?

2. Explain the similarity postulate that you discovered in Exploration 2. What would you call this postulate?

3. Explain the similarity postulate that you discovered in Exploration 3. What would you call this postulate?

4. Use a counterexample to show that if corresponding angles of one quadrilateral are congruent to corresponding angles of a second quadrilateral, then the quadrilaterals are not necessarily similar.

Practice & Apply

Can each pair of triangles be proven similar? Explain why or why not.

5.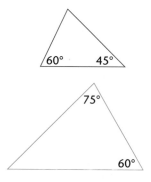

6.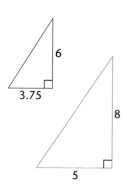

7.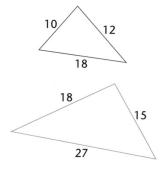

8. $\triangle ABC \sim \triangle DEF$. The perimeter of isosceles triangle ABC is 99 and the perimeter of isosceles triangle DEF is 48. If the length of the base of $\triangle ABC$ is 33, what is the length of the base of $\triangle DEF$?

9. The ratio of the corresponding sides of two triangles is equal to the ratio of the perimeters. If the perimeters of two triangles are 16 cm and 24 cm, what is the ratio of the perimeters? What is the ratio of the sides?

Use △ABC to complete Exercises 10 and 11.

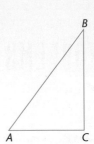

10. Draw a triangle in which each side is twice as long as each side of △ABC. Is this triangle similar to △ABC? Why or why not?

11. Draw a triangle in which each side is 3 cm longer than each side of △ABC. Is this triangle similar to △ABC? Why or why not?

In Exercises 12 and 13, indicate which figures are similar and explain why.

12.

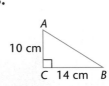

13.

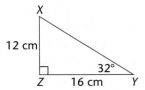

14. Can each set of triangles be proven similar? Why or why not?

a.

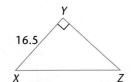

b.

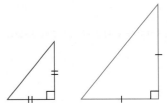

15. △ABC ~ △XYZ. Find the lengths of sides $\overline{XZ}$ and $\overline{YZ}$.

16. What is the ratio of the perimeters of the two similar triangles? What is the ratio of the corresponding sides?

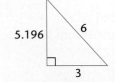

Surveying Surveyors sometimes use similar triangles to measure inaccessible distances. A surveyor could find distance *AB* by setting up similar triangles △*ABC* and △*EDC*. Assuming that all lengths may be directly measured except *AB*, the surveyor can use these measures to set up a proportion and solve for *AB*.

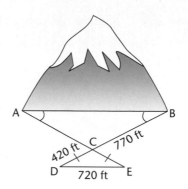

17. How does the surveyor make △*ABC* similar to △*EDC*?

18. Set up a proportion and solve for *AB*.

19. Find the areas of the two triangles in Exercise 17. Compare the ratio of the areas to the ratio of the perimeters. What do you notice?

Look Back

20. In the diagram lines *j* and *k* are parallel; $\overline{AB} \cong \overline{BC}$. Find *v*, *x*, *y*, and *z*. **[Lesson 3.3]**

21. Find the distance between the points (−6, 2) and (3, 8). **[Lesson 5.6]**

Algebra In Exercises 22–25, the fine for speeding in Marshall County is found using the function $f(n) = 40 + 15n$. The variable *n* is the number of miles per hour over the speed limit and *f(n)* is the fine for going *n* miles per hour over the limit.

22. Find the *y*-intercept of the function. What does it mean here?

23. Find the slope of the function. What does it mean here?

24. How many miles per hour over the limit will result in a $100 fine?

25. Find the equation of the line that contains the points (−6, 2) and (3, 8).

Look Beyond

26. Copy the figures onto a piece of paper. In each picture, draw lines connecting *A* to *A'*, *B* to *B'*, and *C* to *C'*. Extend the lines until they meet. Label the point where they meet *O*. Compare the lengths of segments $\overline{OA}$ and $\overline{OA'}$. Compare the lengths of segments $\overline{AB}$ and $\overline{A'B'}$. What do you notice? Compare other lengths. What did you find?

The Side-Splitting Theorem

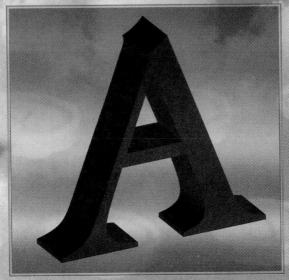

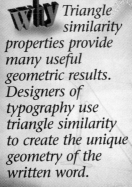

Triangle similarity properties provide many useful geometric results. Designers of typography use triangle similarity to create the unique geometry of the written word.

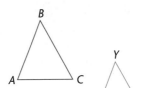

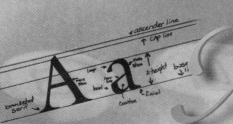

In the last lesson you investigated three different combinations of sides and angles that allow you to conclude triangle similarity from limited information. Formal statements of each combination are given below as postulates.

AA (ANGLE-ANGLE) SIMILARITY POSTULATE

If two angles of one triangle are equal in measure to two angles of another triangle, then the triangles are similar.   **8.4.1**

In $\triangle ABC$ and $\triangle XYZ$, if m$\angle A$ = m$\angle X$ and m$\angle B$ = m$\angle Y$, then $\triangle ABC \sim \triangle XYZ$.

SSS (SIDE-SIDE-SIDE) SIMILARITY POSTULATE

If the measures of pairs of corresponding sides of two triangles are proportional, then the two triangles are similar.   **8.4.2**

In $\triangle ABC$ and $\triangle XYZ$, if $\frac{AB}{XY} = \frac{AC}{XZ} = \frac{BC}{YZ}$, then $\triangle ABC \sim \triangle XYZ$.

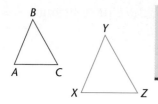

In $\triangle ABC$ and $\triangle XYZ$, if $\frac{AB}{XY} = \frac{AC}{XZ}$ and $m\angle A = m\angle X$, then $\triangle ABC \sim \triangle XYZ$.

Can the proportion above be written $\frac{XY}{AB} = \frac{AC}{XZ}$? Explain.

In Lesson 3.7 you learned a theorem about the midsegment of a triangle. Are there proportional relationships between segments created by drawing other segments parallel to the side of a triangle?

SIDE-SPLITTING THEOREM

A line parallel to one side of a triangle divides the other two sides proportionally.   8.4.4

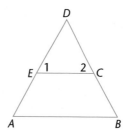

Given: $\overline{EC} \parallel \overline{AB}$

Prove: $\dfrac{AE}{ED} = \dfrac{BC}{CD}$

Proof:

STATEMENTS	REASONS
1. $\overline{EC} \parallel \overline{AB}$	Given
2. $m\angle A = m\angle 1$ and $m\angle B = m\angle 2$	If $\parallel$ segments are cut by a transversal, corresponding angles are congruent.
3. $\triangle ADB \sim \triangle EDC$	AA Similarity Postulate
4. $\dfrac{AD}{ED} = \dfrac{BD}{CD}$	Definition of Similar Triangles
5. $AD = AE + ED$, $BD = BC + CD$	Segment Addition Postulate
6. $\dfrac{AE + ED}{ED} = \dfrac{BC + CD}{CD}$	Substitution
7. $\dfrac{AE}{ED} = \dfrac{BC}{CD}$	"Add-One" Property of Proportions

Lesson 8.4 The Side-Splitting Theorem   **437**

What other proportions can you find in △*ADB* using the Side-Splitting Theorem?

The diagram at right suggests an easier way to remember the proportions that arise from the Side-Splitting Theorem. Using the diagram, the following proportions result:

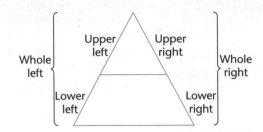

$$\frac{\text{Upper left}}{\text{Whole left}} = \frac{\text{Upper right}}{\text{Whole right}}$$

$$\frac{\text{Upper left}}{\text{Lower left}} = \frac{\text{Upper right}}{\text{Lower right}}$$

$$\frac{\text{Lower left}}{\text{Whole left}} = \frac{\text{Lower right}}{\text{Whole right}}$$

EXAMPLE 1

Solve for *x* using Theorem 8.4.4.

A

B

C

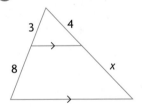

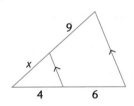

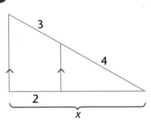

Solution ➤

A $\frac{3}{8} = \frac{4}{x}$

$3x = 32$

$x = 10\frac{2}{3}$

B $\frac{4}{6} = \frac{x}{9}$

$36 = 6x$

$x = 6$

C $\frac{x}{2} = \frac{7}{3}$

$3x = 14$

$x = 4\frac{2}{3}$ ❖

A corollary is a theorem that is easily proven from another theorem. A corollary that follows directly from the Side-Splitting Theorem is given on the next page.

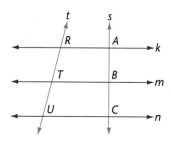

If $k \parallel m \parallel n$, with transversals s and t as shown, then $\frac{AB}{BC} = \frac{RT}{TU}$.

Use the words *upper*, *lower*, and *whole* to describe the proportions in Corollary 8.4.5.

EXAMPLE 2

Cultural Connection: Africa
Students in ancient Egypt studied geometry to solve practical problems (such as the following problem) in building the pyramids.

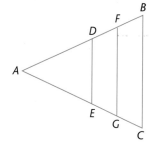

Segments $\overline{DE}$, $\overline{FG}$, and $\overline{BC}$ are parallel to each other.

$AD = AE = 7$ cubits

$DF = FB = 3\frac{1}{2}$ cubits

$DE = 2\frac{1}{2}$ cubits

Solution ➤

$\dfrac{AD}{DF} = \dfrac{AE}{EG}$	$\dfrac{7}{3.5} = \dfrac{7}{EG}$	$EG = 3.5$
$\dfrac{DF}{FB} = \dfrac{EG}{GC}$	$\dfrac{3.5}{3.5} = \dfrac{3.5}{GC}$	$GC = 3.5$
$\dfrac{AD}{AF} = \dfrac{DE}{GF}$	$\dfrac{7}{10.5} = \dfrac{2.5}{GF}$	$GF = 3.75$
$\dfrac{AD}{AB} = \dfrac{DE}{BC}$	$\dfrac{7}{14} = \dfrac{2.5}{BC}$	$BC = 5$ ❖

EXERCISES & PROBLEMS

Communicate

1. Why isn't there an AAA Postulate for similarity?

2. What is the relationship between the ratios of the sides and the ratios of the perimeters of two similar triangles? Do you think the same relationship will occur for similar quadrilaterals? Explain.

3. Are all isosceles triangles similar? Explain or give a counterexample.

4. Are all equilateral triangles similar? Explain or give a counterexample.

5. Are all right triangles similar? Explain or give a counterexample.

Practice & Apply

State which triangles are similar in Exercises 6–8. Which similarity postulate supports your answer? Determine the scale factor where possible.

6.

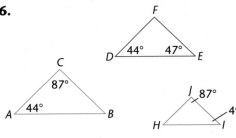

7. $\overline{AB} \parallel \overline{DE}$

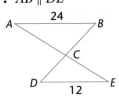

8.

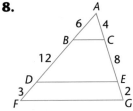

9. $\overline{TS} \parallel \overline{QR}$. Find x.

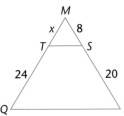

10. $\overline{DC} \parallel \overline{EB}$. Find x and y.

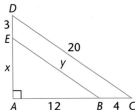

11. $\triangle ABC \sim \triangle AED$. Find x.

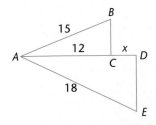

12. $QRST$ is a parallelogram. Find x and y.

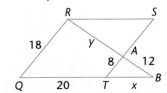

Indirect Measurement You can estimate the height of an object by placing a mirror on level ground so that you can see the top of the object in the mirror. By the laws of optics, light bounces off the mirror at an angle equal to the angle at which it strikes it, so ∠*HRE* ~ ∠*TRB*.

13. By which postulate is △*TBR* ~ △*HER*?

14. A person whose eyes are 5 feet above the ground places a mirror so that the dinosaur's nose is visible in it. The mirror is 12 feet from the point directly below the dinosaur's nose and 3 feet from the person. How high is the dinosaur's nose above the ground?

Height of person

Height of object

T

H

B   *R*   *E*

Mirror

Don't try this in a real museum. This photo is a computer trick!

15. A designer wants to divide segment $\overline{AB}$ into three equal parts.

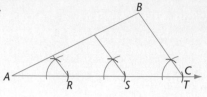

a. First he draws ray $\overrightarrow{AC}$ and marks three equal segments on the ray using his compass.

b. He then makes triangle $\triangle ABT$.

c. Finally, the designer constructs parallels through R and S. Explain why the construction works.

16. A rectangular strip of cloth $6\frac{1}{4}$ inches long must be divided into 7 congruent parts, and $6\frac{1}{4} \div 7$ involves a fraction that does not appear on a ruler. How can the Side-Splitting Theorem be applied to divide the cloth.

Look Back

17. Find the height of the stack of pipes if the diameter of the outside of each pipe is 8 feet. (Hint: Assume that segments connecting the centers of the circles form an equilateral triangle whose sides measure 8 feet.)
[Lesson 5.4]

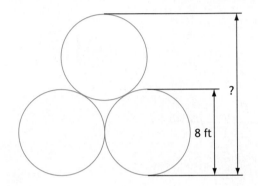

Use $\triangle ABC$, with A(3, 5), B(6, 1) and C(−2, −7) for Exercises 18–24. [Lesson 5.6]

18. Graph $\triangle ABC$. **19.** Find the perimeter of $\triangle ABC$.

20. Find the midpoints of each side. Label them D, E, and F.

21. Find the perimeter of $\triangle DEF$.

22. Draw the median of $\triangle ABC$ from A, and find its length using the Distance Formula.

23. Draw the median of $\triangle ABC$ from B, and find its length using the Distance Formula.

24. Draw the median of $\triangle ABC$ from C, and find its length using the Distance Formula.

Look Beyond

Use $\triangle XYZ$ and $\triangle XJK$ for Exercises 25–27.

25. Find the ratios of the corresponding sides of the two triangles. Are they equal?

26. Find the midpoints of sides $\overline{XZ}$ and $\overline{XK}$. Find the lengths of the medians of $\triangle XYZ$ and $\triangle XJK$.

27. Find the ratio of the corresponding medians in Exercise 26.

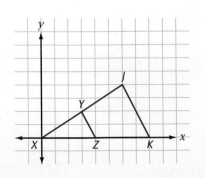

LESSON 8.5

Indirect Measurement, Additional Similarity Theorems

Why *If you needed to measure the breadth of a lake or the height of a mountain, you certainly could not do it directly with a tape measure. Instead, you would need to use a method that could approximate the mountain's height using indirect measurement.*

Using Similar Triangles to Measure Distances

Civil Engineering Inaccessible distances can often be measured by using similar triangles. For example, a military engineer must build a temporary bridge across a river. To do this he must find the distance across the river.

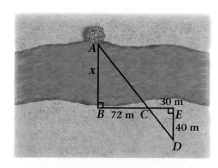

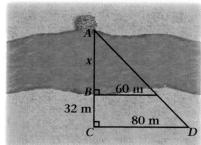

Sighting a tree across the river, the engineer sets up right triangles along the bank of the river.

Method 1 Using the right angles and their vertical angles, the engineer knows that $\triangle ABC \sim \triangle DEC$ because of AA similarity.
Therefore, $\frac{AB}{DE} = \frac{BC}{EC}$.

Then $\frac{x}{40} = \frac{72}{30}$.

$$30x = 2880$$
$$x = 96 \text{ m}$$

Method 2 Since $\angle ABE \cong \angle ACD$ and $\angle A \cong \angle A$, we know that $\triangle ABE \sim \triangle ACD$ because of AA similarity.
So, $\frac{x}{x + 32} = \frac{60}{80}$.

$$80x = 60x + 1920$$
$$20x = 1920$$
$$x = 96 \text{ m}$$

CRITICAL *Thinking*

Why do you think the engineer chose 90° angles? Could other angles have been used in each case?

EXAMPLE 1

Recreation   Kim wants to know whether the basketball hoop in the new playground is the correct height, which is 10 feet. She notices the shadow of the pole that holds the hoop and wonders whether the shadow could be used to check the height of the hoop. Kim is 5 feet 10 inches (70 inches) tall.

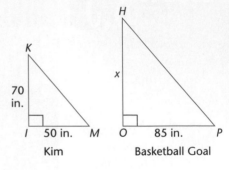

Solution ➤

ALGEBRA
Connection

Kim measures her shadow and the shadow of the pole. Notice that $\angle I \cong \angle O$, and since the sun's rays travel parallel to each other, $\angle M \cong \angle P$. Thus, $\triangle KIM \sim \triangle HOP$ because of AA similarity.

$$\frac{x}{70} = \frac{85}{50}$$

$$50x = 5950$$

$$x = 119 \text{ in.} = 9 \text{ ft } 11 \text{ in. The hoop is a little low.} \ \diamond$$

EXAMPLE 2

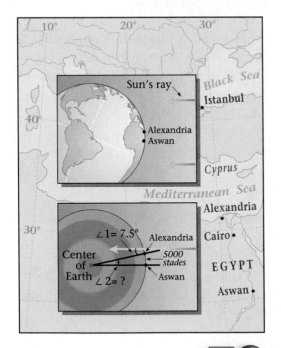

Cultural Connection: Africa   About 200 B.C.E., Eratosthenes, a Greek astronomer who lived in North Africa, estimated the circumference of the Earth using indirect measurement. He knew that a vertical rod in the city of Aswan would not cast a shadow at noon at the time of the summer solstice. In Alexandria, which is 5000 stades (575 miles) from Aswan, a vertical rod would cast a shadow at the same time. He measured the angle the Sun's rays formed with the vertical rod as 7.5°.

Ⓐ Explain why $\angle 1 \cong \angle 2$. Remember that the Sun's rays are considered to be parallel.

Ⓑ Find what Eratosthenes estimated the Earth's circumference to be.

Ⓒ Find the relative error of Eratosthenes' measurement. (The circumference of the Earth is approximately 24,900 miles.)

Solution ➤

ALGEBRA
Connection

Ⓐ $\angle 1 \cong \angle 2$ because they are alternate interior angles.

Ⓑ $\dfrac{7.5}{360} = \dfrac{575}{c}$ $c = 27,600 \text{ mi}$

Ⓒ $E_R = \dfrac{|24,900 - 27,600|}{24,900} \times 100\%$

$\approx 10.8\% \ \diamond$

Additional Similarity Theorems

The sides of similar triangles are proportional. But, as you will see, other parts of similar triangles are also in proportion.

> **PROPORTIONAL ALTITUDES THEOREM**
> If two triangles are similar, then their corresponding altitudes have the same ratio as the corresponding sides.
>
> **8.5.1**

For example, if $\triangle ABC \sim \triangle XYZ$, $\overline{AD} \perp \overline{BC}$, and $XM \perp YZ$ then $\frac{AD}{XM} = \frac{AB}{XY}$.

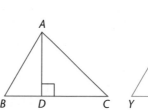

If an altitude of a triangle is not a side of the triangle, then it may be either inside or outside of the triangle. Therefore, a proof of the Proportional Altitudes Theorem must cover both possibilities. The proof below considers only the case of an altitude inside a triangle.

> **Given:** $\triangle ABC \sim \triangle XYZ$; $\overline{AD}$ is an altitude of $\triangle ABC$; $\overline{XM}$ is an altitude of $\triangle XYZ$.
>
> **Prove:** $\frac{AD}{XM} = \frac{AB}{XY}$
>
> **Proof:**

Case 1: The altitude is inside the triangle. $\angle B \cong \angle Y$, by definition of similar triangles. $\angle BDA$ and $\angle YMX$ are right angles, by definition of an altitude of a triangle, and so $\angle BDA \cong \angle YMX$. Then $\triangle BDA \sim \triangle YMX$ by AA similarity, and so $\frac{AD}{XM} = \frac{AB}{XY}$.

Given: $\triangle ABC \sim \triangle XYZ$; $\overline{AD}$ is an altitude of $\triangle ABC$; $\overline{XM}$ is an altitude of $\triangle XYZ$. Write a plan to prove Case 2 of the Proportional Altitudes Theorem, in which the altitude is outside the triangle.

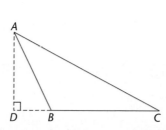
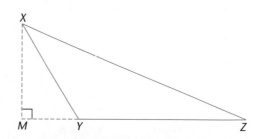

EXAMPLE 3

Solve for *x*.

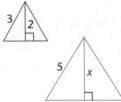

A

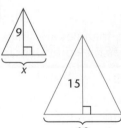

B

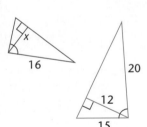

C

Solution ➤

A $\dfrac{2}{x} = \dfrac{3}{5}$

$10 = 3x$

$x = 3\dfrac{1}{3}$

B $\dfrac{9}{15} = \dfrac{x}{10}$

$90 = 15x$

$x = 6$

C $\dfrac{x}{12} = \dfrac{16}{20}$

$20x = 192$

$x = 9.6$ ❖

ALGEBRA
Connection

PROPORTIONAL MEDIANS THEOREM

If two triangles are similar, then their corresponding medians have the same ratio as the corresponding sides.   **8.5.2**

For example, if $\triangle ABC \sim \triangle XYZ$, *D* is the midpoint of $\overline{BC}$, and *M* is the midpoint of $\overline{YZ}$ then $\dfrac{AD}{XM} = \dfrac{AB}{XY}$.

You will be asked to complete a proof of this theorem in the exercise set.

EXAMPLE 4

Solve for *x*.

A

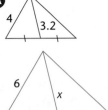

B

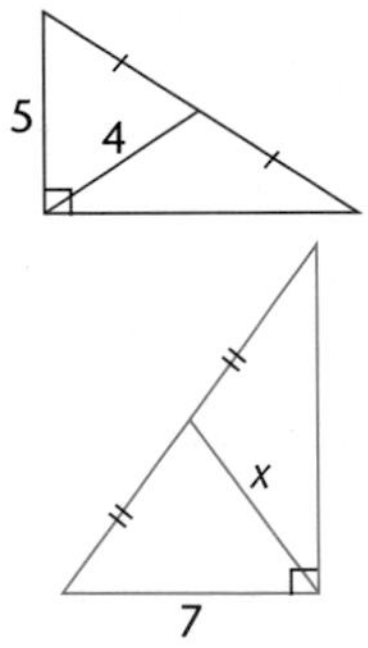

C

Solution ➤

ALGEBRA
Connection

A $\dfrac{4}{6} = \dfrac{3.2}{x}$

$4x = 19.2$

$x = 4.8$

B $\dfrac{5}{7} = \dfrac{4}{x}$

$5x = 28$

$x = 5.6$

C $\dfrac{6}{x} = \dfrac{5}{10}$

$60 = 5x$

$x = 12$ ❖

EXERCISES & PROBLEMS

Communicate

1. Explain why it is convenient to use similarity for measuring the height of a mountain.

2. If △ABC ~ △DEF, will the medians of the two triangles be proportional? Why?

3. If △ABC ~ △DEF, will the altitudes of the two triangles be proportional? Why? If the triangles were isosceles, how would that affect your answer? If the triangles were equilateral, how would that affect your answer?

4. Explain how Eratosthenes found the approximate circumference of the Earth. What postulate or theorem did he apply?

Practice & Apply

Algebra Solve for *x* in Exercises 5 and 6.

5.

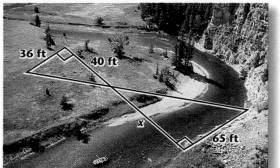

6.

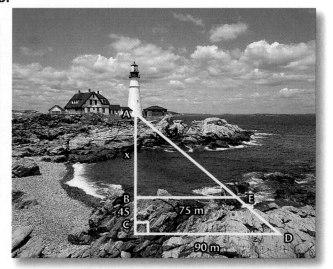

7. Mal wants to know the height of a light pole in the neighborhood. He measures his shadow and the pole's shadow at the same time of day. Find the height of the pole if Mal is 170 centimeters tall.

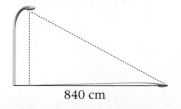

In Exercises 8–10, △*ADE* ~ △*RVW*,
AC = *CD*, **and** *RT* = *TV*.

8. Find *RW*.

9. Find *EB*.

10. Find *WV*.

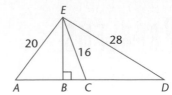
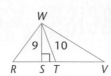
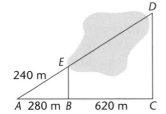

In Exercises 11 and 12, find the distance across the lake. Justify your solution.

11. $\overline{AB}$ is parallel to $\overline{DE}$.

12. $\overline{EB}$ is parallel to $\overline{DC}$.

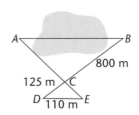
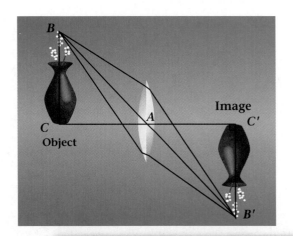

Astronomy Astronomers usually specialize in one of the many branches of their science, such as instruments and techniques. This includes devices used in the formation of images.

The figures at the right show that a convex lens (a lens that is thicker at the middle than at the edges) can bring parallel light rays together to form an image on the opposite side of the lens from the object. Rays that strike the edge of a lens bend by a certain amount. Rays that pass through the center of a lens are not bent. Thus, two similar triangles △*ABC* and △*AB′C′* are formed, and

$$\frac{BC}{B'C'} = \frac{AC}{AC'} \text{ or}$$

$$\frac{\text{Object size}}{\text{Image size}} = \frac{\text{Object distance}}{\text{Image distance}}$$

13. If an object 6 centimeters from a lens forms an image 15 centimeters on the other side of the lens, what is the ratio of the object size to the image size?

14. Rosa placed a lens 25 centimeters from an object 10 centimeters tall. An image was formed at a distance of 5 centimeters from the lens. How tall was the image?

15. In the given figures, prove that △*ABC* ~ △*AB′C′*.

16. How could you arrange an object, a lens, and an image so that the image is 20 times taller than the object to which it corresponds?

17. How would you arrange an object, a lens, and an image, so that the object and the image are the same size?

Surveying A county wants to connect two towns on opposite sides of a forest with a road. Surveyors have laid out the map shown. The road can be built through the forest or around the forest through Point *F*. The line joining Town *A* to Town *B* is parallel to $\overline{CD}$.

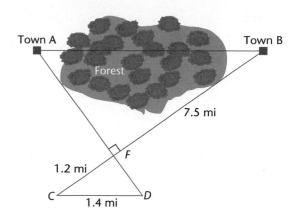

18. Find the distance between the towns through the forest.

19. Find the distance from Town *A* to Town *B* through Point *F*.

20. If it costs $4.5 million per mile to build the road around the forest and $6.2 million per mile to build the road through the forest, which road would be cheaper to build?

21. **Cultural Connection: Africa** These pyramids were built over 2200 years ago in the African Kingdom of Kush, in what is now the Sudan. Over time they have lost their tops. Describe a method that uses similar triangles to find how high they were when they were built.

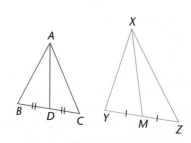

According to the Proportional Medians Theorem, If two triangles are similar, then the corresponding medians are proportional to the corresponding sides.

Complete the proof of the Proportional Medians Theorem on the following page.

Given: $\triangle ABC$ with median $\overline{AD}$, $\triangle XYZ$ with median $\overline{XM}$, $\triangle ABC \sim \triangle XYZ$

Prove: $\dfrac{AD}{XM} = \dfrac{BA}{YX} = \dfrac{BD}{YM} = \dfrac{BC}{YZ}$

Proof:

STATEMENTS	REASONS
_____(22)_____	Given
D is the midpoint of $\overline{BC}$. M is the midpoint of $\overline{YZ}$.	_____(23)_____
BD = DC, YM = MZ	_____(24)_____
_____(25)_____	Segment Addition
BC = 2(BD), YZ = 2(YM)	_____(26)_____
$\dfrac{BC}{YZ} = \dfrac{2(BD)}{2(YM)} = \dfrac{BD}{YM}$	_____(27)_____
$\dfrac{BC}{YZ} = \dfrac{BA}{YX},\ \angle B \cong \angle Y$	_____(28)_____
$\dfrac{BD}{YM} = \dfrac{BA}{YX}$	_____(29)_____
_____(30)_____	SAS Similarity Postulate
_____(31)_____	Definition of Similar Polygons
$\dfrac{AD}{XM} = \dfrac{BA}{YX} = \dfrac{BD}{YM} = \dfrac{BC}{YZ}$	_____(32)_____

Complete the proof of the following theorem:

THE PROPORTIONAL ANGLE BISECTORS THEOREM

If two triangles are similar then the corresponding angle bisectors are proportional to the corresponding sides.

8.5.3

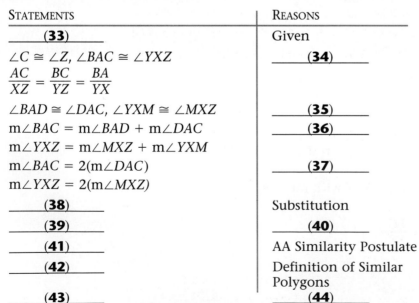

Given: $\triangle ABC \sim \triangle XYZ$, $\overline{AD}$ bisects $\angle BAC$, and $\overline{XM}$ bisects $\angle YXZ$

Prove: $\dfrac{AD}{XM} = \dfrac{AC}{XZ} = \dfrac{AB}{XY} = \dfrac{BC}{YZ}$

Proof:

STATEMENTS	REASONS
_____(33)_____	Given
$\angle C \cong \angle Z$, $\angle BAC \cong \angle YXZ$ $\dfrac{AC}{XZ} = \dfrac{BC}{YZ} = \dfrac{BA}{YX}$	_____(34)_____
$\angle BAD \cong \angle DAC$, $\angle YXM \cong \angle MXZ$	_____(35)_____
$m\angle BAC = m\angle BAD + m\angle DAC$ $m\angle YXZ = m\angle MXZ + m\angle YXM$	_____(36)_____
$m\angle BAC = 2(m\angle DAC)$ $m\angle YXZ = 2(m\angle MXZ)$	_____(37)_____
_____(38)_____	Substitution
_____(39)_____	_____(40)_____
_____(41)_____	AA Similarity Postulate
_____(42)_____	Definition of Similar Polygons
_____(43)_____	_____(44)_____

THE PROPORTIONAL SEGMENTS THEOREM

The angle bisector of a triangle divides the opposite side into segments proportional to the other two sides of the triangle.   **8.5.4**

Given: $\triangle ABC$, $\overline{AD}$ bisects $\angle CAB$

Prove: $\dfrac{DC}{DB} = \dfrac{AC}{AB}$

Proof: To prove this theorem, first construct $\overleftrightarrow{CE}$ parallel to $\overline{AD}$, and extend $\overrightarrow{BA}$ to intersect with $\overleftrightarrow{CE}$ at point E.

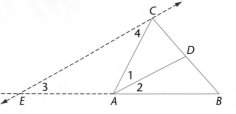

STATEMENTS	REASONS
_____(45)_____	Given
Draw $\overleftrightarrow{CE}$ parallel to $\overline{AD}$.	Parallel Postulate
Extend $\overrightarrow{BA}$ to intersect $\overleftrightarrow{CE}$ at point E.	2 lines intersect in 1 point.
In $\triangle CBE$, $\dfrac{DC}{DB} = \dfrac{AE}{AB}$.	_____(46)_____
_____(47)_____	If ∥ lines are intersected by a transversal, corresponding angles are congruent.
_____(48)_____	If ∥ lines are intersected by a transversal, alternate interior angles are congruent.
$\angle 1 \cong \angle 2$	_____(49)_____
$\angle 3 \cong \angle 4$	_____(50)_____
$\overline{AE} \cong \overline{AC}$	_____(51)_____
_____(52)_____	_____(53)_____

Look Back

54. Find the slope of the line containing (4, 5) and (6, 1).   **[Lesson 3.8]**

55. Find the equation of the line containing (0, 2) and (−3, 0).   **[Lesson 3.8]**

**For Exercises 56–59, write Sometimes, Always, or Never.
[Lesson 4.6]**

56. A rhombus is a square.

57. Consecutive angles of a trapezoid are supplementary.

58. Consecutive angles of a rectangle are complementary.

59. The diagonals of a kite bisect each other.

Look Beyond

60. Are the two squares similar? Explain.

61. What is the ratio of their sides? their perimeters? their areas?

62. What is the relationship between the ratio of the perimeters and the ratio of the areas?

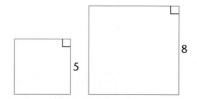

LESSON 8.6

Exploring

Area and Volume Ratios

Why *A town's water-treatment facility uses a spherical container with a radius of 4 meters. The town council wants to install a new container with double the volume. They can find the dimensions of the new container using ratios.*

•Exploration 1 *Ratios of Areas of Similar Figures*

Calculator

You will need
Calculator

For each pair of similar figures below, the ratio of a pair of corresponding segments is given. Find the ratio of the areas for each.

1 Two squares

$$\frac{\text{Side of square A}}{\text{Side of square B}} = \frac{3}{1}$$

$$\frac{\text{Area of square A}}{\text{Area of square B}} = \frac{?}{?}$$

2 Two triangles

$$\frac{\text{Side of triangle A}}{\text{Side of triangle B}} = \frac{2}{1}$$

$$\frac{\text{Area of triangle A}}{\text{Area of triangle B}} = \frac{?}{?}$$

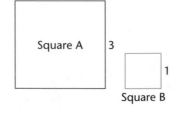

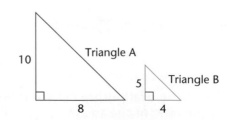

452 CHAPTER 8

3 Two rectangles

$$\frac{\text{Side of rectangle A}}{\text{Side of rectangle B}} = \frac{3}{2}$$

$$\frac{\text{Area of rectangle A}}{\text{Area of rectangle B}} = \frac{?}{?}$$

4 Two circles

$$\frac{\text{Radius of circle A}}{\text{Radius of circle B}} = \frac{4}{3}$$

$$\frac{\text{Area of circle A}}{\text{Area of circle B}} = \frac{?}{?}$$

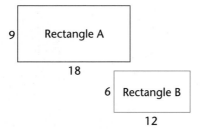

9 | Rectangle A

18

6 | Rectangle B

12

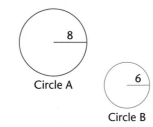

8

Circle A

6

Circle B

5 Use the results above to complete the following conjecture: Two similar figures with corresponding sides in the ratio $\frac{a}{b}$ have areas in the ratio __?__ . ❖

Exploration 2 *Ratios of Volume*

Calculator

You will need
Calculator

Solids are similar if they are the same shape and all the corresponding dimensions are proportional. For example, two rectangular prisms are similar if the dimensions for each corresponding face and base are proportional. For each pair of similar solids below, the ratio of a pair of corresponding segments is given. Find the ratio of the volumes for each.

1 Two cubes

$$\frac{\text{Side of cube A}}{\text{Side of cube B}} = \frac{3}{1}$$

$$\frac{\text{Volume of cube A}}{\text{Volume of cube B}} = \frac{?}{?}$$

2 Two rectangular prisms

$$\frac{\text{Side of prism A}}{\text{Side of prism B}} = \frac{3}{2}$$

$$\frac{\text{Volume of prism A}}{\text{Volume of prism B}} = \frac{?}{?}$$

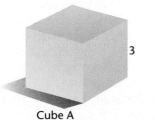

3

Cube A

1

Cube B

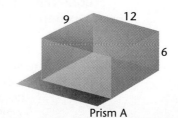

9   12

6

Prism A

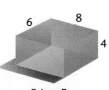

6   8

4

Prism B

3 Two spheres

$$\frac{\text{Radius of sphere A}}{\text{Radius of sphere B}} = \frac{2}{1}$$

$$\frac{\text{Volume of sphere A}}{\text{Volume of sphere B}} = \frac{?}{?}$$

4 Two cylinders

$$\frac{\text{Radius of cylinder A}}{\text{Radius of cylinder B}} = \frac{2}{1}$$

$$\frac{\text{Volume of cylinder A}}{\text{Volume of cylinder B}} = \frac{?}{?}$$

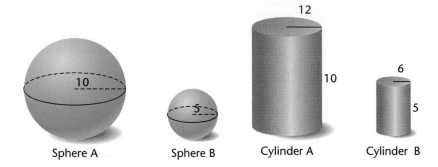

Sphere A   Sphere B   Cylinder A   Cylinder B

5 Use the results above to complete the following conjecture: Two similar solids with corresponding sides in the ratio $\frac{a}{b}$ have volumes in the ratio ___?___. ❖

CRITICAL *Thinking* Are all cubes similar? Why or why not? Are all spheres similar? Why or why not? Are all cylinders similar? Why or why not?

Try This 1. The corresponding sides of two similar triangles are in the ratio $\frac{5}{9}$. What is the ratio of their areas?

2. The corresponding areas of two similar rectangular prisms are in the ratio $\frac{9}{4}$. What is the ratio of their sides?

3. One sphere has a radius of 5 meters. Another sphere has a radius of 15 meters. What is the ratio of their volumes?

Cross-Sectional Areas, Weight, and Height

Many folk tales feature accounts of human giants. However, some mathematical investigation will show that there is a limit to just how tall a human or an animal can be. The amount of weight that a bone can support is proportional to its cross-sectional area. For example, the larger bone has three times the volume of the smaller bone, but only one-fourth more cross-sectional area. The larger bone must support three times more weight than the smaller bone because of its volume. Relative to the small bone, however, the large bone is not as strong because its cross-sectional area has not increased in its proportion to the size.

A young elephant that is 4 ft tall at the shoulder has leg bones with a cross-sectional radius of 4 cm. How much thicker would the leg bones need to be to support an elephant 8 ft tall at the shoulder? By what scale factor does the original radius need to be multiplied, to provide an adequate cross-sectional area?

The two-dimensional size of the elephant increases by a factor of 2. However, the volume of the elephant increases by a factor of 8. Thus, the larger elephant would need a leg bone with a cross-sectional area 8 times larger than the original elephant.

So, $8\pi r^2$ gives the new cross-sectional area.

ALGEBRA
Connection

Let R represent the radius of the larger elephant. Then $\pi R^2 = 8\pi r^2$, so $R^2 = 8r^2$, or $R = \sqrt{8}r$. Thus the radius of the larger bone is $\sqrt{8}$ or about 2.8 times the radius of the smaller bone. ❖

•Exploration 3 *Increasing Height and Volume*

Have you read stories of giants hundreds of feet tall? Do you think they could be realistic? The elephant example shows the effect that increasing size (linear) dimensions has on volume and on the dimensions of the structures that must support the increased bulk. Think about the question of giants as you do this exploration.

 Complete the chart to determine the necessary scale factors for the cross-sectional radii of leg bones of giant horses.

Height scale factor	Volume scale factor	Cross-sectional radius scale factor
2	8	$\sqrt{8}$
3	27	$\sqrt{27}$
4	64	?
5	?	?
100	?	?

2 If a normal-sized horse requires a leg bone with a cross-sectional radius of 2.5 centimeters for support, what must the cross-sectional radius of a leg bone be for a horse 20 times taller? 50 times taller? 100 times taller?

3 What effect does increasing the size of the horse have on the relative proportions of the legs? Would the legs be proportionally thicker or thinner than those of a normal-sized horse? ❖

Exercises & Problems

Communicate

1. What is the relationship between the ratio of the sides of two cubes and the ratio of their volumes?

2. Why are all spheres and cubes similar? Is this true for other three-dimensional figures? If so, name or describe them.

3. If you know the ratio of the areas of two similar prisms, explain how you can find the ratio of the volumes.

4. What happens to the volume of a cylinder if the radius is doubled but the height stays the same?

5. How does the cross-sectional area of a bone relate to an animal's weight and height?

Practice & Apply

6. A 12 ounce box of noodles is 15 cm wide, 20 cm high, and 4 cm deep. If the noodle company would like to sell a 24 ounce box, would it have to double all the dimensions of the 12 ounce box? Explain.

7. A new pizzeria sells an 8-inch diameter pizza for $4 and a 16-inch diameter pizza for $6. Which pizza is the better deal? Explain.

The ratio between the sides of two similar pyramids is $\frac{7}{5}$. Find the ratio between the following.

8. the perimeters of their bases

9. the areas of their bases

10. their volumes

Two spheres have radii 5 cm and 7 cm. Find the ratio between the following.

11. the circumferences of their great circles

12. their volumes

The ratio between the areas of the bases of two similar cones is $\frac{16}{25}$. Find the ratio between the following.

13. their heights

14. the circumferences of their bases

15. their volumes

In the figure, $\overline{EC} \parallel \overline{AB}$.

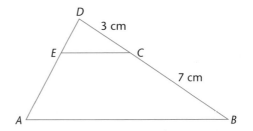

16. Find the ratio between the perimeters of $\triangle ABD$ and $\triangle ECD$.

17. Find the ratio between the areas of $\triangle ABD$ and $\triangle ECD$.

$ABCDE \sim QRSTU$. $ABCDE$ has perimeter 24 meters and area 50 square meters.

18. Find the perimeter of $QRSTU$.

19. Find the area of $QRSTU$.

20. Two similar cylinders have bases with areas 16 cm² and 49 cm². If the larger cylinder has height 21 cm, find the height of the smaller cylinder.

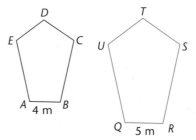

If the ratio between the heights of two similar cones is $\frac{7}{9}$, find the ratio between the following.

21. the circumferences of their bases

22. their volumes

23. the areas of their bases

The ratio between the surface areas of two spheres is 144:169. Find the ratio between the following.

24. their radii

25. their volumes

26. the circumferences of their great circles

If the ratio between the volumes of two similar prisms is $\frac{64}{125}$, find the ratio between the following.

27. their surface areas

28. the perimeters of their bases

29. the diagonals of their bases

30. the areas of their bases

31. *ABCDEF ~ PQRSTU. ABCDEF* has perimeter 42 meters and area 96 square meters. Find the perimeter and area of *PQRSTU*.

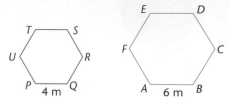

Two similar cones have surface areas of 225 cm² and 441 cm².

32. If the height of the larger cone is 12 cm, find the height of the smaller cone.

33. If the volume of the smaller cone is 250 cm³, find the volume of the larger cone.

34. A leg bone of a horse has a cross-sectional area of 19.6 cm². What is the diameter of the leg bone?

35. How much longer must the diameter of the leg bone in Exercise 34 be to support a horse twice as tall?

36. A 100-foot cylindrical tower has a cross-sectional radius of 26 feet. What would be the radius of a 350-foot tower if it were proportional to the 100-foot tower?

Cone *EDC* is formed by cutting off the bottom of cone *EAB* parallel to the base.

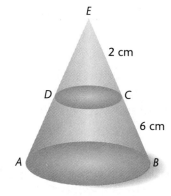

37. If cone *EAB* has volume 288 cm³, find the volume of cone *EDC*.

38. If the base of cone *EDC* has area 3.6 cm², find the area of the base of cone *EAB*.

39. If the Earth and the Moon were perfect spheres, their equators would be approximately 40,200 km and 10,000 km respectively. Compare the diameters and the volumes of the Earth and the Moon.

40. Packaging A toothpaste company packages its toothpaste in a tube with a circular opening with radius 2 mm. The company increases the radius to 3 mm. Predict what will happen to the amount of toothpaste used if the same number of people continue to use the same length of toothpaste. Explain your reasoning.

41. The area of the parking lot at Jerome's Restaurant is 400 m². Jerome buys some land and is able to make the parking lot 1.5 times as wide and 1.75 times as long. Find the area of the expanded lot.

42. Sports The diameter of a standard basketball is approximately 9.5 inches. A company that makes standard-sized basketballs contracts to make a basketball with diameter 5 inches for a restaurant promotion. If the materials to make a standard-sized ball cost the company $1.40, how much will it cost the company to make the promotional basketball if it uses the same materials?

43. A city stores rock salt in a dome-shaped building that is 82 feet in diameter at the base. The building is rated to hold 3,366 tons of salt. Because the city is growing, the city planners decide to add a second, smaller dome. The linear dimensions of the new dome will be three-fourths of the original dome. Estimate the storage capacity of the new dome.

44. One can of beans has twice the linear dimensions of another. The volume is how many times greater?

Look Back

45. Each interior angle of a regular polygon measures 165°. How many sides does the polygon have? **[Lesson 3.6]**

46. The exterior angles of a regular polygon each measure 40°. How many sides does the polygon have? **[Lesson 3.6]**

47. Which of these lines are parallel? **[Lesson 3.8]**

  a. $5x + 4y = 18$ **b.** $-10x + 8y = 21$ **c.** $30x - 24y = 45$

48. Write the equation of the line parallel to $5x + 4y = 18$ and through the point (8, 3). **[Lesson 3.8]**

49. **Algebra** Graph: $y = x^2 - 6x - 4$ **[Lesson 3.8]**

50. **Algebra** Solve for x: $8x^2 + 5x - 8 = 5x^2 - 2x + 4$

Look Beyond

Algebra *ABCD* **is a square with side length 1 and perimeter 4. The squares inside *ABCD* are formed by connecting the midpoints of the bigger square.**

51. What happens to the perimeter of each successive square? Using a calculator, add the perimeters of the first eight squares together. Keep track of the sum after each number is added. Do you think the sum of all the squares generated this way will ever reach 14? Explain.

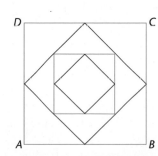

CHAPTER 8
PROJECT

INDIRECT
MEASUREMENT AND
SCALE MODELS

Activity 1: Measuring Buildings

1. Use several different methods, including photographs, shadows, and mirrors, to indirectly find the dimensions of buildings or other structures at your school or in your neighborhood.

2. Draw a diagram of the building with the direct measurements labeled. Include the photograph if you used a photograph.

3. Explain how you made the measurements.

4. Explain how you calculated the dimensions you did not measure directly.

5. If available, find the known dimensions of the structures you measured. Explain how you can account for any differences between what you calculated and the known dimensions.

Activity 2: Scale Models

1. Obtain information on the dimensions of a structure near your school or in your neighborhood. Or, use one of the structures from Activity 1.

2. Build a scale model of the building using cardboard or some other suitable modeling material.

3. Explain the process you used to determine the dimensions of your model and how you constructed the model.

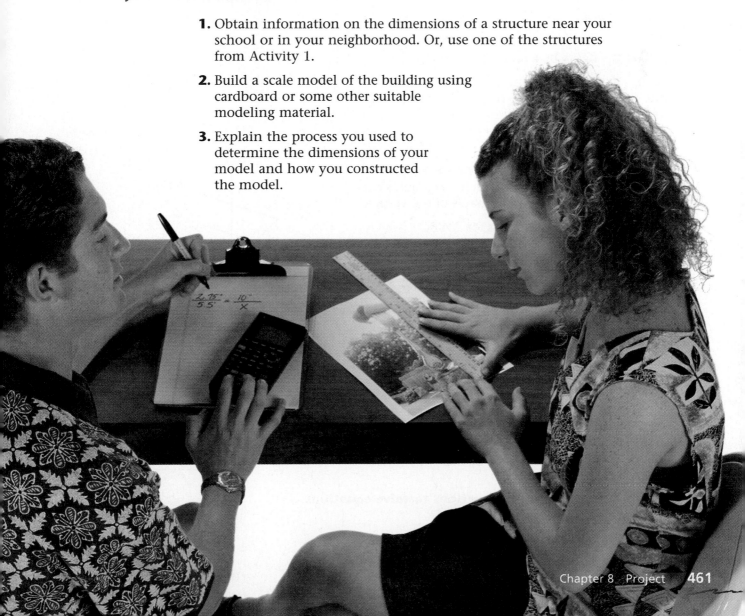

Chapter 8 Review

Vocabulary

contraction	414	proportional	420 ✓
dilation	414	scale factor	412
properties of proportion	422	similar figures	420 ✓

Key Skills and Exercises

Lesson 8.1

➤ **Key Skills**

Given a scale factor, draw a dilation.

Dilate the figure by a scale factor of 0.5.

Multiply the coordinates of the vertices of the figure by 0.5, and sketch the dilation. In this case it is a contraction.

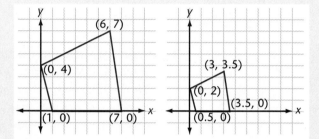

➤ **Exercises**

In Exercises 1–2, refer to the figure at the right.

1. Dilate $\triangle ABC$ by a scale factor of 3 and sketch the result, giving coordinates of the vertices.

2. What is the scale factor for $\triangle A'BC'$?

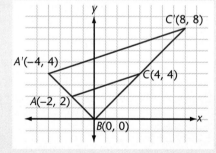

Lesson 8.2

➤ **Key Skills**

Determine if polygons are similar.

In the pair of triangles, $EA = \frac{2}{3} AC$, $DA = \frac{2}{3} BA$, $m\angle D = m\angle B$, and $\angle DAE$ and $\angle BAE$ are right angles. Are the triangles similar?

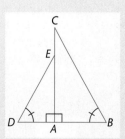

You are given that two angles are congruent, so by the Triangle Sum Theorem the third angles are congruent. Two sets of corresponding sides are proportional with a ratio of 2:3. Use the Pythagorean Theorem to show that the hypotenuses are in the same proportion, thus showing that corresponding sides are all proportional.

Use the properties of proportions to solve equations.

Solve for x.

$$\frac{11}{x + 3} = \frac{3}{x - 5}$$

By the Cross-Multiplication Property,

$$11(x - 5) = 3(x + 3)$$
$$11x - 55 = 3x + 9$$
$$8x = 64$$
$$x = 8$$

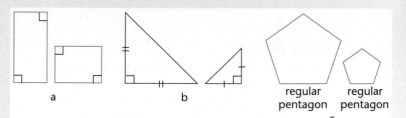

regular pentagon   regular pentagon
c

➤ **Exercises**

3. Which of the pairs of polygons above are similar? Justify your answer.

In Exercises 4–5, solve for x.

4. $\dfrac{26}{x} = \dfrac{2x}{13}$ **5.** $\dfrac{3 + 5x}{7} = \dfrac{3x - 4}{5}$

Lesson 8.3

➤ *Key Skills*

Show that triangles are similar.

Which triangles can be shown to be similar?

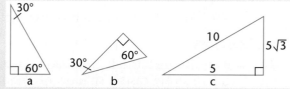

Using the Angle-Angle Similarity Postulate, △A and △B are similar because they have two equal angles. What about △C? The proportions of the lengths of the sides of the triangle tell us that *a triangle with one leg equal to half the longest side and one leg equal to √3 times half the longest side is a 30-60-90 right triangle.* Thus △C is similar to △A and △B.

➤ *Exercises*

In Exercises 6–7, are the triangles similar? Justify your answer.

6.

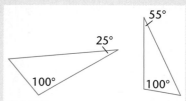

7.

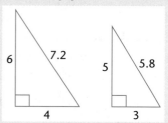

Lesson 8.4

➤ *Key Skills*

Show that triangles are similar using the similarity postulates.

In the triangle below, $\overline{DE}$ is parallel to $\overline{AB}$. Show that △DEC is similar to △ABC.

By the Side-Splitting Theorem, we know that sides $\overline{AC}$ and $\overline{BC}$ are divided proportionally. Thus, $\overline{AC}$ is proportional to $\overline{DC}$, and $\overline{BC}$ is proportional to $\overline{EC}$. ∠C is equal to itself, so the SAS Similarity postulate applies, and △DEC is similar to △ABC.

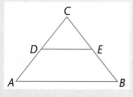

➤ *Exercises*

In Exercises 8–9, which triangles are similar? Justify your answer.

8.

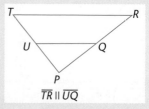

$\overline{TR} \parallel \overline{UQ}$

9.

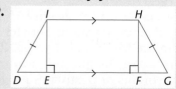

Lesson 8.5

➤ Key Skills

Use similar triangles to measure distance.

A right rectangular pyramid has a base length of 40 feet. If the pyramid casts a shadow 60 feet long and a yardstick casts a shadow 6 feet long, what is the height of the pyramid?

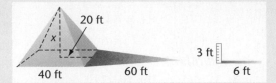

We use similar triangles to find the proportions of the two heights. The horizontal side of the small triangle is six feet; of the larger, 60 feet + 20 feet, or 80 feet. Set up the proportion:

$$\frac{6}{80} = \frac{3}{x}$$
$$6x = 240$$
$$x = 40 \qquad \text{Thus the height of the pyramid is 40 feet.}$$

➤ Exercises

10. You are 1.6 meters tall and cast a shadow of 3.5 meters. If a house simultaneously casts a shadow of 17.5 meters, how tall is the house?

Lesson 8.6

➤ Key Skills

Given similar figures, find ratios between their volume, area, etc.

The two cylinders to the right are similar. What is the ratio of their lateral areas?

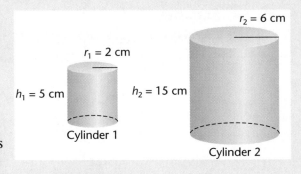

Cylinder 1

Cylinder 2

Method 1: Calculate the lateral areas L_1 and L_2. $L_1 \approx 62.5 \text{ cm}^2$, and $L_2 \approx 565.5 \text{ cm}^2$. The ratio of L_1 to L_2 is $\frac{62.5}{565.5}$, which simplifies to approximately $\frac{1}{9}$.

Method 2: Notice that $r_2 = 3r_1$ and $h_2 = 3h_1$. Thus,
$$\frac{L_1}{L_2} = \frac{2\pi r_1 h_1}{2\pi r_2 h_2} = \frac{2\pi r_1 h_1}{2\pi (3r_1)(3h_1)} = \frac{1}{9}$$

➤ Exercises

11. The two prisms shown are similar. What is the ratio of their volumes? of their heights?

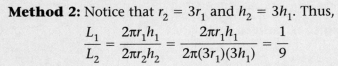

Applications

12. Suppose you put a mirror on the ground 15 meters from the base of a tree. If you stand 1 meter from the mirror, you can see the top of the tree. If your eyes are 1.8 meters above the ground, how tall is the tree? Which theorems or postulates prove that this method of indirect measurement works?

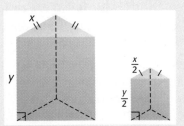

Drawing is not to scale.

Chapter 8 Assessment

In Exercises 1–2, refer to the figure.

1. What is the scale factor for *F'G'H'I'*?

2. Dilate *FGHI* by a scale factor of 1.5 and sketch the result, giving coordinates of the vertices.

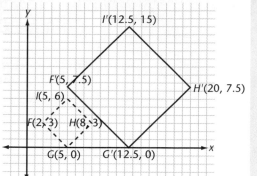

3. Which of the pairs of polygons are similar? Justify your answer.

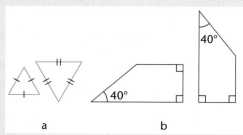

a   b   c

In Exercises 4–5, solve for x.

4. $\dfrac{x}{18} = \dfrac{21}{4}$

5. $\dfrac{10}{2 - 3x} = \dfrac{6}{x + 6}$

In Exercises 6–7, state which triangles are similar. Justify your answer.

6.

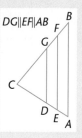

7.

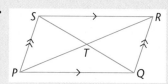

8. A surveyor made the measurements shown below. $\overline{ZY}$ is parallel to $\overline{WX}$. Find the distance across the base of the hill.

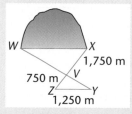

9. The two right cones below are similar. What is the ratio of their lateral areas?

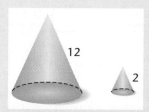

10. The surface areas of two spheres have the ratio 169:900. If the smaller sphere has a volume of 4,599.05, what is the volume of the larger?

Chapters 1–8 Cumulative Assessment

College Entrance Exam Practice

Quantitative Comparison Exercises 1–3 consist of two quantities, one in Column A and one in Column B, which you are to compare as follows:

A. The quantity in Column A is greater.

B. The quantity in Column B is greater.

C. The two quantities are equal.

D. The relationship cannot be determined from the information given.

	Column A	Column B	Answers
1.	Assume there are no hidden cubes Exterior surface area of A	Exterior surface area of B	(A) (B) (C) (D) **[Lesson 6.1]**
2.	Ratio of A's area to B's area	Ratio of B's area to A's area	(A) (B) (C) (D) **[Lesson 5.3]**
3.	$\overline{EF}$ is congruent to $\overline{CA}$. m∠A	m ∠D	(A) (B) (C) (D) **[Lessons 5.4, 5.5]**

4. Which are consecutive interior angles? **[Lesson 3.3]**

a. 1 and 5 **b.** 2 and 3

c. 3 and 6 **d.** 4 and 6

5. Which angle is not an angle of rotational symmetry of a regular octagon? **[Lesson 3.6]**

  a. 135° **b.** 45° **c.** 60° **d.** 90°

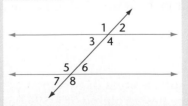

6. Two similar pyramids have slant heights in the ratio of 4:7. The ratio of their surface areas is **[Lesson 7.1]**

    a. 4:7   **b.** 8:14   **c.** 16:28   **d.** 16:49

7. Which property of proportion justifies this conditional? **[Lesson 8.2]**

If $\frac{x}{2y} = \frac{a}{b}$, then $\frac{(x + 2y)}{2y} = \frac{(a + b)}{b}$.

    a. "Add-One" Property   **b.** Cross-Multiplication Property

    c. Exchange Property   **d.** Reciprocal Property

8. Is this a true conditional? "If a quadrilateral is equilateral, then it is a square." If false, provide a counterexample. **[Lesson 2.2]**

9. Mon, Tues, and Wed are towns on a road that runs straight from Mon to Wed. The distance from Mon to Tues is 20 kilometers. The distance from Mon to Wed is 5 kilometers less than three times the distance from Tues to Wed. What is the distance between Tues and Wed? **[Lesson 1.4]**

10. In the figure at the right, name all the pairs of triangles that can be shown to be congruent. **[Lessons 4.2, 4.3]**

$\overline{MO} \cong \overline{NO}$

11. Describe the intersection of plane *FGH* and plane *PQR*. **[Lesson 1.2]**

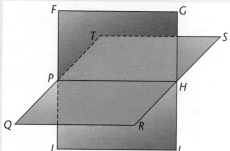

12. Find the volume of the sphere below. **[Lesson 7.6]**

3 in.

13. An equilateral triangle with a side length of 5.2 units is inscribed in a circle with a radius of 3 units. If a point is picked at random anywhere inside the circle, what is the probability that the point will not be inside the triangle? **[Lesson 5.7]**

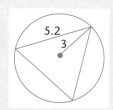

5.2

3

Free-Response Grid Exercises 14–17 may be answered using a free-response grid commonly used by standardized test services.

14. Find the slope of a line that passes through coordinates (2, 6) and (9, 12). **[Lesson 3.8]**

15. *AC* = 12 m, and *DB* = 10 m. What is the area of the rhombus? **[Lessons 5.2, 5.4]**

16. What is the area of the regular octagon? **[Lesson 5.5]**

6 in.

5 in.

17. The dimensions of a right rectangular prism are length = 14 mm, width = 3 mm, and height = 11 mm. Find the length of the prism's diagonal. **[Lesson 6.3]**

CHAPTER 9

Circles

LESSONS

9.1 *Exploring* Chords and Arcs

9.2 *Exploring* Tangents to Circles

9.3 *Exploring* Inscribed Angles and Arcs

9.4 *Exploring* Circles and Special Angles

9.5 *Exploring* Circles and Special Segments

9.6 Circles in the Coordinate Plane

Chapter Project
The Olympic Symbol

Greek geometers at the time of Euclid believed that circles have a special perfection. With the rediscovery of Euclid's Elements by the English philosopher Adelard (12th century), this way of thinking made its way into the European world. The designs in many early churches were based on geometric principles learned from Euclid.

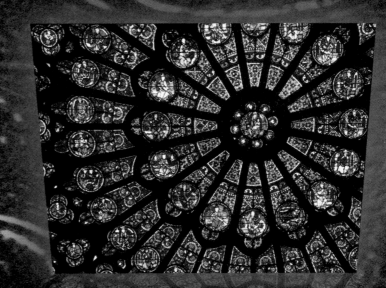

Among the most famous works of art based on the geometry of the circle is the north rose window in the Notre Dame Cathedral in Paris (thirteenth century C.E.). As you can see, tangents to inner circles form regular polygons whose vertices are the centers of the outer circles.

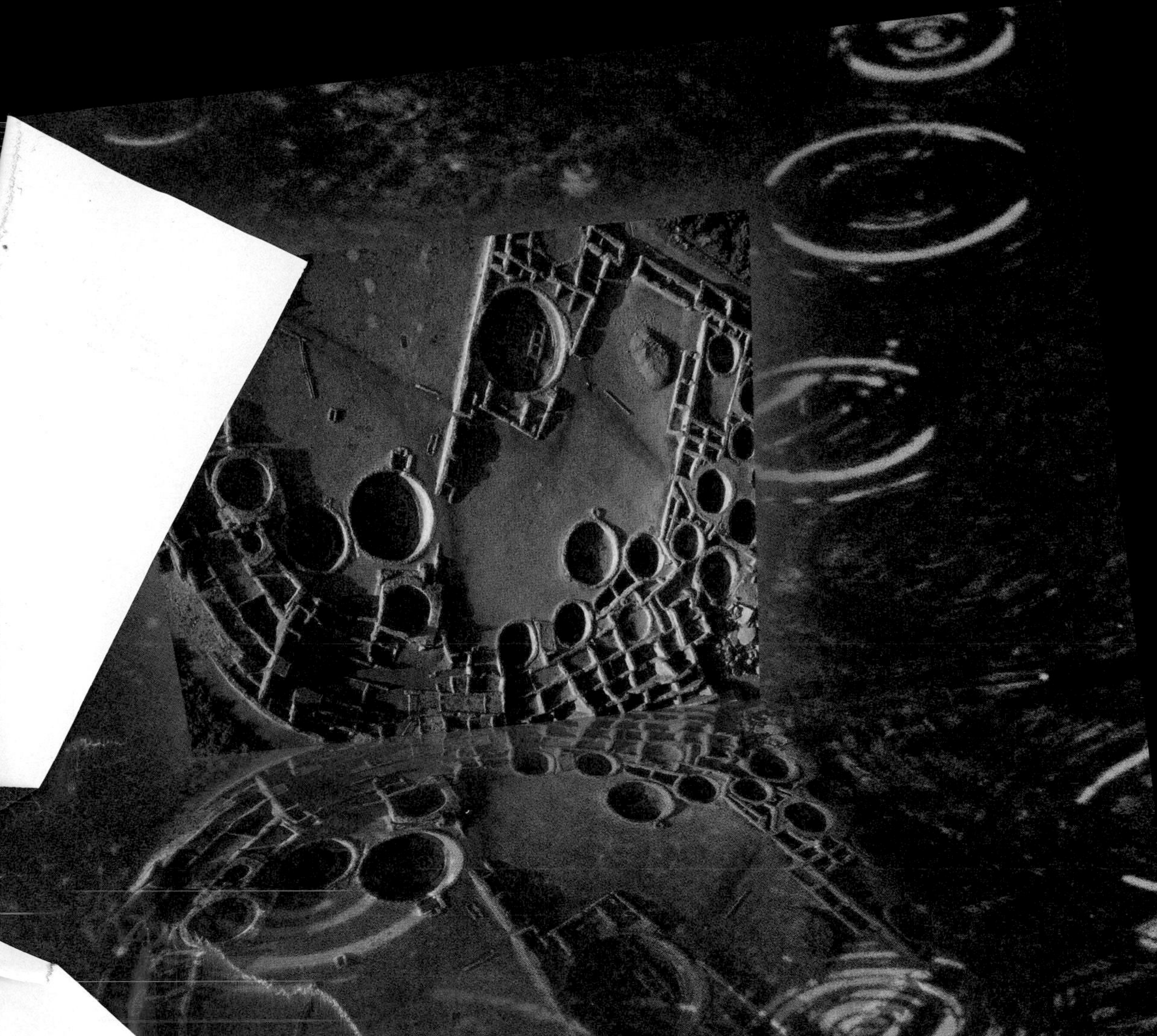

...cular structures in this
...the Pueblo ruins at Chaco
...s *kivas*. The circular design
...ctures reflects the Pueblo
...recurrence of celestial
...e is circular in

center of
...ment from the
...A *chord* is a
...A **diameter** is
...of a circle. **9.1.1**

...nstructions, you
...congruent circles have
...t follows from the

...ng the symbol ⊙ and
...cle.

PORTFOLIO ACTIVITY

...ia According to the Chinese philosophy
...s of an interplay of opposite forces yin
...t symbolizes the tao (meaning "the
...y of the two principles geometrically.
...r striking geometric
...scover in this chapter.

...figure prominently in many
...ols. Find examples of these to
...nclude in your portfolio.

Exploring

Chords and Arcs

why *Mathematical relationships of circles and segments make many interesting compass-and-straight edge constructions possible.*

Defining the Parts of a Circle

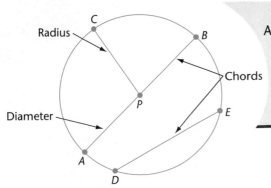

Circle *P*, or ⊙*P*

CIRCLE

A **circle** consists of the points in a plane that equidistant from a given point known as th the circle. A **radius** (plural, radii) is a seg center of a circle to a point on the circle segment whose endpoints lie on a circl a chord that passes through the cente

When you use circles in c make use of the fact that congruent radii. This fa definition of a circle.

A circle is named usi the center of the ci

•Exploration 1 Constructing a Regular Hexagon

You will need
Compass and straightedge

1 Draw a circle with a compass. Label the center *P*. Choose a point on the circle and label it *A*. (Figure a.)

2 Without changing your radius setting, set the point of your compass on *A*. Draw an arc that intersects the circle at a new point. Label the new point *B*. (Figure b.)

3 Draw chord $\overline{AB}$. Draw radii $\overline{PA}$ and $\overline{PB}$. (Figure c.)

4 What kind of triangle is △*ABP*? What are the measures of its angles? Explain your reasoning.

5 Without changing your radius setting, set the point of your compass on *B*. Draw an arc that intersects the circle at a new point. Label the new point *C*. (Figure d.)

6 Draw chord $\overline{BC}$ and the new radius $\overline{PC}$.

7 Continue drawing new points, chords, and radii until you have completed a figure like the one shown. (Figure e.)

8 Is the polygon *ABCDEF* a regular hexagon? Explain your answer.

9 An angle such as ∠*APB* is known as a **central angle** of a circle (and also of a regular polygon). Are the central angles of the circle (and of *ABCDEF*) congruent? Does the sum of their measures equal 360°? Explain your reasoning. ❖

a.

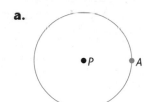

b.

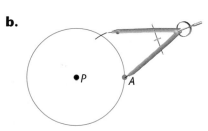

c.

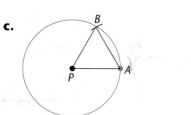

d.

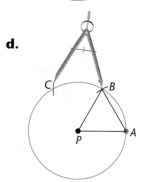

e.

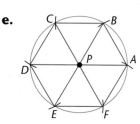

If two central angles of a circle (such as ∠*APB* and ∠*CPD*) are congruent, what can you conclude about their chords? Does the size of the central angles matter? State your answer as a theorem and explain your reasoning.

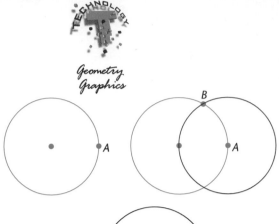

Exploration 2 Constructing a "Circle Flower"

Geometry
Graphics

You will need
Geometry technology or
Compass

1 Draw a circle with a compass or geometry technology. Choose a point on the circle and label it *A*.

2 Using *A* as the center point, draw another circle congruent to the first.

3 Pick one of the points where circle *A* intersects the original circle. Label this point *B*. Draw circle *B* congruent to circle *A*.

4 Repeat Step 3, going around the circle in order, until you have completed the flower.

5 Discuss the mathematical features of the flower. For example, do all six of the outer circles meet at a single point? How is the flower related to the hexagon in Exploration 1? ❖

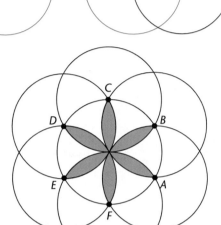

Major and Minor Arcs

An **arc** is a part of a circle. An arc is named using its endpoints and an arc symbol.

Notice that there are two different arcs that have their endpoints at *Q* and *R*. One arc (known as the **minor arc**) is less than a half-circle. The other arc (known as the **major arc**) is greater than a half-circle. When an arc is designated by only two letters, it is assumed to be a minor arc.

To designate a major arc, use the two endpoints and a third point that lies on the arc. Write the third point between the endpoints.

CRITICAL Thinking A **semicircle** is an arc that is exactly half of a circle. Would you need a third point in your notation of a semicircle? Explain your reasoning.

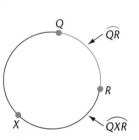

Measurement of Arcs: Degrees

Arcs may be measured in two different ways: by degree measure or by linear units.

To find the degree measure of arc $\overset{\frown}{RQ}$, draw a central angle whose rays pass through the endpoints of the arc and whose vertex is at the center of the circle. Measure the angle.

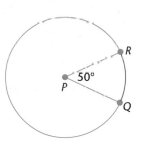

Imagine rotating an endpoint of the arc, such as Q in the illustration, about the center of the circle until it coincides with the other endpoint R. The amount of rotation, which is the measure of the central angle, is the degree measure of the arc:

$$m\overset{\frown}{RQ} = 50°.$$

A complete rotation is 360°, so the measure of the major arc is

$$m\overset{\frown}{RXQ} = 360° - m\overset{\frown}{RQ}$$

$$= 360° - 50° = 310°.$$

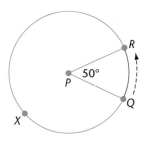

Measurement of Arcs: Length

ALGEBRA
Connection

Notice that arc $\overset{\frown}{XY}$ is part of the circumference of circle P. Since the measure of the central angle is 90° (one-fourth of a complete rotation), the measure of the arc is one-fourth the circumference C of the circle ($C = 2\pi r$):

$$\text{Length of } XY = \frac{90°}{360°} (2)(\pi)(6)$$

$$= \frac{1}{4} (2)(\pi)(6) = 3\pi \text{ units.}$$

In general,

$$\text{arc length} = \frac{A}{360°} \times \text{circumference,}$$

where A is the measure of the central angle.

Is it possible for two arcs to have the same degree measures but different lengths? Explain why or why not.

·Exploration 3 · Chords and Arcs Theorem

You will need
Compass and straightedge

1 In the figure, chords $\overline{AB}$ and $\overline{CD}$ are congruent. Do you think that arcs $\overparen{AB}$ and $\overparen{CD}$ have equal measures? To find the answer, copy the figure and construct central angles $\angle APB$ and $\angle CPD$.

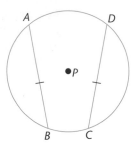

2 Prove that triangles $\triangle APB$ and $\triangle CPD$ are congruent. What can you conclude about central angles $\angle APB$ and $\angle CPD$? about arcs $\overparen{AB}$ and $\overparen{CD}$?

3 In a circle, if two chords are congruent, what may you conclude about their arcs? State your conclusion as a theorem. **(Theorem 9.2.1)** ❖

EXERCISES & PROBLEMS

Communicate

In Exercises 1–5, classify each statement as true or false and explain your reasoning.

1. Every diameter of a circle is also a chord of the circle.

2. Every radius of a circle is also a chord of the circle.

3. Every chord of the circle contains exactly two points of the circle.

4. If two chords of a circle are congruent, then their arcs are congruent.

5. If two arcs of a circle are congruent, then their chords are also congruent.

Practice & Apply

Use the figure for Exercises 6–12.

6. Name the center of the circle.

7. Name a radius of the circle.

8. Name a chord of the circle.

9. Name a diameter of the circle.

10. Name a central angle of the circle.

11. Name a minor arc and a major arc of the circle.

12. Describe the conditions necessary for two arcs to be congruent.

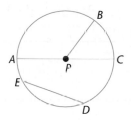

13. 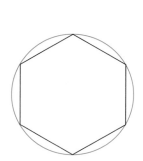 **Portfolio Activity**   In Exploration 1 you constructed a regular hexagon. Construct a regular 12-gon by finding the perpendicular bisectors of each side of the hexagon and marking the intersection of the bisectors with the circle. Connect the new points and the original hexagon vertices to create a regular 12-gon.

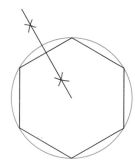

14. 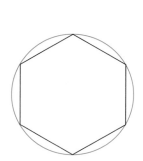 **Portfolio Activity**   Use the 12-gon from Exercise 13 to construct a 12-petal circle flower.

Find the measure of each arc using the central angle measures given on circle Q.

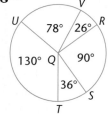

15. $\overarc{RT}$

16. $\overarc{UR}$

17. $\overarc{VS}$

18. $\overarc{VTR}$

19. $\overarc{US}$

20. $\overarc{SUV}$

21. **Civil Engineering**   An engineer needs to calculate the length of a circular section of road. Use the measurements given to determine the length.

22. 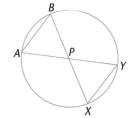 **Statistics** At Smith High School, 450 students are freshmen, 375 students are sophomores, 400 students are juniors, and 325 students are seniors. Create a pie chart showing the distribution of the students. First, find the percentage for each class of the total student body. Multiply each percentage by 360°, and then find the measure of the central angle.

23. In a circle, if two arcs are congruent, what can you conclude about their chords? State your conclusion as a theorem and prove it.

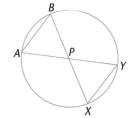 **Algebra** For each arc length measure and radius given below, find the measure of the central angle.

24. 14π; $r = 70$

25. 20π; $r = 100$

26. 3π; $r = 15$

27. 5π; $r = 25$

28. In $\odot P$, if $m\angle APB = 43°$ and $AB = 5$ cm, find XY.

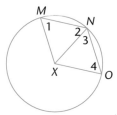

Use circle X, with $\angle MXN \cong \angle NXO$, for Exercises 29–31.

29. Explain why $\overline{MN} \cong \overline{NO}$.

30. Explain why $\overline{MX} \cong \overline{NX}$ and $\overline{NX} \cong \overline{OX}$.

31. Explain why $\triangle MNX \cong \triangle NOX$.

32. **Portfolio Activity**
Look in magazines for examples of circles used as symbols or product logos to add to your portfolio. Write a short description that identifies a unique feature of each symbol.

33. What effect does doubling the radius of a sphere have on the volume? **[Lessons 7.1, 7.6]**

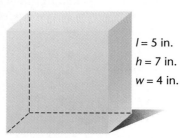 **Algebra** Find the volume and surface area for each solid given below. **[Lessons 7.2, 7.6]**

34.

47 cm

35.

l = 5 in.
h = 7 in.
w = 4 in.

Are the triangles similar? Why or why not? **[Lessons 8.3, 8.4]**

36.

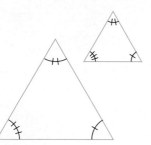

37.

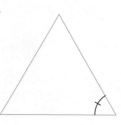

38.

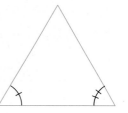

Look Beyond

39. Imagine a line and a circle in the same plane. Describe three ways that the line and circle can intersect.

40. A line that touches a circle in only one point is called a tangent. The perpendicular to the tangent at the point of intersection passes through which point?

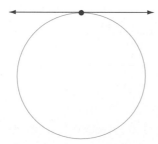

Exploring

Momentum of Shuttle   Shuttle's Orbit

Pull of
the Earth

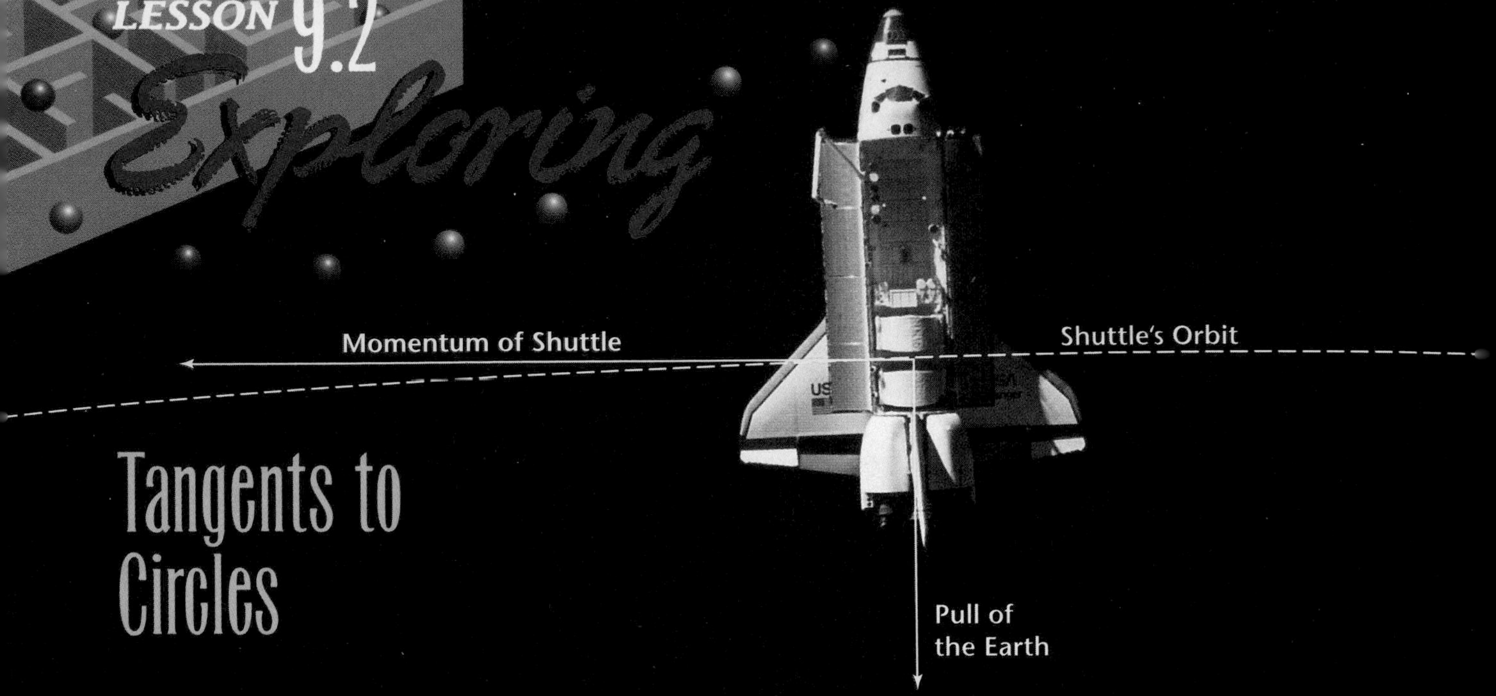

Tangents to
Circles

Why *The momentum of an
object orbiting the
Earth is in the direction of a
line tangent to the orbit of the
object. You will encounter such
lines frequently, both in the
arts and the sciences.*

Secants and Tangents

A line can intersect a circle in three ways.

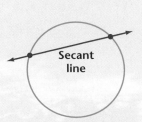

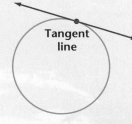

Secant
line

Tangent
line

2 points of intersection   1 point of intersection   0 points of intersection

The word *tangent* comes from the Latin word meaning "to touch." The word *secant* comes from the Latin word meaning "to cut." Why are these words appropriate names for the lines they describe?

Exploration 1 *Radii Perpendicular to Chords*

You will need

Geometry technology or Ruler, compass, and protractor

Geometry Graphics

 1 Draw ⊙*P* with chord $\overline{AB}$.

 2 Construct a radius that is perpendicular to the chord. Label the point of intersection *X*.

 3 Measure $\overline{AX}$ and $\overline{BX}$. What seems to be true of the measures of the two segments?

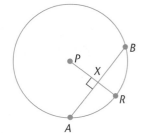

 4 Repeat the above steps using different circles and different chords. If you are using geometry technology, experiment by changing the size of the circle and by dragging the chords to different locations. Make a conjecture about radii that are perpendicular to chords in circles. **(Thm 9.2.2)**

5 Draw segments $\overline{PA}$ and $\overline{PB}$ in one of your circles. Using this diagram, write out a paragraph proof of your conjecture. ❖

EXTENSION

ALGEBRA
Connection

⊙*P* has a radius of 5 inches and *PX* is 3 inches. $\overline{PR}$ is perpendicular to $\overline{AB}$ at point *X*. Find *AB*.

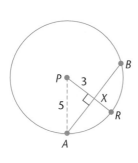

By the "Pythagorean" Right-Triangle Theorem,

$$(AX)^2 + 3^2 = 5^2$$

$$(AX)^2 = 5^2 - 3^2$$

$$(AX)^2 = 16$$

$$AX = 4.$$

By the conjecture you proved in Exploration 1, $\overline{PR}$ bisects $\overline{AB}$, so *BX* = 4. Therefore, *AB* = *AX* + *BX* = 4 + 4 = 8. ❖

Exploration 2 *Radii and Tangents*

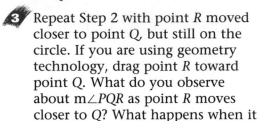

You will need
Geometry technology or
Ruler, compass, and protractor

 1 Draw ⊙*P* with radius $\overline{PQ}$.

 2 Locate a point *R* on the circle and draw a line $\overleftrightarrow{QR}$. Measure ∠*PQR*.

3 Repeat Step 2 with point *R* moved closer to point *Q*, but still on the circle. If you are using geometry technology, drag point *R* toward point *Q*. What do you observe about m∠*PQR* as point *R* moves closer to *Q*? What happens when it coincides with *Q*?

4 Make a conjecture about the relationship between a tangent to a circle and a radius drawn to the point of tangency. ❖

CRITICAL
Thinking

In the diagram, which is like the one from Exploration 1, imagine moving point *X* toward point *R*. What happens to intersection points *A* and *B* as *X* gets closer and closer to *R*? What happens when *X* touches *R*? What conjecture does this suggest to you? How is it different from the conjecture you made in Exploration 2?

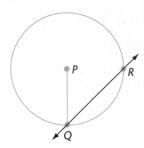

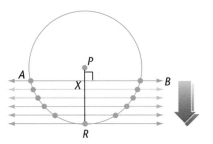

An Unusual Proof

The following proof is related to the conjecture you made in Exploration 2. It uses the fact that the hypotenuse is the longest side of a triangle. Its method is unlike the proofs you have studied so far in this book.

Given: Point *P* is on ⊙*O*, and $\overline{OP}$ is perpendicular to $\overline{AB}$.

Prove: $\overline{AB}$ is tangent to ⊙*O* at *P*.

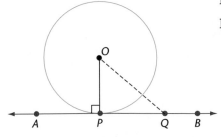

Proof: Choose any point on $\overline{AB}$ other than *P* and label it *Q*. Draw right triangle *OPQ*. Since $\overline{OQ}$ is the hypotenuse of the triangle, it is longer than $\overline{PO}$, which is a radius of the circle. Therefore, the point *Q* does not lie on the circle. This is true for all points on $\overline{AB}$ except point *P*, so $\overline{AB}$ touches the circle just at the one point. By definition, $\overline{AB}$ is tangent to ⊙*O* at *P*.

The result proven here can be stated as a theorem.

> ### TANGENT THEOREM
> If a line is perpendicular to a radius of a circle at its endpoint, then the line is tangent to the circle. **9.2.3**

The converse of the theorem, which is also true, will be proven in Look Beyond, Exercises 22–25.

An artist wants to draw the largest circle that will fit into a square.

1. She connects the midpoints of each of the sides of the square by segments parallel to the sides.

2. She measures the distance from the intersection P of the new segments to the sides of the square. Using this distance as the setting of her compass, she draws a circle with its center at the point P. ❖

CRITICAL Thinking

How can you prove that the artist's method gives the desired result? How can you show that no part of the circle lies outside of the square?

Exercises & Problems

Communicate

1. Explain the three ways in which a line and a circle intersect in a plane.

2. Explain how a secant intersects a circle.

3. Explain how a tangent intersects a circle.

4. Explain what "point of tangency" is.

Practice & Apply

Use ⊙R, with $\overline{RY} \perp \overline{XZ}$ at W, for Exercises 5–7.

5. $\overline{XW} \cong$? .

6. If $RY = 7$ and $RW = 2$, what is the length of $\overline{XW}$? of $\overline{WZ}$?

7. If $RY = 3$ and $RW = 2$, what is the length of $\overline{XW}$? of $\overline{WZ}$?

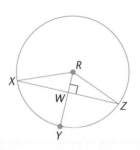

8. **Communications** A radio station installs a VHF radio tower that stands 1500 feet tall. What is the maximum effective signal range for the tower? The diagram below suggests a way to utilize tangents to solve the problem. Use the "Pythagorean" Right-Triangle Theorem to find d.

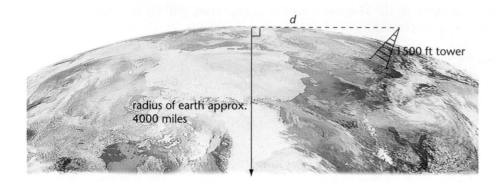

9. How many tangent lines can a circle have? Explain your reasoning.

10. How many lines are tangent to a circle at a given point on the circle? Explain your reasoning.

11. **Space Flight** The space shuttle orbits at 155 miles above the Earth. If the diameter of the Earth is approximately 8000 miles, how far is it from the shuttle to the horizon?

Space shuttle

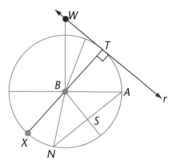

Algebra Use the diagram to find the indicated lengths. Line r is tangent to $\odot B$ at T. $BT = 2$, $BS = 1$, and $WT = 5$.

12. $BA = ?$ 13. $SA = ?$

14. $SN = ?$ 15. $BW = ?$

16. $XT = ?$

17. In Exploration 1 you proved a conjecture about radii that are perpendicular to a chord. Prove the following converse.

THEOREM
The perpendicular bisector of a chord passes through the center of the circle.   **9.2.4**

18. **Algebra** A triangle has a perimeter of 24 cm and an area of 24 sq cm. What is the perimeter and area of a larger similar triangle if the scale factor is $\frac{2}{1}$? **[Lessons 8.1, 8.6]**

19. **Algebra** A rectangle has a perimeter of 22 ft and an area of 22 sq ft. What is the perimeter and area of a larger similar rectangle if the scale factor is $\frac{8}{3}$? **[Lessons 8.1, 8.6]**

20. **Algebra** A rectangular prism has dimensions $l = 12$ in., $w = 8$ in., and $h = 15$ in. What is the volume of a larger similar rectangular prism if the scale factor is $\frac{5}{3}$? **[Lessons 8.1, 8.6]**

21. **Algebra** A cylinder has a radius of 4 ft and a height of 27 ft. What is the surface area of a larger similar cylinder if the scale factor is $\frac{4}{1}$? **[Lessons 8.1, 8.6]**

Look Beyond

Answer the questions below to prove the following theorem.

THE CONVERSE OF THE TANGENT THEOREM
If a line is tangent to a circle, then it is perpendicular to a radius of the circle at the point of tangency.   **9.2.5**

22. Suppose that the theorem is *false*. That is, suppose that m is tangent to circle O at point A, but m is *not* perpendicular to $\overline{OA}$. If this is true, then there *is* some segment, different from $\overline{OA}$, that is perpendicular to m. Call that segment $\overline{OB}$. Then $\triangle OBA$ is a right triangle. (What is the hypotenuse of $\triangle OBA$?)

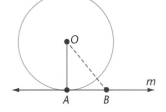

Which segment would be longer, $\overline{OA}$ or $\overline{OB}$?

23. Point B must be in the exterior of the circle, since m is a tangent line. What does this imply about the relative lengths of $\overline{OA}$ and $\overline{OB}$? Explain your answer.

24. Compare your answers to Exercises 22 and 23. What do you observe?

25. When an assumption leads to a contradiction, then the assumption must be rejected. This is the basis for a type of proof known as an **indirect proof**. Explain how the argument above leads to the desired conclusion.

Exploring Inscribed Angles and Arcs

A carpenter's square can be used to find the center of a circle.

Inscribed Angles

An **inscribed angle** is a special type of angle whose vertex lies on the circle and whose rays each extend from the vertex and intersect the circle again. The arc from one of these intersection points to the other is called the intercepted arc of the angle. (The vertex of the angle does not lie on the intercepted arc.)

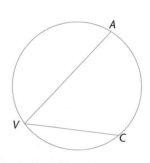

∠AVC is inscribed in $\overarc{AVC}$ *and is said to* **intercept** *(cut off)* $\overarc{AC}$. $\overarc{AC}$ *is the* **intercepted arc** *of* ∠AVC.

A special relationship exists between the measure of an inscribed angle and the degree measure of its intercepted arc.

Exploration 1 Intercepted Arcs

Geometry Graphics

You will need
Geometry technology or
Ruler, compass, and protractor

1 Draw three different figures in which inscribed angle *AVC* intercepts an arc of the circle. Include one major arc, one minor arc, and one semicircle.

2 Measure the inscribed angle and the intercepted arc in each figure. (You will need to add central angles to determine the measures of the arcs.)

3 Compare the measure of the inscribed angle and its intercepted arc in each case.

4 If m$\overset{\frown}{AC}$ = 148°, what is m∠*AVC*? If m∠*AVC* = 53°, what is m$\overset{\frown}{AC}$?

5 Make a conjecture about the relationship between the measure of an inscribed angle and its intercepted arc. ❖

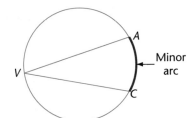

Minor arc

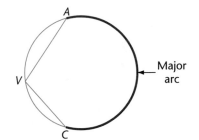

Major arc

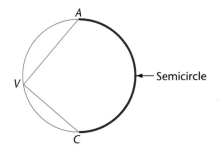

Semicircle

Exploration 2 Proving the Relationship

You will need
No special tools

Part I

1 In the figure, one ray of the inscribed angle contains the center of the circle. What is the relationship between m∠1 and m∠2?

2 Notice that ∠3 is an exterior angle in △*AVP*. What is the relationship among m∠3, m∠1, and m∠2?

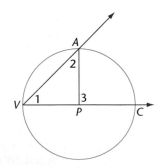

3 Fill in a table like the one below. For each entry in the last row of the table, give a reason.

m∠1	m∠2	m∠3	m$\overset{\frown}{AC}$
20°	?	?	?
30°	?	?	?
40°	?	?	?
$x°$	?	?	?

4 What does your table show about the relationship between m∠1 and m$\overset{\frown}{AC}$?

Part II

1 In the figure at right, the center of the circle is in the interior of the inscribed angle. What is the relationship between m∠1 and m$\overset{\frown}{AX}$? between m∠4 and m$\overset{\frown}{CX}$?

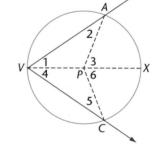

2 Fill in a table like the one below. For each entry in the last row of the table, give a reason.

m∠1	m$\overset{\frown}{AX}$	m∠4	m$\overset{\frown}{CX}$	m∠AVC	m$\overset{\frown}{AXC}$
20°	?	20°	?	?	?
30°	?	20°	?	?	?
40°	?	50°	?	?	?
$x°$	?	$y°$	?	?	?

3 What does your table show about the relationship between m∠AVC and m$\overset{\frown}{AC}$? ❖

CRITICAL
Thinking

How can you prove the case in which the inscribed angle does not encompass the center of the circle?

The results of the explorations are stated in the following theorem.

INSCRIBED ANGLE THEOREM
An angle inscribed in an arc has a measure equal to one-half the measure of the intercepted arc.

9.3.1

EXTENSION

Find the measure of $\angle A$.

$\angle A$ is inscribed in a half-circle; therefore, the arc it intercepts is also a half-circle. The measure of the intercepted arc is thus 180° (why?), and so the measure of $\angle A$ is $\frac{1}{2} \times 180°$, or 90°. ❖

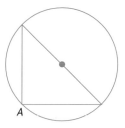

The extension above illustrates the following corollary.

INSCRIBED ANGLE COROLLARY
An angle inscribed in a half-circle is a right angle.

9.3.2

EXTENSION

Find the measure of $\angle B$.

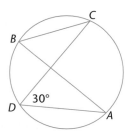

The measure of $\overset{\frown}{AC}$ is twice the measure of $\angle D$, or $2 \times 30° = 60°$. The measure of $\angle B$ is one-half the measure of $\overset{\frown}{AC}$, or $\frac{1}{2} \times 60° = 30°$. ❖

The extension above illustrates an important principle. Notice that $\angle D$ and $\angle B$ intercept the same arc. As you saw, they have the same angle measure. This leads to the following theorem.

THEOREM
If two inscribed angles intercept the same arc, then they have the same measure.

9.3.3

Lesson 9.3 Exploring Inscribed Angles and Arcs **487**

Given ⊙P with diameter $\overline{AB}$, m$\widehat{CB}$ = 110°, and m$\widehat{BD}$ = 130°. Find the measures of ∠1, ∠2, ∠3, ∠4, ∠APC, ∠ADB, and ∠CAD.

Copy the figure and add information as you work through the solution.

Begin by filling in as many arc measures as you can. Since $\overline{AB}$ is a diameter, $\widehat{ACB}$ and $\widehat{ADB}$ each measure 180°. Therefore, m$\widehat{AC}$ = 70° and m$\widehat{AD}$ = 50°.

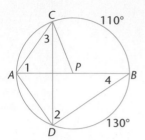

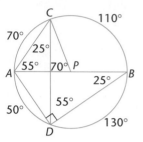

Angles 1 and 2 intercept $\widehat{CB}$. Thus m∠1 = m∠2 = $\frac{1}{2}$(110°) = 55°.

Angles 3 and 4 intercept $\widehat{AD}$. Thus m∠3 = m∠4 = $\frac{1}{2}$(50°) = 25°.

∠APC is a central angle. Thus m∠APC = 70°.

∠ADB is inscribed in a semicircle. Thus m∠ADB = 90°.

∠CAD intercepts $\widehat{CBD}$. Thus m∠CAD = $\frac{1}{2}$(110° + 130°) = 120°. ❖

Carpentry   A student needs to find the center of a small table top. How can she do this using a carpenter's square? The student inscribes two right angles, ∠ABC and ∠DEF, inside the circle. She then adds the hypotenuses of the triangles by connecting points A and C and points D and F. The intersection of $\overline{AC}$ and $\overline{DF}$ is the center of the table. ❖

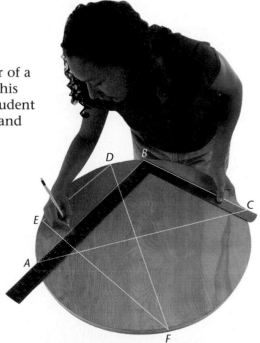

CRITICAL
Thinking

How can the student check the accuracy of her procedure by carrying the process a step further? Discuss.

EXERCISES & PROBLEMS

Communicate

Refer to circle *O* in the photo for Exercises 1–5.

1. Explain how to find m∠1.

2. Explain how to find m$\widehat{AC}$.

3. Name an inscribed angle on circle *O*.

4. Explain how to find m∠2.

5. Explain how to find m$\widehat{BD}$.

6. Explain why two angles inscribed in the same arc have the same measure.

Practice & Apply

 Algebra For Exercises 7–10, refer to the circle at right.

7. m$\widehat{AB}$ = 68°   m∠C = _?_   m∠D = _?_

8. m∠D = 30°   m$\widehat{AB}$ = _?_   m∠C = _?_

9. m$\widehat{CD}$ = 87°   m∠B = _?_   m∠A = _?_

10. m∠B = a°   m$\widehat{DC}$ = _?_   m∠A = _?_

 Algebra For Exercises 11–18, refer to ⊙*P* with diameter $\overline{AC}$. Find the following:

11. m∠A

12. m∠B

13. m∠BCA

14. m$\widehat{AB}$

15. m∠PCD

16. m∠CPD

17. m$\widehat{DC}$

18. m$\widehat{AD}$

 Algebra Quadrilateral *QUAD* is inscribed in the circle at right. Given m∠U = 100°, find the following:

19. m$\widehat{QDA}$

20. m$\widehat{QUA}$

21. m∠D

22. m∠Q + m∠A

23. Stained Glass An artist is creating a circular stained-glass window with the design shown. The artist wishes for the arc intercepted by ∠A and ∠B to measure 80°. What are the measures of ∠A and ∠B?

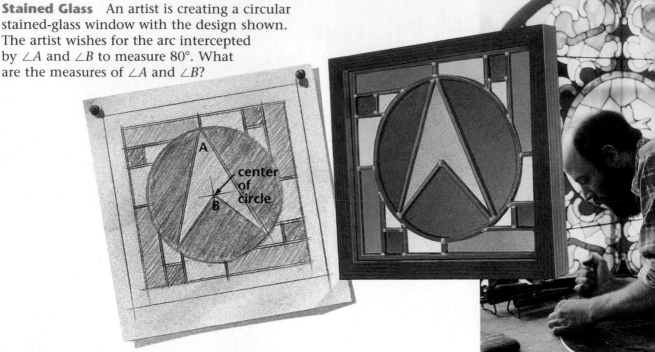

center of circle

A

B

Look Back

24. Given: 1. If the bees are happy, then it pours.
2. If flowers grow, then the bees are happy.
3. If it rains, then flowers grow.

Prove: "If it rains, then it pours" by arranging the three given statements according to the If-Then Transitive Property. **[Lesson 2.2]**

25. Given: *ABCD* is a rhombus with *AC* = 6 and *BD* = 8. Find the perimeter of *ABCD*. **[Lesson 5.4]**

26. Given: In ⊙P, *AB* = 12; *PX* = 4. Find the length of the radius $\overline{PQ}$. **[Lesson 5.4]**

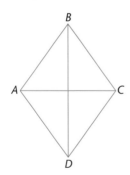

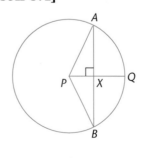

Look Beyond

27. Consider the following argument: *If an object consists of a set of points equidistant from a given point, then the object is a circle. The object at the right is not a circle. Therefore, the object at the right does not consist of all points equidistant from a given point.* Is the argument valid? Explain your reasoning.

LESSON 9.4

Exploring

Circles and Special Angles

why *Navigators at sea must use principles of geometry in many ways. For example, the "horizontal angle of danger" enables a navigator to stay a safe distance from a dangerous region.*

Lighthouses are landmarks for navigators at night, and they often serve as warnings of dangerous shoals or reefs.

Angles or intersecting lines that touch a circle in two or more places can be studied systematically. There are three cases to consider, according to the placement of the vertex of the angles.

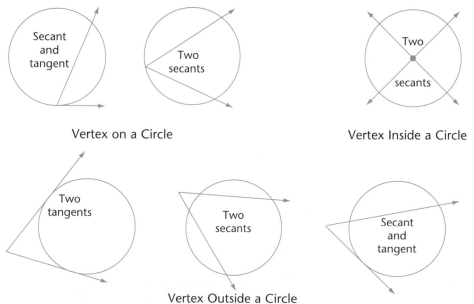

Secant and tangent

Two secants

Vertex on a Circle

Two secants

Vertex Inside a Circle

Two tangents

Two secants

Secant and tangent

Vertex Outside a Circle

Exploration 1 *Vertex on a Circle, Secant-Tangent Angles*

You will need
No special tools

In this exploration you will examine
three special cases of secant-tangent
angles.

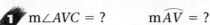

Secant
and
tangent

Case 1: The secant-tangent angle
is a right angle. The secant line
contains the center of the circle.

1 $m\angle AVC = ?$ $m\widehat{AV} = ?$

2 How does this relationship compare
with the one between an inscribed
angle and its intercepted arc?

Case 2: The secant-tangent angle is acute.

1 Copy and complete the
following table.

m$\widehat{AV}$	m∠1	m∠2	m∠PVC	m∠AVC
120°	120°	30°	?	60°
100°	?	?	?	?
80°	?	?	?	?
60°	?	?	?	?
x°	?	?	?	?

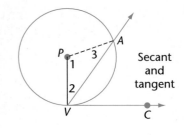

Secant
and
tangent

2 Complete the following statement:

*The measure of an acute secant-tangent angle with its vertex on a circle is
 ? its intercepted arc.*

Case 3: The secant-tangent
angle is obtuse.

1 Copy and complete the
following table.

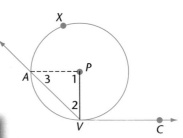

m$\widehat{AXV}$	m∠1	m∠2	m∠PVC	m∠AVC
200°	160°	10°	?	100°
220°	?	?	?	?
240°	?	?	?	?
260°	?	?	?	?
x°	?	?	?	?

2 Complete the following statement:

The measure of an obtuse secant-tangent angle with its vertex on a circle is __?__ *its intercepted arc.*

How does this relationship compare with that of an inscribed angle and its intercepted arc?

Complete the theorem below. ❖

> **THEOREM**
>
> If a tangent and a secant (or a chord) intersect on a circle at the point of tangency, then the measure of the angle formed is __?__ the measure of its intercepted arc.
>
> **9.4.1**

•Exploration 2 *Lines Intersect Inside a Circle*

You will need

No special tools

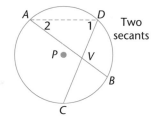

Two secants

1 In the given figure, note that ∠AVC is an exterior angle of △ADV. What is the relationship between the measure of ∠AVC and the measures of ∠1 and ∠2?

2 Copy and complete the following table.

m$\widehat{AC}$	m$\widehat{BD}$	m∠1	m∠2	m∠AVC	m∠DVB
160°	40°	80°	20°	100°	100°
180°	70°	?	?	?	?
x_1°	x_2°	?	?	?	?

3 Complete the theorem below. ❖

> **THEOREM**
>
> The measure of an angle formed by two secants or chords intersecting in the interior of a circle is __?__ of the __?__ of the measures of the arcs intercepted by the angle and its vertical angle.
>
> **9.4.2**

•Exploration 3 *Vertex of an Angle Is Outside a Circle*

You will need

No special tools

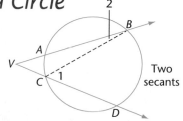

Two secants

1 In the given figure, ∠1 is an exterior angle of △BVC. What is the relationship between the measure of ∠1 and the measures of ∠2 and ∠AVC?

2 Copy and complete the following table.

m$\widehat{BD}$	m$\widehat{AC}$	m∠1	m∠2	m∠AVC
200°	40°	100°	20°	80°
250°	60°	?	?	?
100°	50°	?	?	?
80°	20°	?	?	?
$x_1°$	$x_2°$	?	?	?

3 Complete the theorem below. ❖

THEOREM

The measure of an angle formed by two secants intersecting in the exterior of a circle is __?__ of the __?__ of the measures of the intercepted arcs.

9.4.3

You will explore the two remaining configurations in the exercises.

APPLICATION

Navigation The illustration shows a ship at point *F* and two lighthouses at points *A* and *B*. A circle has been drawn that encloses a region of dangerous rocks. The captain wants to avoid the danger by staying outside the circle.

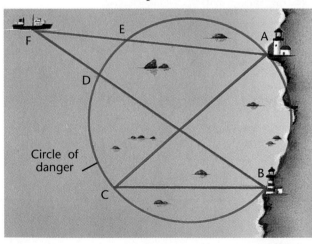

The ship's navigator measures ∠F, the angle between his lines of sight to the two lighthouses. He knows that if m∠F is less than m∠C, then the ship is outside the circle of danger. ∠C is known as the "horizontal angle of danger." Why does this method work?

Notice that ∠C is an inscribed angle and that it intercepts arc $\widehat{AGB}$. So if the ship is on the circle, then ∠F ≅ ∠C. It should be obvious that if m∠F < m∠C, the ship is outside the circle. ❖

CRITICAL *Thinking*

How can you use theorems from this lesson to prove this "obvious" fact? How can you prove that if ∠F > ∠C, then the ship is inside the circle?

The following extension illustrates how to use the angle-arc relationships explored in this lesson.

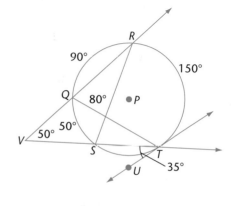

Given: $\overline{TU}$ is tangent to $\odot P$ at T.

$m\widehat{QR} = 90°$;

$m\widehat{RT} = 150°$;

$m\widehat{QS} = 50°$.

Find the following:

A $m\angle STU$ **B** $m\angle 1$ **C** $m\angle 2$

Make a sketch of the figure. As you obtain new information, add it to the figure.

Solution ➤

ALGEBRA
Connection

A Since the vertex of $\angle STU$ is on $\odot P$, and is formed by a tangent and a secant line, $m\angle STU = \frac{1}{2} m\widehat{ST}$. To find $m\widehat{ST}$, note that $50° + 90° + 150° + m\widehat{ST} = 360°$. Thus, $m\widehat{ST} = 70°$ and $m\angle STU = \frac{1}{2}(70°) = 35°$.

B Since the vertex of $\angle 1$ is in the interior of $\odot P$ and is formed by two intersecting chords,
$m\angle 1 = \frac{1}{2} (m\widehat{QR} + m\widehat{ST}) = \frac{1}{2}(90° + 70°) = 80°$.

C Since the vertex of $\angle 2$ is in the exterior of the circle and is formed by two secants, $m\angle 2 = \frac{1}{2}(m\widehat{RT} - m\widehat{QS}) = \frac{1}{2}(150° - 50°) = 50°$. ❖

EXERCISES & PROBLEMS

Communicate

1. The measure of an acute secant-tangent angle with its vertex on a circle is __?__ .

2. The measure of an obtuse secant-tangent angle with its vertex on a circle is __?__ .

3. If two secants (or chords) intersect in the interior of a circle, the measure of any angle formed is __?__ .

4. If two secants intersect in the exterior of a circle, the measure of the angle formed is __?__ .

Practice & Apply

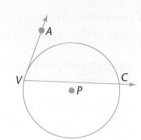

Algebra In the figure, $\overline{VA}$ is tangent to ⊙P at V.

5. If m$\widehat{VC}$ = 150°, find m∠AVC.

6. If m∠AVC = 80°, find $\widehat{VC}$.

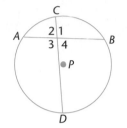

Algebra In the figure, $\overline{AB}$ and $\overline{CD}$ are chords, m$\widehat{CB}$ = 60°, and m$\widehat{AD}$ = 110°. Find each of the following:

7. m∠1   **8.** m∠2

9. m∠3   **10.** m∠4

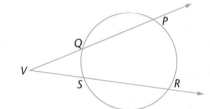

Algebra In the figure, m$\widehat{PR}$ = 100° and m$\widehat{QS}$ = 20°.

11. Find m∠QVS.

Exercises 12–22 explore two additional angle-circle cases.

Case 1: A secant and a tangent intersect with a vertex that is outside the circle. Use ⊙P for Exercises 12–15. $\overrightarrow{VC}$ is tangent to ⊙P at C.

12. What is the relationship between m∠1 and the measures of ∠2 and ∠AVC?

13. Copy and complete the table.

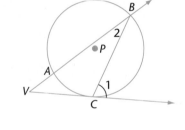

m$\widehat{BC}$	m$\widehat{AC}$	m∠1	m∠2	m∠AVC
250°	60°	(a)	(b)	(c)
200°	40°	(d)	(e)	(f)
130°	40°	(g)	(h)	(i)
70°	30°	(j)	(k)	(l)
x_1°	x_2°	(m)	(n)	(o)

14. Write an equation that describes m∠AVC in terms of m$\widehat{BC}$ and m$\widehat{AC}$.

15. Complete the following statement.

THEOREM

The measure of a secant-tangent angle with its vertex outside the circle is ___?___.

9.4.4

Case 2: Two tangents intersect with a vertex that is outside the circle.

Use ⊙M for Exercises 16–22. $\overrightarrow{VA}$ and $\overrightarrow{VC}$ are tangent to ⊙M at A and C, respectively.

16. What is the relationship between m∠1 and the measures of ∠2 and ∠AVC?

17. The measure of ∠1 is half the measure of its intercepted arc. Name the arc.

18. The measure of ∠2 is half the measure of its intercepted arc. Name the arc.

19. Suppose m$\widehat{AXC}$ = 260°; find m$\widehat{AC}$.

20. Copy and complete the table.

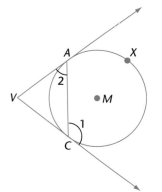

m$\widehat{AXC}$	m$\widehat{AC}$	m∠1	m∠2	m∠AVC
300°	(a)	(b)	(c)	(d)
250°	(e)	(f)	(g)	(h)
220°	(i)	(j)	(k)	(l)
200°	(m)	(n)	(o)	(p)
x°	(q)	(r)	(s)	(t)

21. Write an equation that describes m∠AVC in terms of m$\widehat{AXC}$ and m$\widehat{AC}$.

22. Complete the following statement.

THEOREM

If two tangents intersect in the exterior of a circle, the measure of the angle formed is ___?___.

9.4.5

---

23. Communications The maximum distance a radio signal can reach directly is the length of the segment tangent to the earth's curve. If the tangent radio signals from a tower form angles with the tower to the horizon of 89.5°, what is the measure of the intercepted arc of the Earth? Use your answer and the information in the photo to estimate the area of coverage of the signal.

Note: Drawing is not to scale.

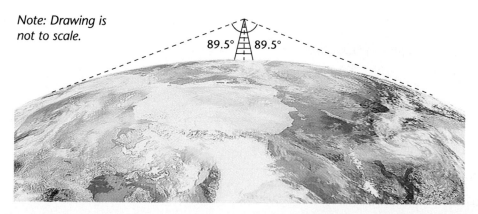

89.5°   89.5°

Exercises 12–22 completed the investigations of angle-arc relationships in circles. Exercises 24–30 summarize the angles studied.

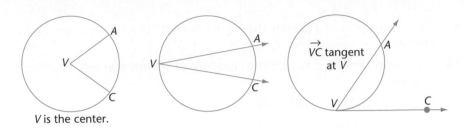

V is the center.

Description of angle   Central angle   **24.** _____   **25.** _____

formula: m∠AVC = ?   m∠AVC = m⌢AC   _____   _____

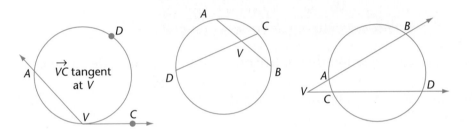

Description of angle   26. _____   **27.** _____   **28.** _____

formula: m∠AVC = ?   _____   _____   _____

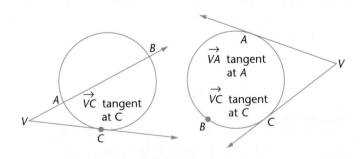

Description of angle   29. _____   **30.** _____

formula: m∠AVC = ?   _____   _____

Use ⊙O for Exercises 31 and 32. $\overleftrightarrow{AF}$
is tangent at A, m$\widehat{CD}$ = 105° and
m$\widehat{BC}$ = 47°. △BDC ≅ △CAB.
m$\widehat{AB}$ = m$\widehat{DC}$.

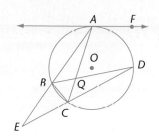

31. Find m∠CQD.

32. Find m∠BQC.

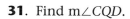 **Look Back**

ABCD is a parallelogram. [Lessons 5.2, 5.5]

33. Find the perimeter.

34. Find the area.

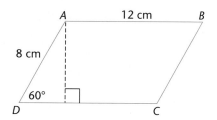

**A photocopier was used to reduce △ABC by a
factor of 0.75.**

35. Are triangles ABC and A′B′C′ similar? Explain
your reasoning. [Lesson 8.2]

36. Find the value of the following ratios: $\frac{A'B'}{AB}$, $\frac{A'C'}{AC}$,
and $\frac{C'B'}{CB}$. [Lesson 8.6]

37. Find A′C′. [Lesson 8.6]

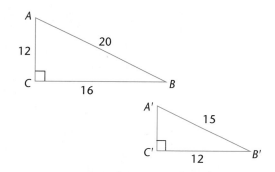

Look Beyond

A **conic section** is the intersection of a right
double cone and a cutting plane. A circle, for
example, can be described as a conic section.
Other geometric figures result when the angle
at which the plane cuts the double cone is
adjusted.

**Make a sketch showing how each of the
following geometric figures can be
produced as a conic section.**

38. ellipse

39. parabola

40. point

41. two intersecting lines

42. line

43. hyperbola

Point of Disaster

Severe Earthquake Hits Los Angeles
Collapsed Freeways Cripple City

Imagine that you are a geologist in the Los Angeles area. You have records from three different stations (A, B, and C) of waves produced by the earthquake. From these seismograms you must pinpoint the epicenter, the place on the earth's surface directly above the origin of the earthquake.

To find the epicenter, you will use two types of seismic waves, **S** waves and **P** waves. Because **S** waves and **P** waves travel at different speeds, you can use the difference in their travel times to determine how far they have gone. You are now ready to begin.

COOPERATIVE LEARNING

1. First, determine how the time between the arrival of the **S** waves and **P** waves depends on how far you are from the origin of the earthquake.

   a. Copy and complete the table. Use the wave speed values shown. What happens to D_t as **d** increases? Why?

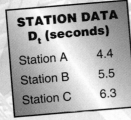

WAVE SPEED VALUES

Speed of **P** wave
$v_p = 6$ km/s

Speed of **S** wave
$v_s = 3.5$ km/s

Distance in kilometers	Travel time in seconds for S wave (t_s)	Travel time in seconds for P wave (t_p)	Time difference between S and P waves (D_t)
10.0	?	?	?
20.0	?	?	?
30.0	?	?	?
40.0	?	?	?

STATION DATA
D_t **(seconds)**

Station A	4.4
Station B	5.5
Station C	6.3

   b. Write an equation for D_t in terms of v_s, v_p, and **d**. (**Hint:** First write equations for t_s and t_p.)

   c. Solve your equation for **d**.

2. Use your equation from Part (c) to find the distances in kilometers from the earthquake origin.

   a. Distance from station A (d_A).

   b. Distance from station B (d_B).

   c. Distance from station C (d_C).

*This seismogram shows **s** waves and **p** waves.*

3. Now you are ready to locate the epicenter. Copy the map of stations A, B, and C.

   Draw three circles to scale:

   Radius = d_A, center at A
   Radius = d_B, center at B
   Radius = d_C, center at C.

   The epicenter is in the region where the three circles overlap.

   For more precision, draw chords connecting the intersections of each pair of circles. The epicenter is where the three chords intersect. How far and in what direction from station B is the epicenter?

   The seismograph readings used to locate the epicenter of the 1994 Los Angeles earthquake were actually from stations all southeast of the epicenter. How would that make finding the location of the epicenter more difficult?

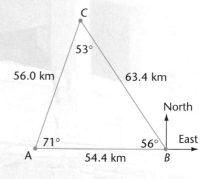

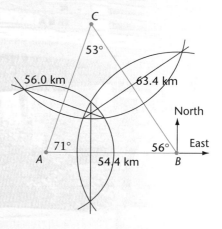

Exploring

Circles and Special Segments

Temple of Kukulkan at Chichén Itzá

The concentration of cenotes is greatest along a semicircular arc.

Chichén Itzá

Chichén Itzá cenote

YUCATAN

Why *An apparent impact crater on the Yucatan Peninsula may explain the extinction of the dinosaurs. An interesting theorem in this lesson will extend your list of techniques for finding the center of such circular objects.*

A semicircular arc of natural wells (cenotes) such as this one at Chichén Itzá, stretches across the Yucatan Peninsula. Some scientists believe that the cenote ring is evidence of an impact crater from an asteroid or comet.

In previous lessons you investigated special angles and arcs formed by secants and tangents to circles. Do you think segments formed by secants and tangents might also have special properties? The following terms will be used in this lesson.

If $\overleftrightarrow{XA}$ is a tangent line and $\overleftrightarrow{XB}$ is a secant line, then

$\overline{XA}$ is a **tangent segment**,

$\overline{XB}$ is a **secant segment**,

$\overline{XC}$ is an **external secant segment**,

and $\overline{BC}$ is a **chord**.

•Exploration 1 *Lines Formed by Tangents*

You will need

Geometry Graphics

Geometry technology or
Ruler and compass

1 You are given ⊙P with tangent lines $\overleftrightarrow{XA}$ and $\overleftrightarrow{XB}$. Construct the figure.

2 Measure the lengths of $\overleftrightarrow{XA}$ and $\overleftrightarrow{XB}$.

3 Write a conjecture about the lengths of two segments that are tangent to a circle from the same external point.

4 Add segments $\overline{AP}$, $\overline{BP}$, and $\overline{XP}$ to your figure. Use the resulting figure to prove your conjecture. ❖

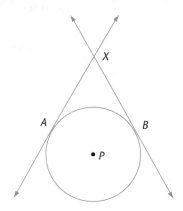

•Exploration 2 *Lines Formed by Secants*

You will need

Geometry Graphics

Geometry technology or
Ruler and compass

Part I

1 You are given a circle with secant lines $\overleftrightarrow{XA}$ and $\overleftrightarrow{XB}$. Construct the figure.

2 Construct segments $\overline{AD}$ and $\overline{BC}$ that intersect at a point O. Name the two triangles that are formed inside the circle.

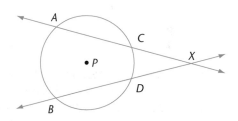

3 Name two angles of the triangles that intercept the same arc of the circle. What can you conclude about these angles?

4 What other angles of the triangle can you show to be congruent? What can you conclude about the triangles inside the circle?

5 Name the two large triangles in your figure that each have a vertex at X. What can you conclude about them? Complete the similarity statement:

$$\Delta AXD \sim \underline{\quad?\quad}$$

6 Complete the proportion by relating two sides of one triangle to two sides of the other triangle:

$$\frac{AX}{?} = \frac{XD}{?}$$

7 Cross-multiply and state your result.

Based on your result, complete the following theorem.

> **THEOREM**
>
> If two secants intersect outside of a circle, then the product of the lengths of one secant segment and its external segment equals __?__ .
> (Whole × Outside = Whole × Outside.) **9.5.1**

8 Measure segments $\overline{XA}$, $\overline{XC}$, $\overline{XB}$, and $\overline{XD}$. Calculate the products $XA \cdot XC$ and $XB \cdot XD$. Do your results agree with the theorem?

Part II

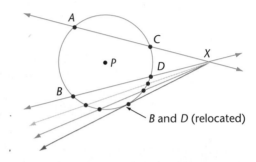

1 Imagine moving point B along the circle so that $\overleftrightarrow{XB}$ becomes a tangent line. You can think of points B and D as coinciding at the point of tangency. What is the relationship between $\overline{XB}$ and $\overline{XD}$?

2 Substitute $\overleftrightarrow{XB}$ for $\overleftrightarrow{XD}$ in your result from Part 1, Step 6. Complete the following theorem.

> **THEOREM**
>
> If a tangent and a secant intersect outside of a circle, then the product of the length of the secant segment and its external segment equals __?__ .
> (Whole × Outside = Tangent Squared) **9.5.2**

3 Construct the figure with point B moved. Measure segments $\overline{XA}$, $\overline{XC}$, and $\overline{XB}$. Calculate the product $XA \cdot XC$ and XB^2. Do your results agree with the theorem? ❖

·Exploration 3 *Intersection of Chords Inside a Circle*

Geometry Graphics

You will need
Geometry technology or
Ruler and compass

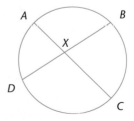

1 You are given a figure with chords $\overline{AC}$ and $\overline{DB}$ intersecting at point X. Construct the figure.

2 Draw segments $\overline{AD}$ and $\overline{BC}$. Name the two triangles that are formed.

3 Name two angles of the triangles that intercept the same arc of the circle. What can you conclude about these angles?

4 What other angle of the triangles can you show to be congruent? What can you conclude about the triangles?

5 Complete the proportion by relating two sides of one triangle to two sides of the other triangle.

$$\frac{DX}{XA} = \frac{?}{?}$$

Cross-multiply and state your result.

6 Based on your result, complete the following theorem.

THEOREM

If two chords intersect inside a circle, then the product of the lengths of the segments of one chord equals __?__ .

9.5.3

---

7 Measure segments $\overline{BX}$, $\overline{DX}$, $\overline{AX}$, and $\overline{CX}$. Calculate the product $BX \cdot DX$ and $AX \cdot CX$. Do your results agree with the theorem? ❖

APPLICATION

Geology On this map of the *cenote* ring, the distance from point *A* on the coast to point *B* near Chichén Itzá measures 3.8 cm. You can use the theorem you have just learned to find the center and diameter of the *cenote* ring.

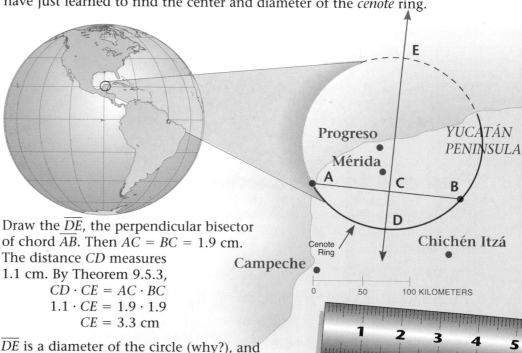

Draw the $\overline{DE}$, the perpendicular bisector of chord $\overline{AB}$. Then $AC = BC = 1.9$ cm. The distance CD measures 1.1 cm. By Theorem 9.5.3,

$$CD \cdot CE = AC \cdot BC$$
$$1.1 \cdot CE = 1.9 \cdot 1.9$$
$$CE = 3.3 \text{ cm}$$

$\overline{DE}$ is a diameter of the circle (why?), and

$$DE = 1.1 + 3.3 = 4.4 \text{ cm}$$

Using the map scale, the actual diameter is about 170 km.
Using a radius of 2.2 cm for the *cenote* ring, you can locate the center point on $\overline{DE}$. This point is 0.7 cm east of Progreso on the map. Using the map scale, the actual distance is a little over 15 km east of Progreso.

EXERCISES & PROBLEMS

Communicate

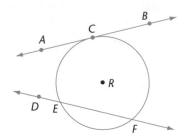

$\overleftrightarrow{AB}$ is tangent at point *C*. Identify each of the following in ⊙*R*.

1. a tangent segment
2. a secant segment
3. an external secant segment

Identify each of the following in ⊙*S*.

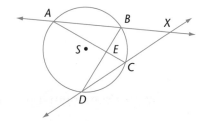

4. a pair of similar triangles
5. 2 pairs of congruent angles

 Algebra Find *x* in each of the following.

6.

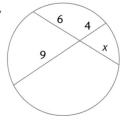

7.

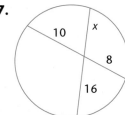

Practice & Apply

 Algebra $\overrightarrow{VA}$ and $\overrightarrow{VB}$ are tangent to ⊙*P*. *VA* = 6 cm; radius of ⊙*P* = 3 cm. Find each of the following.

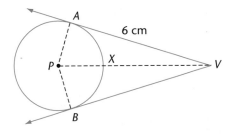

8. *VB*   9. *AP*   10. *BP*   11. *PV*

12. Name an angle that is congruent to ∠*AVP*.

13. Name an angle that is congruent to ∠*APV*.

14. Name an arc that is congruent to $\widehat{AX}$.

 Algebra Use the figure for Exercises 15–17. (You may need to use the quadratic formula.)

15. Given: *VA* = 4; *VB* = 10; *VC* = 5. Find *CD*.

16. Given: *VB* = *x*; *VA* = 6; *VD* = *x* + 3; *VC* = 5. Find *x*.

17. Given: *VB* = *x*; *VA* = *x* − 16; *VD* = 8; *VC* = 5. Find *x*.

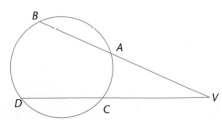

Algebra Use the figure for Exercises 18 and 19.

18. Given: $VY = 16$; $VX = 4$.
Find VW.

19. Given: $VW = 10$; $VX = 8$.
Find VY.

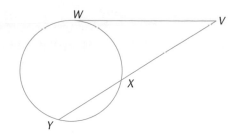

Algebra Use the figure for Exercises 20 and 21.

20. Given: $AX = x$; $XB = x - 2$; $XC = 3$; $XD = 8$.
Find x.

21. Given: $AB = 10$; $CX = 2$; $CD = 12$.
Find AX.

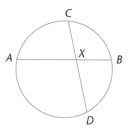

22. Machining Jeff is restoring a clock and needs a new gear drive to replace the broken one shown. To machine a new gear, he must determine the diameter of the original. In the picture, F is the midpoint of $\overline{BD}$. Use the product of chord segments to find the original diameter.

$BD = 10$ cm

$EF = 2.5$ cm

BD = 10.1 cm
EF = 2.5 cm

Draw a net for each solid.   [Lesson 6.1]

23.

24.

25.

**Given: $m\widehat{AD} = 140°$, $m\widehat{BC} = 40°$ and
$m\widehat{CD} = 120°$. Find the following: [Lesson 9.4]**

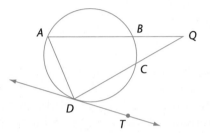

26. m∠AQD

27. m∠QDT

28. m∠QAD

29. m∠ADQ

30. Are there any perpendicular segments in the figure? Explain.

31. Are there any parallel segments in the figure? Explain.

Look Beyond ~~~~

32. **Algebra**   The drawing below suggests a visual proof of the "Pythagorean" Right-Triangle Theorem. Use chord-segment products to explain how the proof works.

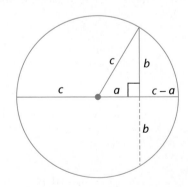

 Why *Computer graphics software must create a variety of geometric objects, such as points, lines, and circles. To do this, the software has subroutines that make use of mathematical representations of these objects—that is, equations.*

Graphing a Circle From an Equation

ALGEBRA
Connection

In your work in algebra, you may have investigated graphs of equations such as $y = 2x - 3$ (a line), $y = x^2 - 3$ (a parabola), and $y = 3 \cdot 2^x$ (an exponential curve). In this lesson you will investigate equations in which both x and y are squared.

EXAMPLE 1

Given: $x^2 + y^2 = 25$

Graphics Calculator

Sketch and describe the graph by finding ordered pairs that satisfy the equation. Use a graphics calculator to verify your sketch.

Solution ➤

When sketching the graph of an equation, it is often helpful to investigate its intercepts, which is where the graph crosses the x- and y-axes. To find the x-intercept(s), find the value(s) of x when $y = 0$.

$$x^2 + 0^2 = 25$$

$$x^2 = 25$$

$$x = \pm 5$$

Thus the graph has two x-intercepts, $(5, 0)$ and $(-5, 0)$.

To find the y-intercept(s), find the value(s) of y when $x = 0$.

$$0^2 + y^2 = 25$$
$$y^2 = 25$$
$$y = \pm 5$$

Thus the graph has two y-intercepts, $(0, 5)$ and $(0, -5)$.

Make a table. Next, set x equal to some other value, such as 3.

$$3^2 + y^2 = 25$$
$$y^2 = 16$$
$$y = \pm 4$$

Thus there are two points with an x-value of 3: $(3, 4)$ and $(3, -4)$.

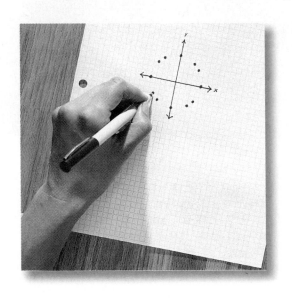

Similarly, by choosing other convenient values for x, a table like the one below is obtained.

x	y	Points on Graph
3	± 4	$(3, 4), (3, -4)$
-3	± 4	$(-3, 4), (-3, -4)$
4	± 3	$(4, 3), (4, -3)$
-4	± 3	$(-4, 3), (-4, -3)$

Add these new points to the graph. A circle with radius 5 and its center at the origin $(0, 0)$ begins to appear. Sketch in the curve. ❖

 CRITICAL Thinking How does the graph change if 25 is changed to 49? to 81? to 51?

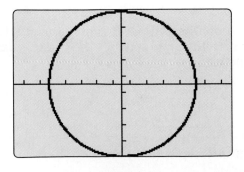

You can also graph the curve using a graphics calculator or a computer with graphics software. You can use the trace function of the computer software or calculator to find the coordinates of individual points on the graph. Note: Some graphing technology requires that you write your equation in the form $y = $ __?__ . If your technology requires this, you will need to solve your equation for y.

Deriving the Equation of a Circle

In a circle, all the points are a certain distance r from a fixed point. In the simplest case, that fixed point is the origin, as shown.

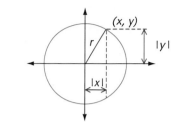

For any point (x, y) on the circle that is not on the x- or y-axis, you can draw a right triangle whose legs have the values $|x|$ and $|y|$. The length of the hypotenuse is the distance r of the point from the origin. So for any such point,

$$x^2 + y^2 = r^2. \quad \text{(Equation 1)}$$

By substituting $y = 0$ or $x = 0$ into the equation, you can see that it is also true for points (x, y) that are on the x- or y-axis.

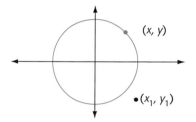

If point (x_1, y_1) is not on the circle, its distance from the origin is some value not equal to r (why?), so the equation will not be true. That is, for the same value of r,

$$x_1{}^2 + y_1{}^2 \neq r^2.$$

Explain why this is true.

Notice that Equation 1 satisfies the following two conditions:

1. It is true of all points (x, y) that are on the circle.

2. It is not true of any point (x, y) that is not on the circle.

Thus Equation 1 is the equation of the circle.

Moving the Center of the Circle

To find the general equation of a circle centered at a point (h, k) that is not at the origin, study the diagram at right. For such a circle,

$$(x - h)^2 + (y - k)^2 = r^2.$$
(Equation 2)

It should also be clear that the equation is not true for a point that does not lie on the equation of the circle.

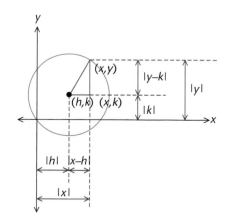

CRITICAL Thinking

How can you show that the relationship given in the diagram holds for points (h, k) in quadrants II, III, and IV?

ALGEBRA
Connection

EXAMPLE 2

Given: $(x - 7)^2 + (y + 3)^2 = 36$

Find the center of the circle and its radius.

Solution ➤

Comparing the given equation with the general equation of the circle, you find the following correspondences:

The general equation		The given equation
$(x - h)^2$	↔	$(x - 7)^2$
$(y - k)^2$	↔	$(y + 3)^2$ or $(y - (-3))^2$
r^2	↔	36

From this you can conclude that

$$h = 7, k = -3, \text{ and } r = 6.$$

That is, the radius of the circle is 6 units, and the center is at $(7, -3)$. ❖

Try This For each of the equations, find the radius and the center of the circle represented. Graph the equation, and compare the graph with your values for the radius and the center of the circle.

a. $(x + 3)^2 + (y - 3)^2 = 49$ **c.** $(x - 3)^2 + (y + 3)^2 + 1 = 50$

b. $(y - 4)^2 + (x - 5)^2 = 30$ **d.** $(x + 2)^2 + (y - 5)^2 = 50$

EXERCISES & PROBLEMS

Communicate

Find the x- and y-intercepts for each equation.

1. $x^2 + y^2 = 100$

2. $x^2 + y^2 = 64$

3. $x^2 + y^2 = 50$

4. State the general form for the equation of a circle.

State the formula for each circle with the center and radius given.

5. radius = 7, center $(4, -5)$

6. radius = 2.5, center $(0, 0)$

7. radius = 10, center $(-1, -7)$

In Exercises 8–11, which equations are circles? How can you tell?

  8. $x^2 = 10 - y^2$ **9.** $x + y^2 = 100$

10. $x^3 + y^3 = 64$ **11.** $x^2 - y^2 = 9$

Practice & Apply

 Algebra Find the center and radius of each circle.

12. $x^2 + y^2 = 100$ **13.** $x^2 + y^2 = 101$

Write an equation of a circle with the given characteristics.

14. center = (0, 0); radius = 6   **15.** center = (0, 0); radius = $\sqrt{13}$

16. Write an equation for the given circle.

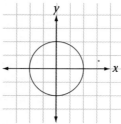

Algebra Find the center and radius of each circle.

17. $(x - 6)^2 + y^2 = 9$ **18.** $x^2 + (y + 3)^2 = 4$

19. $(x + 5)^2 + (y - 2)^2 = 16$ **20.** $(x + 1)^2 + (y + 3)^2 = 19$

21. $x^2 + y^2 = 36$ **22.** $y^2 + (x + 3)^2 = 49$

Algebra Write an equation of a circle with the given characteristics.

23. center = (2, 3); radius = 4   **24.** center = (−1, 5); radius = 7

25. center = (0, 6); radius = 5   **26.** center = (4, −3); radius = $\sqrt{7}$

Algebra Write an equation for the given circle.

27.

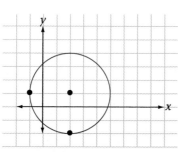

28.

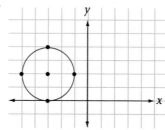

Algebra Use a graphics calculator to find an equation of a circle with the given characteristics, or sketch a graph.

29. center = (5, 3); tangent to the x-axis

30. center = (5, 3); tangent to the y-axis

⌨◇ **Algebra** Use a graphics calculator or graph paper to find the
following:

31. Given: $x^2 + y^2 = 9$

Find the x-intercepts.

Find the y-intercepts.

32. Given: $(x - 2)^2 + y^2 = 9$

Find the x-intercepts.

Find the y-intercepts.

33. Given: $x^2 + (y - 4)^2 = 25$

Find the x-intercepts.

Find the y-intercepts.

34. Given: $(x - 3)^2 + (y - 4)^2 = 25$

Find the x-intercepts.

Find the y-intercepts.

35. Structural Design John
is designing a wheelchair
to use in wheelchair
basketball. He wants the
push rim to be 6.75
inches less in diameter
than the wheel rim. The
wheel diameter is 24.5
inches. Using the wheel
hub as the origin and
inches as the
measurement unit, draw a
graph for the push rim
and wheel rim. Write an
equation for each.

*Wheelchair basketball is a highly competitive, organized sport.
The players' custom-made chairs cost up to $2000.*

You may use geometry technology for Exercises 36–44.

36. Sketch a graph of $(x - 3)^2 + (y - 5)^2 = 4$.

Reflect the graph over the x-axis and sketch the image circle. Write an
equation for the image circle.

37. Sketch a graph of $(x - 4)^2 + (y - 2)^2 = 1$.

Reflect the graph over the y-axis and sketch the graph of the image
circle. Write an equation for the image circle.

38. Sketch a graph of $(x - 2)^2 + y^2 = 9$.

Translate the graph 6 units to the right and sketch a graph of the
image circle. Write an equation for the image circle.

39. Sketch a graph of $(x - 6)^2 + (y - 4)^2 = 9$.

Translate the graph 2 units to the left and then 1 unit down. Sketch a
graph of the image circle and write its equation.

40. Sketch a graph of $(x - 5)^2 + (y - 4)^2 = 9$.

Rotate the graph 180° about the origin. Sketch a graph of the image
circle and write its equation.

41. Find an equation of a circle with center (2, 3) and containing the
point (8, 3).

42. Find an equation of a circle with center (2, 3) and containing the point (8, 11).

43. Find an equation of the tangent to the circle $x^2 + y^2 = 100$ at the point $(-6, 8)$.

44. Show that $x^2 + 6x + y^2 - 4y + 12 = 0$ is a circle.

45. **Portfolio Activity** Draw a design on a coordinate grid consisting of circles and lines. Then list the equations of the geometric figures in the picture on another sheet of paper. Trade lists with a classmate and reproduce the picture by graphing the equations.

Look Back

The wheel of a bicycle has a radius of 14 in. [Lessons 5.3, 9.3]

46. If the wheel makes 3 complete revolutions, how far will the bicycle travel?

47. If the wheel rotates through 45°, how far will the bicycle travel?

48. Find *DE*. **[Lesson 9.5]**   **49.** Find *DE*. **[Lesson 9.5]**

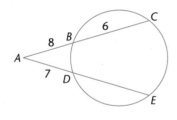 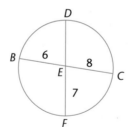

Look Beyond

Unit circle is the name given to the circle with center at the origin and radius 1.

50. Write an equation for the unit circle.

51. What are the coordinates of *A*, *B*, *C*, and *D*?

52. What is the circumference of the unit circle?

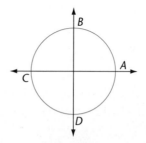

Use the unit circle to find the following.

53. $m\widehat{AB}$

  length of $\widehat{AB}$

54. $m\widehat{ABC}$

  length of $\widehat{ABC}$

55. $m\widehat{ABD}$

  length of $\widehat{ABD}$

THE OLYMPIC SYMBOL

The official symbol of the Olympic Games consists of five interlocking circles representing the continents of Europe, Asia, Africa, Australia, and the Americas. The circles are of different colors—blue, yellow, black, green, and red—and symbolize friendship among peoples.

ACTIVITY 1 The Geometry of the Olympic Symbol

1. Make a copy of the five-circle figure.

2. Use construction techniques to locate the centers of the five circles. Name the centers of the circles *A*, *B*, *C*, *D*, and *E* as shown.

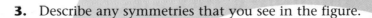

3. Describe any symmetries that you see in the figure.

4. ⊙B can be viewed as a reflection image of ⊙A. Draw the reflection line.

5. Measure the following angles.
   a. m∠ABC   b. m∠CBD
   c. m∠CDE   d. m∠BAC
   e. m∠BCA
   f. the angle with vertex at A and containing the points of intersection of ⊙A and ⊙B

6. Find the following ratios.
   a. $\frac{AB}{BC}$ b. $\frac{AC}{AB}$

7. Describe the interesting features of the figure. Look for congruent angles, segments, and triangles. Look for transformations and special polygons.

8. Suppose that North America and South America each want their own circle. Add a circle to your figure and describe the method that you used.

ACTIVITY 2 A Marching Band Application

The high school band director wants to put on a "Salute to the Olympic Games" at the half-time of the next football game. The band has 120 members. Describe where the members of the band might stand to form an Olympic symbol centered on the 50 yard line.

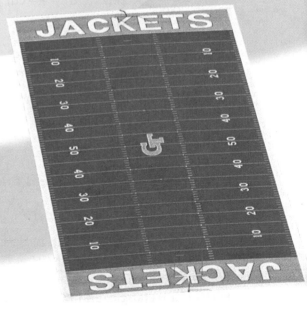

CHAPTER 10

LESSONS

10.1 *Exploring* Tangent Ratios

10.2 *Exploring* Sines and Cosines

10.3 *Exploring* the Unit Circle

10.4 *Exploring* Rotations With Trigonometry

10.5 The Law of Sines

10.6 The Law of Cosines

10.7 Vectors in Geometry

Chapter Project
Plimpton 322 Revisited

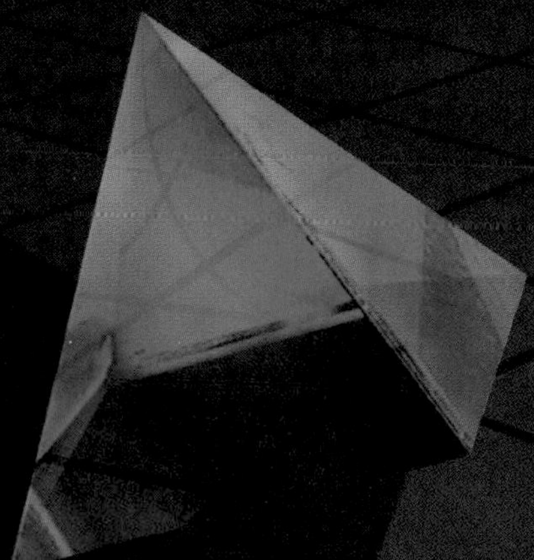

Trigonometry

Have you ever wondered how highway engineers are able to be sure that one section of a freeway or overpass will correctly match up with another section that is under construction a considerable distance away? Accurate measurements and calculations are necessary to ensure success. In this kind of work, trigonometry is an indispensable tool.

Trigonometry, like much of geometry, depends upon triangles. The simple study of relationships between the parts of right triangles quickly leads ultimately to more sophisticated calculation techniques that are well within your grasp at this time in your mathematical studies.

As you progress in your study of trigonometry, you will be introduced to vectors, which are among the most important of all of mathematical concepts for scientists and engineers.

PORTFOLIO ACTIVITY

Choose one, or both, of the following activities.

1. Using a protractor, string, and a small weight, find a way to determine the angle that is formed by your line of sight to the top of a tall structure. Then, by using this angle measure and a trigonometric ratio, determine the height of the structure.

2. The motions of a satellite around a planet, as viewed from the Earth, can be represented by a trigonometric function. Explain this using your own illustrations, and perhaps including your own observational data.

LESSON 10.1

Exploring Tangent Ratios

why *Trigonometry is an essential tool of surveyors. One of the most famous surveys in history was the great Trigonometric Survey of India, which, among other things, determined the height of the world's tallest mountain, Mount Everest.*

•Exploration 1 *A Familiar Ratio*

Geometry Graphics

You will need
Geometry technology or
Ruler and protractor
Calculator

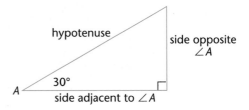

1 Draw a right triangle with a 30° angle. Label this angle *A*.

2 Measure the **side opposite ∠A**. Record your result.

3 One of the sides of ∠A is the hypotenuse of the right triangle. The other side of ∠A is called the **adjacent side**. Measure the adjacent side. Record your result.

4 Divide the measure of the side opposite ∠A by the side adjacent to ∠A. Record your result.

5 Repeat Steps 2–4 for other right triangles with 30° angles. Or compare your result in Step 4 with other class members' results. What do you notice? Make a conjecture about the ratio of the side opposite an acute angle of a right triangle to the side adjacent to the angle. ❖

In an earlier lesson you divided the "rise" of a segment by its "run" to determine its slope. How is the ratio you calculated in the exploration related to the concept of slope? How is it different?

The Trigonometric Ratios

In the right triangles $\triangle ABC$ and $\triangle MNO$, $\angle A$ is congruent to $\angle M$. Therefore, $\triangle ABC$ is similar to $\triangle MNO$. (Explain.)

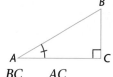

Because the triangles are similar, $\dfrac{BC}{NO} = \dfrac{AC}{MO}$.

Then, by the Exchange Property of Proportions, $\dfrac{BC}{AC} = \dfrac{NO}{MO}$.

In each ratio, the side opposite the angle is the numerator and the side adjacent to the angle is the denominator. The result shows us that if you construct a right triangle with an angle of a given measure, *the ratio of the side opposite the angle to the side adjacent to the angle will always have the same value.* This ratio is the **tangent ratio** of the angle.

The tangent is one of six different ratios that can be formed by the sides of a right triangle. These six ratios form the basis for the study of **trigonometry**. The ideas of trigonometry date back to ancient times. The word itself comes from the Greek word meaning "triangle measurement."

List the six possible side ratios for $\angle A$ in terms of "opposite," "adjacent," and "hypotenuse." Identify the tangent ratio.

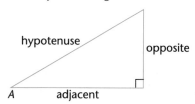

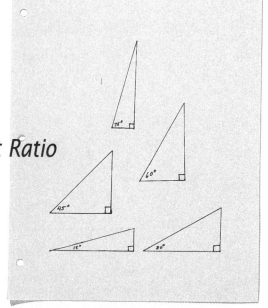

Exploration 2 *Graphing the Tangent Ratio*

You will need
Geometry technology or
Ruler and protractor
Graph paper

1 Draw five different right triangles, as shown, with m$\angle A$ ranging from 15° to 75° at 15° intervals.

Lesson 10.1 Exploring Tangent Ratios **525**

In each of the triangles that you formed in Step 1, measure the following:

$\angle A$,
the side opposite $\angle A$, and
the side adjacent to $\angle A$.

 For each angle, compute its tangent by dividing the measure of the side opposite $\angle A$ by the measure of the side adjacent to $\angle A$. Record your results.

 Plot the ordered pairs (angle measure, tangent) for the different angles you drew. Draw a smooth curve through the points.

ALGEBRA
Connection

Does your graph seem to consistently increase or decrease? If so, does it do so at a steady rate? Describe its behavior. ❖

•Exploration 3 *Using the Tangent Ratio*

You will need

TECHNOLOGY

Calculator

Surveying

A ruler and protractor
A calculator
Your graph from Exploration 2

In this exploration you will simulate or model the work of a surveying crew as they measure the distance across a canyon. To make your answers realistic, you can let 1 centimeter = 10 meters.

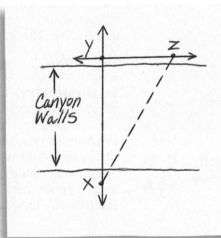

1 Draw curves to represent the sides of a canyon. Draw a line across the canyon as shown. This line represents a line of sight from point X to point Y.

2 Draw a line perpendicular to the line of sight $\overleftrightarrow{XY}$ through point Y. Label a point on the perpendicular as point Z. Connect points X and Z.

3 Measure $\overline{YZ}$ and $\angle ZXY$.

4 The tangent ratio of $\angle ZXY$ is the measure of the side opposite the angle $(\overline{YZ})$ divided by the measure of the side adjacent to the angle $(\overline{XY})$. In mathematical notation,

$$\tan \angle ZXY = \frac{YZ}{XY}.$$

ALGEBRA
Connection

Use your graph from Exploration 2 to find an approximate value for the tangent of $\angle ZXY$. Substitute this value and the measure of $\overline{YZ}$ into your equation. Solve for XY.

5 Measure $\overline{XY}$. Compare your measured value with your calculated value from Step 4. How accurate is your answer? What do you think are the main sources of error in your work? ❖

Find the indicated measure in each figure by using the given equation.

ALGEBRA
Connection

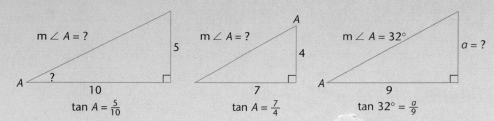

$m \angle A = ?$

5

A ∠ ?

10

$\tan A = \frac{5}{10}$

$m \angle A = ?$

4

7

$\tan A = \frac{7}{4}$

A

$m \angle A = 32°$

$a = ?$

9

A

$\tan 32° = \frac{a}{9}$

APPLICATION

Cultural Connection: Africa   Trigonometry has been used for more than 4,000 years. A certain trigonometric ratio was used as early as 2650 B.C.E. in ancient Egypt.

To build smooth-sided, square pyramids, the ancient Egyptians knew that the ratio of the "rise" and the "run" of the sides must be kept constant.

In modern times, the tangent ratio is generally used for such purposes—that is, the rise divided by the run. The pyramid builders used the reciprocal of the tangent, which is known today as the **cotangent**. The ancient Egyptians called this value the **seked** of the sloping surface. A typical problem from an ancient papyrus reads as follows:

IF A PYRAMID IS 250 CUBITS HIGH AND THE SIDE OF ITS BASE IS 360 CUBITS LONG, WHAT IS ITS SEKED?
—PROBLEM 56 OF THE AHMES PAPYRUS

The pyramid is assumed to be a right square pyramid. One-half the length of a side of the base, or 180 cubits, is the run of the shaded triangle. The rise is 250 cubits.

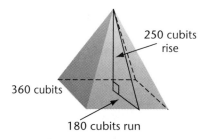

250 cubits
rise

360 cubits

180 cubits run

Thus, seked of pyramid $= \dfrac{\text{run}}{\text{rise}} = \dfrac{180}{250} = \dfrac{18}{25}$.

Unlike our trigonometric ratios, which have no units, the Egyptian ratio was written as though it were a length measure, cubits. A fractional part of a cubit was converted to palms. (There are seven palms to a cubit.) Thus the papyrus reads,

Seked is $5\frac{1}{25}$ palms. ❖

Communicate

1. What is the difference between a tangent ratio and the ratio that the ancient Egyptians called a seked?

2. Does the tangent ratio increase or decrease as an angle gets larger? Explain your answer.

Which sides are opposite and adjacent to ∠A in each right triangle below? Explain your answers. Then find the tangent ratio for ∠A.

3.

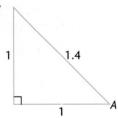

4.

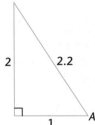

5.

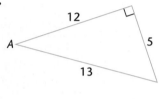

Practice & Apply

Find the tangent ratio for ∠A in each right triangle below.

6.

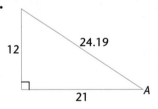

7.

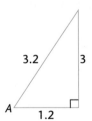

8.

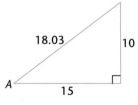

9.

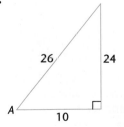

10.

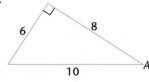

11.

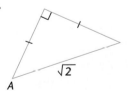

Find the tangent ratio for ∠C in each right triangle below.

12.
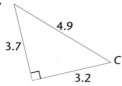
4.9
3.7
C
3.2

13.
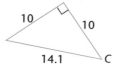
10
10
14.1
C

14.
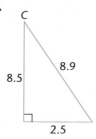
C
8.9
8.5
2.5

15.
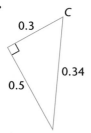
C
0.3
0.34
0.5

16.
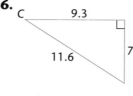
C
9.3
11.6
7

17.
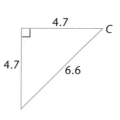
4.7
C
4.7
6.6

Use the graph from Exploration 2 to estimate the angle associated with each tangent ratio.

18. 0.5

19. 1.5

20. 0.85

Technology   Use a calculator with scientific functions to find the tangent ratio or tangent for each angle indicated.

21. 67°

22. 25°

23. 87°

24. 19°

25. 47°

26. 53°

27. 21°

28. 75°

Technology   Use a calculator with scientific functions to find the angle θ for each tangent. Use the inverse tangent function ($\tan^{-1}$). Round to the nearest degree.

29. $\tan \theta = .4663$

30. $\tan \theta = .6009$

31. $\tan \theta = 2.3559$

32. $\tan \theta = 1$

Use a table of trigonometric values or a scientific calculator to answer Exercises 33 and 34.

33. What happens to the tangent values as the angles get larger?

34. What happens to the pattern in Exercise 33 for angles larger than 90°?

Technology   For Exercises 35–39, write an equation and solve for *x*. Use a scientific calculator to find the tangent ratio. An example is given below.

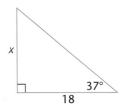

$$\tan 37° = \frac{x}{18}$$

$$0.7536 = \frac{x}{18}$$

$$x = 18(0.7536)$$

$$x = 13.56$$

35.

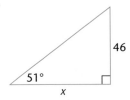

36.

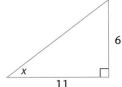

37.

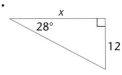

38.

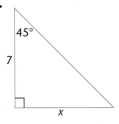

39.

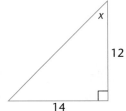

40. Engineering   In order to construct a bridge across a lake, an engineer wishes to determine distance *AB*. A surveyor found that *AC* = 530 meters and m∠*C* = 43°. If ∠*A* is a right angle, find *AB*, the distance across the lake.

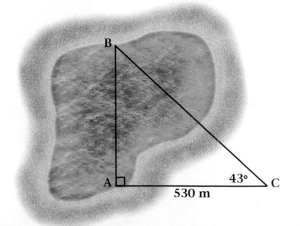

Technology   Use a scientific calculator for Exercises 40–43. Express the answer to the nearest tenth or nearest degree.

41. Engineering   The steepness, or grade, of a highway or railroad is expressed as a percent. A railway that rises 18 feet for 100 feet of horizontal run has a grade of 18 percent, as in the photograph of the cog railway at Pike's Peak, Colorado. What is θ, the angle of inclination of the railway in the photograph?

The maximum grade of the railway at Pike's Peak is 25 percent. What angle is this?

42. Determine the height of the flagpole shown in the figure at the right.

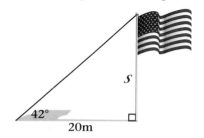

43. Surveying   Points *P* and *Q* are on the north and south rims, respectively, of Glen Canyon, with *Q* directly south of *P*. Point *R* is located 300 feet from *P* and $\overline{QP} \perp \overline{PR}$. The measure of $\angle PRQ$ is 75°. Find *PQ*, the width of the canyon.

Look Back

Find the volume and surface area of each.

44. right prism   **[Lesson 7.2]**   **45.** cylinder   **[Lesson 7.4]**

4

10

6

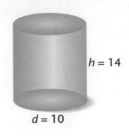

$h = 14$

$d = 10$

46. cone   **[Lesson 7.5]**

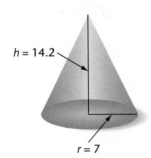

$h = 14.2$

$r = 7$

47. sphere   **[Lesson 7.6]**

$r = 2.5$

Use ⊙P, with $\overline{MN} \perp \overline{RP}$ at Q, for Exercises 48–50.
[Lessons 9.4, 9.5]

48. $\overline{MQ} \cong$ ___?___

49. If $PR = 8$ and $PQ = 3$, what is the length
of $\overline{MQ}$? of $\overline{QN}$?

50. If $PR = 12$ and $PQ = 4$, what is the length
of $\overline{MQ}$? of $\overline{QN}$?

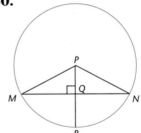

Look Beyond

51. Does tan 20° + tan 60° = tan 80°? Why or why not?

52. If a pyramid is 300 cubits high and the side of its base is 420 cubits
long, what is its seked? (Hint: Use the cotangent as on page 527.)

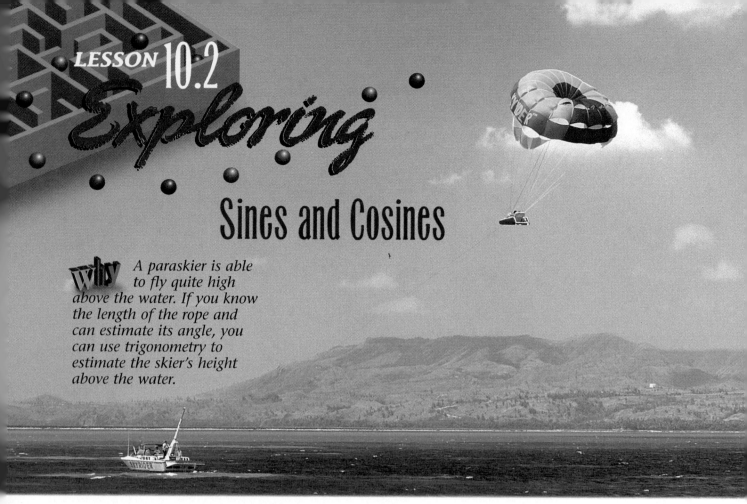

LESSON 10.2

Exploring

Sines and Cosines

Why *A paraskier is able to fly quite high above the water. If you know the length of the rope and can estimate its angle, you can use trigonometry to estimate the skier's height above the water.*

Trigonometric Ratios

In the previous lesson you learned two different ratios for an angle in a right triangle.

$$\text{tangent of } \angle A, \text{ or } \tan A = \frac{\text{length of side opposite } \angle A}{\text{length of side adjacent to } \angle A}$$

$$\text{cotangent of } \angle A, \text{ or } \cot A = \frac{\text{length of side adjacent to } \angle A}{\text{length of side opposite } \angle A}$$

Now consider the following two additional ratios:

$$\textbf{sine of } \angle A, \text{ or } \sin A = \frac{\text{length of side opposite } \angle A}{\text{length of hypotenuse}}$$

$$\textbf{cosine of } \angle A, \text{ or } \cos A = \frac{\text{length of side adjacent to } \angle A}{\text{length of hypotenuse}}$$

These ratios can also be written $\sin \theta$, $\cos \theta$, etc., where θ (theta) is the measure of the angle.

CRITICAL *Thinking*

The cotangent is the reciprocal of the tangent. Is the cosine the reciprocal of the sine? Explain.

Lesson 10.2 Exploring Sines and Cosines **533**

•Exploration 1 *Sines and Cosines*

Scientific Calculator

You will need
Scientific calculator
Graph paper (small grid)

1 As ∠A increases in the illustration below, what happens to the value of sin θ? Does it increase or decrease? Write a conjecture about the sine of 0° and of 90°.

$$\sin \theta = \frac{opp.}{hyp.}$$

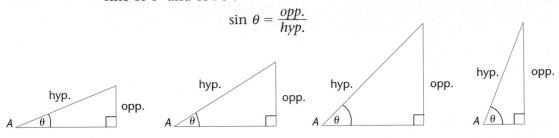

2 As ∠A increases in the illustration below, what happens to the value cos θ? Does it increase or decrease? Write a conjecture about the cosine of 0° and 90°.

$$\cos \theta = \frac{adj.}{hyp.}$$

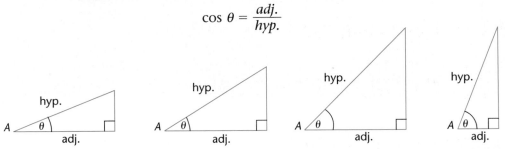

3 Write a statement that contrasts the behavior of the sine and cosine ratios as an angle increases from 0° to 90°.

4 Use a scientific calculator to find the sine and cosine of each angle indicated. Fill in a table like the one below. Three figures to the right of the decimal points is sufficient for this purpose. (Be sure your calculator is set to the "degree" mode.)

θ	sin θ	cos θ
0°	?	?
10°	?	?
20°	?	?
·	·	·
90°	?	?

5 Plot the values of the sine ratios for the angles in the table (angle measure, sine). Draw a smooth curve through the points. Repeat for the cosine ratios. Compare your graphs with your statement in Step 3. ❖

Recreation

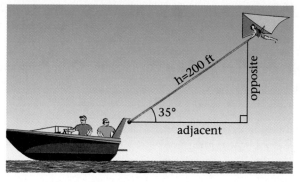

A paraskier is towed behind a boat with a 200-foot rope. The spotter in the boat estimates the angle of the rope to be 35° above the horizontal. Estimate the skier's height above the water.

Label the hypotenuse and the opposite and adjacent sides for the given angle. The hypotenuse is known, and the side you want to find is the opposite side. Use the sine ratio.

$$\sin 35° = \frac{o}{h}$$

Solve for a and substitute the values for the sine ratio and the length of the rope.

$$o = h(\sin 35°)$$
$$= 200(0.5736) \approx 115 \text{ ft } ❖$$

Astronomy

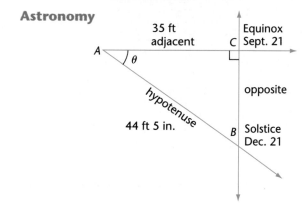

The members of an astronomy club keep records of their observations of the position of the sunrise as viewed from a fixed point A in a parking lot. They set up lines of sight from point A in the directions of sunrise at the autumnal equinox and the winter solstice. Then they set up a line perpendicular to the solstice line as shown. Finally, they measure the distances AB and AC. What is the measure of angle θ?

$$\cos \theta = \frac{AC}{AB} = \frac{35 \text{ ft } 0 \text{ in.}}{44 \text{ ft } 5 \text{ in.}}$$

$$= \frac{35.0}{44.417} \approx 0.788$$

To find the angle whose cosine is 0.788, you can use a trigonometry table. To find the angle using a scientific calculator, use the $\cos^{-1}$ key (the inverse cosine key).

$$\boxed{\cos^{-1}} \ .788 \approx 38° ❖$$

•Exploration 2 *Two Trigonometric "Identities"*

You will need
Scientific calculator

Scientific Calculator

Part I

1 Using the `SIN`, `COS`, and `TAN` keys on a scientific calculator, copy and complete the table.

θ	sin θ	cos θ	tan θ	$\dfrac{\sin \theta}{\cos \theta}$
20°	?	?	?	?
40°	?	?	?	?
60°	?	?	?	?

2 What do you notice about the values in the tangent column? Make a conjecture about a relationship between the tangent of an angle and the sine and cosine of the angle.

ALGEBRA
Connection

3 In the equation below, what happens when you simplify the second expression?

$$\frac{\text{sine of } \angle A}{\text{cosine of } \angle A} = \frac{\dfrac{\text{length of side opposite } \angle A}{\text{length of hypotenuse}}}{\dfrac{\text{length of side adjacent to } \angle A}{\text{length of hypotenuse}}}$$

Does the equation prove your conjecture? Discuss.

Part II

1 Using the `SIN` and `COS` keys on a scientific calculator, fill in a table like the one below.

θ	sin θ	cos θ	$(\sin \theta)^2 + (\cos \theta)^2$
20°	?	?	?
40°	?	?	?
60°	?	?	?

If a graphics calculator is available, graph the following equation.

$$y = (\sin x)^2 + (\cos x)^2$$

 What do you notice about the values of y? Make a conjecture about the value of $(\sin x)^2 + (\cos x)^2$ for any given angle x.

ALGEBRA
Connection

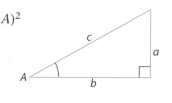 Give a reason for each step in the proof below.

$$(\text{sine of } \angle A)^2 + (\text{cosine of } \angle A)^2$$

$$= \left(\frac{a}{c}\right)^2 + \left(\frac{b}{c}\right)^2$$

$$= \frac{(a^2 + b^2)}{c^2}$$

Compare the numerator and the denominator of the resulting fraction. What do you observe? How does this prove the conjecture? ❖

EXERCISES & PROBLEMS

Communicate

Complete each statement.

1. How is the sine of an angle found in a right triangle?

2. How is the cosine of an angle found in a right triangle?

3. Does the sine ratio increase or decrease as the angle gets larger? Explain your answer.

4. Does the cosine ratio increase or decrease as the angle gets larger? Explain your answer.

Complete each trigonometric identity.

5. Describe two ways of finding the tangent of an angle in a right triangle.

6. Why does $(\sin A)^2 + (\cos A)^2$ equal 1?

Practice & Apply

Find the following for the triangle on the right.

7. sin C 8. cos D 9. sin D

10. cos C 11. tan C 12. tan D

13. m∠C (use a scientific calculator)

14. m∠D

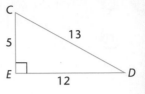

Find the following for the triangle on the right.

15. $\dfrac{15}{17} = \cos$ ____?____

16. $\dfrac{8}{17} = \sin$ ____?____

17. $\dfrac{8}{17} = \cos$ ____?____

18. $\dfrac{15}{17} = \sin$ ____?____

19. $\dfrac{8}{15} = \tan$ ____?____

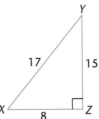

Use a scientific calculator to find the following.

20. sin 35° 21. cos 72° 22. tan 37°

23. sin 57° 24. cos 52°

Use a scientific calculator to find the angles for each of the following. Round to the nearest degree.

25. $\sin^{-1} 0.3$ 26. $\sin^{-1} 0.8775$ 27. $\cos^{-1} 0.56$

28. $\cos^{-1} 0.125$ 29. $\tan^{-1} 2.432$

30. **Recreation** A straight water slide is 25 m long and starts at the top of a tower that is 21 m high. Find the angle between the tower and the slide.

Use a calculator to answer Exercises 31–33.

31. Choose any angle measure between 0° and 90°. Find the sine of your angle and then apply the $\sin^{-1}$ key to the sine value. Repeat this procedure for four more angles. Look for a pattern and write a conjecture based on the pattern.

32. Repeat Exercise 31 but replace the sine with tangent and replace $\sin^{-1}$ with $\tan^{-1}$. What do you notice?

33. Repeat Exercise 31 for cosine and $\cos^{-1}$. What do you notice?

Exercises 34–37 lead through the development of the cofunction identities for sine and cosine. Use a scientific calculator.

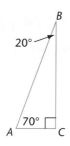

34. In △ABC, what is the sine of ∠A? What is the cosine of ∠B? What do you notice about these two values?

35. Repeat the steps in Exercise 34 for another set of complimentary angles. (Remember that the acute angles of a right triangle sum to 90°.) What do you notice about the relationship between sin A and cos B?

36. Complete the conjecture below.

In a right triangle, the __?__ of one acute angle is __?__ to the __?__ of the other acute angle.

37. Since the acute angles A and B of a right triangle sum to 90°, sin A = cos B can also be written sin A = cos (90° – m∠A). Complete the following cofunction identities. (The first one is done for you.)

sin A = cos (90° – m∠A)

cos B = __?__

sin B = __?__

cos A = __?__

38. Building Codes
For easy access, a sidewalk ramp is supposed to have an angle of 15° or less to the ground. A ramp to a landing 1.5 m high is 5 m long. Is the angle 15° or less? Explain your reasoning.

39. Use trigonometry to find the area of the triangle.

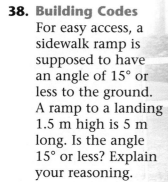

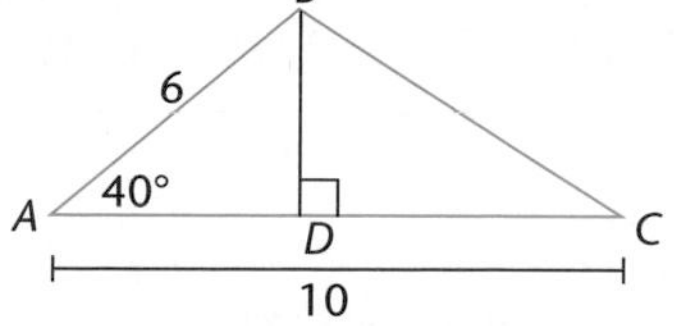

40. Use trigonometry to find the area of the parallelogram.

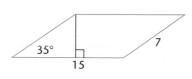

For the given length, find each of the remaining two lengths.
[Lesson 5.5]

41. $x = 7$

42. $z = 13$

43. $y = 14$

44. $q = 3$

45. $p = 1$

46. $r = 16$

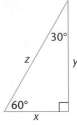

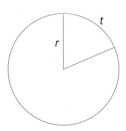

For each arc length and radius given below, find the measure of the central angle.   [Lesson 9.1]

47. $t = 12\pi$; $r = 20$

48. $t = 10\pi$; $r = 100$

49. $t = 3\pi$; $r = 25$

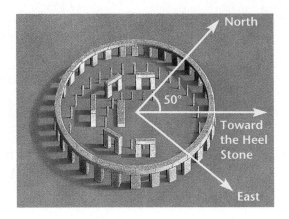

Look Beyond

Portfolio Activity The placement of certain stones at **Stonehenge** suggests that they were markers for the rising and setting positions of the sun and the moon. Most important, the "Heel Stone" marks the position of the rising sun on the day of the summer solstice as seen from the center of the stone circle. At the latitude of Stonehenge in southern England, this angle is about 50° east of true north.

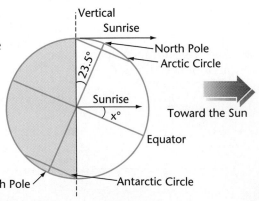

Use the diagram to answer Exercises 50 and 51. A globe may also be helpful.

On the day of the summer solstice, the Earth's North Pole is inclined toward the sun at an angle of 23.5° with the vertical.

50. What is the direction of the summer solstice sunrise for (a) a person on the equator and (b) a person at the Arctic Circle (66.5° N)? (North is the direction toward the North Pole along a line of longitude.)

51. Describe the motion of the sun in the sky on the day of the summer solstice for a location on the Arctic Circle.

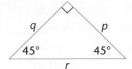

Exploring

The Unit Circle

why *The curve shown on the oscilloscope is a graph of the sound wave generated by the synthesizer. Such a curve is known as a sine curve.*

Are Trigonometric Ratios Limited to Acute Angles?

So far, trigonometric ratios have been defined in terms of acute angles of right triangles. Such definitions seem to limit trigonometric ratios to angles greater than 0° and less than 90°. But you have already seen that a calculator gives values for the sines and cosines of 0° and 90°.

•Exploration 1 Trigonometric Ratios on a Calculator

Scientific Calculator

You will need

Scientific calculator

1 Using a scientific calculator, see what happens if you press the keys for sin 0° and sin 90°. Record your results.

2 Repeat Step 1 for the cosine and tangent ratios. Record your results.

3 Experiment with the sine, cosine, and tangent ratios of angles greater than 90°. Does your calculator give values for these? Record your results.

4 Experiment with the sine, cosine, and tangent ratios of angles with negative measures. Does your calculator give values for these? Record your results.

Lesson 10.3 Exploring the Unit Circle **541**

5 Find the sine, cosine, and tangent ratios of 65° and 115°. Repeat for angles of 70° and 110°. Record your results. Do you observe a pattern? If so, describe it. ❖

Extending the Trigonometric Ratios

Begin on a coordinate plane and focus on the ray consisting of the origin and the positive *x*-axis. Imagine this ray rotating counterclockwise about the origin. As it rotates, it sweeps through a certain number of degrees, say $\theta°$. Although measures of angles are restricted between 0° and 180°, the ray is free to rotate any number of degrees. The illustration shows that a rotation can, in fact, be greater than 360°.

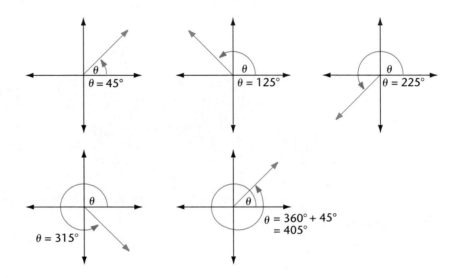

Explain how a rotation can have a negative measure.

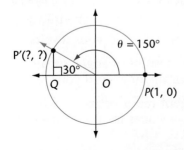

EXTENSION

A **unit circle** is a circle with its center at the origin and radius 1. It consists of all the rotation images of $P(1, 0)$ about the origin. Find the coordinates of P'.

P' is on the 150° rotation image of $P(1, 0)$ about the origin. Find the coordinates of P'.

Since $\angle P'OQ$ is supplementary to $\angle P'OP$, its measure is 30°. Thus, $\triangle P'OQ$ is a 30-60-90 triangle with its hypotenuse equal to 1. $P'Q = \frac{1}{2}$ and $QO = \frac{\sqrt{3}}{2}$. The *x*-coordinate of P' is $-\frac{\sqrt{3}}{2} \approx -0.866$. The *y*-coordinate of P' is $\frac{1}{2} = 0.5$. Thus, the coordinates of P' are $(-0.866, 0.5)$. ❖

•Exploration 2 *Redefining the Trigonometric Ratios*

You will need
Scientific calculator
Graph paper

Scientific Calculator

1 *P'* is the 30° rotation image of *P*(1, 0) about the origin. Use the rules from the 30-60-90 Right-Triangle theorem to find the coordinates of *P'*.

2 *P'* is the 210° rotation image of *P*(1, 0) about the origin. Find the coordinates of *P'*.

3 *P'* is the 330° rotation image of *P*(1, 0) about the origin. Find the coordinates of *P'*.

4 Use the results of Example 1, Exploration 2, and a scientific calculator to complete the table.

θ	x-coordinate of image point	y-coordinate of image point	cos θ	sin θ
30°	?	?	?	?
150°	?	?	?	?
210°	?	?	?	?
330°	?	?	?	?

5 Write definitions for sin θ and cos θ based on the information in the table. Complete the figure to illustrate your definition.

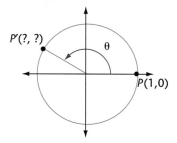

P'(?, ?) θ *P*(1,0)

6 Do your values for sin 30° and cos 30° agree with those obtained using the right-triangle definitions? Explain. ❖

Before giving an official definition for sin θ and cos θ, we need to adopt a convention: a counterclockwise rotation has positive measure and a clockwise rotation has negative measure.

Let θ be the measure of a rotation of *P*(1 ,0) about the origin. Sin θ is the *y*-coordinate of the image point. Cos θ is the *x*-coordinate of the image point.

The definition for tan θ obtained in Lesson 10.2 can be generalized to give the following definition:

Let θ be the measure of a rotation of *P*(1, 0) about the origin.

$$\tan \theta = \frac{\sin \theta}{\cos \theta}$$

•Exploration 3 *Graphing the Trigonometric Ratios*

You will need
Scientific calculator
Graph paper

1 Complete the table; θ is the measure of a rotation of $P(1, 0)$ about the origin.

Quadrant of image point	sign of sin θ	sign of cos θ	sign of tan θ
I	+	?	?
II	+	?	?
III	−	?	?
IV	?	?	?

2 Extend the graphs of each of the trigonometric functions below to $\theta = 360°$ by plotting points at intervals of 30°.

3 Use your graphs to determine the quadrants in which the sine, cosine, and tangent are positive and those in which they are negative. Does this information agree with the table that you completed in Part 1?

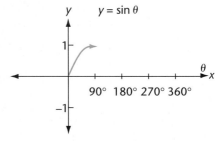

4 In Exploration 1 you found that sin 65° = sin 115° and that sin 70° = sin 110°. Locate these values on your graph of the sine curve. List at least three other pairs of angle measures that share the same sine ratios.

5 Use the cosine curve. List at least three pairs of angle measures that have the same cosine ratios.

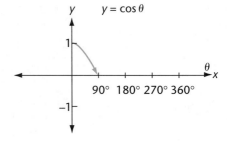

6 Use the tangent curve. List at least three pairs of angle measures that have the same tangent ratios.

7 Consider the following identity:

sin θ = sin (180° − θ), for all values of θ.

Do the results from Step 1 agree with this identity? Explain.

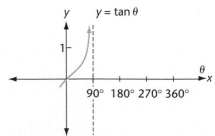

8 Does cos θ = cos (180 − θ) for all values of θ in the table above? If yes, explain why. If no, provide a counterexample and modify the equation to make a true statement. ❖

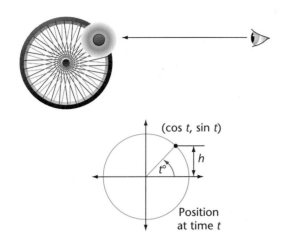

(cos t, sin t)

h

$t°$

Position
at time t

A wheel with a 1-foot radius is turning slowly at a constant velocity of one degree per second and has a light mounted on its rim. A distant observer watching the wheel from the side sees the light moving up and down in a vertical line. Write an equation for the vertical position of the light starting from the horizontal position shown at time $t = 0$. What will be the vertical position of the light after 1 minute? After 5 minutes? After 24 hours?

Imagine a coordinate system with the origin at the center of the wheel as shown. After t seconds have elapsed, the value of θ is $t°$. The coordinates of the light are (cos t, sin t). Notice that sin t, the y coordinate, is the vertical position of the light. Thus, at time t the vertical position h of the light is given by the equation,

$$h = \sin t.$$

At $t = 1$ minute, or 60 seconds, $h = \sin 60° \approx 0.867$ units.

At $t = 5$ minutes, or 300 seconds, $h = \sin 300° \approx -0.867$ units.

At $t = 24$ hours, or 86,400 seconds, $h = \sin 86,400° = 0$ units. ❖

Try This   Write the equation for the vertical position of the light if the wheel is turning at the rate of one complete rotation per second. Write its equation if the diameter of the wheel is 2 units. (Assume t is in seconds.)

EXERCISES & PROBLEMS

Communicate

1. How is the unit circle used to extend the trigonometric ratios beyond right triangles?

2. Explain how to find sin 90°, cos 90°, and tan 90°.

3. Explain how to find sin 180°, cos 180°, and tan 180°.

Explain how to find the sign of the following:

4. cosine in quadrant III   5. sine in quadrant II

6. tangent in quadrant III

7. Explain how a trigonometric function can represent the vertical motion of point X on the oil-rig pump.

Practice & Apply

Use graph paper and a protractor or geometry software to sketch a unit circle and a ray with the given angle θ. **Estimate the sine, cosine, and tangent of θ to the nearest hundredth. Include the sketch with your answers.**

8. θ = 45°   **9.** θ = 110°   **10.** θ = 175°   **11.** θ = 450°

For Exercises 12–15, use a scientific calculator to find the following values.

12. sin 110°   **13.** cos 157°   **14.** tan 230°   **15.** sin 480°

16. If sin θ = .4756, what are all the possible values of θ between 0° and 360° rounded to the nearest degree?

17. If cos θ = −.7500, what are all the possible values of θ between 0° and 360° rounded to the nearest degree?

Find the coordinates of *P′*, the rotation image of *P*(1, 0), for the following angles of rotation.

18. θ = 145°   **19.** θ = 45°   **20.** θ = 60°   **21.** θ = 72°

22. θ = 470°   **23.** θ = 250°   **24.** θ = 840°   **25.** θ = 760°

26. Astronomy An astronomer observes that a satellite moving around a planet moves one degree per hour. Assuming that the radius of the satellite's orbit is one unit, what is the horizontal position, in coordinate form, of the satellite after 2 days? after 5 days?

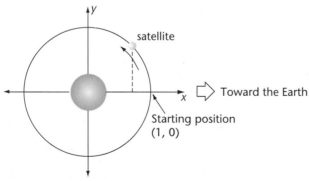

View of planet from above.

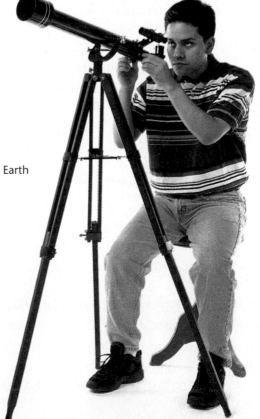

Look Back

Use △ABC for Exercises 27–34. [Lessons 10.1, 10.2]

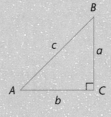

27. $\sin A = \dfrac{?}{c}$

28. $\cos B = \dfrac{?}{c}$

29. $\tan A = \dfrac{a}{?}$

30. $\cos A = \dfrac{?}{c}$

31. $(\sin A)^2 + \underline{\ \ ?\ \ } = 1$

32. $\dfrac{\sin A}{\cos A} = \underline{\ \ ?\ \ }$

33. $\sin A = \cos (\underline{\ \ ?\ \ })$

34. $\cos B = \sin (\underline{\ \ ?\ \ })$

Look Beyond

The graph shows the positions of the four Galilean Moons of Jupiter from the point of view of the Earth at midnight on April 1–16, 1993. The parallel lines in the center of the graph represent the visible width of Jupiter.

35. **Portfolio Activity** Galileo was able to observe Jupiter's four largest satellites. Viewed from the Earth, they appear to move back and forth in an approximate straight line through the center of the planet. Today these four satellites are known as the "Galilean Moons" of Jupiter. Their names are

      I. Io
     II. Europa
   III. Ganymede
   IV. Callisto

Sketch the positions of the planet Jupiter and its four Galilean Moons as they would appear to a person with a small telescope or a pair of binoculars on (a) April 4, (b) April 8, and (c) April 12. Use dots for the moons and a circle for Jupiter.

36. Use the graph to estimate the orbital periods for each of the four Galilean Moons of Jupiter.

37. What kind of curves do the satellite lines on the graph appear to be? Explain why they have these shapes. (Note: The orbits of the four satellites are nearly circular.)

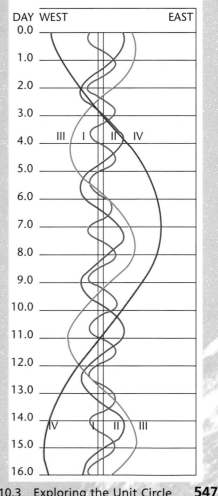

Configurations of satellites I–IV for April 1–16, 1993 at midnight Greenwich Meridian Time (GMT)

Exploring Rotations With Trigonometry

 Why *Computers use geometric transformations to show moving objects. In this lesson you will add rotations to the transformations you can do on a coordinate plane.*

Astronauts prepared for the Hubble Space Telescope repair mission by studying "virtual-reality" simulations of the planned event.

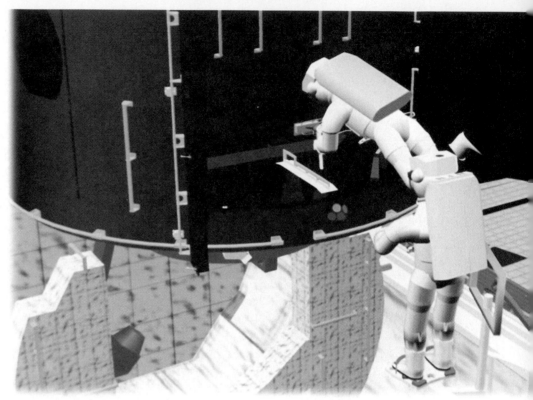

In earlier lessons you studied translations, reflections, and dilations in a coordinate plane or space. You also studied a number of special cases of rotations. Now, with trigonometry as a tool, you are ready to rotate a point in a coordinate system by *any desired amount*.

Transformation Equations

ALGEBRA
Connection

The two equations below are known as transformation equations.

$$x' = x \cos \theta - y \sin \theta$$
$$y' = x \sin \theta + y \cos \theta$$

To use the equations, you need to know the coordinates x and y of the point you want to rotate and also the measure of the amount of rotation θ about the origin. The coordinates x' and y' are the coordinates of the rotated image point.

Exploration I Rotating a Point

You will need
Scientific calculator
Graph paper (small grid)
Ruler and protractor

Scientific Calculator

Part I

1 Set up coordinate axes on a sheet of graph paper. Select a convenient point $P(x, y)$.

2 Determine the coordinates x and y of point P and label the point with these values. Choose a value for θ, an amount of rotation. (For simplicity, choose a value between 0° and 180° for your first example.)

3 Substitute your values for x, y, and θ into the transformation equations on the previous page. Use a calculator to determine x' and y', the coordinates of the image point P'. Plot point P' on your graph and measure the rotation angle. Discuss your results.

Part II

1 Find the values of the sines and cosines in the table below.

θ	0°	90°	180°	270°	360°
Sine	?	?	?	?	?
Cosine	?	?	?	?	?

2 For each value of θ, substitute the pair of sine and cosine values into the transformation equations. Use the resulting equations to state a rule in words for rotations of 0°, 90°, 180°, 270°, and 360°.

3 Experiment with negative values for θ, such as −90°, and values greater than 360°, such as 540°. Describe your results. ❖

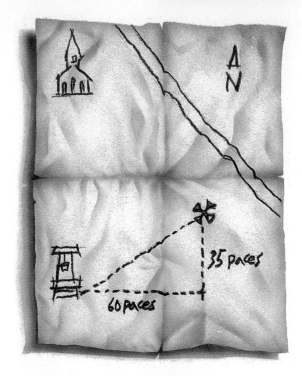

According to an old family story just uncovered, some jewels are buried at a point 60 paces east and 35 paces north of the old water well. Before visiting the site, the family draws a coordinate map of the property with true north as the direction of the positive y-axis.

If magnetic north for the location of the site were used instead of true north, point J (the jewelry location) would need to be rotated $5\frac{1}{2}$ degrees counterclockwise. What would the coordinates of the new point J' be?

Substitute in the rotation equations:

$$x' = 60 \cos 5\frac{1}{2}° - 35 \sin 5\frac{1}{2}°$$
$$\approx 60(0.9954) - 35(0.0958)$$
$$\approx 56 \text{ paces}$$

$$y' = 60 \sin 5\frac{1}{2}° + 35 \cos 5\frac{1}{2}°$$
$$\approx 60(0.0958) + 35(0.9954)$$
$$\approx 41 \text{ paces}$$

The coordinates of J' are (56, 41). ❖

Rotation Matrices

Recall from algebra that a matrix is an effective tool for visually storing data. Matrices can be added, subtracted, and multiplied. Matrices also provide a convenient way to represent the rotation of a point in a coordinate system.

$$\begin{bmatrix} \cos \theta & -\sin \theta \\ \sin \theta & \cos \theta \end{bmatrix}$$

2 × 2 rotation matrix

To rotate a point (x, y) about the origin by an amount θ, place the coordinates of the point in a 2×1 column matrix. Then multiply the column matrix by the 2×2 **rotation matrix** as shown below.

$$\begin{bmatrix} \cos \theta & -\sin \theta \\ \sin \theta & \cos \theta \end{bmatrix} \times \begin{bmatrix} x \\ y \end{bmatrix} = \begin{bmatrix} x\cos \theta - y\sin \theta \\ x\sin \theta + y\cos \theta \end{bmatrix}$$

2 × 2 rotation matrix *2 × 1 matrix* *2 × 1 matrix*

Notice that the expressions in the product matrix agree with the expressions in the rotation equations at the beginning of this lesson.

CRITICAL *Thinking*

Give the rotation matrices for $\theta = 0°$, $90°$, $180°$, $270°$, and $360°$ by filling in the sine and cosine values. Compare and contrast the matrices. Do you see a 2×2 **identity matrix** among them? Discuss.

Rotating Polygons

The convenience of rotation matrices becomes especially apparent when more than one point is rotated at a time. In the matrix multiplication below, the 2×2 matrix rotation matrix moves the three vertices of $\triangle ABC$ $60°$ counterclockwise about the origin.

Create a $60°$ rotation matrix.

$$\begin{bmatrix} \cos 60° & -\sin 60° \\ \sin 60° & \cos 60° \end{bmatrix} = \begin{bmatrix} .5 & -.866 \\ .866 & .5 \end{bmatrix}$$

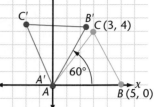

Multiply this matrix times the triangle matrix.

$$\begin{matrix} & A & B & C \end{matrix}$$
$$\begin{bmatrix} .5 & -.866 \\ .866 & .5 \end{bmatrix} \begin{bmatrix} 0 & 5 & 3 \\ 0 & 0 & 4 \end{bmatrix} = \begin{bmatrix} 0 & 2.5 & -1.964 \\ 0 & 4.33 & 4.598 \end{bmatrix} \begin{matrix} A' & B' & C' \end{matrix}$$

Try This   Rotate $\triangle ABC$ $90°$, $180°$, and $270°$ about the origin.

•Exploration 2 Combining Rotations

You will need
Calculator
Graph paper
Ruler and protractor

Calculator

If you rotate point P first $30°$ counterclockwise and then $40°$ counterclockwise about the origin, what will the final result be? In this exploration you will experiment with combined rotations using matrices.

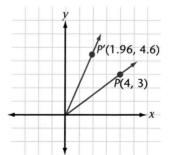

1 Pick a point in the first quadrant of the coordinate plane. Rotate the point $30°$ counterclockwise using the rotation matrix. Record your results. Plot your new point on the coordinate plane and check the angle of rotation with a protractor.

2 Rotate your new point $40°$ counterclockwise using the rotation matrix. Record your results. Plot the resulting point on the coordinate plane and check the angle of rotation with a protractor.

3 Find a single rotation that does the same thing as the combination of rotations. ❖

CRITICAL
Thinking

Does the order of the rotations in Steps 1 and 2 make a difference? How can you demonstrate your answer? When would the order of rotations make a difference?

EXERCISES & PROBLEMS

Communicate

1. What is the identity matrix for 2×2 matrices? What happens when you multiply it by another matrix? Explain why.

2. Give an angle of rotation that produces a 2×2 identity matrix when substituted into a rotation matrix.

3. Give examples of two rotations, one clockwise and one counterclockwise, that produce the same result.

Practice & Apply

Find the image point for each pre-image and rotation angle given.

4. $(-3, 2)$; 270°   5. $(4, -5)$; 90°   6. $(5, 10)$; 180°

7. $(-1, -4)$; 360   8. $(1, -4)$; 270°   9. $(3, 0)$; 90°

Technology For each point and given rotation, name the image point of the rotation. Use a calculator and the transformation equations from Exploration 1.

10. $(4, 2)$; 27°   11. $(5, -10)$; 160°   12. $(-1, -1)$; 15°

13. $(15, 10)$; 230°   14. $(-5, -6)$; 295°   15. $(-2, 3)$; 85°

Given the pre-image and image below, determine the angle of rotation (0°, 90°, 180°, 270°, or 360°).

Pre-image	Image
16. $(4, 2)$	$(-2, 4)$
17. $(3, 7)$	$(-3, -7)$
18. $(-10, -9)$	$(-9, 10)$
19. $(3, -2)$	$(3, -2)$

Find the image point for each pre-image point and rotation angle given.

20. $(-10, 2)$; 270°   21. $(-3, 5)$; 65°

22. $(2, -7)$; 150°   23. $(2, 1)$; 150°

Technology For Exercises 24–27, create the appropriate rotation matrix. First write the matrix using sines and cosines. Then calculate the values and round them to the nearest thousandth.

24. 90°   25. 270°   26. 45°   27. 60°

Technology For Exercises 28–35, the use of a graphics calculator is recommended to perform the matrix multiplication.

28. Rotate $\triangle ABC$ with $A(2, 0)$, $B(4, 4)$, and $C(5, 2)$ 45° about the origin. Call the image triangle $\triangle A'B'C'$ and plot both triangles on the same set of axes.

29. Rotate $\triangle A'B'C'$ 45° about the origin. Call the image triangle $A''B''C''$ and plot it along with $\triangle ABC$ and $\triangle A'B'C'$.

30. Describe the transformation $\triangle ABC \rightarrow \triangle A''B''C''$ as a single rotation. Write the rotation matrix for this rotation. Check your answers by multiplying this rotation matrix times the point matrix for $\triangle ABC$.

31. Let $[R_{60}]$ denote a 60° rotation matrix. Fill in a table like the one below. Plot the points in the last column on graph paper along with the point $P(4, 0)$.

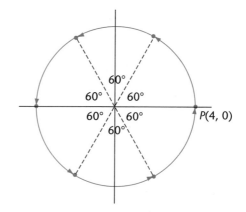

n	$[R_{60}]^n$	$[R_{60}]^n \begin{bmatrix} 4 \\ 0 \end{bmatrix}$
1	?	?
2	?	?
3	?	?
4	?	?
5	?	?
6	?	?

32. Perform the following matrix multiplication. Work from right to left as you multiply.

$$[R_{60}][R_{60}][R_{60}][R_{60}][R_{60}][R_{60}] \begin{bmatrix} 4 \\ 0 \end{bmatrix}$$

33. Repeat Exercise 32, but this time work from left to right as you multiply. Compare the results of Exercises 32 and 33 with $[R_{60}]^6 \begin{bmatrix} 4 \\ 0 \end{bmatrix}$, the last entry in your table in Exercise 31.

34. One vertex of a regular pentagon centered at the origin is (4, 0). Describe a procedure for finding the coordinates of the other four vertices. Carry out your procedure and plot all five vertices.

35. If A and B are 2 × 2 rotation matrices, does AB = BA? For example, is it true that $[R_{30}][R_{60}] = [R_{60}][R_{30}]$? Is the same thing true for 2 × 2 matrices other than rotation matrices? Illustrate your answer with an example.

Translations Recall that to translate the
point $P(2, 1)$ 5 units to the right and 3 units
down, you can add 5 units to the x-coordinate
and -3 units to the y-coordinate:

$$P' = (2 + 5, 1 + -3) = (7, -2)$$

Addition of matrices can be used to perform
this translation.

36. Fill in the missing entries.

$$\begin{bmatrix} ? \\ ? \end{bmatrix} + \begin{bmatrix} 2 \\ 1 \end{bmatrix} = \begin{bmatrix} 7 \\ -2 \end{bmatrix}$$

37. Translate $\triangle PQR$, with $P(2, 1)$, $Q(4, 8)$, and $R(8, 5)$, five units to the
right and three units down. Fill in the missing entries.

$$\begin{matrix} \quad P \;\; Q \;\; R \quad\quad P'Q'R' \\ \begin{bmatrix} ? \; ? \; ? \\ ? \; ? \; ? \end{bmatrix} + \begin{bmatrix} 2 \; 4 \; 8 \\ 1 \; 8 \; 5 \end{bmatrix} = \begin{bmatrix} ? \; ? \; ? \\ ? \; ? \; ? \end{bmatrix} \end{matrix}$$

Look Back

**Find the angle θ between 0° and 90° for each trigonometric
ratio given. Round to the nearest degree. [Lesson 10.1]**

38. $\sin \theta = 0.767$ **39.** $\cos \theta = 0.3565$ **40.** $\tan \theta = 2.676$ **41.** $\sin \theta = 0.3565$

**Find the sine, cosine, and tangent for each angle below.
Round to the nearest hundredth. [Lesson 10.1]**

42. 37°   **43.** 187°   **44.** 90°

45. 250°   **46.** 400°   **47.** 1800°

Look Beyond

48. A triangle has a right angle C, two other angles A and B, and sin
$A = \cos A$. What are the measures of $\angle A$ and $\angle B$? Classified by its side
lengths, what type of triangle is the triangle? Expressed in terms of
the hypotenuse c, what are the lengths of sides a and b?

The Law of Sines

Satellite photo of the San Francisco Bay area

Why *Although the triangle formed by the three points on the photo is not a right triangle, its measures can be found using trigonometry.*

•Exploration Law of Sines

You will need

Geometry technology or Centimeter ruler, protractor, and scientific calculator

Geometry Graphics

1 Construct an acute triangle, a right triangle, and an obtuse triangle. Measure the sides and angles of each triangle.

2 Use the measurements from Step 1 to complete a table like the one below.

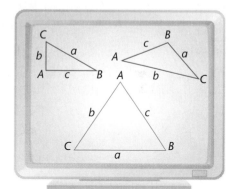

	m∠A	m∠B	m∠C	a	b	c	$\frac{\sin A}{a}$	$\frac{\sin B}{b}$	$\frac{\sin C}{c}$
Acute Triangle	?	?	?	?	?	?	?	?	?
Right Triangle	?	?	?	?	?	?	?	?	?
Obtuse Triangle	?	?	?	?	?	?	?	?	?

3 Write a conjecture using the data in the last three columns of the table. ❖

Your conjecture should be equivalent to the following theorem.

THE LAW OF SINES

For any triangle ABC $\dfrac{\sin A}{a} = \dfrac{\sin B}{b} = \dfrac{\sin C}{c}$.

10.5.1

Proof: Acute case

Each angle of $\triangle ABC$ is an acute angle. The measures of the altitudes from C and B are h_1 and h_2, respectively.

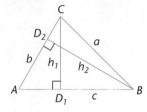

In right triangles $\triangle ACD_1$ and $\triangle BCD_1$,

$$\sin A = \frac{h_1}{b} \text{ and } \sin B = \frac{h_1}{a}.$$

Thus, $b \sin A = h_1$ and $a \sin B = h_1$.

Since h_1 can be written in two different ways, $b \sin A = a \sin B$.

Dividing both sides of the equation by ab gives

$$\frac{\sin A}{a} = \frac{\sin B}{b}.$$

Similarly, in right triangles $\triangle ABD_2$ and $\triangle CBD_2$,

$$\sin A = \frac{h_2}{c} \text{ and } \sin C = \frac{h_2}{a}.$$

Thus, $\dfrac{\sin A}{a} = \dfrac{\sin C}{c}$.

Therefore, $\dfrac{\sin A}{a} = \dfrac{\sin B}{b} = \dfrac{\sin C}{c}$.

The proof for right triangles and for obtuse triangles is left for you to try in the exercises.

The Law of Sines can be used to find the measures of sides and angles in any triangle, as long as you know certain combinations of measures in the triangle.

EXAMPLE 1

Find b and c.

Solution ➤

Set up a proportion based on the Law of Sines in which 3 of the 4 terms are known. Look for a known angle whose opposite side is also known, in this case $\angle A$ and a.

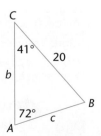

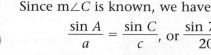
Since m∠C is known, we have

$$\frac{\sin A}{a} = \frac{\sin C}{c}, \text{ or } \frac{\sin 72°}{20} = \frac{\sin 41°}{c}.$$

Cross-multiplication gives

$$c = \frac{20 \sin 41°}{\sin 72°} \approx 13.8.$$

To find b, we must first find m ∠B: $180 - (72 + 41) = 67°$. Again using the known angle-with-side pair,

$$\frac{\sin A}{a} = \frac{\sin B}{b}, \text{ or } \frac{\sin 72°}{20} = \frac{\sin 67°}{b}.$$

Cross-multiplication gives

$$b \sin 72° = 20 (\sin 67°).$$

$$b = \frac{20 \sin 67°}{\sin 72°} \approx 19.36. ❖$$

The Law of Sines can also be used to find angles.

EXAMPLE 2

Find m∠Q.

Solution ➤

$$\frac{\sin 82°}{34} = \frac{\sin x}{27}$$

$$\sin x = \frac{27 \sin 82°}{34}$$

$$\sin x \approx 0.78639$$

$$x = \sin^{-1} (0.78639)$$

$$x \approx 52°$$

Caution: In solving for an angle using the Law of Sines, you will need to know in advance whether the angle is acute or obtuse. This is because an acute and an obtuse angle may have the same sine ratio. For example, $\sin 50° = \sin 130° = 0.7660$. However, your calculator will give only one value for the angle whose sine is 0.7660. (Which one does it give?)

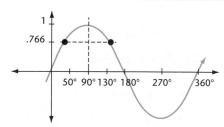

In general, the Law of Sines can be used in any triangle if you know the measures of

1. two angles and a side of a triangle or

2. two sides and an angle that is opposite one of the sides of a triangle. ❖

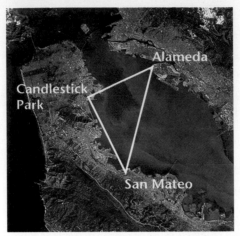

A boater regularly travels across San Francisco Bay from San Mateo to Alameda, a distance of approximately 12.8 miles. If the boater decides to travel first to Candlestick Park, what is the distance to Alameda?

The angle formed at Candlestick Park is 95°, at Alameda 53°. Setting up the proportion:

$$\frac{\sin 95°}{12.8} = \frac{\sin 53°}{b} = \frac{\sin 32°}{c}$$

Solve the ratios in pairs:

$$\frac{\sin 95°}{12.8} = \frac{\sin 53°}{b}$$

$$b\sin 95° = 12.8 \sin 53°$$

$$b = \frac{12.8 \sin 53°}{\sin 95°}$$

$$= \frac{12.8(0.7986)}{0.9962}$$

$$\approx 10.26$$

$$\frac{\sin 95°}{12.8} = \frac{\sin 32°}{c}$$

$$c \sin 95° = 12.8 \sin 32°$$

$$c = \frac{12.8 \sin 32°}{\sin 95°}$$

$$= \frac{12.8(0.5299)}{0.9962}$$

$$\approx 6.81$$

Total distance = 17.1 miles

The total boating distance from San Mateo to Alameda via Candlestick Park is approximately 17.1 miles. ❖

EXERCISES & PROBLEMS

Communicate

1. State the Law of Sines. How is it used?

2. In solving for an angle using the Law of Sines, why must you know in advance if the angle is acute or obtuse?

3. Give two examples of triangle measures that can be used to apply the Law of Sines.

4. The length of the side and the angle opposite it must be known to apply the Law of Sines. Why is this true?

5. Draw and label a triangle for which the Law of Sines cannot be applied. Use your diagram to give an explanation.

Practice & Apply

For Exercises 6–13, sketch triangle ABC roughly to scale using the given triangle measures. Then find the measure of the indicated triangle part.

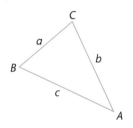

6. Find side c given m∠B = 24°, m∠A = 56°, and b = 1.22 cm.

7. Find side b given m∠B = 73°, m∠C = 85°, and a = 3.14 cm.

8. Find side a given m∠A = 35°, m∠B = 44°, and c = 24 mm.

9. Find side b given m∠C = 72°, m∠A = 53°, and c = 2.34 cm.

10. Find ∠A given a = 3.12, b = 3.28, and m∠B = 29°.

11. Find side c in Exercise 10.

12. Find the measure of ∠C given c = 2.13 cm, b = 7.36 cm, and m∠B = 67°.

13. Find side a in Exercise 12.

Surveying   You are driving toward a mountain. At a rest stop, you use a clinometer (a pocket-size device that can measure angles of elevation) to measure the angle of inclination from your eye to the top of the mountain. The angle measures 11.5°. You then drive toward the mountain on level ground for 5 miles. You stop and measure the angle of inclination to be 27.5°.

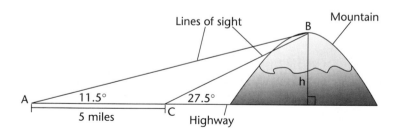

14. Determine the measure of ∠ABC.

15. Find the distance BC.

16. Using the results from Exercises 14 and 15, what is the height, h, of the mountain.

17. **Architecture**   A real-estate developer wants to build an office tower on a triangular lot between Oak Street and Third Avenue. The lot measures 42 m along Oak Street and 39 m along Third Avenue. Oak Street and Third Avenue meet at a 75° angle and the third side of the lot is 48.8 m long. The architects want to know the other angles formed by the sides of the lot before they start drawing the blueprints. Find the other angles.

18. **Wildlife Management** Scientists are tracking several polar bears that have been fitted with radio collars. The scientists have two stations that are connected by a straight road that is 9 km long. The scientists find that the line between Station 1 and a polar bear forms a 49° angle with the road, while the line between Station 2 and the polar bear forms a 65° angle with the road.

a. How far is the polar bear from Station 2.

b. Find the shortest distance from the polar bear to the road.

Complete the proof for the obtuse case of the Law of Sines in Exercises 19–33.

Given: In $\triangle ABC$, $\angle ABC$ is obtuse. The measures of the altitudes from C and B are h_1 and h_2, respectively.

Prove: $\dfrac{\sin A}{a} = \dfrac{\sin B}{b} = \dfrac{\sin C}{c}$

Proof:

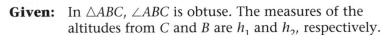

STATEMENTS	REASONS
In $\triangle BCD_1$ $\sin B = \dfrac{h_1}{a}$ and in $\triangle ACD_1$ $\sin A = \dfrac{h_1}{b}$	Definition of sine
$h_1 = $ __(19)__ and $h_1 = $ __(20)__	__(21)__
__(22)__	Substitution property of equality
$\dfrac{\sin B}{b} = \dfrac{\sin A}{a}$	__(23)__
In $\triangle CD_2B$ $\sin C = $ __(24)__ and in $\triangle BD_2A$ $\sin A = $ __(25)__	__(26)__
$h_2 = $ __(27)__ and $h_2 = $ __(28)__	__(29)__
$c \sin A = a \sin C$	__(30)__
__(31)__	Division property of equality
__(32)__	__(33)__

Use the diagram to prove the Law of Sines for right triangles for Exercises 34–36.

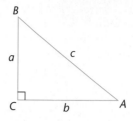

34. Write the sine of angles A, B, and C in terms of a, b, and c.

35. Form the three ratios and simplify $\frac{\sin A}{a}$, $\frac{\sin B}{b}$, and $\frac{\sin C}{c}$.

36. What can you conclude about $\frac{\sin A}{a}$, $\frac{\sin B}{b}$, and $\frac{\sin C}{c}$?

Look Back

37. Find the surface area and volume of a sphere with a radius of 15 m. What would happen to the volume and surface area of the sphere if the radius were tripled?   **[Lesson 7.6]**

38. Find x.   **[Lesson 8.3]**

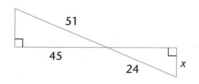

39. $\overline{BC}$ is parallel to $\overline{DE}$. Find x.
    [Lesson 8.4]

Find these without using a calculator.   [Lesson 10.2]

40. $\sin^{-1}\left(\dfrac{\sqrt{3}}{2}\right)$

41. $\tan^{-1}\sqrt{3}$

42. $\cos^{-1}\left(\dfrac{\sqrt{2}}{2}\right)$

Look Beyond

43. The image of the rabbit is distorted. It can be seen in proper perspective if you hold the book so that your line of sight forms a certain angle to the page. Determine the approximate angle for your line of sight in order to see the image properly.

44. Explain how the image might be considered a "projection."

Eyeglasses In Space

Astronauts Snare Hubble Telescope for Vital Repairs

By John Noble Wilford
Special to the *New York Times*

HOUSTON, Dec. 4—Arriving for a house call 357 miles above Earth, the astronauts of the space shuttle Endeavor today reached out with a mechanical grappling arm and easily snared the Hubble space telescope and prepared to treat the crippled spacecraft in five days of the most complex orbital repairs yet attempted.

The Hubble telescope was deployed in orbit by another shuttle in April, 1990. The fanfare of early expectations faded in the following weeks as astronomers were dismayed to see the blurry pictures. Tests traced the problem to an improperly ground mirror.

The flawed curvature meant that the mirror could not focus incoming light to a precise point. As a result, observations were not as clear as desired and the effective range was limited to about 4 billion light years; astronomers had been promised views out to 10 to 14 billion light years.

The Big Fix

By Ron Cowen
Science News, Nov. 6, 1993

Astronomers had long awaited the ultrasharp images that the Hubble Space Telescope seemed poised to offer. But the fuzzy pictures that the craft began radioing back to Earth dashed their hopes. The telescopes' primary mirror was hopelessly flawed. Instead of focusing most of the light from faint stars and galaxies into a tight circle with a radius of 0.1 arc second, the mirror spread the concentration of incoming rays. Scientists later learned that the flaw stemmed from a spherical defect introduced when a contractor ground the mirror's surface to its final shape.

With that bad news came one ray of hope: The flaw was so precise that additional mirrors could be introduced into the telescope's light path might correct for the error. Researchers suggested that if the removed one of Hubble's four instruments–phone-booth-size devices that lie parallel to Hubble's optical axis–they could replace it with a similarly shaped compartment mounted with corrective mirrors. These mirrors would intercept the fuzzily focused light bounced from the flawed primary mirror before it had a chance to enter any of the deflectors.

1. According to NASA, the main mirror in the Hubble telescope ground too flat by about 2 microns, or 2 millionths of a meter. *Newsweek* Magazine reported the amount as 0.000039 inch. Is discrepency between these figures? Explain.

2. In the NASA description below, is the last sentence consistent with the data in the description on the telescope's stability? Explain your reasoning. Use the data on the right.

> 1 arcminute = 1/60°
> 1 arcsecond = 1/3600°
>
> Los Angeles to San Francisco is about 560 km

A unique Pointing Control Subsystem aligns the spacecraft to point to and remain locked on any target. The pointing subsystem is designed to hold the Telescope with 0.007 arcsec stability for up to 24 hours while the Telescope continues to to orbit the Earth at 17,000 mph. If the Hubble Telescope were in Los Angeles, it could hold a beam of light on a dime in San Francisco, without the beam straying from the coin's diameter.

——from NASA Media Reference Guide

3. To get an idea of how the shape of a mirror affects the way it works, examine the diagram below. A vertical ray of light $\overrightarrow{FG}$ reflects off a tilted mirror $\overline{AC}$. If $\overleftrightarrow{DE} \perp \overleftrightarrow{AC}$, $\overrightarrow{FG} \parallel \overline{BC}$, and $m\angle BAC = a$, find $m\angle FGD$ and $m\angle DGH$. (Hint: Extend $\overrightarrow{FG}$.)

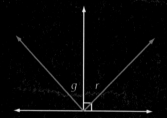

Light reflects at the same angle as its approach. In the picture, $\angle g \cong \angle r$.

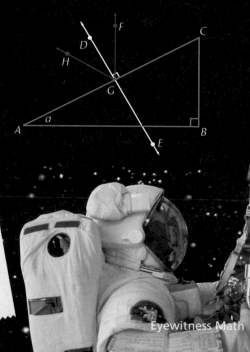

Shuttle's Crew Ends Repairs to Telescope

By John Noble Wilford
Albuquerque Journal, Dec. 1993

HOUSTON, Dec, 8—Wrapping up the most critical repairs to the Hubble Space Telescope with surprising ease and alacrity, the spacewalking shuttle astronauts have now given astronomers everything they asked for: two new sets of corrective mirrors designed to overcome the telescope's flawed vision and give clear sight to its two cameras and other scientific instruments.

why Will a parking attendant at Station 1 be able to talk directly to a parking attendant at Station 3? The Law of Cosines will enable you to determine the answer.

At a football stadium there are three parking attendants directing traffic into the parking areas. When one area is full, the attendant at one station notifies the attendants at the other two stations. The parking attendants are equipped with walkie-talkies that have a range of 1,320 feet.

In the triangle formed by the three parking attendants, two sides and the included angle are known (SAS). But the Law of Sines will not work in this situation because any proportion in the Law of Sines would have two unknowns.

The Law of Cosines provides a way to find measures of sides and angles of triangles when two sides and an included angle or all three sides of a triangle are known. (A proof of the acute case of Law of Cosines is included at the end of the lesson.)

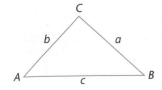

THE LAW OF COSINES

For any triangle ABC,

$$a^2 = b^2 + c^2 - 2bc \cos A$$
$$b^2 = a^2 + c^2 - 2ac \cos B$$
$$c^2 = a^2 + b^2 - 2ab \cos C$$

10.6.1

The important thing to notice about the Law of Cosines is that the length of the side used on the left side of the equation is opposite the angle used in the equation.

EXAMPLE 1

Find *b*.

Solution ➤

ALGEBRA
Connection

Using the Law of Cosines,

$$b^2 = a^2 + c^2 - 2ac \cos B$$

$$b^2 = 28^2 + 32^2 - 2(28)(32) \cos 43°$$

$$b^2 \approx 1808 - 1310.586$$

$$b^2 \approx 497.414$$

$$b \approx 22.303$$

So *b* is 22.303 m long. ❖

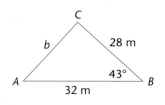

CRITICAL
Thinking

Would the Law of Sines have worked in Example 1? Explain your reasoning.

EXAMPLE 2

Find m∠*D*, m∠*E*, and m∠*F*.

Solution ➤

ALGEBRA
Connection

Part I

The Law of Sines will not work initially because at least one angle is required. So we use the Law of Cosines to find the largest angle, which must be opposite the largest side, *e*.

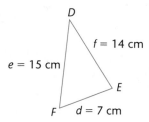

Thus, we will use $e^2 = d^2 + f^2 - 2df \cos E$.

$$15^2 = 7^2 + 14^2 - 2(7)(14) \cos E$$

$$225 = 49 + 196 - 196 \cos E$$

$$-20 = -196 \cos E$$

$$E = \cos^{-1}(0.10204) \approx 84°$$

Part II

Since we now know m∠*E*, we can use the Law of Sines to find m∠*F*.

$$\frac{\sin F}{14} = \frac{\sin 84°}{15}$$

$$\sin F = \frac{14 \sin 84°}{15}$$

$$\sin F \approx 0.9282$$

$$F = \sin^{-1}(0.9282) \approx 68°$$

Part III

$$m∠D = 180 - (m∠E + m∠F) \approx 180 - (84 + 68) \approx 28° ❖$$

In general, the Law of Cosines is used in triangles when you know the measures of

1. two sides and the included angle of the triangle (SAS) or

2. all three sides of the triangle (SSS).

APPLICATION

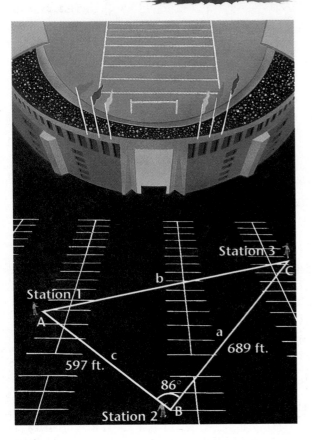

Let's take another look at the question posed at the beginning of the lesson. The parking attendant at Station 1 is 597 feet away from the attendant at Station 2. The parking attendant at Station 2 is 689 feet away from the attendant at Station 3. The angle between these segments is 86°, as shown. The parking attendants communicate with walkie-talkies that have a range of 1,320 feet. Will the parking attendant at Station 1 be able to talk directly to the parking attendant at Station 3?

Label the figure and then apply the Law of Cosines:

$$b^2 = a^2 + c^2 - 2ac \cos B$$

$$b^2 = 689^2 + 597^2 - 2(689)(597) \cos 86°$$

$$b^2 = 474{,}721 + 356{,}409 - 822{,}666 \cos 86°$$

$$b^2 = 831{,}130 - 822{,}666 \cos 86°$$

$$b^2 \approx 831{,}130 - 57{,}386.28$$

$$b^2 \approx 773{,}743.72$$

$$b \approx 880 \text{ ft}$$

Yes, the parking attendant will be able to communicate directly with the walkie-talkies. ❖

Proof of the Law of Cosines: The Acute Case

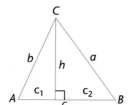

ALGEBRA
Connection

To prove the Law of Cosines, first consider whether the triangle in the proof is acute or obtuse. The following proof is of the acute case. A proof of the obtuse case appears in the exercises.

Each angle of $\triangle ABC$ is an acute angle, and the measure of the altitude from C is h. Segment $\overline{AB}$ is divided into two segments, with length c_1 and c_2 as indicated in the figure. From the definition of cosine, you know that

$$\cos A = \frac{c_1}{b}.$$

Since multiplying both sides of the equation by the same thing preserves the equality, you multiply both sides by $2bc$ to produce a helpful equality.

$$(2bc) \cos A = \frac{(2bc)c_1}{b}$$

$$2bc \cos A = 2cc_1$$

From the "Pythagorean" Right-Triangle Theorem,

$$a^2 = h^2 + c_2{}^2.$$

Since $c_2 = c - c_1$, you can substitute

$$a^2 = h^2 + (c - c_1)^2$$

$$a^2 = h^2 + (c_1{}^2 + c^2 - 2cc_1)$$

$$a^2 = (h^2 + c_1{}^2) + c^2 - 2cc_1.$$

Since $b^2 = h^2 + c_1{}^2$, you can substitute

$$a^2 = b^2 + c^2 - 2cc_1.$$

From above you know that $2cc_1 = 2bc \cos A$. Substituting produces

$$a^2 = b^2 + c^2 - 2bc \cos A,$$

which is the Law of Cosines.

EXERCISES & PROBLEMS

Communicate

Refer to the general triangle shown in Exercises 1–4.

1. Summarize the Law of Cosines.

2. Explain why the Law of Sines will not work to solve a triangle when two sides and an included angle are known (SAS). Will it work when two angles and the side between them are known (ASA)?

3. What parts of a triangle need to be known in order for the Law of Cosines to be applied?

4. Explain how right-triangle trigonometry and the "Pythagorean" Right-Triangle Theorem are used in the proof of the Law of Cosines for acute triangles.

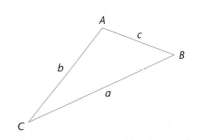

Practice & Apply ~~~~~

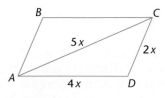 **Algebra** For Exercises 5–9, sketch △*ABC* roughly to scale using the information given. Then find the indicated side or angle measure.

5. Find side *b* given *a* = 12, *c* = 17, and *B* = 33°.

6. Find the measures of angles ∠*C* and ∠*A* in Exercise 5.

7. Find side *c* given *a* = 2.2, *b* = 4.3, and m∠*C* = 52°.

8. Find the measures of angles ∠*A* and ∠*B* in Exercise 7.

9. Find side *a* given *b* = 68.2, *c* = 23.6, and m∠*A* = 87°.

10. Mark and Stephen walk into the woods along lines that form a 72° angle. If Mark walks at 2.8 miles per hour and Stephen walks at 4.2 miles per hour, how far apart will they be after 3 hours?

11. On a standard baseball field, the bases form a square with sides 90 ft long. The pitcher's mound is 60.5 ft from home plate on a diagonal to second base. Find the distance from the pitcher's mound to first base.

12. On a standard softball field, the bases form a square with sides 60 ft long, and the pitcher stands 40 ft from home plate. Find the distance from the pitcher to first base.

13. The distance from Greenfield to Brownsville is 37 km, and the distance from Greenfield to Red River is 25 km. The angle between the road going from Greenfield to Brownsville and the road going from Greenfield to Red River is 42°. The state decides to build a road directly from Brownsville to Red River.

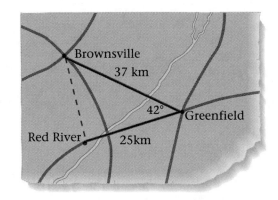

 a. How long will the road be if it is straight?

 b. At what angle to the road from Greenfield to Brownsville should the new road be built?

14. 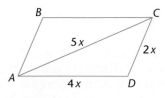 Given parallelogram *ABCD* as shown, use the Law of Cosines to find the measures of the interior angles.

15. How does this diagram differ from the diagram shown in the proof of the Law of Cosines given in the lesson?

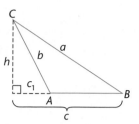

16. Explain why the proof given in the lesson does not work for this case.

In Exercises 17–19, you will prove the obtuse case of the Law of Cosines. Use the diagram in Exercises 15 and 16.

17. Use the "Pythagorean" Right-Triangle Theorem to show that $a^2 = h^2 + (c + c_1)^2$. Then use algebra to show that $a^2 = h^2 + c^2 + 2cc_1 + c_1^2$.

18. By right-triangle trigonometry, show that $h = b \sin A$ and $c_1 = -b \cos A$. Hint: $\cos (180 - \theta) = -\cos \theta$.

19. By substituting the results of Exercise 18 into the result of Exercise 17, prove that $a^2 = b^2 + c^2 - 2bc \cos A$ for the obtuse triangle ABC. Hint: Use the identity $(\sin A)^2 + (\cos A)^2 = 1$.

Look Back

For Exercises 20–22, draw the image of $\triangle ABC$ under the given mapping.   [Lesson 8.1]

20. $(x, y) \rightarrow (3x, 3y)$

21. $(x, y) \rightarrow (x - 2, y - 2)$

22. $(x, y) \rightarrow \left(\dfrac{3x}{4}, \dfrac{3y}{4}\right)$

23. The perimeter of a rectangle is 80 cm. The ratio of the length to the width of the rectangle is 1 to 5. Find the length and width.   **[Lesson 8.2]**

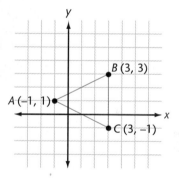

Determine if the triangles in each pair are similar. Explain why or why not.   [Lesson 8.3]

24.

25.

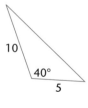

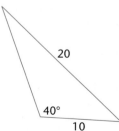

Look Beyond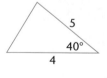

26. Can you draw this figure without lifting your pencil or retracing? Why or why not?

Vectors in Geometry

why *A goalie attempting to kick the ball to a forward downfield must take the wind into account as she takes aim. A stadium flag gives her a visual indication of the wind's **direction** and its **magnitude**, or strength. Because the wind has these two properties, it can be represented as a vector.*

A **vector** is a mathematical "object" that has both **magnitude** (a certain numerical measure) and **direction**. Arrows are used to represent vectors, because arrows have both magnitude and direction. The length of a vector arrow represents the magnitude of the vector.

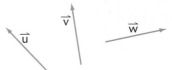

$\vec{u}$, $\vec{v}$, and $\vec{w}$ are vectors

Physics Anything that has both magnitude and direction can be modeled by a vector. In the pictorial examples on this and the following pages, vectors are used to represent a velocity, a force, and a displacement (a relocation). Describe the magnitude and direction of the vector in each.

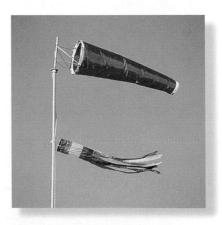

(**Velocity**) The wind sock indicates the strength of the wind.

(**Force**) *A tugboat applies a force of 2000 kg on a barge in the direction the barge is pointed.*

(**Displacement**) *The car speeds down the quarter-mile track.*

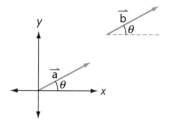

To describe the direction of a vector, you need a direction to use as a reference. On a coordinate grid this is usually the positive *x*-axis direction. (What are the reference directions in the three pictorial examples?)

Vectors with the same direction, such as vectors $\vec{a}$ and $\vec{b}$ in the illustration, are said to be **parallel**. Vectors at right angles are perpendicular.

Vector Sums

In many situations it is appropriate to combine vectors to get a new vector called the **resultant**. The process of combining vectors represented in the examples is called **vector addition**. The resultant is called the **vector sum**. For example,

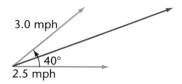

A hiker travels 3 miles northeast, then 1 mile east. The resultant vector (red) is the total distance traveled in a specific direction from the starting point.

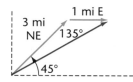

The vector of a current in a lake is 2.5 mph east. The vector of a swimmer is 3.0 mph (relative to the water) at an angle of 40° with the direction of the current. The actual velocity of the swimmer in the direction shown is the resultant vector.

Two tractors pull on a tree stump with the forces and directions shown. The combined force exerted by the tractors in the direction shown is the resultant vector.

The Head-to-Tail Method of Vector Addition

Physics In the hiker example, each vector represents a displacement—that is, a relocation of the hiker from a given point. In such cases, the **head-to-tail method** is a natural way to do vector addition.

To use the head-to-tail method of vector addition, place the *tail* of one vector at the *head* of the other. The vector sum is a new **displacement vector** from the tail of the first vector to the head of the second vector.

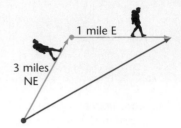

1 mile E

3 miles NE

The Parallelogram Method of Vector Addition

Physics In many cases the **parallelogram method** of vector addition is a natural way to think of combining vectors. This is especially true when two forces act on the same point.

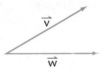

$\vec{v}$

$\vec{w}$

Magnitude of $\vec{v} = |v|$
Magnitude of $\vec{w} = |w|$

A lowercase letter with a vector arrow over it may be used to indicate a vector. The same symbol between double bars represents the magnitude of the vector.

To find the sum of vectors $\vec{a}$ and $\vec{b}$, complete a parallelogram by adding two segments to the figure. The vector sum is a vector along the diagonal of the parallelogram starting at the common point.

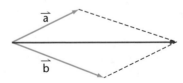

$\vec{a}$

$\vec{b}$

Try This What is the vector sum if two equal forces are applied to a point in opposite directions? What happens to the sum if two such forces are unequal?

CRITICAL
Thinking

The parallelogram method and the head-to-tail method are equivalent since they produce the same resultant. How does the diagram show that the methods are equivalent?

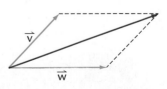

$\vec{v}$

$\vec{w}$

EXAMPLE 1

A swimmer swims
perpendicular to a
3-mph current in a
lake. Her speed in still
water is 2.5 mph. Find
the actual speed and
actual direction of the
swimmer.

ALGEBRA
Connection

Solution ➤

The parallelogram method for
adding the two vectors gives a
rectangle in this case. Thus, you can
solve for x using the "Pythagorean"
Right-Triangle Theorem.

$$x^2 = 3^2 + 2.5^2$$

$$x^2 = 15.25$$

$$x \approx 3.9 \text{ mph (actual speed of the swimmer)}$$

There is more than one way to find θ. For example,

$$\tan \theta = \frac{3}{2.5} = 1.2$$

$\theta = \tan^{-1}(1.2) \approx 50°$. The swimmer actually swims at an angle
50° to the current. ❖

EXAMPLE 2

The swimmer in Example 1 changes direction so that she is swimming
against the current at an angle of 40° with the direction of the current.
What is her actual speed and direction?

ALGEBRA
Connection

Solution ➤

One simple and often quite effective way to solve such problems is to
draw an accurate vector diagram and measure the resultant vector.

If you wish to find x mathematically, use the law of
cosines.

$$x^2 = 3^2 + 2.5^2 - 2(3)(2.5)\cos 40°$$

$$x^2 \approx 3.759$$

$$x \approx 1.94 \text{ mph}$$

Use the Law of Sines to find θ.

$$\frac{\sin \theta}{3.0} = \frac{\sin 40°}{1.94}$$

$$\sin \theta = \frac{3.0 \sin 40°}{1.94} \approx 0.9940$$

$\theta \approx \sin^{-1}(0.9940) \approx 84°$ with the current. ❖

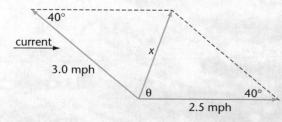

Exercises & Problems

Communicate

1. What is the "magnitude" of a vector. Explain.

2. What is the "direction" of a vector. Explain.

Describe the magnitude and direction of the vector or vectors in each of the following.

3. an airplane flying northwest at 175 knots

4. a boat going 15 knots upstream against a current of 3 knots

5. two equal tug-of-war teams pulling on opposite ends of a rope

Practice & Apply

Copy these vectors and draw a⃗ + b⃗ using the parallelogram method. You may have to translate one of the vectors.

6.
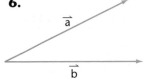

7.
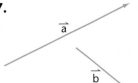

8.
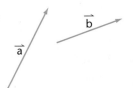

9.

Copy these vectors and draw a⃗ + b⃗ using the head-to-tail method. You may need to translate one of the vectors.

10.
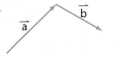

11.

12.

13.

Use the diagram below for Exercises 14–18.

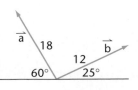

14. Draw a parallelogram using a⃗ and b⃗.

15. Find the angles of the parallelogram.

16. Draw vector c⃗, the resultant of a⃗ and b⃗.

17. Find the magnitude of c⃗.

18. Find the angle that c⃗ makes with b⃗.

19. On a coordinate grid, vector a⃗ has an initial point of (3, 14) and a terminal point (where the arrow goes) of (8, 6). Vector b⃗ has an initial point of (−2, 3) and a terminal point of (3, −5). Are the magnitude of a⃗ and b⃗ equal? Explain your reasoning.

For Exercises 20–22, draw a vector model and solve.

20. In football, receivers run pass routes that can be thought of as vectors. Ahmed's pass route calls for him to start behind the line of scrimmage and run on a line parallel to the line of scrimmage at 3.5 meters per second for 4 seconds. When the ball is snapped, Ahmed must spin around sprint downfield on a line that forms a 60° angle with his original line at 8.2 meters per second for 1.5 seconds. How far from his original position will Ahmed be?

21. Dan is a Navy SEAL investigating a boat sunk at the mouth of a river. To reach the wreck, Dan must swim against the current of 2.7 miles per hour. Suppose Dan dives and starts swimming at 4.1 miles per hour (still water speed) at a 15° angle with the water's surface. Find the speed Dan will swim as a result of the current.

22. A fishing boat leaves port and sails 5.6 miles per hour for 3.5 hours. The boat then turns at a 57° angle and sails at 4.9 miles per hour for 4.25 hours.

    a. If the boat sails for port at 6 miles per hour, how long will it take the boat to reach port?

    b. At what angle will the boat have to turn to head directly back to port?

Look Back

Use the Law of Sines and △ABC for Exercises 23–25. [Lesson 10.5]

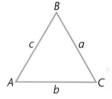

23. Find side c given m∠B = 27°, m∠A = 40°, and b = 142.

24. Find side b given m∠B = 79°, m∠C = 81°, and a = 3.14.

25. Find side c given a = 7.4, b = 9.5, and m∠B = 79°.

Use the Law of Cosines and △DEF for Exercises 26–28. [Lesson 10.6]

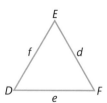

26. Find side d given e = 13, f = 19, and m∠D = 40°.

27. Find side f given d = 3.2, e = 4.5, and m∠F = 55°.

28. Find the measures of angles ∠D and ∠E in Exercise 27.

Look Beyond

Describe the conditions necessary to make each statement true.

29. It is a sunny day, and my car has a flat tire.

30. It is a sunny day, or my car has a flat tire.

31. Today is Tuesday and the lawn needs mowing, or cats have kittens and dogs have fleas.

PLIMPTON 322 REVISITED

The arrangement of the numbers in the tablet known as Plimpton 322 (see Lesson 5.4) may seem to be quite haphazard. Large numbers are mixed with smaller ones. But there is an underlying method to the arrangement.

Cuneiform Tablet, ca. 1900-1600 B.C

"Pythagorean" Triples

Leg	Hyp.	Leg.	Tan θ	θ
119	169	?	?	?
3367	4825	?	?	?
4601	6649	?	?	?
12709	18541	?	?	?
65	97	?	?	?
319	481	?	?	?
2291	3541	?	?	?
799	1249	?	?	?
481	769	?	?	?
4961	8161	?	?	?
45	75	?	?	?
1679	2929	?	?	?
161	289	?	?	?
1771	3229	?	?	?
56	106	?	?	?

To discover the key to the arrangement of the numbers, fill in the numbers in the third of the "Pythagorean" triples columns. Which column contains the hypotenuses of the right triangles? Which one contains the shorter legs? The longer legs?

Divide the length of the shorter leg by the length of the longer leg in each of the triangles and place these numbers in the column with the heading Tan θ. Then find the values of θ and add them to your table.

What do you notice about the arrangement of the values of the angles. What does the average difference between the sizes of the angles seem to be?

The Babylonian number system seems to have originally been a base 10 system like our own, as the **cuneiform** ("wedge-shaped") numbers below illustrate.

1 ⊤	6 ⊤⊤⊤	11 ◁⊤	16 ◁⊤⊤⊤	20 ◁◁	60 ◁◁◁	100 ⊤►	1000 ◁⊤►
2 ⊤⊤	7 ⊤⊤⊤	12 ◁⊤⊤	17 ◁⊤⊤⊤	30 ◁◁◁	70 ◁⊤	200 ⊤⊤⊤►	2000 ⊤⊤◁⊤►
3 ⊤⊤⊤	8 ⊤⊤⊤⊤	13 ◁⊤⊤⊤	18 ◁⊤⊤⊤⊤	40 ◁◁◁	80 ◁◁⊤		
4 ⊤⊤⊤	9 ⊤⊤⊤	14 ◁⊤⊤	19 ◁⊤⊤⊤⊤	50 ◁◁◁	90 ◁◁◁⊤		
5 ⊤⊤⊤	10 ◁	15 ◁⊤⊤⊤					

The numbers on Plimpton 322 are in base 60. The first three entries in Column II are:

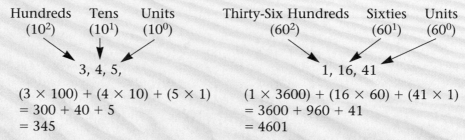

｜⟨⟨⟨‖‖	⟨⟨‖‖‖	｜⟨‖‖⟨⟨｜
1, 59	56, 7	1, 16, 41

To understand the base 60 number system, compare a base 10 number with a base 60 number.

Hundreds (10^2)	Tens (10^1)	Units (10^0)		Thirty-Six Hundreds (60^2)	Sixties (60^1)	Units (60^0)

$$3, 4, 5,$$

$$1, 16, 41$$

$(3 \times 100) + (4 \times 10) + (5 \times 1)$ $(1 \times 3600) + (16 \times 60) + (41 \times 1)$
$= 300 + 40 + 5$ $= 3600 + 960 + 41$
$= 345$ $= 4601$

The transcription of the cuneiform numbers on Plimpton 322 is shown. (Some of the damaged numbers have been supplied by researchers.) Compare the numbers in the transcription with the table on the previous page. Four of the entries in columns II and III are errors. Can you find them?

I	II	III	IV
1,59,15	1,59	2,49	ki-1
1,56,56,58,14,50,6,15	56,7	3,12,1	ki-2
1,55,7,41,15,33,45	1,16,41	1,50,49	ki-3
1,5,3,10,29,32,52,16	3,31,49	5,9,1	ki-4
1,48,54,1,40	1,5	1,37	ki-5
1,47,6,41,40	5,19	8,1	ki-6
1,43,11,56,28,26,40	38,11	59,1	ki-7
1,41,33,59,3,45	13,19	20,49	ki-8
1,38,33,36,36	9,1	12,49	ki-9
1,35,10,2,28,27,24,26,40	1,22,41	2,16,1	ki-10
1,33,45	45	1,15	ki-11
1,29,21,54,2,15	27,59	48,49	ki-12
1,27,3,45	7,12,1	4,49	ki-13
1,25,48,51,35,6,40	29,31	53,49	ki-14
1,23,13,46,40	56	53	ki-15

The numbers in the first column are fractional values. Notice for example, the number 1,38,33,36,36 in line 9. The decimal value of this number can be found as shown.

$$1 + \frac{38}{60} + \frac{33}{3600} + \frac{36}{216,000} + \frac{36}{12,960,000} = 1.6266944444\ldots$$

Compare this value with the following result:

$$\left(\frac{769}{600}\right)^2 = 1.64266944444\ldots$$

What does the first column in the Babylonian table seem to represent? Make your own conversions like the one above to test your idea. Do the numbers exactly match your conversions? Or are there errors in the table? If so, which entries are in error?

Chapter 10 Review

Vocabulary

cosine	533	magnitude	570	tangent ratio	525
cotangent	527	parallelogram method	572	unit circle	542
direction	570	resultant	571	vector	570
displacement vector	572	rotation matrix	550	vector addition	571
head-to-tail method	572	sine	533	vector sum	571

Key Skills and Exercises

Lesson 10.1

➤ **Key Skills**

Find the tangent ratio of a triangle.

In $\triangle RST$ the tangent ratio of $\angle T$ is the length of the opposite side, RS, divided by the length of the adjacent side, TS: $\frac{3}{(3\sqrt{3})} = \frac{1}{\sqrt{3}} \approx 0.577$.

➤ **Exercises**

1. Find the tangent ratio for $\angle A$.

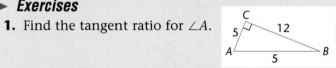

Lesson 10.2

➤ **Key Skills**

Find the sine and cosine of a triangle.

In $\triangle ABC$ the sine of $\angle B$ is the length of the opposite side, CA, divided by the length of the hypotenuse, CB: $\frac{5}{6.4} \approx 0.781$. The cosine of $\angle B$ is the length of the adjacent side, BA, divided by the length of the hypotenuse, CB: $\frac{4}{6.4} \approx 0.625$.

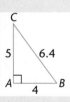

Find the measure of an angle.

In $\triangle ABC$ above, find m$\angle B$ from either the sine, 0.781, or the cosine, 0.625: m$\angle B \approx 51.3°$.

➤ **Exercises**

In Exercises 2–4, refer to the figure at the right.

2. Find cos $\angle D$.

3. Find sin $\angle F$.

4. Find m$\angle F$.

Lesson 10.3

> ## *Key Skills*

P′ is on the 60° rotation image of P(1, 0) about the origin. Find the coordinates of P′.

The coordinates of P′ are (0.5, 0.866).

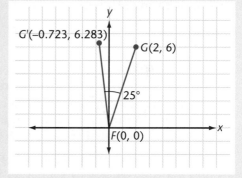

> ## *Exercises*

In Exercises 5–7, find the ratio.

5. P′ is the 420° rotation image of P(1, 0) about the origin. Find the coordinates of P′.

6. Estimate the ratio cos 390° to the nearest hundredth.

7. Estimate the ratio sin 235° to the nearest hundredth.

Lesson 10.4

> ## *Key Skills*

Name the image point of a rotation.

Find the vertices of the image of line $\overleftrightarrow{FG}$ rotated 25° clockwise about the origin.

Create a 25° rotation matrix.

$$\begin{bmatrix} \cos 25° & -\sin 25° \\ \sin 25° & \cos 25° \end{bmatrix} \approx \begin{bmatrix} .906 & -.423 \\ .423 & .906 \end{bmatrix}$$

Multiply this matrix by the line matrix.

$$\begin{bmatrix} .906 & -.423 \\ .423 & .906 \end{bmatrix} \times \begin{bmatrix} 0 & 2 \\ 0 & 6 \end{bmatrix} = \begin{bmatrix} 0 & -0.723 \\ 0 & 6.283 \end{bmatrix}$$

Point *F* is at (0, 0), and point *G′* is at (−0.723, 6.283).

> ## *Exercises*

In Exercises 8–9, name the image point of the rotation.

8. Point (5, 8) rotated 270°

9. Point (−2, 3) rotated 34°

Lesson 10.5

> ## *Key Skills*

Apply the Law of Sines to find measures in a triangle.

$$\frac{\sin A}{a} = \frac{\sin B}{b} = \frac{\sin C}{c}$$

Find m∠*I*.

Set up a proportion with the sine of an angle and the length of the side opposite: $\frac{\sin I}{30} = \frac{\sin 53°}{24}$.

$$\sin I = 30\left(\frac{\sin 53°}{24}\right) = .998$$

We find m∠*I* ≈ 86.37°.

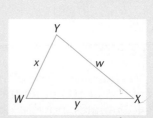

> ## *Exercises*

In Exercises 10–11, refer to the figure.

10. Find side *y* when ∠*X* = 64°, ∠*W* = 41°, and *w* = 100.

11. Find m∠*X* when *y* = 412, *w* = 533, and m∠*W* = 39°.

Lesson 10.6

➤ Key Skills

Apply the Law of Cosines to find measures in a triangle.

$$c^2 = a^2 + b^2 - 2ab\cos C$$

Find the length of side q.

We know the length of two sides and the measure of the included angle, so we can use the Law of Cosines:

$$q^2 = r^2 + s^2 - 2rs(\cos \angle Q)$$

$$q^2 = 61^2 + 87^2 - 2(61)(87)\cos 45°$$

$$q^2 \approx 3{,}784.8 \text{ m}^2$$

$$q \approx 61.5 \text{ m}$$

➤ Exercises

In Exercises 12–13, refer to the figure.

12. Find side f when $d = 23$, $e = 41$, and $\angle F = 25°$.

13. Find m$\angle D$ when $e = 321$, $f = 233$, and $d = 300$.

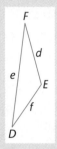

Lesson 10.7

➤ Key Skills

Add vectors using two methods.

A swimmer is moving 30° west of north at 2 miles per hour (relative to the water) through a current that flows north at 1.3 miles per hour. Find the actual speed and direction of the swimmer.

You can use vectors to represent current and swimming velocity and add the vectors to get actual speed and direction. Using the head-to-tail method, draw an accurate vector diagram and measure the resultant.

Using the parallelogram method, sketch a parallelogram and use the Law of Cosines to find the magnitude and the Law of Sines to find the direction.

$$x^2 = 2^2 + 1.3^2 - 2(2)(1.3)\cos 150°; \quad x \approx 3.2 \text{ miles}$$

$$\frac{\sin y}{1.3} = \frac{\sin 150°}{3.2}; \quad y \approx 11.7°$$

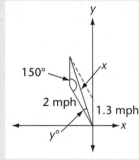

➤ Exercises

In Exercises 14–15, add the vectors below, drawing your solution.

14. Use the head-to-tail method.   **15.** Use the parallelogram method.

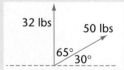

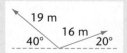

Application

16. Navigation   Two planes set off from an airport, one with a heading of 045 at 100 mph, the other with a heading of 150 at 115 mph. After 40 minutes, how far apart will the planes be?

Chapter 10 Assessment

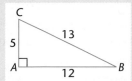

1. Which ratio does $\frac{5}{12}$ represent?

C

13

5

A

12

B

    a. tan B **b.** cot B

    c. cos C **d.** sin A

2. Find m$\angle R$.

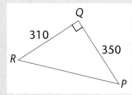

3. Find the cosine of 140°.

In Exercises 4–5, name the image point of the rotation.

  4. Point $(7, -9)$ rotated 450°

  5. Point $(-8, 3)$ rotated 73°

In Exercises 6–9, refer to the figure below.

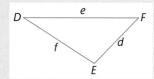

  6. Find side e when $\angle E = 86°$, $\angle D = 28°$, and $d = 42$.

  7. Find m$\angle D$ when $e = 211$, $f = 263$, and m$\angle F = 75°$.

  8. Find side f when $d = 1.7$, $e = 3.1$, and $\angle F = 125°$.

  9. Find m$\angle F$ when $e = 440$, $f = 240$, and $d = 340$.

10. Add the vectors below, drawing your solution.

Chapters 1-10   Cumulative Assessment

College Entrance Exam Practice

Quantitative Comparison   Exercises 6–9 consist of two quantities, one in Column A and one in Column B, which you are to compare as follows:

A. The quantity in Column A is greater.
B. The quantity in Column B is greater.
C. The two quantities are equal.
D. The relationship cannot be determined from the information given.

	Column A	Column B	Answers
1.	$\triangle DEF$ is similar to $\triangle GHI$. $\boxed{\tan \angle E}$	$\boxed{\tan \angle H}$	Ⓐ Ⓑ Ⓒ Ⓓ **[Lesson 10.1]**
2.	Angle $\theta > 45°$ $\cos \theta$	$\sin \theta$	Ⓐ Ⓑ Ⓒ Ⓓ **[Lesson 10.2]**
3.	$\boxed{\text{Slope of } l}$	$\boxed{\text{Slope of } m}$	Ⓐ Ⓑ Ⓒ Ⓓ **[Lesson 3.8]**
4.	$\overline{CB} \cong \overline{DB}$ $\boxed{\text{Area of } \triangle ACB}$	$\boxed{\text{Area of } \triangle BDE}$	Ⓐ Ⓑ Ⓒ Ⓓ **[Lesson 5.2]**

5. Choose the most complete, accurate description of the two polygons. **[Lesson 4.6]**

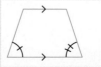

a. quadrilaterals
b. trapezoids
c. similar trapezoids
d. congruent trapezoids

6. A surveyor has taken the measures shown below. What method can he use to find x? **[Lessons 10.5, 10.6]**

   a. Cross-Multiplication Property   **b.** Exchange Property

   c. Law of Sines   **d.** Law of Cosines

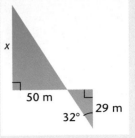

7. Which ratio does $\frac{20}{13}$ represent? **[Lessons 10.1, 10.2]**

   a. tan B   **b.** cot B

   c. cos C   **d.** sin A

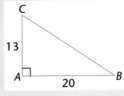

8. In circle O, which angle or arc measures 60°? **[Lesson 9.3]**

   a. ∠ABD   **b.** ∠BDC

   c. $\widehat{AD}$ **d.** $\widehat{BC}$

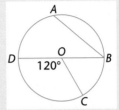

9. Find the volume of the oblique pyramid. **[Lesson 7.3]**

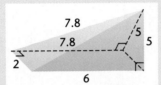

10. The ratio between the volumes of two spheres is 27:1. If the smaller sphere has a radius of 15 inches, what is the radius of the larger sphere? **[Lessons 7.1, 7.6]**

In Exercises 11–12, refer to the figure at the right.

11. Construct a rotation of the segment below. Rotate the segment 30° counterclockwise about its endpoint (0, 0). **[Lesson 4.9]**

12. Give the coordinates of the endpoints of the rotated segment. **[Lesson 10.4]**

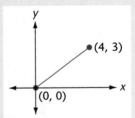

13. Write a paragraph proof that △PYW and △PYX in circle P are congruent. **[Lessons 4.2, 4.3]**

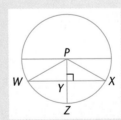

Free-Response Grid   Exercises 14–17 may be answered using a free-response grid commonly used by standardized test services.

14. Find the volume of a cylinder with a radius of 3 cm and a height of 10 cm. **[Lesson 7.4]**

15. What is the cosine of 240°? **[Lesson 10.2]**

16. Find the area of the parallelogram. **[Lesson 5.2]**

17. What is the area of a regular pentagon whose sides are 1 inch long? **[Lesson 5.5]**

CHAPTER
11

LESSONS

11.1 Golden Connections

11.2 Taxicab Geometry

11.3 Networks

11.4 Topology: Twisted Geometry

11.5 Euclid Unparalleled

11.6 *Exploring* Projective Geometry

11.7 Fractal Geometry

Chapter Project
Two Random Process Games

Taxicabs, Fractals, and More

Can you imagine a geometry in which a coffee cup is equivalent to a donut? Or one in which there is more than one "shortest path" between two points? There are many such strange ideas in this chapter.

By questioning commonly held assumptions, or by taking unusual imaginative leaps, mathematicians create entirely new branches of mathematics that often prove to be rewarding fields of study.

Before branching out to more recent discoveries in mathematics, you will first study an idea that goes back to classical times—the "golden ratio." This is an idea so powerful that it appears again and again in mathematics and nature.

PORTFOLIO ACTIVITY

Choose one or more of the following activities.

1. Draw a "Star of Pythagoras" and calculate the measures of its parts. See how many occurrences of the golden ratio you can find in it.

2. Construct a Möbius strip. Then, cut it in "half" and record what happens. What happens if you cut it again? What happens if you give a Möbius strip one extra twist before cutting it? two extra twists?

3. Draw a fractal such as the Koch Snowflake or the Sierpenski Gasket. Give it as much detail as you can. Then design and draw your own fractal.

Why *What do the Parthenon, a seashell, and a pentagon have in common? As you will learn, they each involve a mathematical principle known as the golden ratio, a concept that appears in many seemingly unrelated fields of study.*

Golden Rectangles

The golden rectangle, which is considered to have pleasing proportions, has been used by artists and architects for centuries.

To test whether a rectangle is golden, you can construct a square inside it as shown. A rectangle is considered golden if the large and small rectangles in the resulting diagram are similar.

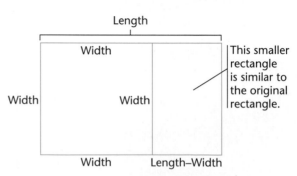

This smaller rectangle is similar to the original rectangle.

The similarity of the two rectangles can be expressed by the proportion below. The resulting value is known as the **golden ratio**.

$$\frac{\text{length}}{\text{width}} = \frac{\text{width}}{\text{length} - \text{width}}$$

Exploration 1 *The Dimensions of a Golden Rectangle*

You will need
Ruler
Graphics calculator

Graphics Calculator

1 Draw a vertical line segment 16 cm long. This segment will be the width of the golden rectangle you will draw.

2 For your rectangle to be "golden," it must satisfy the proportion on the previous page. Using a table like the one below, experiment with different values for *l* until the third and fourth columns match closely. Your final value for *l* should be accurate to the nearest tenth of a centimeter.

l	*w*	$\frac{l}{w}$	$\frac{w}{l-w}$
32	16	2	1
?	16	?	?
?	16	?	?
?	16	?	?
?	16	?	?

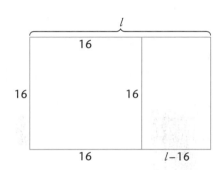

3 Draw the rest of the golden rectangle with the correct dimension *l*.

4 Complete the following statement:
The ratio $\frac{\text{length}}{\text{width}}$ in a golden rectangle is ___?___. ❖

Exploration 2 *Seashells*

You will need
Geometry technology or
Ruler
Compass
Your golden rectangle from Exploration 1

Geometry Graphics

1 Label your golden rectangle *ABCD*.

2 Form square *EBCF*.

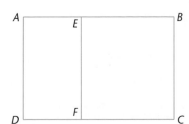

3 Form square *AEGH*.

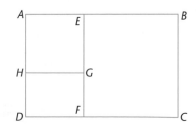

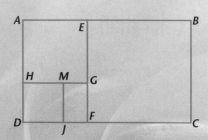

 4 Continue this process until you have five nested rectangles. Do the rectangles seem to be golden? Measure the length and width of each. What is the ratio $\frac{l}{w}$ in each case?

 5 Use a compass to make quarter-circles in each square. The resulting equiangular or logarithmic spiral models the growth pattern of seashells such as the chambered nautilus. ❖

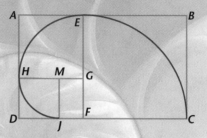

Computing the Golden Ratio

The proportions of the golden ratio can be solved algebraically to find the numerical value for $\frac{l}{w}$. First recognize that in a golden rectangle, the proportion must be true regardless of the value of w. For example, if $w = 1$, we can find a value for l that satisfies the proportion.

$$\frac{l}{1} = \frac{1}{(l-1)}$$
$$l(l-1) = 1$$
$$l^2 - l - 1 = 0$$

Notice that the resulting equation is a quadratic equation of the form

$$Ax^2 + Bx + C = 0, \text{ where } A = 1, B = -1, \text{ and } C = -1.$$

Substitute these values into the quadratic formula,

$$\frac{-B \pm \sqrt{B^2 - 4AC}}{2A}.$$

The result is

$$l = \frac{-(-1) \pm \sqrt{(-1)^2 - 4(1)(-1)}}{2(1)} = \frac{1 \pm \sqrt{5}}{2} \approx 1.618$$

(for length, use the positive value only).

Since $w = 1$,

$$\frac{l}{w} = \frac{l}{1} = \frac{1.618}{1} = 1.618.$$

This ratio is called the golden ratio. It is often represented by the Greek letter ϕ (phi):

$$\phi = \frac{1 + \sqrt{5}}{2} = 1.618033989\ldots$$

CRITICAL
Thinking

Do you think you would get the same result for $\frac{l}{w}$ if a value for w other than 1 is used? Try other values for w. What do you discover?

Exploration 3   *Constructing a Golden Rectangle*

You will need

Geometry
Graphics

Geometry technology or
Compass
Straightedge

1 Draw a vertical segment and label points $\overline{AB}$ as shown. Find the midpoint M. Label the distance AB 2 units.

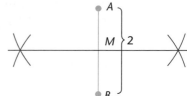

2 Construct a perpendicular at point A. Use your compass to locate point C one unit from A on the perpendicular line.

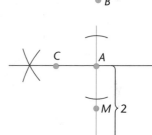

3 Construct the hypotenuse of the right triangle ABC. Use the "Pythagorean" Right-Triangle Theorem to find the length of the hypotenuse.

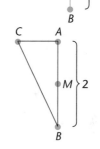

4 Use AB (2 units) as the width of your rectangle. Measure a distance on the perpendicular line that is $1 + \sqrt{5}$ units.

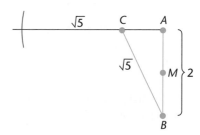

5 You now have two neighboring sides of a rectangle. Complete the rectangle by constructing perpendicular and parallel lines.

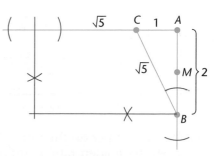

6 How do you know that the rectangle you have constructed is golden? Write an expression for the ratio of the sides of the rectangle. ❖

Lesson 11.1   Golden Connections   **589**

EXERCISES & PROBLEMS

Communicate

1. Explain how to construct a golden rectangle. How is the "Pythagorean" Right Triangle Theorem important to this construction?

2. Describe places in art, architecture, and nature where the golden ratio can be seen.

3. Describe the geometric relationship between the sides of a golden rectangle. How does this relationship relate to what you know about similar figures?

4. How is the golden rectangle self-replicating?

5. The golden rectangle is supposed to be pleasing to the eye. Do you agree or disagree? Explain your answer.

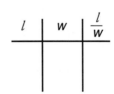

Practice & Apply

Use the five nested golden rectangles you constructed in Exploration 2 for Exercises 6–8.

6. Using a ruler, measure the dimensions of each of the five rectangles. Put your measurements in a chart like the one shown.

l	w	$\dfrac{l}{w}$

7. Compute the ratio $\dfrac{l}{w}$ for each of the five rectangles.

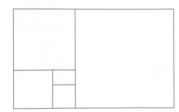

8. What number do the ratios approach? How do your ratios compare with the golden ratio computed in the lesson?

In Exercises 9–12, you will construct a regular pentagon and determine its relationship to the golden ratio.

9. Construct a regular pentagon using geometry technology or a compass and straightedge.

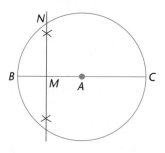

    a. Draw circle A and diameter $\overline{BC}$.

    b. Construct the perpendicular bisector of $\overline{AB}$ in order to mark point M, the midpoint of $\overline{AB}$.

    c. Label N, a point where the perpendicular bisector intersects the circle.

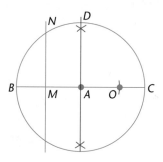

    d. Construct the perpendicular bisector of $\overline{BC}$ and label D, a point where it intersects the circle.

    e. Starting at M, mark off distance MD on $\overline{BC}$. Label the point O.

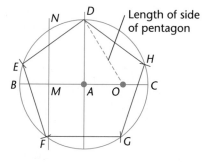

Length of side of pentagon

    f. DO is the length of the pentagon's side. Starting at D, mark off length DO around the circle. Draw segments connecting the arcs.

    g. The result is regular pentagon $DEFGH$.

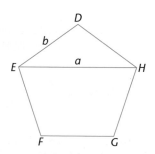

10. Draw $\overline{EH}$. Measure a and b to the nearest tenth of a centimeter.

11. Compute $\frac{a}{b}$ to the nearest tenth.

12. Write a conjecture about the result in Exercise 11.

Confirm the conjecture you made in Exercise 12 using trigonometry. Exercises 13–16 lead you through the steps.

13. Use the pentagon you constructed in Exercise 9. Find the measure of ∠D by recalling the formula from Lesson 3.6 for the interior angles of a regular polygon.

14. Since the sides of a regular polygon are congruent, △EDH is isosceles. Use this fact and the measure of ∠D to compute m∠DEH and m∠DHE.

15. Use the Law of Sines and properties of proportions to calculate $\frac{a}{b}$ to as many places as your calculator will display.

16. How does the ratio from Exercise 15 compare with the one you computed in the lesson?

Star of Pythagoras Extending the sides of a regular pentagon creates a figure known as the Star of Pythagoras. The five small triangles in the star are isosceles and congruent.

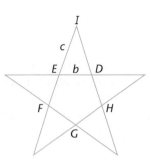

17. Use a protractor and find the measures of the angles of △DEI.

18. Use the Law of Sines to compute $\frac{c}{b}$.

19. Complete the conjecture for a triangle from a star of Pythagoras.

The ratio of the __?__ to the base is equal to the __?__.

Fibonacci Sequence The Fibonacci Sequence is a series of numbers that appear in many places in nature. Each term is the sum of the two terms preceding it. The following is a Fibonacci sequence.

1 1 2 3 5 8 13 21 . . .

20. Find the next seven terms of the sequence shown above.

21. Divide each number in the sequence by the number that immediately precedes it. Record your results to as many decimal places as your calculator will display. What do you notice?

22. The 28th term in the Fibonacci sequence is 317,811, and the 29th term is 514,229. Find the ratio of these numbers and write a conjecture.

23. In the lesson, you constructed a golden rectangle from the ratio $\frac{(1 + \sqrt{5})}{2}$. Here is an alternative construction.

   a. Begin by constructing square $ABCD$. Construct the midpoint of $\overline{AB}$. Call the midpoint E.

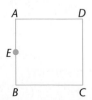

b. Extend the segment on $\overleftrightarrow{AB}$ as shown. Using E as the center of a circle and $\overline{EC}$ as the radius, construct an arc that extends the shorter distance from point C to $\overleftrightarrow{AB}$. Call the point of intersection of the arc with $\overleftrightarrow{AB}$ point F.

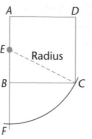

c. Construct the perpendicular through F. Extend the perpendicular and extend the segment on $\overleftrightarrow{DC}$ until they intersect. Call the intersection G.

24. Explain why $AFGD$ is a golden rectangle.

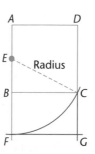

 **Look Back**

Technology Use geometry technology or graph paper and a protractor to sketch a unit circle and a radius with the given angle θ. Estimate the sine, cosine, and tangent of θ to the nearest hundredth. Include the sketch with your answers. **[Lesson 10.3]**

25. $\theta = 17°$ **26.** $\theta = 387°$ **27.** $\theta = 65°$ **28.** $\theta = 122°$

Look Beyond

29. Look closely at the picture of the sunflower seed head. The seeds form two sets of spirals in opposite directions. Count the number of spirals to the right and the number of spirals to the left. Both numbers should be Fibonacci numbers. What are they?

30. **Portfolio**

Activity Certain flowers, leaf and stem patterns, and fruits such as pineapples all have mathematical structures related to the Fibonacci sequence. Write a short research report about the Fibonacci sequence in nature.

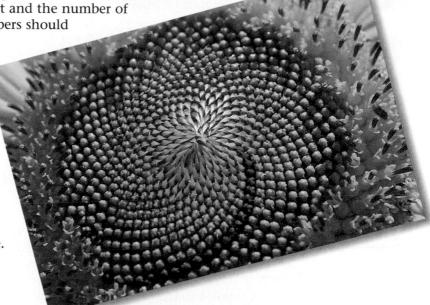

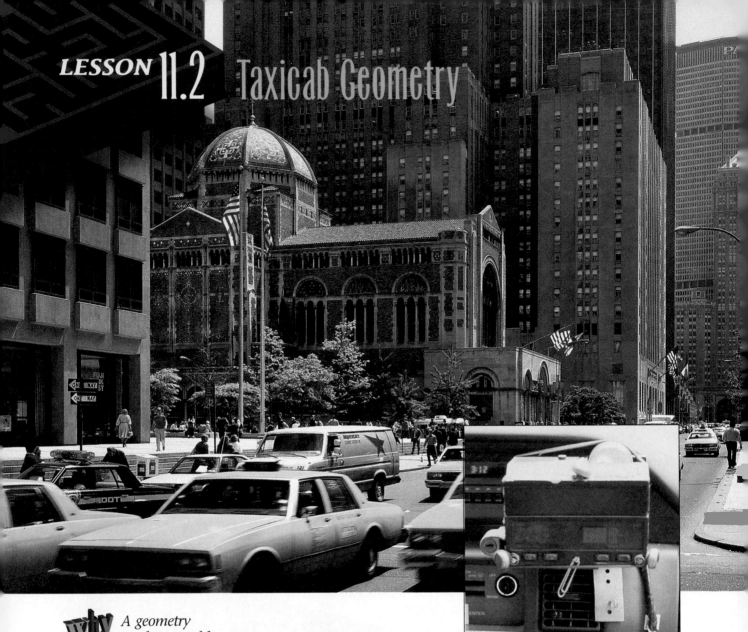

LESSON 11.2 Taxicab Geometry

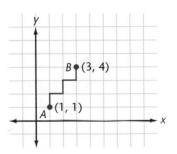

Why *A geometry student would probably use the "Pythagorean" Right-Triangle Theorem to find the distance between two points on a city map. However, a taxicab driver might have a very different idea. As you will see, it is possible to develop a logically consistent geometry around a taxicab driver definition of the distance between two points.*

Taxidistance

In "taxicab" geometry, points are located on a special kind of map or coordinate grid. The horizontal and vertical lines of the grid represent streets. But unlike points on a traditional coordinate plane, points on a taxicab grid can be only at intersections of two "streets." So the coordinates are always integers.

In taxicab geometry, the distance between two points, known as the **taxidistance**, is the smallest number of grid units, or "blocks," a taxi must travel to go from one point to the other. On the map shown, the taxidistance between the two points is 5.

How many other ways can you travel five "blocks" from point A to point B?

•Exploration 1 *Exploring Taxidistances*

Transportation

You will need
Graph paper (large grid)

Part I: The taxidistance from Central Dispatch

Assume that all taxis leave for their destinations from a central terminal at point $O(0, 0)$.

1 Draw the six destination points on a taxicab grid as shown in the diagram. Label the points A through F and their coordinates.

2 Find the taxidistances from $(0, 0)$ to each of the six destination points. (Make sure that you have found the shortest taxidistance in each case.) Arrange your information in a chart.

3 Write a conjecture about the taxidistance between the point $(0, 0)$ and a point (x, y) on a taxicab grid.

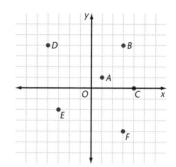

Destination point coordinates	Taxi distance from O
A (?, ?)	
B	
C	
D	
E	
F	

Part II: The taxidistance between any two points

1 Using the diagram from Part 1, find the taxidistance between at least six pairs of points in addition to the pair illustrated. Use different combinations of points in the four quadrants to

x_1, y_1	x_2, y_2	x_1	x_2	y_1	y_2	Taxi distance
A (1, 1)	B (3, 4)	1	3	1	4	5

make sure all cases are covered. Arrange your information in a chart.

2 Write a conjecture about the taxidistance between a point (x_1, y_1) and a point (x_2, y_2) on a taxicab grid. ❖

Two Points Determine . . . ?

For two points that are a given taxidistance apart, how many minimum-distance pathways are there between them?

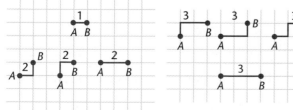

If the taxidistance between the points is just one unit, there is just one minimum pathway.

If the taxidistance between the points is two units, there are three minimum pathways.

If the taxidistance between the points is three units, there are four minimum pathways.

Lesson 11.2 Taxicab Geometry **595**

Try This This problem quickly gets complicated as the taxidistance between the points increases. Make sketches to show all of the minimum paths between points separated by a distance of 4. There are 11!

Exploration 2 *Taxicab Circles*

You will need
Grid paper

In Euclidean geometry, a circle consists of points that are a given distance from a given point. What happens if this definition is applied to points in taxicab geometry?

1 Plot a point *P* on graph paper. Then plot all the points that are located 1 taxidistance from point *P*. The very uncircular-looking result is a **taxicab circle** with a **taxicab radius** of 1.

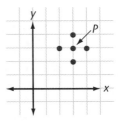

A taxicab circle with radius 1

2 Draw additional taxicab circles with taxicab radii of 3, 4, 5, and 6. Count the number of points each circle contains. Find each circle's circumference by finding the *taxidistances* between all the points as you trace a path "around" the circle.

3 Complete the chart to find a formula that predicts the number of points a given taxicab circle will contain and a formula for its circumference based on its radius.

Radius	Number of points in circle	Circumference
1	5	8
2	?	?
3	?	?
4	?	?
5	?	?
r	?	?

4 Use the information in the chart to determine a taxicab equivalent for π. (Hint: $\pi = \frac{\text{circumference}}{\text{diameter}}$) ❖

EXERCISES & PROBLEMS

Communicate

1. Why is the geometry studied in this lesson called "taxicab" geometry?

2. Explain what a taxicab grid is.

3. Explain how the distance between points is determined in taxicab geometry.

4. What are some practical applications of taxicab geometry?

How does taxicab geometry affect the final bill for the ride?

Practice & Apply

5. Draw a grid that could be used in taxicab geometry and identify the points A(5, −3) and B(−2, 4). What is the distance between these points? How would you find the distance without counting movement around the squares?

6. Without plotting, find the taxidistance between the points (−11, 4) and (−3, 9).

7. Find the taxidistance between (1, 7) and (−2, −5).

8. Draw a grid that could be used in taxicab geometry. On that grid, find a "circle" with its center at (0, 0) and a radius of 3 units.

9. Identify two points on a taxicab grid that have a taxidistance of 4. Verify, using the formula, that the taxidistance is 4.

10. Without plotting, find the taxidistance between the points (−129, 43) and (152, 236).

11. Draw a grid that could be used in taxicab geometry. On the grid, find the "circle" with its center at (5, −2) and a radius of 4 units.

Though π is not used in taxicab geometry, there is a number that serves a similar purpose in finding the circumference of a taxicab circle. **Exercises 12–21 explore taxicab circles and help you to determine what that number is.**

12. Why is π not used in taxicab geometry?

13. Plot any point P on a taxicab grid. Now find all the points that are located a taxidistance of 1 from this point.

14. Using different colored pencils, draw taxicab circles with radii of 2, 3, 4, 5, and 6.

15. Using the following table as a way to organize your results, count the number of points each circle contains. Then find the circumference of the taxicab circle by finding the taxidistances between all the points as you trace a path "around" the circle. The first two are done for you.

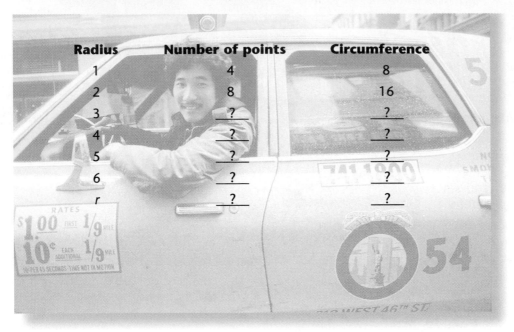

Radius	Number of points	Circumference
1	4	8
2	8	16
3	?	?
4	?	?
5	?	?
6	?	?
r	?	?

16. Making observations about the pattern in your table, predict the number of points and the circumference for a taxicab circle of radius *r*.

17. The number π is a ratio of what two parts of a circle?

18. How could you define the diameter of a taxicab circle?

19. In Euclidean geometry, the relationship between the radius and the diameter of a circle is $2r = d$. Is this also true in a taxicab circle? Why or why not?

20. In terms of *d* (diameter), what is the circumference of a taxicab circle?

21. What is the taxicab geometry equivalent of π?

22. In Chapter 4, you learned that a point is equidistant from a segment's endpoints if and only if it lies on the segment's perpendicular bisector. You can use this theorem to discover what a taxicab perpendicular bisector might look like by doing the following steps.

    a. On a taxicab grid, plot the points $A(0, 0)$ and $B(4, 2)$. Locate all the points that are a taxidistance of 2 from point *A* and from point *B*.

    b. On the same diagram, locate all the points that are a taxidistance of 3 from both *A* and *B*.

    c. Locate points that are a taxidistance of 4 from both *A* and *B*.

    d. Continue locating points that are the same taxidistance from both *A* and *B* until you have constructed the "perpendicular."

23. How is the perpendicular you constructed in Exercise 22 similar to a perpendicular in Euclidean geometry? How is it different?

Find _y_ for Exercises 24 and 25.   [Lesson 9.5]

24.

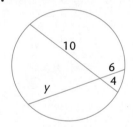

25.

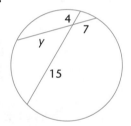

26. _BC_ = 5; _AC_ = 22;
_CD_ = 6. Find _DE._
[Lesson 9.5]

27. _MP_ = 18; _MN_ = 7.
Find _MR._
[Lesson 9.5]

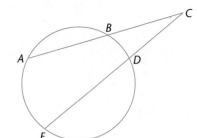

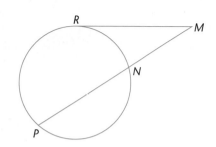

Look Beyond ~~~~~

Consider the following statements:

> _If today is February 30, then dogs can drive taxicabs._
>
> _Today is February 30._
>
> _Therefore, dogs can drive taxicabs._

28. Does this argument follow a proper order of reasoning?
Explain your answer.

29. Can you determine if the argument is true or false?
Explain your answer.

LESSON 11.3 Networks

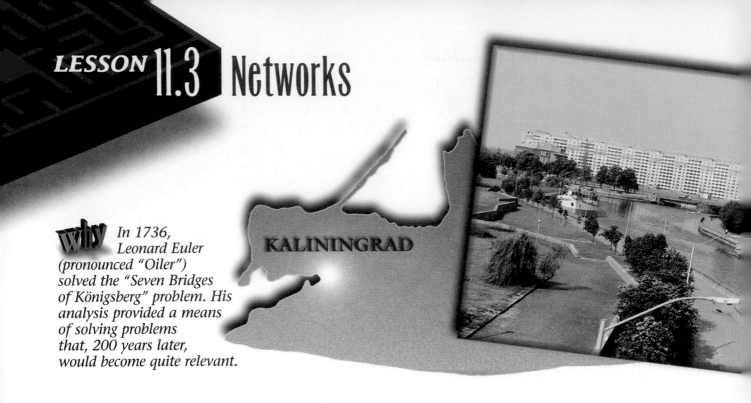

why In 1736, Leonard Euler (pronounced "Oiler") solved the "Seven Bridges of Königsberg" problem. His analysis provided a means of solving problems that, 200 years later, would become quite relevant.

KALININGRAD

The Bridge Problem

On Sunday afternoons in the 1700s, people in the old city of Königsberg (now Kaliningrad, Russia) developed an interesting mathematical pastime. As people took their Sunday strolls, they wondered if it was possible to cross each of the seven bridges of the city once and only once. The diagram below represents the city. It shows the two islands in the river that runs through the city, as well as the seven connecting bridges.

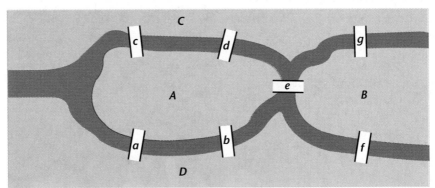

Euler represented the problem by a diagram called a **network**. The points represent the land masses, and the line segments and arcs represent the bridges. The bridge problem becomes that of traversing the network— that is, of going over every path once and only once—without lifting your pencil from the paper. The points can be crossed as many times as you like.

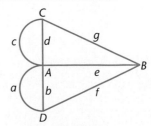

The shapes and sizes of the land masses are irrelevant.

Odd and Even Vertices

You may have tried to traverse the "Seven Bridges" network and decided that it was impossible. The people of Königsberg certainly thought so, but no one could prove it with logical argument until Euler presented his proof.

As Euler discovered, the traversability of a network can be determined by classifying its points, or vertices, as **even** or **odd**. Odd vertices have an odd number of paths (segments or arcs) going to them. Even vertices have an even number of paths going to them.

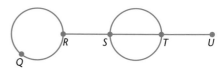

In the diagram, *U* has one path going to it, so it is an odd vertex. *R* is also an odd vertex, with three segments going to it. *Q*, *S*, and *T* are even vertices.

The network is traversable. One possible route begins at *U* and traces a path represented by the letter sequence *UTSTSRQR*. What other paths are possible?

 *Networks*

You will need
No special tools

Which of the networks seem to be untraversable? Make a table like the one at the bottom of the page and see if you can discover Euler's proof.

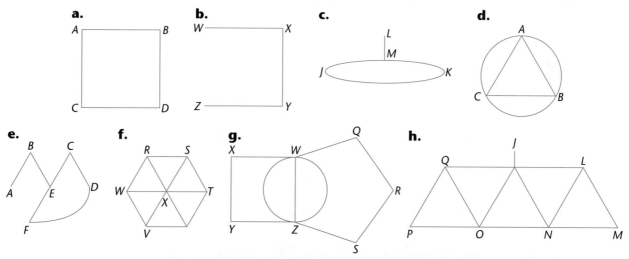

Number of vertices	Number of odd vertices	Number of even vertices	Can the network be traversed?
a. ___?___	___?___	___?___	___?___
b. ___?___	___?___	___?___	___?___
c. ___?___	___?___	___?___	___?___
•	•	•	•

Make networks from the plans of the two houses, indicating all possible routes between the rooms.

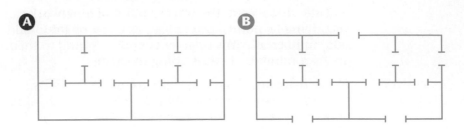

A

B

Solution ➤

A Draw and label five points to represent each of the rooms. Connect the points to indicate doorways from one room to the next. Notice that there is no connection between *D* and *E* in the network because there is no doorway between the rooms.

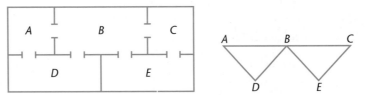

B Two possibilities for representing the space outside the house are shown below, along with their networks. ❖

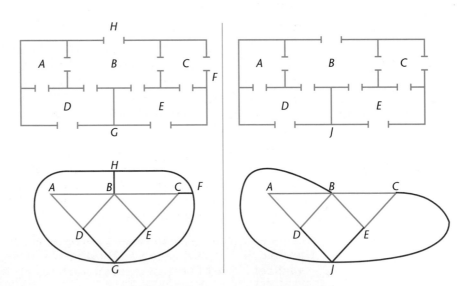

CRITICAL *Thinking*

Which of the two possibilities for representing the space outside the house do you prefer? Which do you think is the better abstraction of the problem? Give reasons for your answer. Does the "shape" of the space outside the house matter? Is there more than one "space" outside the house?

Starting and Stopping

The following problem may help you understand Euler's proof.

While hiking one snowy day, you startle a rabbit. It scampers out of the woods and runs from bush to bush, hiding under each one. It covers all the paths between the bushes without going over the same path more than once. Then it stops. Which bush is the rabbit hiding under?

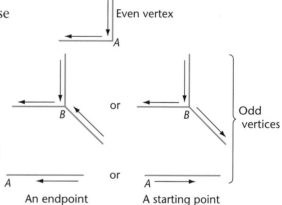

After a few tries you should see that the rabbit always ends up under bush *E,* an odd vertex. The only other odd vertex is the woods, where the rabbit's journey begins.

Consider the difference between an odd and even vertex when traversing a network: *An even vertex always allows a path into the point and a different path out of it.*

If you are attempting to traverse a network, one of the paths of an odd vertex will eventually lead into a vertex with no unused path leading out.

This means that an odd vertex can only be the beginning or ending point of a route in a network. Otherwise a retracing will occur, or a part of the network will remain untraversed.

This insight about odd vertices provided Euler with a convincing explanation that solved the problem of whether a given network can be traversed. From the data in your exploration, it appears that networks with only even vertices can always be traversed. This is in fact true, since there will always be an unused path out of an even vertex if it is needed to complete a route.

Networks with odd vertices can be traversed, but such points must either begin or end the route. Otherwise a retracing will occur, or a part of the network will remain untraversed. When a network contains exactly two odd vertices, as in the rabbit problem, one odd vertex must begin the journey and the other must end it.

Networks with more than two odd vertices are impossible to traverse. Do you think it is possible for a network to have just one odd vertex?

Further Developments From Euler's Result

Euler's analysis of the "Seven Bridges" problem led to the development of two new areas of mathematics: topology and graph theory. You will learn about topology in the next lesson.

EXERCISES & PROBLEMS

Communicate

1. What was the inspiration for Euler's method of analyzing networks?

2. Describe when a network can be traversed.

3. If the soccer ball traverses the network, which player must be the starting point? the ending point? Explain why.

4. Can the seven bridges of Königsberg be traversed? Explain your reasoning.

5. Describe some current applications of network theory.

Practice & Apply

6. Draw a network illustrating all possible routes between rooms in Figure 1.

7. Can the network in Figure 1 be traversed? Explain your reasoning.

8. Draw a network illustrating all possible routes between rooms in Figure 2. Notice that outside doors have been added. Can this network be traversed? Explain your reasoning.

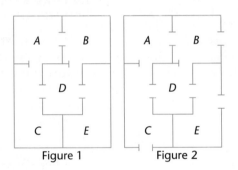

Figure 1   Figure 2

Below is a drawing adapted from Euler's original paper. He used this as an example to help illustrate the concept behind the bridges problem. Use this illustration for Exercises 9–11.

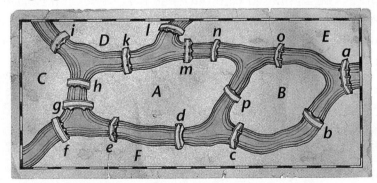

9. Draw Euler's diagram as a simplified network.

10. Determine if this network can be traversed. Explain your reasoning.

11. If you were to traverse this network, where would you begin and where would you end?

12. The following is a floor plan of a house. Draw a network that represents the circulation in the house.

13. Is it possible to walk through the house in Exercise 12 and pass through each door exactly once? Must you start or end at a particular place, or can you start anywhere? Explain your reasoning.

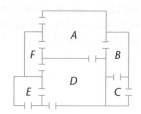

14. The following is a floor plan of a house. Draw a network that represents the circulation in the house.

15. Is it possible to walk through the house in Exercise 14 and pass through each door exactly once? Must you start or end at a particular place, or can you start anywhere? Explain your reasoning.

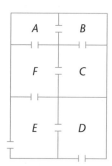

16. An **Euler circuit** is a network that has only even vertices. What is the benefit of an Euler circuit over a network that can simply be traversed?

Law Enforcement A member of the police force is collecting money from the parking meters on the streets diagramed below. The dots represent the meters. If meters occur on both sides of the street, the police officer must go down the street twice. Use this diagram to complete Exercises 17–19.

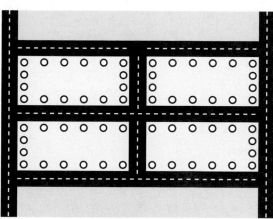

17. Draw a network that represents the parking meter problem.

18. Is this network traversable? Is it an Euler circuit? Explain your reasoning.

19. Where should the officer park in order to start and end at the same point, and to take the most efficient route possible? Explain your reasoning.

You may be familiar with a game where the challenge is to draw a figure in one stroke without lifting the pencil from the page or retracing steps. Use the figures for Exercises 20–22.

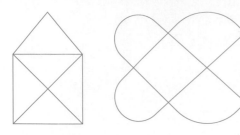

20. Which, if any, are traversable? How can you tell?

21. Which, if any, are Euler circuits? How can you tell?

22. If any of the above are traversable but are not Euler circuits, where must you start and end? Why?

 **Look Back**

Copy and reflect each figure with respect to the given line. [Lesson 1.6]

23.

24.

Write each statement below as a conditional.   [Lesson 2.2]

25. All people who live in California live in the United States.

26. The measures of the angles of a triangle sum to 180°.

27. A square is a parallelogram with four congruent sides and four congruent angles.

Look Beyond

28. Suppose you draw a triangle on a non-inflated balloon and then add air to the balloon. Will the angles of the triangle still sum to 180°? Why or why not?

LESSON 11.4 Topology: Twisted Geometry

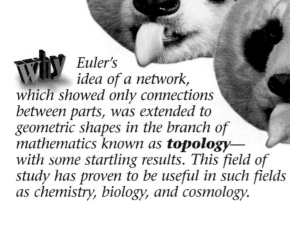

Are the faces of the panda and the tiger topologically equivalent?

why *Euler's idea of a network, which showed only connections between parts, was extended to geometric shapes in the branch of mathematics known as* **topology**—*with some startling results. This field of study has proven to be useful in such fields as chemistry, biology, and cosmology.*

Knots, Pretzels, Molecules, and DNA

Genetics

DNA

Möbius molecule

The shape at right is known as a **Möbius strip**, a topological classic with intriguing properties. (You will explore Möbius strips in the exercises.) The mathematical analysis of such objects has proven useful in other fields, such as chemistry and biology.

David Walba and his co-workers at the University of Colorado have synthesized a molecule in the shape of a Möbius strip. Structures like this help them understand why certain drugs with the same molecular structure can have vastly different effects.

Researchers have also used a branch of topology called **knot theory** to study the structures of complex, tangled structures such as DNA molecules.

Imagine that the coffee cup is made of extremely elastic material so that it can be easily stretched or squashed. As you see, it can be deformed into a doughnut shape.

Do you think it is possible to unlink the rings of the "pretzel" without tearing or cutting in any way?

Topological Equivalence

In topology, two shapes are equivalent if one of them can be stretched, shrunk, or otherwise distorted into the other without cutting or tearing. In a plane, all of the shapes below are topologically equivalent.

Such shapes are called **simple closed curves**. They enclose a distinct region, and their segments or curves do not cross themselves.

Curves or shapes whose lines cross over each other form other categories of shapes.

The two shapes on the left are topologically equivalent to each other but not to the shape on the right, since the latter contains two intersections instead of one.

CRITICAL
Thinking

Do you think that there is a categorization, involving "loops" and intersections, that occurs for 3-dimensional objects similar to the 2-dimensional objects above? Explain.

Jordan's Theorem

A fundamental theorem in topology was first stated by the French mathematician Camille Jordan (1838–1922) in the nineteenth century.

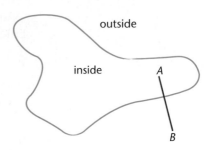

outside

inside

A

B

> ## JORDAN'S THEOREM
> Every simple closed curve divides the plane into two distinct regions. To connect a point inside the region with a point outside, you must "cut" the curve.
>
> **11.4.1**

The theorem, which is very difficult to prove mathematically, seems obvious for the curve shown above. But for some curves it is not so simple. Are points *P* and *Q* inside or outside this "simple" closed curve?

Jordan's Theorem is also true for surfaces other than planes. For example, the equator divides the Earth into two regions. To go from one region to the other you must cross or "cut" the equator.

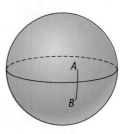

On the surface of a donut shape, however, Jordan's Theorem does not hold. Use the illustration to explain why.

A   B

Topologists call a doughnut-shaped surface a **torus**. Because Jordan's Theorem is true for a sphere but not for a torus, topologists can conclude that the two shapes are not topologically equivalent without trying to see if they can "deform" one into the other.

Invariants

Properties that stay the same no matter how a figure is deformed are called **invariants**. If a pentagon is distorted into a curve, the distances between the vertices may change, but the order of the points around the circle stays the same. The order of the points is an invariant.

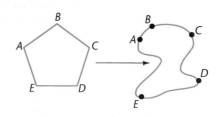

One of the most important invariants in topology comes from Euler's formula for solids.

Imagine distorting the pyramid into a sphere. You can see that the sphere is separated into four regions corresponding to the four faces of the pyramid. (The face, *BCD*, is the entire lower hemisphere.)

Next imagine deflating the sphere and flattening it out like a pancake. You can verify that Euler's formula works for the resulting topologically equivalent shape. Notice that the bottom face *BCD* is the region surrounding the circle in the flattened-out shape.

If you draw a number of vertex points on a plane or a sphere and connect them into a single figure, Euler's formula will always yield the number 2. (Connecting a single vertex to itself forms a simple closed curve.) The area outside the figure is considered a face.

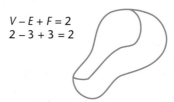

$V - E + F = 2$
$2 - 3 + 3 = 2$

The region outside the shape is counted as the third face.

However, it is possible to draw figures on a torus for which a different formula applies:

$$V - E + F = 0$$

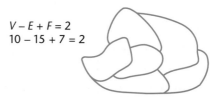

$V - E + F = 2$
$10 - 15 + 7 = 2$

Hence, the Euler number of a torus is said to be 0. Topologists use the number resulting from Euler's formula as an invariant, allowing them to classify different types of surfaces mathematically. This is important, because some shapes that seem to be topologically equivalent in fact are not equivalent.

$V = 1$
$E = 2$
$F = 1$

Finally, after everything you have learned about topology, it may not surprise you to learn that the "pretzel" shown at the beginning of this lesson can indeed be unlinked.

EXERCISES & PROBLEMS

Communicate

1. Explain what is meant by two shapes being topologically equivalent.

Decide if each of the following pairs is topologically equivalent. Explain your reasoning.

2.

3.

4. What is an invariant?

5. Explain Jordan's Theorem.

Practice & Apply

Decide if the following are topologically equivalent. Explain your reasoning.

6.

7.

8. Draw a curve for which it is difficult to determine what is inside and what is outside.

9. Imagine distorting a cube into a sphere the same way that the pyramid in the lesson was distorted. Draw a picture of the cube distorting into a sphere.

10. Use your illustration from Exercise 9 to draw the vertices and edges flattened onto a plane (see p. 610).

11. Count the vertices, edges, and faces of your figure in Exercise 10 (remember to count the "outside" area as a face). Does your calculation verify Euler's Theorem?

Portfolio Activity Exercises 12–19 explore the Möbius strip. You will need paper, tape, scissors, and a pencil.

12. Construct a Möbius strip by following these instructions:

a. Cut a piece of paper about 3 cm wide and 28 cm long (the length of your notebook paper will work well).

b. Connect the ends of the piece of paper together with tape so that your loop has a half-twist in it. Be sure to tape the side completely so that it will stay together.

13. Starting at the tape, use your pencil to draw the path of an imaginary bug that crawls in the center of the strip. Describe your results.

14. Let your bug crawl on the strip starting at the tape again, but this time draw the bug's path close to an edge of the strip. Describe your observations.

15. Materials Handling What do you think would be the advantage of conveyor belts shaped as Möbius strips?

16. Now you are going to construct a double Möbius strip. Take two rectangles of paper, 3 cm by 28 cm, and lay them on top of each other. Put a half-twist on both strips and tape the ends together to produce two nested Möbius strips.

17. Using your nested Möbius strips, put your finger in between the two and run it all the way around the strip to feel the separation. Pay attention to the "ceiling" and the "floor" as if you were a little person walking. What do you observe?

18. Pull the strips apart. What is the result? As a challenge, try to nest the strips back together again.

19. Take a long strip of paper (about 3 cm wide and 50 cm long) and put three half-turns into the strip. Connect the ends with tape. Cut the strip down the middle, lay your shape flat on the table, and draw a picture of the result. This result is a **trefoil knot**.

Look Back

20. Identify the pairs of congruent angles in the diagram below, given $l_1 \parallel l_2$. **[Lesson 3.3]**

21. Find the measures of $\angle ABC$ and $\angle DBE$. **[Lesson 2.6]**

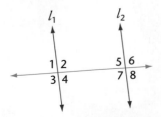

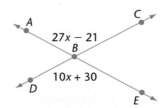

Use the Law of Cosines and △*ABC* for Exercises 22 and 23. [Lesson 10.6]

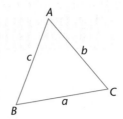

22. Find side *a* given *b* = 25, *c* = 20, and *A* = 55°.

23. Find side *b* given *a* = 104, *c* = 47, and *B* = 92°.

Look Beyond

Klein Bottle   The shape called a Klein bottle is an example of a closed surface (like a sphere) that has only one side (like a Möbius strip). An illustration of a Klein bottle drawn in a traditional way is shown. To understand the Klein bottle takes some imagination. Imagine a tube that is stretched at the bottom and pulled up and twisted into itself. It is not actually possible to construct a "real" Klein bottle in three dimensional space—or even to draw one—but the following paper model will give you a sample of some of its properties.

24. Paper model of a Klein bottle:

   a. Take a rectangular piece of paper 4 in. by 11 in. and fold it lengthwise down the center. Tape the entire length together to produce a flattened tube.

   b. Holding the flattened tube vertically, cut a horizontal slot the width of the paper, about a quarter of the way down the paper, through only the side of the tube nearest you.

   c. Fold the paper up and insert the bottom edge of the tube through the slot. Align both ends of the tube together and tape the two pairs of edges together to produce one "hole" from above. This is your model of a Klein bottle. Study the picture and compare the drawing to your construction.

25. When you think that you have an adequate understanding of your bottle and how it has the same interior as exterior, lay the paper flat and cut up the middle. Without cutting or tearing your paper, unfold the two halves of your bottle as much as you can. You should discover two familiar shapes. What are they?

26. It is possible to cut the paper model into one Möbius strip. Can you discover how it is done?

why *Many revolutionary concepts in math and science begin by questioning traditional assumptions.* Circle Limit X, *a woodcut by M. C. Escher, is an artist's conception of a non-Euclidean geometry. Such geometries reject Euclid's assumption about parallel lines.*

Imagine that you live on this two-dimensional surface. As you move outward from the center, everything gets smaller. Would you ever reach the edge of your "universe"?

The Non-Euclidean Geometries

Euclid's geometry rested upon five assumptions or postulates:

1. A straight line may be drawn between any two points.

2. Any terminated straight line may be extended indefinitely.

3. A circle may be drawn with any given point as center and any given radius.

4. All right angles are equal.

5. If two straight lines lying in a plane are met by another line, and if the sum of the internal angles on one side is less than two right angles, then the straight lines will meet.

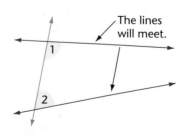

The lines will meet.

$m\angle 1 + m\angle 2 < 180°$

Because the Fifth Postulate seemed less obvious than the four that came before it, many mathematicians wished to prove it in terms of the earlier four. Although they never succeeded in doing this, several statements were found to be equivalent to the Fifth Postulate:

1. If a straight line intersects one of two parallels, it will intersect the other.

2. Straight lines parallel to the same straight line are parallel to each other.

3. Two straight lines that intersect one another cannot be parallel to the same line.

4. If given, in a plane, a line l and a point P not on l, then through P there exists one and only one line parallel to l.

These statements are equivalent to the Fifth Postulate because any one of them (with the help of the first four postulates) can be used to prove the Fifth. Statement 4 is the version most mathematicians now think of as "Euclid's Fifth Postulate" or the "Parallel Postulate."

$m\angle 1 + m\angle 2 + m\angle 3 = 180°$?

For hundreds of years mathematicians tried to prove the Parallel Postulate, until finally it was shown that the task is impossible. Many even wondered whether the Parallel Postulate is actually true.

If the Parallel Postulate is not true, then the theorems that depend on it must be questioned. One such theorem is the Triangle Sum Theorem. (To prove the theorems, you must construct a line through one vertex parallel to the opposite base.)

The great mathematician Karl Friedrich Gauss (1777–1855) went so far as to measure the angles of a triangle formed by points on three different mountaintops to see if their measures added up to 180°. (Within the limits of the accuracy of his measurements, they did.)

Some mathematicians adopted a different attitude. They found that they could develop entirely new systems of geometry *without using the Parallel Postulate.* Systems in which the Parallel Postulate does not hold are called **non-Euclidean geometries**. To their surprise, mathematicians found no major flaws in these new geometries.

Spherical Geometry

Geometry on the Earth's surface is an example of a type of non-Euclidean geometry known as **Riemannian geometry**, after its discoverer, G. F. B. Riemann (1826–1866).

In spherical geometry, a line is defined as a great circle of the sphere—that is, a circle that divides the sphere into two equal halves. In this geometry, as with any Riemannian geometry, there are *no parallel lines* at all, since all great circles intersect.

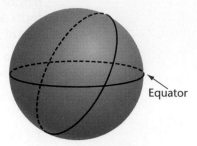

Equator

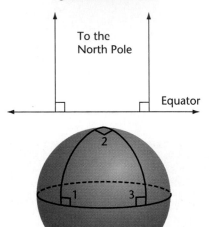

All great circles intersect.

Imagine two superhuman runners who start running at the equator. Their paths form right angles to the equator. What will happen to their paths as they approach the North Pole? How does this differ from the result you would expect on a plane?

Theorems that depend on the Parallel Postulate for their proof may actually be false in spherical geometry. On a sphere, for example, the sum of the measures of the angles of a triangle is greater than 180°, as in the spherical triangle at right.

$m\angle 1 + m\angle 2 + m\angle 3 = 270°$

CRITICAL
Thinking

Do the first four postulates of Euclid seem to be true on the surface of a sphere? If they are, should the theorems that follow them all be true on a sphere?

Hyperbolic Geometry on a "Saddle"

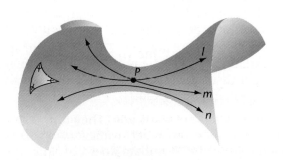

Unlike the surface of a sphere, which curves *outward*, the surface of a "saddle" curves *inward*. On such a surface, there is more than one line through a point that is parallel to a given line. In fact, there are infinitely many.

The geometry of a saddle is an example of **hyperbolic**, or **Lobachevskian**, **geometry**. Lobachevski (1773–1856) was one of two mathematicians who discovered this type of geometry independently. Janos Bolyai (1802–1860) was the other.

In hyperbolic geometry, a line is defined as the shortest path between two points. In the illustration, lines *l* and *m* pass through the point *P*, and both are parallel to line *n* (because they will never intersect line *n*).

Once again, the Triangle Sum Theorem does not hold. The sum of the angles of a triangle on a saddle is less than 180°.

Poincaré's Model of Hyperbolic Geometry

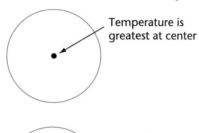

Temperature is greatest at center

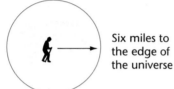

Six miles to the edge of the universe

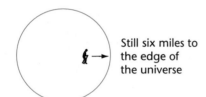

Still six miles to the edge of the universe

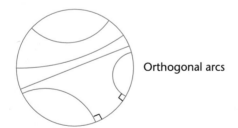

Orthogonal arcs

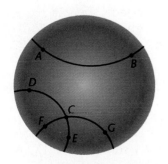

The "universe" in the Escher woodcut on the first page of this lesson is actually a representation of a three-dimensional universe imagined by the French mathematician Henri Poincaré (1854–1912). Poincaré, whose many interests included physics and thermodynamics, imagined physical reasons for the geometric properties of his imagined universe.

The temperature of Poincaré's universe is greatest at the center, and it falls away to absolute zero at the edges. According to Poincaré's rules for his system, no one would be aware of temperature changes.

In Poincaré's universe, the sizes of objects are directly proportional to their temperatures. An object would grow smaller and smaller as it moved away from the center.

Since everything, including rulers, would be shrinking in size, there would be no way of detecting the change. A table that came up to your waist when you were at the center of the circle would still come up to your waist if you moved away from the center. In fact, since distance measures keep shrinking, you would never get any closer to the edge of the universe, no matter how long you traveled.

The arcs inside the diagram represent the curving paths that rays of light are supposed to travel in Poincaré's imagined universe. These are the "lines" of the system. Light rays that stay close to the circumference of the circle have a greater curvature than those that pass close to the center, which are relatively straight. Another feature of the lines in Poincaré's system is that they are **orthogonal**, or "perpendicular," to the outer circle.

In Poincaré's system, a "line segment" is the orthogonal arc connecting two points. This is the shortest distance between the two points. If you wanted to walk from *A* to *B*, the curved line would be shorter than the straight line. This is because your steps would get larger as you approached the center of the circle. You would cover more distance with a single step if you moved toward the center instead of walking directly "to the point."

Notice also on the previous page that there is more than one line through a point drawn parallel to a given line. For example, lines $\overleftrightarrow{DE}$ and $\overleftrightarrow{FG}$ pass through the point *C*, and both are parallel to line $\overleftrightarrow{AB}$ (Why?) Thus the geometry of Poincaré's system is an example of a hyperbolic, or Lobachevskian, geometry.

Finally, notice that the sum of the measures of the angles of a triangle in Poincaré's system is less than 180°, just as it is on the surface of a saddle.

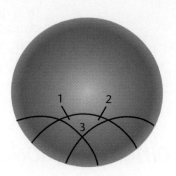

$m\angle 1 + m\angle 2 + m\angle 3 < 180°$

Years Later—Applications

In Einstein's General Theory of Relativity, space is not Euclidean. In fact, owing to the influence of gravity or—equivalently—acceleration, space is curved. The non-Euclidean geometries discovered many years earlier proved both an inspiration and a useful tool to Einstein in formulating fundamental laws of the universe.

Albert Einstein, 1879–1955

EXERCISES & PROBLEMS

Communicate

1. Which postulate of Euclid started the debate? State it in your own words.

2. Euclid's Parallel Postulate is about the existence of a single parallel line under certain conditions. How is the situation different in sphere geometry? How is it different in Poincaré's system?

3. Euclid stated in the parallel postulate that there was one parallel. Why do you think this statement went uncontested for nearly 2,000 years?

4. What are the two options to one parallel that have yielded the two non-Euclidean geometries?

5. In sphere geometry, how does a line have infinite length?

Practice & Apply

6. If you were to cut an orange in half around the "equator," and then cut each half twice at right angles on the poles, you would divide the orange peel into 8 "triangles." If you were to sum the angles of each of these triangles, what would you get? Why does this seem odd?

7. In sphere geometry, a pair of points opposite each other, such as the north and south poles of the Earth, are defined to be a *single point*. Is it true, then, that any two "lines" on a sphere intersect in just a single point? Is it true that there is just one line between any *two* points on a sphere? Illustrate your answers.

8. Draw a two-sided polygon in sphere geometry. Represent it on a sphere. What must be true about the vertices of this shape? How do you know?

9. In sphere geometry, how many degrees can a 2-gon have?

10. Draw a 4-gon on a sphere. Include one diagonal in your illustration.

11. How many degrees does your quadrilateral for Exercise 10 have? How do you know? How might you prove it?

12. Draw an illustration of a triangle in Poincaré's model.

13. In Poincaré's model of geometry, how does a line have infinite length?

14. Construct circle *A*. This circle will represent the "plane" of Poincaré. Pick any point *P* on circle *A* and construct (a) the radius of circle *A* and (b) the tangent to circle *A* at point *P*. Call the tangent line *l*.

15. "Lines" in Poincaré's model are represented by arcs of orthogonal circles. Circles are orthogonal if the radii at the points of intersection are perpendicular. If a radius of one circle is perpendicular to a radius of another, what do you know about the lines that contain the radii of the circles?

16. In your illustration for Exercise 14, the centers of a set of circles that are orthogonal to circle *A* lie on *l*. These circles intersect circle *A* at point *P*. How many circles fit this characteristic? Drawing lightly, construct a circle whose center lies relatively near *P* and one whose center lies relatively far away from *P*. Darken the arcs of the orthogonal circles that fall inside circle *A*. These arcs represent the lines of Poincaré's model.

17. How many points could you have chosen to initially draw the tangent? How many points could you have chosen on the tangent to be the center of the orthogonal circle? How many "lines" are possible in Poincaré's plane?

18. Draw a line *l* and point *Q* not on the line in Poincaré's model. How many lines can you draw through point *Q* that do not intersect line *l*? How is this illustration relevant to Euclid's Parallel Postulate?

19. Draw three lines in Poincaré's model that intersect to form a triangle. Draw one of the midsegments of that triangle. Use this drawing to answer Exercises 20–22.

20. What is the sum of the angles in the triangle you drew in Exercise 19? How does this compare with the sum of the angles in Euclidean and in sphere geometry?

21. What appears to be true about the length of this midsegment?

22. Does the midsegment appear to be parallel to one of the bases of the triangle? Why or why not?

23. Draw a hyperbolic geometry quadrilateral on a Poincaré model.

24. How many degrees does your quadrilateral for Exercise 23 have? How do you know? How might you prove it?

25. Draw a right triangle in Poincaré's model. What do you know about the acute angles of a right triangle in this geometry?

Look Back

26. Draw a floor plan for your house or apartment. Draw a network that represents the circulation in the house. Remember that the rooms are points and the doors represent the segments. **[Lesson 11.3]**

27. Is your network from Exercise 26 traversable? Why or why not? **[Lesson 11.3]**

28. Is a soup bowl topologically equivalent to a coffee cup? to a tennis ball? In each case explain your reasoning. **[Lesson 11.4]**

29. Describe three objects that are topologically equivalent to a coffee cup. **[Lesson 11.4]**

Look Beyond

Hyperbolic Geometry Assume that the two triangles below are similar triangles in a hyperbolic geometry, not necessarily drawn to scale. In hyperbolic geometries, such triangles are congruent! You can prove this using an indirect proof —i.e., by assuming the opposite result.

Given: △ABC ~ △DEF

Prove: △ABC ≅ △DEF

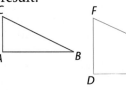

30. What do you know about the corresponding angles?

31. What do you want to temporarily assume?

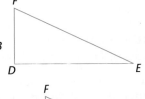

32. Place △ABC onto △DEF such that ∠C lies on top of ∠F.

Explore the relationship of the angles and lines until you reach a contradiction about quadrilateral A'B'ED.

What do you now know?

LESSON 11.6
Exploring

Projective Geometry

why *In spite of the deformities of a Mercator projection, shapes are still recognizable. This is because certain things are invariant when figures are transformed by projections. The study of such invariants is the subject matter of projective geometry.*

The three rigid transformations—reflections, rotations, and translations—preserve the sizes and shapes of objects. Dilations, on the other hand, preserve shape but not size. But there are types of transformations that shrink, expand, or stretch an object in different directions, so that neither shape nor size is preserved.

Affine Transformations

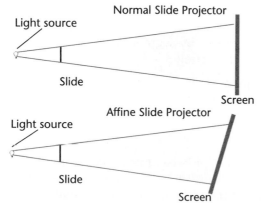

Consider a normal slide projector. The projected image on the screen is geometrically similar to the image on the slide. The original image has been dilated.

Now consider what happens to the image if you tilt the slide or the screen so that the planes of the slide and of the screen are no longer parallel. The image on the screen will be distorted. The resulting transformations are examples of **affine transformations**.

> ### AFFINE TRANSFORMATION
> An affine transformation maps all pre-image points P in a plane to image points P' so that
>
> **1.** collinear points project to collinear points.
>
> **2.** straight lines project to straight lines.
>
> **3.** intersecting lines project to intersecting lines.
>
> **4.** parallel lines project to parallel lines.
>
> 11.6.1

•Exploration 1 *Affine Transformations*

You will need
Graph paper (large grid)
Compass

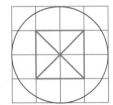

Part I

1 Mark off a 4 × 4 square grid and a 4 × 4 grid of parallelograms as shown. (Use more than one unit square to make the grid large enough to work with.)

2 Inscribe a circle in the square grid. Draw a square with its diagonals inside the circle.

3 Mark the points of intersection of the circle with the grid. Mark the vertices of the square.

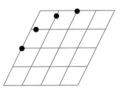

4 Mark the corresponding points of intersection in a parallelogram grid like the one shown. The first few are marked in the diagram.

5 To see how the figure would look under an affine transformation, connect the points of the "square" by segments and draw a smooth curve through the points of the "circle."

6 How does your resulting figure illustrate each of the four points in the definition of an affine transformation?

Part II

Affine transformations can also be represented using coordinates. For example, multiplying the x and y coordinates of the points of a figure by two different numbers is an example of an affine transformation.

Pre-image point	Multiply x by 2, y by 3	Image point
(x, y)	→	(2x, 3y)

1 Draw a square and its diagonals in the first quadrant of a coordinate plane.

2 Multiply the x-coordinates of the vertices of your figure by 2 and the y-coordinates of the vertices by 3. Plot the resulting points and connect them with segments. Describe the resulting figure.

3 Multiply the x-coordinates of the vertices of your figure by 3 and the y-coordinates of the vertices by 2. Plot the resulting points and connect them with segments. Describe the resulting figure.

4 How do your resulting figures illustrate each of the four points in the definition of an affine transformation?

5 How do these affine transformations differ from the one in Part I? How are they alike? ❖

Projections and Projective Geometry

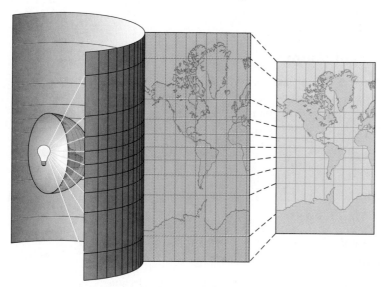

Geography students have long pointed out that Greenland is really much smaller than it appears to be on Mercator maps. Deformities like these result because it is impossible to flatten out the Earth's spherical surface into a rectangle without some distortion of sizes, shapes, distances, and directions.

The diagram shows a projection between two lines in the same plane. The points on line *l* are projected onto line *m* from a center of the projection called point *O*. Rays are drawn from the center of the projection. These lines intersect *l* at points *A*, *B*, and *C*. The rays are extended, and their intersections with line *m* determine the locations of the projected points *A′*, *B′*, and *C′*.

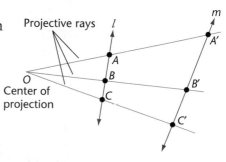

Artists used the concept of projection to give their works depth and realism. Meanwhile, mathematicians used projections to develop an entire system of geometry.

Summary

THE MAIN FEATURES OF PROJECTIVE GEOMETRY

1. It is the study of the properties of figures that do not change under a projection.

2. There is no concept of size, measurement, or congruence.

3. Its theorems state facts about such things as the positions of points and the intersections of lines.

4. An unmarked straightedge is the only tool allowed for drawing figures.

Exploration 2 *Two Projective Geometry Theorems*

Geometry Graphics

You will need
Unlined paper
Straightedge or
Geometry technology

Part I: A Theorem of Pappus

Cultural Connection: Africa Pappus was a mathematician who lived in Alexandria in the third century B.C.E. His work was very important in the development of projective geometry many centuries later. In the steps that follow, you will discover one of his theorems.

1 Orient the paper horizontally. Mark a point *O* toward the left edge of the paper.

2 Draw two rays from point *O*.

3 From point *O*, mark *A*, *B*, and *C* (in that order) on one ray.

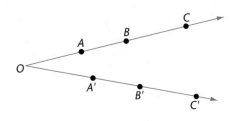

4 In the same manner, mark *A'*, *B'*, and *C'* on the other ray.

5 Draw $\overline{AB'}$ and $\overline{A'B}$. Label their intersection *X*.

Draw $\overline{BC'}$ and $\overline{B'C}$. Label their intersection *Y*.

Draw $\overline{CA'}$ and $\overline{C'A}$. Label their intersection *Z*.

6 **What appears to be true about *X*, *Y*, and *Z*?** Compare your results with your classmates' results. (If you are using geometry technology, drag the rays in various ways to see if your result still holds.) State your results as a conjecture. **(Thm 11.6.2)**

Part II: A Theorem of Desargues

Girard Desargues (1593–1662) was a French mathematician whose ideas are among the most basic in projective geometry. Carefully follow each step to discover one of his most important theorems. (Use as large a piece of paper as is practical.)

1 Draw △*ABC* approximately in the center of the paper. This can be any type of triangle, but it should be small so that the resulting construction will fit on your paper.

2 Mark a point outside △*ABC* and label it *O*. This is the center of projection.

3 Mark a random point *A′* on ray $\overrightarrow{OA}$. (For clarity of construction, it should be farther from point *O* than from point *A*.) *A′* is the projection of *A*.

4 Mark *B′*, the projection of *B*, in a random spot on ray $\overrightarrow{OB}$.

5 Extend $\overleftrightarrow{AB}$ and $\overleftrightarrow{A'B'}$ until they intersect, and label the point of intersection *X*.

6 Mark *C′*, the projection of *C*, in a random spot on ray $\overrightarrow{OC}$.

7 Extend $\overleftrightarrow{AC}$ and $\overleftrightarrow{A'C'}$ until they intersect, and label the point of intersection *Y*.

8 Extend $\overleftrightarrow{BC}$ and $\overleftrightarrow{B'C'}$ until they intersect, and label the point of intersection *Z*. (In some cases the extended segments will intersect at points off the paper. If this happens, reposition *B′* and *C′* so that points *X*, *Y*, and *Z* all fall on the paper.)

9 What appears to be true about points *X*, *Y*, and *Z*? Compare your results with your classmates' results. (If you are using geometry technology, drag the rays and points in various ways to see if your result still holds.) State your result as a conjecture about two triangles when one is the projection of the other. **(Thm 11.6.3)** ❖

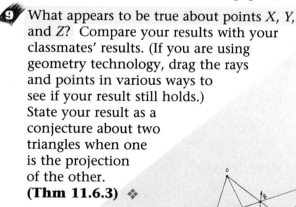

EXERCISES & PROBLEMS

Communicate

1. Describe what an affine transformation is.

2. Which of the following mappings are affine transformations? Explain your reasoning.

    a. $(x, y) \rightarrow (2x, 4y)$

    b. $(x, y) \rightarrow (y, x)$

    c. $(x, y) \rightarrow (-4x, -4y)$

    d. $(x, y) \rightarrow \left(\frac{x}{5}, \frac{y}{3}\right)$

3. Describe the four main features of projective geometry.

4. State the theorems of Pappus and Desargues.

Practice & Apply

Sketch the pre-image and image for each affine transformation below.

5. square $(0, 0)$, $(4, 0)$, $(4, 4)$, $(0, 4)$

    $(x, y) \rightarrow (3x, -2y)$

6. rectangle $(0, 0)$, $(5, 0)$, $(5, 8)$, $(0, 8)$

    $(x, y) \rightarrow (2x, 0.5y)$

7. triangle $(4, 7)$, $(-1, -1)$, $(0, 8)$

    $(x, y) \rightarrow \left(\frac{5}{2}x, 5y\right)$

For Exercises 8–10, identify the following in each of the 2 diagrams:

   a. the center of projection

   b. the projective rays

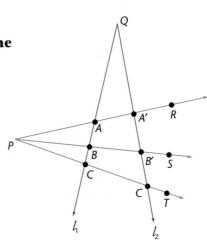

8. Projection of points on l_1 onto l_2. Use the figure above right.

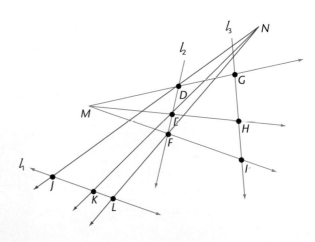

9. Projection of points on l_2 onto l_1. Use the figure at right.

10. Projection of points on l_2 onto l_3.

11. If A is the center of projection, then the projection of B on $\overline{FK}$ is ___?___, the projection of I on $\overline{FK}$ is ___?___, and the projection of J on $\overline{FK}$ is ___?___.

12. If L is the center of projection, then the projection of H on $\overline{MA}$ is ___?___ and the projection of J on $\overline{MA}$ is ___?___.

13. If F is the center of projection, then the projection of ___?___ on $\overline{AG}$ is I, and the projection of ___?___ on $\overline{AG}$ is G.

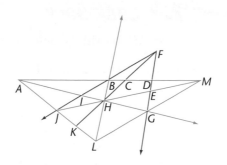

Copy the diagram and draw each projection.

14. Using center of projection O, project points A, B, C, and D onto line l_2. Label the projected points A', B', C', and D'.

15. Using center of projection P, project points A', B', C', and D' onto line l_3. Label the projected points A'', B'', C'', and D''.

16. Draw another line l and label points A, B, and C on it. Draw lines l_2 and l_3, and label a center of projection O.

   a. Sketch the projection of points A, B, and C on l_2. Label the projected points A', B', and C'.

   b. Sketch the projection of points A', B', and C' on l_3. Label the projected points A'', B'', and C''.

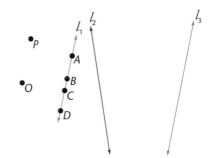

17. Converse of the Desargues Theorem
This construction begins where the activity on the Desargues Theorem ends. Working backward, you are able to locate the original center of projection.

   a. Draw l_1 and label X and Y at arbitrary locations on it.

   b. Draw ray $\overrightarrow{XB}$ and mark a random point A on it.

   c. Draw ray $\overrightarrow{XB'}$ and mark a random point A' on it.

   d. Draw ray $\overrightarrow{YA}$ and mark a random point C on it (see diagram).

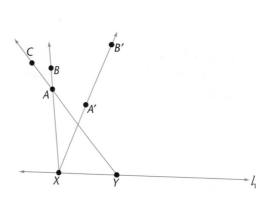

   e. Draw ray $\overrightarrow{CB}$. This will determine the location of point Z on l_1. Label point Z on l_1.

   f. Draw ray $\overrightarrow{YA'}$ and ray $\overrightarrow{ZB'}$. Label their point of intersection C'.

   g. What can you conclude about $\overline{CB}$ and $\overline{C'B'}$? (Hint: compare $\triangle ABC$ to $\triangle A'B'C'$.)

   h. Locate the center of the projection, point O, by drawing $\overline{AA'}$, $\overline{BB'}$, and $\overline{CC'}$.

Look Back

18. Describe what is meant by the "Golden Ratio." **[Lesson 11.1]**

Find the "taxidistance" for each pair of points. [Lesson 11.2]

19. (4, 3), (2, 1)   **20.** (−3, 2), (1, 1)   **21.** (1, 3), (5, 5)

22. Can the figure be traversed? Why or why not? **[Lesson 11.3]**

23. In spherical geometry, how many lines are parallel to another line through a point not on the line? **[Lesson 11.5]**

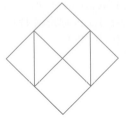

Look Beyond

24. The Nine Coin Puzzle The nine coins are arranged in eight rows of three. Can you rearrange them into ten rows of three? Hint: You can use Pappus's Theorem to solve this puzzle.

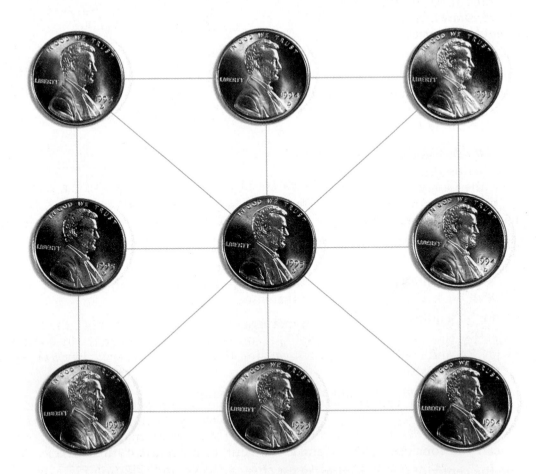

Fractal Geometry

A **fractal**, such as this computer-generated "fern," is a self-similar structure. Notice how each subdivision of the leaves of the fern has basically the same structure as the leaves themselves—all the way down to the curving tips. This self-similarity enables a computer programmer to write a program for drawing such structures in just a few lines of code.

Self-Similarity in Fractals

A fractal, like the **Menger Sponge**, is a geometric object that exhibits some degree of **self-similarity**. This means that the object always looks the same, whether seen in an extremely close view, from middle distance, or from far away. If you were to cut off a small, cube-shaped portion of the Menger Sponge and examine it, you would find it to be a miniature copy of the entire sponge.

In mathematically created fractals like the Menger Sponge, this process can be theoretically continued forever, and the self-similarity will always be evident.

Fractals are created by doing a simple procedure over and over. The Menger Sponge, for example, is created by starting with a certain shape (a cube) and changing it according to a certain rule (removing a part of the cube). This same rule is applied to the newly changed shape. The process is then continued. This repetitive application of the same rule is called **iteration**.

How Long is a Coastline?

A coastline is a good example of a self-similar fractal structure in nature. Jagged irregularities, bays, capes, and inlets can be seen from outer space. The same basic structures are evident over a smaller section of a coastline viewed from a plane, or even a smaller section viewed during a walk or a drive.

How long is a coastline? Unlike ordinary geometric segments or curves, which can often be readily measured, the "length" of a fractal coastline depends on how closely you move in to measure it.

History   During the American Revolution, the British Royal Navy attempted to blockade the American coastline. Although fairly successful, the blockade was not able to prevent shipping in and out of many of the harbors. What does a close examination of the coastline reveal about its "length" to explain the difficulty faced by the British to completely blockade it?

In a theoretical fractal coastline, where the depth of self-similarity is endless, the length is considered to be infinite.

Many objects found in nature also exhibit self-similarity, although to a lesser degree than in mathematical fractals. Notice, for instance, the self-similarity of the broccoli.

•Exploration 1 *Creating a "Cantor Dust"*

One of the simplest fractals was discovered by Georg Cantor (1845–1918) years before fractals were defined and studied. As you will see, Cantor's fractal is a one-dimensional version of the Menger Sponge.

You will need
A ruler

1. Draw a line 27 cm long.

2. Erase the middle third of the segment. You should now have two segments 9 cm long, with a 9-centimeter gap between them.

3. Erase the middle third of each of the two 9-centimeter segments. Now you will have four segments, each 3 cm long.

4. Continue erasing the middle third of each of these segments until you are left with a scattering of pointlike segments.

27 cm			
9 cm		9 cm	Iteration 1
3 cm	3 cm	3 cm	3 cm	Iteration 2
			Iteration 3
			Iteration 4
			Iteration 5

ALGEBRA *Connection*

5. Calculate the number of segments and their combined lengths after each iteration.

Iteration	0	1	2	3	4	5	n
Number of segments	?	?	?	?	?	?	?
Combined length	?	?	?	?	?	?	?

6. As the number of iterations increases, what happens to the number of segments? What happens to the combined lengths of the segments? Do you think there is a limit greater than zero of the combined lengths of the segments as the process continues? ❖

Communication Technology

Communications   When an electric current transmits data over a wire, a certain amount of "noise" occurs, which can cause errors in the transmission. The noise seems to occur in random bursts, with "clean" spaces in between. Benoit Mandelbrot, the discoverer of fractal geometry, showed that the noise patterns closely matched the pattern of a Cantor dust. This mathematical representation allowed the development of new strategies for reducing transmission noise to a minimum.

The Sierpinski Gasket

The Sierpinski Gasket, like the Menger Sponge and the Cantor dust, is created by applying the same rule to an initial shape. The rules are as follows:

1. Start with an isosceles or an equilateral triangle. Find the midpoints of each of the sides.

2. Connect the midpoints of the sides to form four congruent isosceles triangles in the interior of the original triangle. This constitutes one iteration.

3. Each new iteration is performed on the three outer triangles while the triangle in the middle is left alone.

Mathematical ideas often turn out to have surprising connections to seemingly unrelated fields. The following exploration shows a connection between a Sierpinski Gasket and Pascal's Triangle.

Exploration 2   *Pascal and Sierpinski*

You will need
Graph paper

ALGEBRA
Connection

Recall that Pascal's Triangle is a triangular number array beginning with 1 at the top. Each number in each row is the sum of the two numbers directly above it. The pattern is repeated endlessly.

```
          1
         1 1
        1 2 1
       1 3 3 1
      1 4 6 4 1
   1 5 10 10 5 1
```

1 Build a Pascal's Triangle with at least 24 rows in a triangle of squares as indicated in the diagram. The more rows you can complete, the more impressive the result will be.

2 Darken each box occupied by an odd number, and leave each box occupied by an even number blank. Describe your results. ❖

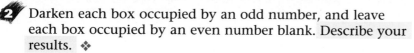

Communicate

1. What is self-similarity? If it helps you to describe an object to explain self-similarity, do so.

2. Name an object from nature that is self-similar, and describe how it is self-similar.

3. Why do you think a set of segments called Cantor dust is called that?

4. How many iterations does any fractal have?

Practice & Apply

5. Mimi, Arnold, and Roberto measure the rocky ocean frontage of a hotel using paces. Arnold's stride is 3 ft long and he counts 24 paces. Roberto's stride is $2\frac{1}{2}$ ft long and he counts 36 paces. Mimi's stride is 2 ft long and she counts 47 paces. How long is the coastline to each? Explain your answer.

6. Explain how a fractal coastline could be considered to have an infinite length.

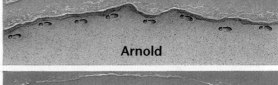

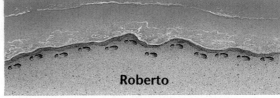

Two different measurements of a "coastline."

Portfolio Activity Exercises 7–11 ask you to construct a fractal called the Koch Snowflake.

7. Construct the first iteration of the Koch Snowflake using the directions below.

   a. Construct an equilateral triangle with 18-cm sides.

   b. Divide each side of the triangle into three 6-cm segments.

   c. Using the middle 6-cm segment on each side of your original triangle, construct an equilateral triangle outside the triangle with the middle segment as the base. Erase the middle segment. You should now have a six-pointed star. This completes the first iteration.

8. Continue the construction of the Koch Snowflake. Repeat the steps by dividing each segment into three congruent segments, constructing new smaller equilateral triangles and erasing the bases. Do at least three iterations.

9. Find the perimeter of the Koch Snowflake for the first two iterations. Remember the snowflake starts with a perimeter of $18 \times 3 = 54$.

10. Is the perimeter increasing or decreasing? If it is increasing, is it increasing by greater leaps each iteration, or by smaller leaps—i.e., is it increasing at an increasing or decreasing rate? What does this tell you about the perimeter?

11. Look at your snowflake and consider the changes it would make if you completed many more iterations. What do you think happens to the area of the snowflake as the iterations increase?

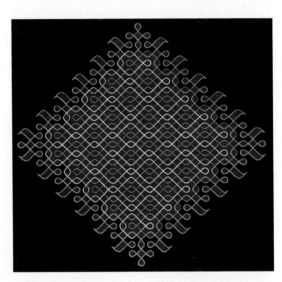

Cultural Connection: Asia People of India have been using fractals as art for many centuries. They are taught to draw a *Kolam* very quickly. A Kolam has an algorithm very similar to a curve known as the Hilbert Curve, which you will construct in Exercises 12–15.

12. Start with a square with no bottom side. Divide the top side into three congruent segments, and construct a square inward from the middle segment. Erase the middle segment of the top of the original square.

Step 1

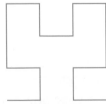
Step 2

13. Divide the sides into three segments. Construct an inward square using the lowest segment on each of the sides. Erase the outside segment from each of these squares. This is the first iteration.

14. Repeat this process on the two squares now extending upward at the top of where your original square was, and on the two squares that turn inward at the bottom of your original square. Repeat this process three times.

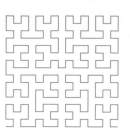

15. This is known as a space-filling curve, or "monster curve." If you were to continue this process indefinitely, what would the result look like?

Look Back

Find the distance between each pair of points.   [Lesson 5.6]

16. $(4, -2), (2, -1)$

17. $(5, -10), (-2, 3)$

18. $(15, 2), (-6, 5)$

Find the distance between each point in a three-dimensional coordinate system.   [Lesson 6.4]

19. $(4, 3, 2), (2, -3, 5)$

20. $(18, 1, 0), (0, -1, 5)$

21. $(5, 1, -5), (2, -12, 0)$

22. Find the measure of the diagonal of a cube with its corner at $(0, 0, 0)$, lying in the top-front-right octant, and with coordinates as shown. **[Lesson 6.4]**

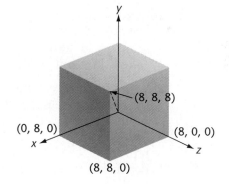

Look Beyond

Candy Making   Dissolve $2\frac{1}{2}$ cups of sugar into a cup of boiling water. Hang a string into the sugar water and leave it untouched for at least a day. Sugar crystals will form on the string, resulting in "rock candy."

23. How is the resultant "rock candy" a fractal?

24. What else do you see in nature that is the same kind of process that results in a fractal?

TWO RANDOM PROCESS GAMES

The Chaos Game

Can there be order in a random process? The following "Chaos Game" may lead you to ask some deep questions!

1. Randomly select a point in the interior of the triangle determined by points A, B, and C. This point is called the **seed point.**

2. Roll a die or use some other method to randomly select either vertex *A*, *B*, or *C*. (You can let 1 or 2 represent A, 3 or 4 represent *C*, and 5 or 6 represent *B*.) Mark a point halfway between the selected vertex and the seed point. This point becomes the **new seed point.**

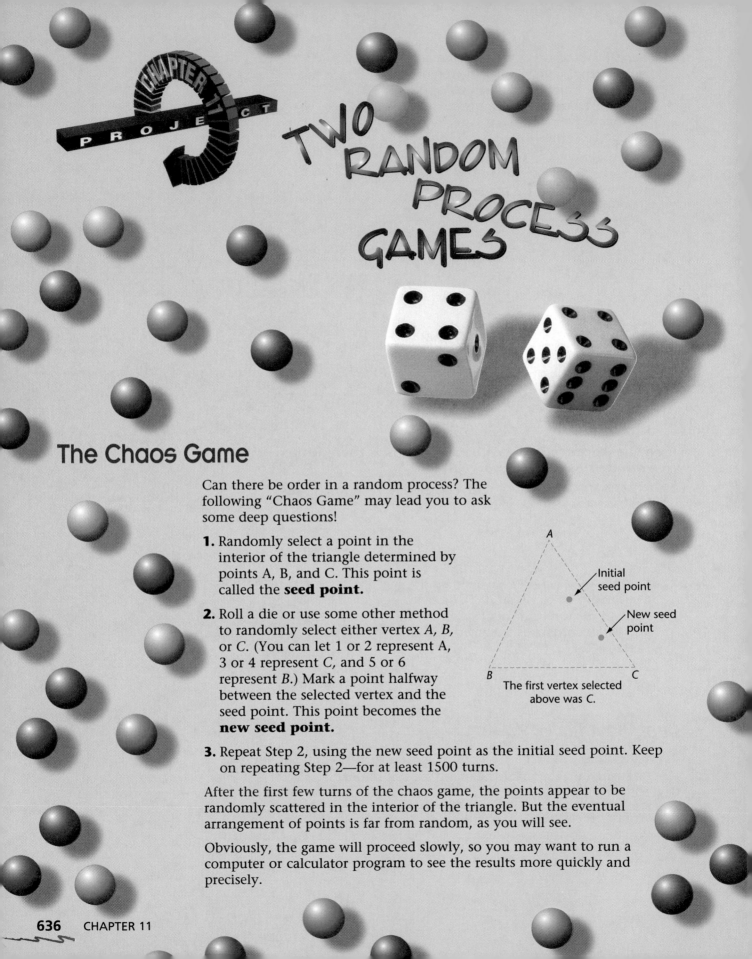

Initial seed point

New seed point

The first vertex selected above was C.

3. Repeat Step 2, using the new seed point as the initial seed point. Keep on repeating Step 2—for at least 1500 turns.

After the first few turns of the chaos game, the points appear to be randomly scattered in the interior of the triangle. But the eventual arrangement of points is far from random, as you will see.

Obviously, the game will proceed slowly, so you may want to run a computer or calculator program to see the results more quickly and precisely.

A Forest Fire Simulation

A forest fire is less likely to spread through a forest where trees are sparse than where they are densely concentrated. But what is the exact density and pattern of trees that guarantees the spread of the fire?

In this simplified model, you will try to discover the critical density of trees required for a forest fire to spread. Follow the steps below. (It will be necessary to pool your data with the rest of the class to obtain valid results.)

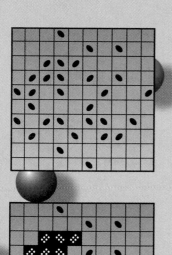

1. Your forest will be a 10 × 10 grid, which you will randomly plant with a certain percentage of trees. To have a forest with 30% trees, you will need to fill 30 of the 100 grid units with trees. One way to do this randomly is to number small squares of paper from 1 to 100 and mix them up. Pick a number for each square on the grid. If the number picked is 1–30, place a "tree" in that grid unit.

2. To simulate a fire's spread, start the "fire" with all the trees on the left side of the grid. The fire can spread to a neighboring tree directly above, below, to the right, or to the left of a burning tree—not diagonally.

3. See if the fire spreads from the left side of the grid to the right side. In the diagram, the fire dies out before it reaches the right side of the grid.

4. Each person or group in the class should simulate fires with 10%, 20%, 30%, 40%, 50%, 60%, 70%, 80%, 90% tree concentrations (0% and 100% tree concentrations are obvious).

5. Collect and tabulate the data of the class. Does there appear to be a critical percentage that assures the spread of the fire across the grid?

Chapter 11 Review

Vocabulary

affine transformation	621	Lobachevskian geometry	616	self-similarity	629
even vertices	601	Möbius strip	607	simple closed curve	608
fractal	629	network	600	taxicab circle	596
golden ratio	586	non-Euclidean geometry	615	taxicab radius	596
hyperbolic geometry	616	odd vertices	601	taxidistance	594
invariant	609	orthogonal	617	torus	609
knot theory	607	Riemannian geometry	616	trefoil knot	607

Key Skills and Exercises

Lesson 11.1

> **Key Skills**

Define and construct golden rectangles.

Is the quadrilateral a golden rectangle?

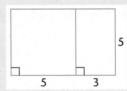

A rectangle is golden if it is constructed from a square and a similar rectangle. Set up a proportion to see if the rectangle is golden.

$$\frac{8}{5} \overset{?}{=} \frac{5}{3}$$

$$1.6 \neq 1.67$$

Or you can compare the ratio of length to width with the golden ratio, $\frac{1 \pm \sqrt{5}}{2}$.

$$\frac{8}{5} \overset{?}{=} \frac{(1 + \sqrt{5})}{2}$$

$$1.6 \neq 1.62$$

Both methods give an inequality, so the rectangle is not golden.

> **Exercises**

1. The figure at the right is a sand dollar. In what ways does a sand dollar's shape reflect the golden ratio?

Lesson 11.2

> **Key Skills**

Using taxicab geometry, find the distance between two points.

Find the taxidistance between points (9, 5) and (3, −4).

To find taxidistance, you calculate the differences between the x values and between the y values and add the absolute value of the differences.

$$|9 - 3| = 6 \qquad |5 - (-4)| = 9$$

$6 + 9 = 15$. The taxidistance is 15.

Create taxicab circles.

The taxicab circle has a radius of 2; that is, the points shown are all the possible ways of going two units from the center of the circle. The circumference is 16.

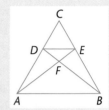

➤ *Exercises*

2. Draw a taxicab circle with a radius of 3. Find the circumference of the circle.

Lesson 11.3

➤ *Key Skills*

Determine whether a network is traversable.

Can you traverse the network at the right?

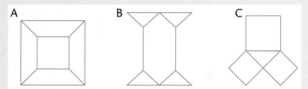

In order to be traversable, a network may have any number of even vertices, but a maximum of two odd vertices. We classify the vertices: *C*, *D*, *E*, and *F* connect an even number of paths; *A* and *B* connect an odd number of paths. This network has only two odd vertices, so it is traversable.

➤ *Exercises*

3. Which of these networks can be traversed?

Lesson 11.4

➤ *Key Skills*

Determine the topological equivalence of two figures.

Are the two closed loops topologically equivalent?

To be equivalent, the shapes must be transformable, one to the other, without tearing or cutting them. The left-hand shape can be pulled into a circle, but it will have no knot in it. The right-hand shape cannot lose its knot without being cut. So the two shapes are not equivalent.

➤ *Exercises*

4. Which figures are topologically equivalent? Explain your answer.

5. Draw a topological equivalent to a water hose.

Lesson 11.5

➤ *Key Skills*

Find the consequences of omitting the Parallel Postulate from a geometry.

In Riemannian geometry, which is one example of a non-Euclidean geometry, a line is a great circle on a sphere. Since all great circles intersect, there are no parallel lines in this geometry. Thus any theorem whose proof depends on the Parallel Postulate may be false in Riemannian geometry.

➤ *Exercises*

6. In Reimannian geometry, the sum of the angles in a triangle is greater than what number?

Lesson 11.6

➤ Key Skills

Make an affine transformation of an image.

Sketch the pre-image and image for the affine transformation $(x, y) \rightarrow (3x, 4y)$ of rectangle $(0, 0)$, $(0, 6)$, $(3, 6)$, $(3, 0)$.

Plot the original points and draw the rectangle. Multiply the x-coordinates by 3 and the y-coordinates by 4, and plot the resulting points for the image.

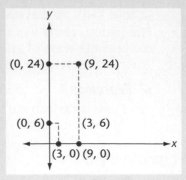

Sketch the projection of points.

Using center of projection O, project points A and B onto line m.

Use a straightedge to draw rays from the center of projection through A and B on line l to line m. The new points A' and B' lie at the intersection of the rays and line m.

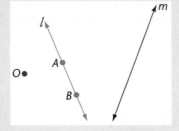
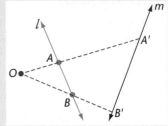

➤ Exercises

7. Sketch the pre-image and image for the affine transformation $(x, y) \rightarrow (2x, -y)$ of triangle $(0, 0)$, $(3, 5)$, $(-1, 4)$.

8. Using center of projection O, project points A, B, and C onto line m.

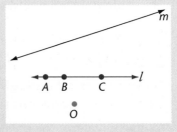

Lesson 11.7

➤ Key Skills

Use fractal concepts to solve problems.

The pilot of a small airplane going over the Fractal Mountains reports that it is 2.5 miles from point A to point B. A hiker traveling from point A to point B reports the distance as 3.5 miles. Explain the difference in reports.

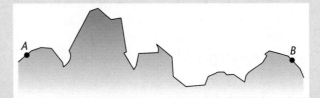

The reports depend on how closely the measurements follow the contours of mountains. The hiker follows the contours very closely. The airplane does not.

➤ Exercises

9. You and I would like to know how long the coast of Oregon is. You find a satellite photo of the coast and find it is 490 kilometers long. I use a series of aerial photos of the coast and find it is 560 kilometers long. Explain the difference in our results.

Application

10. Make a perspective drawing of the checkered floor.

Chapter 11 Assessment

1. Which quadrilaterals are golden rectangles, given that *ACEH* is a golden rectangle?

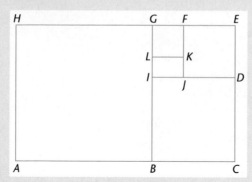

2. Find the taxidistance between points $(7, -3)$ and $(-1, -6)$.

3. Draw a taxicab circle with a radius of 4.

4. Which of these networks are traversable?

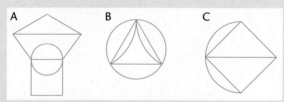

5. Which figures at the right are topologically equivalent? Explain your answer.

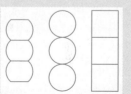

6. Draw a topological equivalent to a paper clip.

7. How many lines are parallel to another line through one point in a hyperbolic geometry? Explain.

8. Sketch the pre-image and image for the affine transformation $(x, y) \rightarrow (3x, 2y)$ of triangle $(-1, -2)$, $(-5, -3)$, $(1, 1)$.

9. Using center of projection O, project points A, B, and C onto line m.

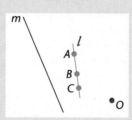

10. Draw the second iteration of the Sierpinski carpet, shown below.

CHAPTER 12

A Closer Look at Proof and Logic

How can a machine be intelligent? Some people believe that computers cannot actually think—and never will be able to. However, machines can certainly follow the rules of logic and make decisions based on given conditions. The decisions made by computer chess players and other "smart" machines are made possible by formal logic.

Formal logic reduces logical procedures to their essential elements, which can, in turn, be implemented by computers or other machines.

LESSONS

12.1 Truth and Validity in Logical Arguments

12.2 "And," "Or," and "Not" in Logic

12.3 A Closer Look at If-Then Statements

12.4 Indirect Proof

12.5 Computer Logic

12.6 *Exploring* Proofs Using Coordinate Geometry

Chapter Project
Two Famous Theorems

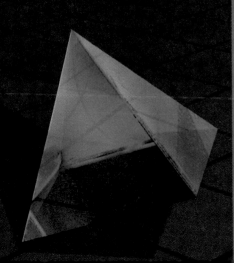

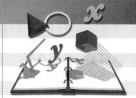

PORTFOLIO ACTIVITY

Imagine your own dream home or apartment. Design a logically linked system to control the lights, air conditioning, security system, appliances, etc. Reduce the system's operation to a set of logical statements. Then design a circuit of logic gates and switches for the system.

If you play chess, you can complete the following rule for a chess-playing computer.

If my King is in check, then I must

- move out of check, or

- capture the attacking piece, or

- _____?_____ or

- _____?_____ .

The student's King is in check. He can capture the attacking piece, but the computer will win on the next move.

Truth and Validity in Logical Arguments

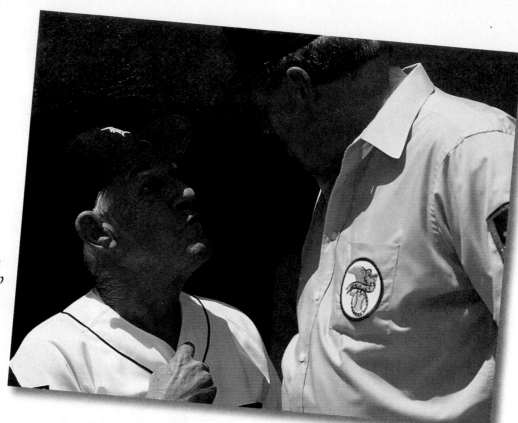

why *Newspaper reporters, politicians, lawyers, and even baseball managers may use logic to convince others of their views. Recognizing valid arguments as well as invalid ones will help you to think clearly in confusing situations.*

Arguments: Valid and Invalid

In logic, an **argument** consists of a sequence of statements. The final statement of the argument is called the **conclusion**, and the statements that come before it are known as **premises**. The following is an example of a logical argument:

> If an animal is an amphibian, then it is a vertebrate.   ← premises
> Frogs are amphibians.
> Therefore, frogs are vertebrates.   ← conclusion

In this argument, the conclusion is said to *follow logically* from the premises. The premises force the conclusion. An argument of this kind is known as a valid argument, and the conclusion of such an argument is said to be a valid conclusion.

A valid argument makes the following "guarantee": *If the premises are all true, then the conclusion is true as well.* In the valid argument above, both of the premises are true. Therefore, the conclusion must be true.

Now consider a different argument:

Some vertebrates are warmblooded. ⎫ ← premises
Frogs are vertebrates. ⎭
Therefore, frogs are warmblooded. ← conclusion

This *argument is invalid*. Both of the premises are true, but the conclusion is false. *This can never happen in a valid argument.*

CRITICAL *Thinking*

Suppose that you did not know that a frog is a coldblooded animal. Would you have questioned the conclusion of the second argument anyway? Is there something basically "wrong" with the argument? If so, try to describe what it is.

Try This Write your own examples of valid and invalid arguments.

The Form of an Argument

Logicians can tell whether an argument is valid or invalid without knowing anything about the validity of its premises or of its conclusions. They are able to do this by analyzing its form. The valid argument on the previous page, for example, has the following form:

If p then q. ⎫
p ⎬ ← premises
Therefore, q. ← conclusion

This argument form is sometimes referred to by its medieval Latin name, **modus ponens**—the "proposing mode." Any argument that has this form is valid, regardless of the statements that are substituted for p and q. The following nonsense argument is valid:

If flivvers twiddle then bokes malk. ⎫ ← premises
Flivvers twiddle. ⎭
Therefore, bokes malk. ← conclusion

This argument's form (*modus ponens*) guarantees that if the first two statements should somehow turn out to be true, then the third statement (the conclusion) will be true as well.

False Premises

If the premises of an argument are false, then there can be no "guarantee" that the conclusion is true, even though the argument might have a valid form. The following *modus ponens* argument is valid, but its conclusion is false.

If an animal is an amphibian, then it can fly. ⎫ ← premises
A frog is an amphibian. ⎭
Therefore, a frog can fly. ← conclusion

Notice that *the conclusion, though false, is a valid conclusion* because the argument has a valid form. But since one of the premises is false, the conclusion is not guaranteed to be true. Remember, a valid argument guarantees that a conclusion is true *if the premises are true*. There is no guarantee if one or more of the premises are false.

CRITICAL *Thinking*

The last argument has a false premise and a false conclusion. Do you think it is possible for a valid argument to have a false premise and a true conclusion? If so, give an example. If not, explain why.

The Law of Indirect Reasoning

Consider the following argument.

If a shirt is a De Morgan, then it has a blackbird logo. } ← premises
This shirt does not have a blackbird logo.
Therefore, this shirt is not a De Morgan.   ← conclusion

Does this argument seem valid to you? If you knew the premises to be true, would you be convinced that the conclusion was true? The argument has the following form:

If p then q. } ← premises
Not q.
Therefore, not p.   ← conclusion

This is, in fact, a valid argument form. It is sometimes referred to by its medieval Latin name, **modus tollens**—the "removing mode." In more recent times it has come to be known as the Law of Indirect Reasoning.

You should be careful not to confuse the *modus tollens* form with the following form:

If p then q. } ← premises
Not p.
Therefore, not q.   ← conclusion

This is the form of a common logical mistake, or **fallacy**, known as "denying the antecedent." The conclusion does not follow logically from the premises even *if it is true*. Be sure you understand the difference between this form and the *modus tollens* form.

Try This   Give two examples of an argument that has the "denying the antecedent" form. In your first example, make the premises true and the conclusion false. In your second example, make the premises true and the conclusion true. Explain why the form is a logical fallacy, or mistake. (Does it make a guarantee about the conclusion?)

EXERCISES & PROBLEMS

Communicate

For Exercises 1–4, determine if the argument is valid. Explain why or why not.

1. If today is Wednesday, then the cafeteria is serving beef stew.

Today is Wednesday.

Therefore, the cafeteria is serving beef stew.

2. If pigs fly, then today is February 30.

Today is February 30.

Therefore, pigs fly.

3. If Jon is a man, then Jon is mortal.

Jon is not mortal.

Therefore, Jon is not a man.

4. If $y > x$, then $a > b$.

$y > x$.

Therefore, $a > b$.

5. Is it possible for a valid argument to have a false conclusion? Explain your reasoning.

Practice & Apply

In Exercises 6–9, write a valid conclusion from the given premises.

6. If a person belongs to Party c, then the person is a conservative.

Maria is a member of Party c.

7. If the team wins on Saturday, the team will be in the playoffs.

The team did not make the playoffs.

8. If Evan beats Rob, then Evan plays Mario.

If Evan plays Mario, then Mario will win the tournament.

Evan beats Rob.

9. If a then b.

Given: a

The If-Then Transitive Property is a form of argument used in proofs throughout the text. It was introduced in Lesson 2.2.

If-Then Transitive Property

If *p* then *q*.
If *q* then *r*.
Therefore, if *p* then *r*.

For Exercises 10 and 11, list all valid conclusions that can be drawn from the given premises.

10. If *x* then *y*.

    If *y* then *k*.

    Given: not *k*.

11. If *n* then *m*.

    If *q* then *r*.

    If *m* then *q*.

    Given: *n*.

12. Given premises:

    If you study, then you will succeed.

    Eleanor studied.

    Tamara did not study.

    José succeeded.

    Mary did not succeed.

    Which of the following conclusions are valid? Explain your reasoning.

    a. Eleanor will succeed.

    b. Eleanor will not succeed.

    c. Tamara will succeed.

    d. Tamara will not succeed.

    e. José studied.

    f. José will not study.

    g. Mary studied.

    h. Mary did not study.

13. Consider the following argument:

    If a quadrilateral is a parallelogram, then its diagonals are congruent.

    PQRS is a parallelogram.

    Therefore, the diagonals of *PQRS* are congruent.

    a. Is the conclusion valid? Explain your reasoning.

    b. Is the first premise true or false?

    c. If *PQRS* is actually a square, is the conclusion true or false? Explain your reasoning.

    d. If *PQRS* is a parallelogram that is not a rectangle or a square, is the conclusion false? Explain your reasoning.

14. A valid argument has premises *a*, *b*, *c*, and *d*, and conclusion *r*.

    a. Does validity guarantee that all four premises and *r* are true? Explain your reasoning.

    b. If the four premises are true, must *r* be true? Explain your reasoning.

    c. If the argument is valid, under what circumstances might *r* be false?

Look Back

Determine if each pair of triangles is congruent. Explain why or why not. [Lessons 4.2, 4.3]

15.

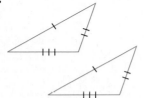

16.

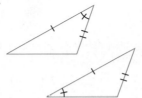

17.

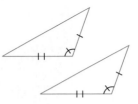

Use the circle at right for Exercises 18–20. [Lessons 9.3, 9.4]

$m\angle 1 = 20°$; $m\angle 2 = 35°$; $m\widehat{SR} = 80°$

Find each of the following.

18. $m\widehat{QR}$

19. $m\widehat{PS}$

20. $m\widehat{PQ}$

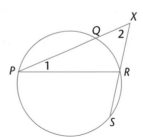

Look Beyond

The **contrapositive** of an if-then statement is written by negating the two parts of the statement and by switching the *if* and *then* parts.

If-then statement: If Joe goes home, then Joe will watch TV.

Contrapositive: If Joe does not watch TV, then Joe did not go home.

Write the contrapositive for each statement below.

21. If I make an A on the geometry test, then my father will give me $5.

22. If Bill is from France, then he can speak French.

23. If the piano has been moved, then the piano will be badly out of tune.

Why The words "and," "or," and "not" are used frequently in everyday conversation. These words have precise meaning in logical arguments.

In logic, a **statement** is a sentence that is either true or false. For example, "Belinda orders pepperoni on her pizza" is a statement, because it is either true or false. A **compound statement** in logic is formed when two statements are connected by **and** or by **or**.

Conjunctions

A compound statement that uses the word **and** is called a **conjunction**.

Sentence p:	*Today is Tuesday.*
Sentence q:	*Tonight is the first varsity basketball game.*
Conjunction p AND q:	*Today is Tuesday **and** tonight is the first varsity basketball game.*

A conjunction is true when both of its statements are true. If one or both of its statements are false, the conjunction is false. The four possibilities for a conjunction can be illustrated with a **truth table**.

All possible combinations of truth values for the two statements that form the conjunction are placed in the first two columns. The last column indicates the truth values for the conjunction. In the first line, for example, both of the statements that make up the conjunction are true. In this case, the conjunction is true.

p	q	p and q
T	T	T
T	F	F
F	T	F
F	F	F

EXAMPLE 1

Determine if the following conjunctions are true:

A George Washington was the first president of the United States, and John Adams was the second.

B The sum of the measures of the angles of a triangle is 200°, and blue is a color.

Solution ➤

A The conjunction is true because both of its statements are true.

B The conjunction is false because one of its statements is false. ❖

Disjunctions

Two statements may also be combined into a single statement by the word **or**. In logic, **OR** statements are known as **disjunctions**. When used in everyday language, **OR** often means "one or the other, but not both." For example, if a waitress says,

"You may have soup **or** salad with your pizza,"

she means that you may choose just one. This kind of *or* is known as the **exclusive OR**.

However, in mathematics and logic, **OR** means "one *or* the other, *or* both." This kind of *or* is known as the **inclusive OR**. If someone asks how John spends his Saturday afternoons, the answer might be,

"He goes swimming **or** bowling."

The sentence can be written in logical form as *p* **OR** *q*, where *p* and *q* are identified as shown:

He goes swimming **or** [he goes] bowling.

$$p \qquad\qquad q$$

The statement is false only if John does neither one. Notice that in the truth table, only the fourth combination gives a value of false for the conjunction. If he does either or both it is true, as the values for the first three combinations show.

p	*q*	*p* OR *q*
T	T	T
T	F	T
F	T	T
F	F	F

EXAMPLE 2

Determine whether each of the following disjunctions is true:

A Teddy Roosevelt liked to go hunting or horseback riding.

B Dogs can fly or 5 − 3 = 2.

Solution ➤

A The disjunction is true because both statements are true. (Teddy Roosevelt liked to go hunting. He also liked to go horseback riding).

B The disjunction is true because one of the statements is true. ❖

Negation

It is raining outside.   *It is **not** raining outside.*

One of the statements above is the **negation** of the other. If *p* is a statement, then **not** *p* is its negation. The negation of *p* is written ~*p*.

It is raining outside.   *It is **not** raining outside.*

p   **~p**

When a statement *p* is true, its negation ~*p* is false.

When a statement *p* is false, its negation ~*p* is true.

CRITICAL
Thinking

How do you write the negation of a conjunction, *p* **AND** *q*, or of a disjunction, *p* **OR** *q*? Hint: You will need to use parentheses.

•Exploration *The Negation of a Conjunction*

You will need
No special tools

1 Copy and complete the truth table for the negation of a conjunction. Remember that the fourth column represents a negation of the third column.

p	q	p AND q	~(p AND q)
T	T	T	F
T	F	?	?
F	T	?	?
F	F	?	?

2 Copy and complete the truth table for a disjunction formed by two negations. The first row is completed for you. Remember that the values for ~*p* and ~*q* are used to determine the truth tables for ~*p* OR ~*q*.

p	*q*	~*p*	~*q*	~*p* OR ~*q*
T	T	F	F	F
T	F	?	?	?
F	T	?	?	?
F	F	?	?	?

3 Compare the last column from Step 1 with the last column from Step 2. Explain what you observe.

4 When two logic statements have the same truth values they are said to be **logically equivalent**. Complete the following statement, which is one of **De Morgan's Laws**.

~(*p* AND *q*) is logically equivalent to __?__. ❖

CRITICAL *Thinking*

In the two truth tables in the exploration, why is it important that the combinations of T and F be listed in exactly the same way in the first two columns of each? If you list T and F values for the three statements *p*, *q*, and *r*, how many different combinations of T and F will there be? What is a good order for these combinations?

EXERCISES & PROBLEMS

Communicate ⌁⌁

1. Explain the conditions necessary for a conjunction to be true.

2. Explain the conditions necessary for a disjunction to be true.

Indicate whether the statement is true or false. Explain your reasoning.

3. $4 + 5 = 9$ and $4 \cdot 5 = 9$.

4. All triangles have three angles or all triangles have four angles.

5. Three noncollinear points determine a unique plane, and a segment has two endpoints.

6. All squares are hexagons or all triangles are squares.

Team strategy meeting before a debate.

State the negation of each statement.

7. The state is Alaska.

8. The weather is not rainy.

Practice & Apply

Write a conjunction for each pair of statements. Determine whether the conjunction is true or false.

9. A carrot is a vegetable.
Florida is a state.

10. A ray has only one endpoint. Kangaroos can fly.

11. The sum of the measures of the angles of a triangle is 180°. Two points determine a line.

Write a disjunction for each pair of statements. Determine whether the disjunction is true or false.

12. An orange is a fruit.
Cows have kittens.

13. Triangles are circles.
Squares are parallelograms.

14. Points equidistant from a given point form a circle. The sides of an equilateral triangle are congruent.

Write the negation of each statement.

15. The figure is a rectangle.

16. My client is not guilty.

17. Rain makes the road slippery.

18. a. Copy and complete the truth table for $\sim(\sim p)$.

p	$\sim p$	$\sim(\sim p)$
T	?	?
F	?	?

b. What is the statement $\sim(\sim p)$ logically equivalent to? Explain your reasoning.

LESSON 12.3

A Closer Look at If-Then Statements

why *In courtrooms, as in everyday life, if-then statements are a very important part of our language. They are also used in mathematical reasoning. The statement, "If a rhombus has four right angles, then it is a square" is just one of many examples of if-then statements to be found in this book.*

Lawyers use logic to convince a jury: "If the defendant committed a crime, then he must have been at 4th Avenue and Crescent between 10 P.M. and 11 P.M."

Conditionals

Various forms of if-then statements have been used throughout this book. You may recall from Lesson 2.2 that if-then statements are called conditionals.

Suppose your neighbor makes the following promise:

If you mow his lawn, then he will give you $10.

p q

Four possible situations can occur:

1. *You mow the lawn and your neighbor gives you $10.*

Since your neighbor kept his promise, we agree that the conditional statement is true.

2. *You mow the lawn. Your neighbor does not give you $10.*

The promise is broken. Therefore, we agree that the conditional statement is false.

For Exercises 19–26, write the statement expressed by the symbols. Use the following simple statements.

p: △ABC is isosceles.   *q:* △ABC has two equal angles.

r: ∠1 and ∠2 are adjacent.   *s:* ∠1 and ∠2 are acute angles.

19. ~*p*   **20.** *q* OR *p*   **21.** *p* AND *q*   **22.** ~*q*

23. ~*s*   **24.** *r* OR *s*   **25.** *s* AND ~*r*   **26.** *q* OR ~*s*

27. Explain all the logical possibilities that make the following statement true:

This weekend we will go camping and hiking, or it will rain and we will cancel the trip.

28. Construct a truth table for (**p** AND **q**) OR (**r** AND **s**). When is (**p** AND **q**) OR (**r** AND **s**) false?

29. Construct a truth table for (**p** OR **q**) OR **r**. When is (**p** OR **q**) OR **r** false?

30. Construct a truth table for (**p** AND **q**) AND **r**. When is (**p** AND **q**) AND **r** true?

Look Back

Is each pair of triangles similar? Explain why or why not.
[Lessons 8.3, 8.4]

31.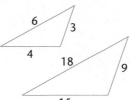

32.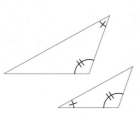

33.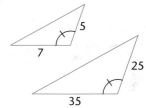

Find the following for the triangle below.
[Lessons 10.1, 10.2]

34. sin *A*

35. cos *B*

36. tan *B*

37. cos *A*

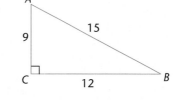

Look Beyond

38. Nonroutine Can you determine if the following sentence is true or false? Explain your reasoning.

I never tell the truth.

For Exercises 19–26, write the statement expressed by the symbols. Use the following simple statements.

p: △ABC is isosceles.   *q:* △ABC has two equal angles.

r: ∠1 and ∠2 are adjacent.   *s:* ∠1 and ∠2 are acute angles.

19. ~*p*   **20.** *q* OR *p*   **21.** *p* AND *q*   **22.** ~*q*

23. ~*s*   **24.** *r* OR *s*   **25.** *s* AND ~*r*   **26.** *q* OR ~*s*

27. Explain all the logical possibilities that make the following statement true:

This weekend we will go camping and hiking, or it will rain and we will cancel the trip.

28. Construct a truth table for (*p* AND *q*) OR (*r* AND *s*). When is (*p* AND *q*) OR (*r* AND *s*) false?

29. Construct a truth table for (*p* OR *q*) OR *r*. When is (*p* OR *q*) OR *r* false?

30. Construct a truth table for (*p* AND *q*) AND *r*. When is (*p* AND *q*) AND *r* true?

Look Back

Is each pair of triangles similar? Explain why or why not. [Lessons 8.3, 8.4]

31.   **32.**   **33.**

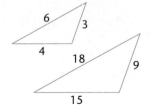   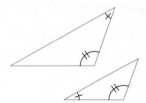   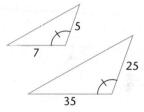

Find the following for the triangle below. [Lessons 10.1, 10.2]

34. sin *A*

35. cos *B*

36. tan *B*

37. cos *A*

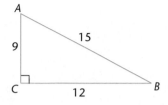

Look Beyond

38. Nonroutine   Can you determine if the following sentence is true or false? Explain your reasoning.

I never tell the truth.

THE ENDS OF TIME

Message of the Maya in Modern Times

By Greg Stec
Special to the Christian Science Monitor

Discovering the titles of Mayan royalty, the names of their gods and their food, the dates of important events, and all the other things great and small that make up an advanced society has taken more than a hundred years of digging and probing. At last, though, a computer program has been developed to calculate Mayan dates.

The Mayans worshiped time and numbers. The current thinking holds that by dating the elite's birth, accomplishments, and death with unimpeachable accuracy, the person's position and rank would be permanently established.

Computer programs wade through a sea of Mayan dates, saving investigators effort that can be used to examine other translations.

Imagine having two calendars on your wall, one of them 260 days long and the other 365 days long. Each day would have two dates, usually different. Your friend's birthday, for example, might be August 4 and April 29.

The Mayan calendar was similar in that it had two different years simultaneously. To get a better idea of how it worked, you can use diagrams of wheels and gears. To keep track of dates, you'll use letters for months instead of the actual Mayan names.

COOPERATIVE LEARNING

1. The Mayan 260-day year, called the Sacred Round, can be represented by the two smaller wheels in the diagram. Each day, the wheels move one notch to the next section.

   a. The date shown in the diagram is 7G. What will the date be 1 day later? How is that different from our calendar?

   b. What will the date be 4 days after 7G? 13 days after 7G? 20 days after 7G?

   c. If the wheels keep turning, will every one of the 260 possible dates occur? How do you know?

   d. Suppose the outer wheel had only 6 letters and the inner wheel had only 4 numbers. Would all 24 pairs of letters and numbers occur? How do you know?

   e. For every date to occur, what must be true about the number of sections on the wheels? Hint: Experiment with small numbers. Make a table of your results. Look for common factors.

Every date occurs	Every date does not occur
13, 20	4, 6

2. The Mayan 365-day year can be represented by a single large wheel. It consists of 18 "months" of 20 days each and a special 5-day "month." In the diagram, the date in the 365-day year is 13a. The date in the 260-day year is 7G. The combined date is 7G13a.

   a. The first day of the 365-day year is 0a. The 21st day is 0b. What is the 365th day of the year?

   b. What will the date be 1 day after 7G13a? 20 days after 7G13a? 260 days after 7G13a? 365 days after 7G13a?

   c. Does every possible date occur? Explain? Hint: Use your answer to question 1d.

The logic may be simple, but the Mayan round calendar was not. It employed two inter-meshing years, one 260 days long, and the other 365 days long...

SUPER CHALLENGE

3. How many times will the large wheel turn before the date 7G13a occurs again? How many days is that?

A Closer Look at If-Then Statements

why *In courtrooms, as in everyday life, if-then statements are a very important part of our language. They are also used in mathematical reasoning. The statement, "If a rhombus has four right angles, then it is a square" is just one of many examples of if-then statements to be found in this book.*

Lawyers use logic to convince a jury: "If the defendant committed a crime, then he must have been at 4th Avenue and Crescent between 10 P.M. and 11 P.M."

Conditionals

Various forms of if-then statements have been used throughout this book. You may recall from Lesson 2.2 that if-then statements are called conditionals.

Suppose your neighbor makes the following promise:

If you mow his lawn, then he will give you $10.

p q

Four possible situations can occur:

1. *You mow the lawn and your neighbor gives you $10.*

Since your neighbor kept his promise, we agree that the conditional statement is true.

2. *You mow the lawn. Your neighbor does not give you $10.*

The promise is broken. Therefore, we agree that the conditional statement is false.

3. *You do not mow the lawn. Your neighbor gives you $10.*

The promise is not broken, so we agree that the conditional is true.

4. *You do not mow the lawn. Your neighbor does not give you $10.*

The promise is not broken, so we agree that the conditional is true.

You can think of a conditional as a promise. In logic, if the "promise" is broken, the conditional is said to be false. Otherwise, it is said to be true.

The truth table below summarizes the truth values for the conditional. Recall that the logical notation for "if p then q" is $p \Rightarrow q$ (read: "p implies q").

p	q	$p \Rightarrow q$
T	T	T
T	F	F
F	T	T
F	F	T

The first two columns list all possible combinations of T and F for the two statements p and q. Notice that the only time $p \Rightarrow q$ is false is when the promise of $10 for mowing the lawn is broken.

The Converse

Recall from Lesson 2.2 that the converse of a conditional results from interchanging its "if" and "then" parts. Consider the following conditional:

If Tamika lives in Montana, then she lives in the United States.

(Is this statement true?)

The converse of the conditional is:

If Tamika lives in the United States, then she lives in Montana.

(Is this statement true?)

The truth table below summarizes the truth values for the conditional and its converse. When studying the table, notice that q and p are reversed in the last column.

p	q	$p \Rightarrow q$	$q \Rightarrow p$
T	T	T	T
T	F	F	T
F	T	T	F
F	F	T	T

Two statements are logically equivalent if they have the same truth values. Compare the truth values for the conditional and the converse. Are they logically equivalent? Explain your reasoning.

The Inverse

The **inverse** of a conditional is formed by negating both the hypothesis and the conclusion. The inverse of the original conditional is as follows:

If Tamika does not live in Montana, then she does not live in the United States. (Is this statement true?)

The truth table below summarizes the truth values for the inverse. Notice that extra rows are required for the negations of **p** and **q**.

p	q	~p	~q	~p ⇒ ~q
T	T	F	F	T
T	F	F	T	T
F	T	T	F	F
F	F	T	T	T

Are a conditional and its inverse logically equivalent? Explain your reasoning.

The Contrapositive

The **contrapositive** of a conditional is formed by interchanging the "if" and "then" parts and negating each part. The contrapositive of the original conditional is as follows:

If Tamika does not live in the United States, then she does not live in Montana. (Is this statement true?)

The truth table below summarizes the truth values for the contrapositive of a conditional. When studying the table, notice that **q** and **p** are reversed and negated in the last column.

p	q	~q	~p	~q ⇒ ~p
T	T	F	F	T
T	F	T	F	F
F	T	F	T	T
F	F	T	T	T

Notice that the final columns for the truth tables of the original conditional and those of its contrapositive are the same. Recall from Lesson 12.2 that this means the two statements are logically equivalent. This is a very useful piece of information. Whenever a conditional is true, its contrapositive must also be true. Thus, the contrapositive of every postulate and theorem that can be written in if-then form must also be true!

CRITICAL *Thinking*  In addition to a conditional and its contrapositive, what other two forms of a conditional are logically equivalent?

Summary

An affine if-then statement has three related forms.

If-then statement:	If p then q.	$p \Rightarrow q$
Converse:	If q then p.	$q \Rightarrow p$
Inverse:	If ~p then ~q.	$\sim p \Rightarrow \sim q$
Contrapositive:	If ~q then ~p.	$\sim q \Rightarrow \sim p$

Try This Write the converse, inverse, and contrapositive of each conditional below. State whether each new statement is true or false.

a. If a triangle is equilateral, then it is an isosceles triangle.

b. If a quadrilateral is a rhombus, then it is a square.

EXERCISES & PROBLEMS

Communicate

For Exercises 1–4, state the converse, inverse, and contrapositive for each conditional.

1. If today is February 30, then the moon is made of green cheese.

2. If all the sides of a triangle are congruent, then the triangle is equilateral.

3. If I do not go to the market, then I will not buy cereal.

4. If the car starts, then I will not be late to school.

5. Describe the circumstances that would make a statement "If *a* then *b*" false.

What is *green cheese, anyway?*

Practice & Apply

For each conditional in Exercises 6–10, write the converse, inverse, and contrapositive. Decide whether each conditional, converse, inverse, and contrapositive is true or false and explain your reasoning.

6. If a figure is a square, then it is a rectangle.

7. If $a = b$, then $a^2 = b^2$.

8. If three angles of a triangle are congruent to three angles of another triangle, then the triangles are congruent.

9. If p and q are even numbers, then $p + q$ is an even number.

10. If water freezes, then its temperature is less than or equal to 32°F.

11. Given: *If p then q.* Write the contrapositive of this statement. Then write the contrapositive of the contrapositive. What may you conclude about the contrapositive of the contrapositive of an if-then statement?

12. Suppose the following statement is true:

    If the snow exceeds 6 inches, then school will be canceled.

    Which of the following statements must also be true? Explain your reasoning.

    a. If the snow does not exceed 6 inches, then school will not be canceled.

    b. If school is not canceled, then the snow does not exceed 6 inches.

    c. If school is canceled, then the snow exceeds 6 inches.

13. State the "Pythagorean" Right-Triangle Theorem, its converse, inverse, and contrapositive. Determine whether each is true or false and explain your reasoning.

14. Choose a postulate or theorem from Chapter 3 that is written in if-then form. Write its converse, inverse, and contrapositive and decide whether these statements are true or false.

15. Choose a postulate or theorem from Chapter 9 that is written in if-then form. Write its converse, inverse, and contrapositive and decide whether these statements are true or false.

Some statements that are not written in if-then form can be rewritten in if-then form. For example,

> *Every rectangle is a parallelogram*

can be rewritten as follows:

> *If a figure is a rectangle, then it is a parallelogram.*

For Exercises 16–19, write each statement in if-then form.

16. All seniors must report to the auditorium.

17. A point on the perpendicular bisector of a segment is equidistant from the endpoints of the segment.

18. Call me if you expect to be late.

19. Doing mathematics homework every night will improve your grade in mathematics.

20. Look for three if-then statements (or statements that can be written in if-then form) in a newspaper, in a magazine, or on TV. Write the statement, its converse, inverse, and its contrapositive. Determine if each is true or false and explain your reasoning.

21. Consider the following statement:

You will make the honor roll only if you get at least a B in mathematics.

Which of the following statements appear to convey the same meaning as the original statement? Explain your reasoning.

   a. If you made the honor roll, then you must have gotten at least a B in mathematics.

   b. If you get at least a B in mathematics, then you will make the honor roll.

   c. If you do not make the honor roll then you did not get at least a B in mathematics.

   d. If you do not get at least a B in mathematics then you will not make the honor roll.

22. The statement p **if and only if** q, written p **iff** q, is equivalent to two statements:

$$p \text{ if } q \text{ and } p \text{ only if } q.$$

Suppose the statement p **iff** q is true. Which of the following must be true? Explain your reasoning.

   a. If p then q.

   b. If q then p.

   c. If $\sim p$ then $\sim q$.

   d. If $\sim q$ then $\sim p$.

23. Statements using "if and only if" are sometimes called **biconditionals**. Based on the results from Exercise 22, what can you conclude about many of the theorems in this book? Explain your reasoning and provide examples from the text.

IF YOU CAN READ THIS
YOU'RE TOO CLOSE

**Use the triangle at right and the Law of Sines to determine
the missing measure.   [Lesson 10.5]**

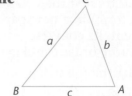

24. Find side c given m$\angle B$ = 37°,
m$\angle A$ = 50°, and b = 100.

25. Find side b given m$\angle C$ = 65°,
m$\angle A$ = 47°, and c = 3.45.

**Copy the vectors below and draw the resultant
using the head-to-tail method. You may have
to translate one of the vectors.   [Lesson 10.7]**

26.

27.

28.

Look Beyond

29. You are on an island and are trying to determine whether you should
go east or west in order to get back to the boat dock. Two different
groups of people live on the island. One group always tells the truth.
The other group always lies. The groups dress differently, but you
have not been able to determine which is which. You approach two
strangers who are dressed differently to ask directions to the dock.
What one question can you ask that will provide you with the correct
direction?

LESSON 12.4

Indirect Proof

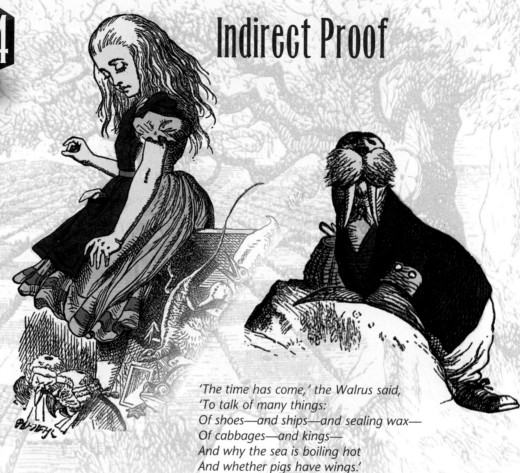

why Lewis Carroll, the author of Alice in Wonderland and Through the Looking-Glass, was a logician who was fond of absurdity as a form of entertainment. But does absurdity have any real place in logic or mathematics? In this lesson you will see that it can, in fact, be quite relevant.

'The time has come,' the Walrus said,
'To talk of many things:
Of shoes—and ships—and sealing wax—
Of cabbages—and kings—
And why the sea is boiling hot
And whether pigs have wings.'

— Lewis Carroll

Have you ever heard an expression like,

"If you are twenty-one, then pigs have wings!"

The speaker, perhaps without realizing it, is inviting you to use the Law of Indirect Reasoning, which you studied in Lesson 12.1. Since it is certainly not true that pigs have wings, the statement in question (in the "if" part) must be false.

If you are twenty-one, then pigs have wings.	(If p then q.)
Pigs do not have wings.	(Not q.)
Therefore, you are not twenty-one.	(Therefore, not p.)

Indirect Proofs

A closely related form of argument is known by its Latin name **reductio ad absurdum**—literally, "reduction to absurdity." In this type of argument, an assumption is shown to lead to an absurd or impossible conclusion. In this case, the assumption must be rejected.

Impossible spatial configurations

In formal logic and mathematics, certain proofs use a *reductio ad absurdum* strategy, but with an important twist. In such proofs, you assume the *opposite*, or in logical terms, the *negation*, of the statement that you want to prove. If this assumption leads to an impossible result, then you can conclude that the assumption was false. (Then you know that the original statement was true.) Such proofs are known as **proofs by contradiction**.

Proofs by Contradiction

What is meant by an "absurd" or "impossible" result? In logic, a **contradiction** is such a result. A contradiction has the following form:

> *p* **and** ~*p*.

That is, a contradiction asserts that a statement and its negation are both true. The following compound statement is a contradiction:

A horse is a vegetarian, and a horse is not a vegetarian.

In formal logic and in mathematics, any assumption that leads to a contradiction must be rejected. Contradictions turn out to be very useful in indirect proofs.

PROOF BY CONTRADICTION

To prove **s**, assume ~**s**. Then the following argument form is valid:

If ~**s** then (**t** and ~**t**).
Therefore, **s**.   **12.4.1**

Corresponding Angles Revisited

The following proof uses a contradiction to prove the converse of the Corresponding Angles Postulate. The converse, which is itself a theorem, states:

> *If two lines are cut by a transversal in such a way that corresponding angles are congruent, then the two lines are parallel.* **(Thm 3.4.1)**

In the proof, the "if" part of the theorem is the given.

Given: Line *l* is a transversal that passes through lines *m* and *n*. ∠1 ≅ ∠2

Prove: *m* ∥ *n*

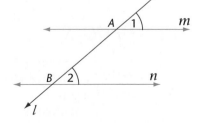

Proof:

Assume that *m* is *not* parallel to *n*.

Since, by assumption, *m* is not parallel to *n*, the two lines will meet at some point *C*, as shown in the redrawn figure.

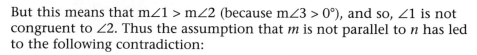

Since ∠1 is an exterior angle of △*ABC*,

m∠1 = m∠2 + m∠3.

But this means that m∠1 > m∠2 (because m∠3 > 0°), and so, ∠1 is not congruent to ∠2. Thus the assumption that *m* is not parallel to *n* has led to the following contradiction:

∠1 ≅ ∠2 and not (∠1 ≅ ∠2).

The assumption must therefore be false. Thus, the conclusion is

m ∥ *n*.

Alibis and Indirect Proof

Arguments using the Law of Indirect Reasoning are more common than you might think. In a court of law, for example, a lawyer might want to show that a claim on the part of the prosecution contradicts the evidence—the "given." Arguments like the following are quite common:

"If the defendant set the fire, then he would have been at the restaurant between 7:30 P.M. and 11:00 P.M. But three witnesses have testified that the defendant was not at the restaurant during those hours—he was in fact at a party on the other side of town. Therefore, the defendant did not set the fire."

The form of the argument is as follows:

1. If the defendant set the fire, then he was at the restaurant between 7:30 P.M. and 11:00 P.M. **(If *p* then *q*.)**

2. The defendant was not at the restaurant between 7:30 P.M. and 11:00 P.M. **(~*q*.)**

3. Therefore, the defendant did not set the fire. **(Therefore, ~*p*.)**

Thinking

Show how the argument above could be made into a proof by contradiction. Let the first two statements be the given. Then assume the opposite of what you want to prove. What contradiction emerges?

EXERCISES & PROBLEMS

Communicate

1. What is a contradiction?

2. Give two real contradictory statements. Explain why they are contradictory.

3. Give a mathematical example of a contradiction.

4. Describe the form of argument known as *reducto ad absurdum*.

5. Summarize the steps for writing an indirect proof.

The artist René Magritte (1898–1967), like other Surrealists, loved to create images that seem to defy logic and common sense. The words in this famous painting (1929) mean, "This is not a pipe." Do you feel a sense of contradiction as you look at the words and the picture?

Practice & Apply

For Exercises 6–10, state whether the given argument is an example of an indirect argument. Explain why or why not.

6. Statement: It is raining.

   Argument: If it were not raining, there would be no puddles on the ground. But I see puddles on the ground. Therefore, it is raining.

7. Statement: You are ill.

   Argument: If you were not ill, then you would eat a large dinner. But you did not eat a large dinner. Therefore, you must be ill.

8. Statement: The sun is shining.

   Argument: I see my shadow. If I see my shadow then the sun must be shining. Therefore the sun is shining.

9. Statement: I am in New York.

   Argument: If I am not in the United States, then I am not in New York. Therefore, I am in New York.

10. Statement: My client is innocent.

    Argument: If my client were guilty then he would look guilty. Since my client does not look guilty, then he must be innocent.

Fill in the blanks to prove that if ∠1 is not congruent to ∠2, then ∠1 and ∠2 are not vertical angles.

Given: __(11)__

Prove: __(12)__

Assume that ∠1 and ∠2 are   __(13)__ . Then ∠1 __(14)__ ∠2. But the given states that ∠1 __(15)__ ∠2, which is a __(16)__ . Therefore, the assumption that ∠1 and ∠2 are vertical angles is false, and ∠1 and ∠2 are __(17)__ .

Fill in the blanks to prove that if m∠1 ≠ m∠2, then line *l* is not perpendicular to line *m*.

Given: __(18)__

Prove: __(19)__

Assume that __(20)__ . Then ∠1 ≅ ∠2 because __(21)__ . If ∠1 ≅ ∠2, then m∠1 __(22)__ m∠2 because __(23)__ . But the given states that __(24)__ , which is a contradiction. Therefore, __(25)__ .

Write an indirect proof for Exercises 26–28.

26. Given: $\overline{BD}$ bisects ∠ABC; $\overline{BD}$ is not a median.

   Prove: $\overline{AB}$ is not congruent to $\overline{BC}$.

27. Given: $\overline{CT}$ is not congruent to $\overline{BK}$.

   Prove: $\overline{BC}$ and $\overline{KT}$ do not bisect each other.

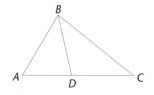

28. Given: ∠1 is not congruent to ∠2.

   Prove: m∠1 ≠ 90°

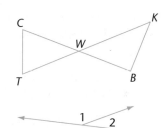

Construct an indirect argument in support of the given statement from the evidence provided.

29. Statement: My client is innocent.

   Evidence: My client was miles away from the scene of the crime at the time it happened.

30. Statement: This is not Elm Street.

   Evidence: Elm Street has a brick house on the corner. All the houses on this street are made from wood.

31. Statement: The temperature must be 32°F or higher.

   Evidence: The water on the sidewalk is not frozen.

32. Statement: If a line intersects a plane not containing it, then the intersection contains exactly one point.

Evidence: If a plane contains two points of a line, then the plane contains the whole line.

33. Statement: If two lines are cut by a transversal so that alternate interior angles are congruent, then the lines are parallel.

Evidence: Use the given "strange" figure and postulates and theorems found earlier in this textbook.

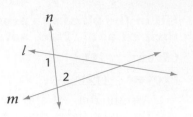

 **Look Back**

Compute the taxidistance between two points on a taxicab grid. [Lesson 11.2]

34. (0, 0) and (3, 3)   **35.** (1, 2) and (3, 4)   **36.** (1, 1) and (3, 4)

Use grid paper for the following exercises. Draw a taxicab circle on a grid with the indicated taxicab radius. [Lesson 11.2]

37. taxicab radius = 4   **38.** taxicab radius = 7

39. List two items that are topologically equivalent to a thumbtack. **[Lesson 11.4]**

40. List two items that are topologically equivalent to a drinking straw. **[Lesson 11.4]**

Look Beyond

41. **Portfolio**
Activity You are designing a system for your dream home that will do the following: turn on the air conditioner if the outside temperature is greater than 85°, turn the lights on in the living room from 2 A.M. to 6 A.M., turn the lights on in the entryway when the door opens, and arm the security system when the front door is locked. Write a series of if-then statements logically describing the system's operation.

LESSON 12.5 Computer Logic

Why *Logic provides the foundation for the decision-making and arithmetic processes of many "smart" electronic devices. The fundamental units of logical circuits are logic gates, which function like logical operators such as AND, OR, and NOT.*

A single computer chip in this "mother board" contains hundreds of thousands of logic gates like the one shown here, greatly magnified.

A computer is an example of a device that uses electric impulses and the **binary number system** to operate. **Binary** means "having two parts," and the binary number system is based on two numbers, 1 and 0. Think of a computer as a series of electrical switches that exist in one of two states, *on* or *off*. When a switch is *on*, it can be represented by 1. When a switch is *off*, it can be represented by 0.

Exploration *On-Off Tables*

You will need
No special tools

The tables in this exploration are a simulation of how computers work. The individual columns of each table represent electrical switches that work together. Let 1 = ON and 0 = OFF. Work in pairs and complete each table.

1 The power button on a TV remote control will turn the TV on if it is off, and vice-versa. Determine whether the TV will be on or off after pressing the power button by filling in the second column of the table.

TV	TV after pressing POWER button
1	?
0	?

2 To record on a video recorder, press both PLAY and RECORD. Determine whether the video recorder will record or not by filling in the last column of the table.

PLAY button	RECORD button	Video Recorder
1	1	?
1	0	?
0	1	?
0	0	?

3 The Student Driver car at Dover High School has been equipped with two brake pedals so that either the student driver or the instructor can stop the car. Complete the last column of the table, determining when the brake is on or off.

Student pedal	Instructor pedal	Brake
1	1	?
1	0	?
0	1	?
0	0	?

4 What did you consider as you filled in the last column of each table? Explain why Steps 2 and 3 have different answers even though the tables are set up in the same way.

5 How can the situation described in Step 3 be changed to get the same answers as in Step 2? Why is this new situation impractical in real life? ❖

CRITICAL
Thinking

How do the tables in the exploration compare with the truth tables you have constructed before?

Logic Gates

Each table in the exploration corresponds to a particular type of electronic circuitry called a **logic gate**. Logic gates are the building blocks for all "smart" electronic devices. Each logic gate has a special symbol representing **NOT**, **AND**, or **OR**.

In the diagram, p represents the electric input pulse. The **input-output table**, which is like a truth table, records what happens to the input as it passes through the logic gate. If input p has a value of 1, when it passes through a **NOT** gate it will have the opposite value, 0. This input-output table corresponds to the table in Step 1 of the exploration.

NOT Logic Gate

Input   Gate   Output

Input-Output Table	
Input	**Output**
p	NOT p
1	0
0	1

An **AND** logic gate needs two input pulses, which are represented in the diagram by p and q. Notice in the table that in order to get an output value of 1, both pulses must work together. Both p and q must have an input value of 1 in order for the output to be 1. This input-output table corresponds to the table in Step 2 of the exploration.

AND Logic Gate

Input   Gate   Output

Input-Output Table		
Input		**Output**
p	q	p AND q
1	1	1
1	0	0
0	1	0
0	0	0

An **OR** logic gate works differently from an AND gate because the input pulses do not have to work together. In order to get an output value of 1, *either p or q* has to have an input value of 1. Notice that to have an output value of 0, *neither p nor q* can have a value of 1. The input-output table on the following page corresponds to the table in Step 3 of the exploration.

OR Logic Gate

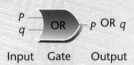

P
q
Input   Gate   Output

p OR q

Input-Output Table		
Input		Output
p	q	p OR q
1	1	1
1	0	1
0	1	1
0	0	0

Networks

Logic gates can be combined to form networks. To determine how a network operates, you can use input-output tables.

EXAMPLE 1

Construct an input-output table for the following network.

Solution ➤

Read from left to right. The first gate is *p* or *q*. Use parentheses to capture the output from this gate: (*p* OR *q*). Then perform the NOT operation: NOT (*p* OR *q*). To construct the input-output table, all possible input combinations must be considered. The numbers are filled in as you would fill in a truth table, where 1 is *true* and 0 is *false*. ❖

p	q	p OR q	NOT (p OR q)
1	1	1	0
1	0	1	0
0	1	1	0
0	0	0	1

EXAMPLE 2

Create a logical expression that corresponds to the following network.

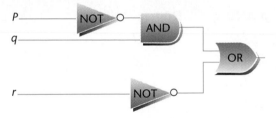

Solution ➤

Read from left to right, using parentheses to capture output.

NOT appears first: (NOT p)

The AND gate gives ((NOT p) AND q)

The bottom branch gives (NOT r)

The OR gate takes the output from the three preceding steps:

((NOT p) AND q) OR (NOT r) ❖

CRITICAL *Thinking*

How many rows and how many columns will an input-output table require for the network in Example 2? Explain your reasoning.

EXERCISES & PROBLEMS

Communicate ～～

Draw the symbol of the gate that corresponds to each of the following.

1. NOT   **2.** AND   **3.** OR

Use the logic gates to answer each question.

4. If $p = 1$ and $q = 0$, what will the output be?

5. If $p = 1$ and $q = 0$, what will the output be?

6. If $p = 1$, what will the output be?

7. If $p = 1$, what will the output be?

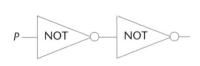

Practice & Apply

For Exercises 8–11, complete the given input-output table for the following network.

	p	q	p **AND** q	**NOT** (p **AND** q)
8.	1	1	?	?
9.	1	0	?	?
10.	0	1	?	?
11.	0	0	?	?

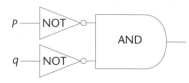

12. Complete an input-output table for the network.

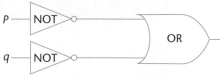

p	q	**NOT** p	**NOT** q	(**NOT** p) **AND** (**NOT** q)
?	?	?	?	?
?	?	?	?	?
?	?	?	?	?
?	?	?	?	?

13. Complete an input-output table for the following network.

14. Two of the logical expressions shown are functionally equivalent. That is, when given the same input they will produce the same output. Identify the two functionally equivalent logical expressions and complete the equivalence statement.

  NOT (p AND q)

  (NOT p) OR (NOT q)

  p AND (NOT q)

  ___?___ ≡ ___?___

15. Decide which two logical expressions are functionally equivalent and write an equivalence statement.

  p OR (NOT q)

  NOT (p OR q)

  (NOT p) AND (NOT q)

Charles Babbage (1791–1871) invented this early computer, which he called the Difference Engine.

For Exercises 16–23, complete an input-output table for the following network.

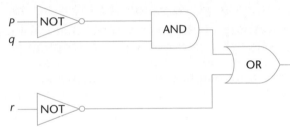

	p	q	r	NOT p	(NOT p) AND q	NOT r	((NOT p) AND q) OR (NOT r)
16.	1	1	1	?	?	?	?
17.	1	1	0	?	?	?	?
18.	1	0	1	?	?	?	?
19.	1	0	0	?	?	?	?
20.	0	1	1	?	?	?	?
21.	0	1	0	?	?	?	?
22.	0	0	1	?	?	?	?
23.	0	0	0	?	?	?	?

Construct a network diagram for the following expressions:

24. p AND (NOT q)   **25.** NOT (p OR q)

Construct a logical expression and an input-output table for the following networks.

26.

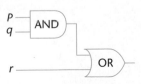

27.

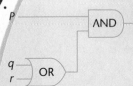

28. Are the two networks in Exercises 26 and 27 functionally equivalent? Explain your reasoning.

In Exercises 29–32, decide whether the situation is best described with a NOT, AND, or OR gate. Create a logical expression, a network diagram, and an input-output table to describe the given situation.

29. To run a particular dishwasher, the door must be locked and the power button must be on.

30. The living-room lights can be turned on by either a switch in the living room or a switch in the hallway.

31. A CD is placed in a CD player. To operate the player, you must first hit the POWER button, and then you must hit the PLAY button to start the music.

32. Using a certain word processing program, if the BOLD icon is selected, the print style will change from regular to bold or vice versa.

33.  **Portfolio Activity** Design a "machine" that must use logic in order to operate. Draw a picture and write a description of your machine. Then draw a network diagram of the machine's circuitry and complete an input-output table for the network.

Look Back

In Exercises 34 and 35, tell whether the proportion is true for all values of the variables. If the proportion is false, give a numerical counterexample. [Lesson 8.2]

34. If $\dfrac{x}{y} = \dfrac{r}{s}$, then $\dfrac{(x + c)}{y} = \dfrac{(r + c)}{s}$.

35. If $\dfrac{x}{y} = \dfrac{r}{s}$, then $\dfrac{x}{y} = \dfrac{(x + r + m)}{(y + s + m)}$.

36. The ratio of the corresponding sides of two similar triangles is equal to the ratio of the perimeters. If the perimeters of two similar triangles are 18 cm and 25 cm, what is the ratio of the perimeters? What is the ratio of the sides? **[Lesson 8.6]**

Look Beyond

37. A 50 ft pipe carries water from a well to a house. The pipe has sprung a leak. What is the probability that the leak will be within 5 ft of the house?

38. A speck of dust lands on the diagram. What is the probability that the speck of dust will land inside the triangle?

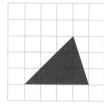

Environmental Science An aerial photograph shows workers cleaning up an oil spill.

39. How can you use objects in the photo to estimate the area of the spill? (Hint: First estimate the size of the squares in the yellow grid.)

40. Estimate the area of the spill.

Exploring

Proofs Using Coordinate Geometry

Meteorologists use latitude and longitude readings to track hurricanes on a map of the Earth. Near the Equator, these coordinates are very similar to x-y coordinates.

why *Properties of geometric figures can be represented in a coordinate geometry drawing. For this reason, coordinate geometry can be used to prove geometry theorems.*

To prove a theorem using coordinate geometry, begin by drawing a figure in a coordinate plane. If the figure is a polygon, it is usually convenient to place one vertex at the origin and at least one side along an axis.

•Exploration 1 *Triangle Midsegment Theorem Revisited*

Geometry Graphics

You will need
Graph paper or
Geometry technology

1 Use the coordinates of the three triangles given in the table to prove the following: *The segment joining the midpoints of two sides of a triangle is half the length of the third side.* Draw the first two triangles in a coordinate plane. Use a different *xy*-axis for each triangle.

Vertices of triangle	Coordinates of M (on $\overline{AB}$)	Coordinates of S (on $\overline{BC}$)	Slope of $\overline{MS}$	Slope of $\overline{AC}$	Length of $\overline{MS}$	Length of $\overline{AC}$
1. $A(0, 0)$ $B(2, 6)$ $C(8, 0)$	?	?	?	?	?	?
2. $A(0, 0)$ $B(6, -8)$ $C(10, 0)$	?	?	?	?	?	?
3. $A(0, 0)$ $B(2p, 2q)$ $C(2r, 0)$	?	?	?	?	?	?

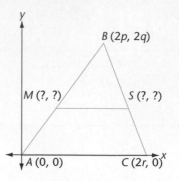

2 Label the vertices of each triangle, indicating the coordinates. For each triangle, find the midpoints of $\overline{AB}$ and $\overline{BC}$, labeling these points M and S, respectively. Draw the midsegment $\overline{MS}$ for each triangle and record the coordinates of the midpoint in a table like the one on the previous page.

3 Copy and complete the table for the first two triangles. Then use the figure shown to state the general case of the Triangle Midsegment Theorem in row three of the table. The general case, along with the completed table, is the proof of the theorem.

4 Based on the information in the last row of your table, what can you conclude about the relationship between $\overline{MS}$ and $\overline{AC}$? ❖

CRITICAL
Thinking

Compare this proof of the Triangle Midsegment Theorem with the one that appears in Lesson 3.7. Which proof appears easier? Explain why?

•Exploration 2   *The Diagonals of a Parallelogram*

You will need
Graph paper or
Geometry technology

*Geometry
Graphics*

1 Use the coordinates given in the table to prove the following: *The diagonals of a parallelogram bisect each other.* Draw the first two parallelograms in a coordinate plane. Use a different *xy*-axis for each parallelogram. Three vertices are given. Find the fourth vertex to complete each figure.

3 Vertices of parallelogram	Fourth vertex	Midpoint of $\overline{BD}$	Midpoint $\overline{AC}$
1. $A(0, 0)$ $B(2, 6)$ $D(10, 0)$	$C(?, ?)$	?	?
2. $A(0, 0)$ $B(4, -8)$ $D(10, 0)$	$C(?, ?)$	?	?
3. $A(0, 0)$ $B(2p, 2q)$ $D(2r, 0)$	$C(?, ?)$	?	?

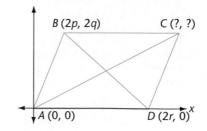

2 For each parallelogram, draw diagonals $\overline{BD}$ and $\overline{AC}$, and determine their midpoints.

3 Copy and complete the table for the first two parallelograms. Then use the figure shown to state a general case for the theorem in row three of the table. The general case, along with the completed table, is the proof of the theorem.

4 Based on the information in the last row of your table, what may you conclude about the diagonals of a parallelogram? ❖

CRITICAL *Thinking*
How can you prove this theorem without using coordinate geometry? Which kind of proof appears easier? Explain why?

•Exploration 3 *Reflection About the Line y = x*

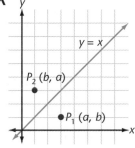

You will need
Graph paper
Calculator

Calculator

You may recall that the effect of reversing the coordinates of a point was to reflect the point through the line $y = x$. You can use coordinate geometry to prove this result.

1 If you know the x-coordinate of a point on the line $y = x$, what can you conclude about the y-coordinate? (Filling in a table like the one on the right will reveal the pattern.)

x	y
1	?
2	?
0	?
−1	?
P	?

2 Find the midpoint between the point $P_1(a, b)$ and $P_2(b, a)$ using the midpoint formula. Does this point lie on the line $y = x$? Explain your reasoning.

3 Pick two points on the line $y = x$ and use them to find the slope of the line $y = x$. Record your result.

4 Find the slope of the line that passes through the points $P_1(a, b)$ and $P_2(b, a)$. Record your result. Hint: You can rewrite $b - a$ as $-(a - b)$.

5 Compare your results in Steps 3 and 4. What can you conclude about the relationship between the two lines?

6 Explain how your results prove that the effect of reversing the coordinates of a point is to reflect the point through the line $y = x$. (Recall the definition of a reflection.) ❖

EXERCISES & PROBLEMS

Communicate

1. In proofs using coordinate geometry, why is it best to place one vertex of the figure at the origin and one side along the *x*-axis?

2. How could you use coordinate geometry to prove that opposite sides of a parallelogram are equal in length?

3. How could you use coordinate geometry to prove that the diagonals of a rhombus are perpendicular?

4. Explain how you would prove that the diagonals of a rhombus are perpendicular without using coordinate geometry.

Practice & Apply

Use graph paper or geometry technology for Exercises 5–9.

5. *ABCD* is a rectangle.

   a. Given: $A(0, 0)$, $B(0, 3)$, $D(7, 0)$

   Find the coordinates of *C*.

   b. Given: $A(0, 0)$, $B(0, p)$, $D(q, 0)$

   Find the coordinates of *C*.

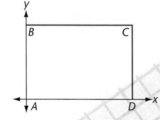

6. *ABC* is an isosceles triangle with $\overline{AB} \cong \overline{BC}$.

   a. Given: $A(0, 0)$, $B(4, 3)$

   Find the coordinates of *C*.

   b. Given: $A(0, 0)$, $B(p, q)$

   Find the coordinates of *C*.

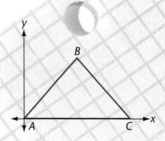

7. *ABCD* is a parallelogram.

   a. Given: $A(0, 0)$, $B(3, 5)$, $D(7, 0)$

   Find the coordinates of *C*.

   b. Given: $A(0, 0)$, $B(p, q)$, $D(r, 0)$

   Find the coordinates of *C*.

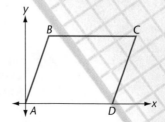

8. *ABCD* is a rhombus.

   a. Given: $A(-2, 0)$, $B(0, 5)$

   Find the coordinates of C and D.

   b. Given: $A(-p, 0)$, $B(0, r)$

   Find the coordinates of C and D.

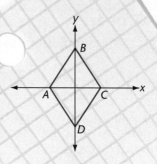

9. *ABCD* is a square.

   a. Given: $A(-3, 0)$

   Find the coordinates of B, C, and D.

   b. Given: $A(-p, 0)$

   Find the coordinates of B, C, and D.

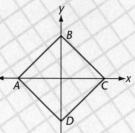

For Exercises 10–13, *ABCD* is a trapezoid with coordinates $A(0, 0)$, $B(2, 6)$, $C(8, ?)$, and $D(12, ?)$.

10. Find the missing coordinates for C and D.

11. Find the coordinates of M and S, the midpoints of $\overline{AB}$ and $\overline{CD}$.

12. Find the lengths of $\overline{AD}$, $\overline{BC}$, and the midsegment $\overline{MS}$.

13. How is the length of a midsegment of a trapezoid related to the length of the two bases of the trapezoid?

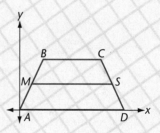

For Exercises 14–17, *ABCD* is a trapezoid with coordinates $A(0, 0)$, $B(2p, 2q)$, $C(2r, ?)$, and $D(2s, ?)$.

14. Find the missing coordinates for C and D in terms of p, q, r, and s.

15. Find the coordinates of M and S, the midpoints of $\overline{AB}$ and $\overline{CD}$, in terms of p, q, r, and s.

16. Prove that the length of the midsegment of a trapezoid is equal to the average of the lengths of the two bases.

17. Prove that the midsegment of a trapezoid is parallel to the bases.

18. What type of quadrilateral is given in the figure below? Justify your answer.

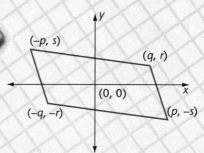

Use △ABC for Exercises 19–21.

19. Find the coordinates of the midpoint of the hypotenuse.

20. Show that the midpoint of the hypotenuse is equidistant from the three vertices.

21. Use a general case of △ABC to prove that the midpoint of the hypotenuse of a right triangle is equidistant from the three vertices.

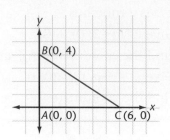

Use quadrilateral ABCD for Exercises 22–24.

22. Find the coordinates of the midpoints of the four sides.

23. Connect the midpoints to form a quadrilateral. What type of quadrilateral is formed? Justify your answer.

24. Generalize the results of Exercise 23 for quadrilateral ABCD with $A(0, 0)$, $B(2p, 2q)$, $C(2r, 2s)$, and $D(2t, 0)$.

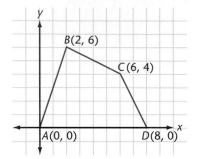

Look Back

Find the volume of each prism. [Lesson 7.2]

25. $B = 6 \text{ cm}^2$, $h = 5$ cm

26. $B = 12 \text{ cm}^2$, $h = 17$ cm

27. Find the volume of a semicircular cylinder formed by cutting a cylinder in half on its diameter. The diameter is 5 in. and the height of the cylinder is 14 in. **[Lesson 7.4]**

Find the surface area of each right cone. [Lesson 7.5]

28.

14 ft

8 ft

29.

8 cm

6 cm

Look Beyond

30. Use coordinate geometry to prove that an angle inscribed in a semicircle is 90°.

31. Using three-dimensional coordinate geometry, show a relationship among the diagonals of the cube.

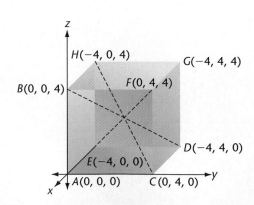

Two Famous Theorems

Two of the most famous theorems in mathematics come to us from classical times. Both of them involve indirect proof, which you studied in Lesson 4 of this chapter. Both of them also involve the following theorem, which is known as the "Fundamental Theorem of Number Theory." It is not difficult to prove, but here it will be assumed.

Every number has a unique prime factorization.

The Infinity of the Primes

Euclid's proof that there are infinitely many primes has captured the imagination of people over the ages. To help you understand the proof, which is actually very brief, a numerical example is given first.

The proof will assume that there is only a finite number of primes. Suppose, for example, that there are only 5 primes, so that 11 is the largest prime number. Then the list of all the prime numbers would read:

   2, 3, 5, 7, 11.

Now form a new number m by multiplying the prime numbers together and adding 1.

   $m = (2)(3)(5)(7)(11) + 1 = 2311$

The new number m must be composite, according to the assumption, which states that 11 is the largest prime number. Therefore, m must be factorable into a combination of prime numbers, which by the assumption range from 2 to 11. But none of the numbers in the list of primes will divide the number evenly, because you will always get a remainder of 1.

For example, $\dfrac{462}{5\overline{)2311}}$ r. 1

But if 2311 is not divisible by one of the primes in the list, then it must be divisible by some other prime greater than 11, which contradicts the assumption that 11 is the largest prime.

Once you understand the example, you should be able to follow the generalization of it, which is given below.

Prove: There are infinitely many primes.
Proof: Assume that there are only finitely many primes, say, n of them. Then the list of the primes would be

$$p_1, p_2, p_3, \cdots p_n.$$

Form a new number $m = (p_1)(p_2)(p_3) \cdots (p_n) + 1$.

By assumption, m must be composite. But m is not divisible by any of the numbers in the list of primes, so there must be a larger prime than p, which contradicts the assumption. Thus there are infinitely many primes. ❖

The "Incommensurability" of the Square Root of 2

In the early history of mathematics it seems to have been widely believed that any number could be represented as a fraction—that is, as a ratio of two integers. This was certainly true with the early Pythagoreans. Thus the proof that the square root of two is *not* such a ratio (i.e., that it is *irrational*) came as a profound shock to them, which shook the foundations of their beliefs.

Before studying the theorem, you will need to know three simple results from number theory. *The following three theorems use the prime factorization theorem.*

Prove: **The square of an even number is even.**
Proof: If a number is even, then it has 2 as one of its prime factors. Otherwise, it would not be divisible by 2. (This is because 2 is the only even prime number.) When the number is squared, then, the number 2 will appear at least twice in the prime factorization of the result. Therefore, the result is divisible by 2.

Prove: **The square of an even number is divisible by 4.**
Proof: The proof is left as an exercise for the reader.

Prove: **The square of an odd number is odd.**
Proof: The proof is left as an exercise for the reader.

You are now ready to tackle the "incommensurability" theorem.

Prove: The square root of 2 cannot be written as a ratio of two integers.
Proof: Assume that the square root of 2 can be represented as the ratio of two numbers—that is, as a fraction. If this fraction is not in lowest terms then there is a fraction in lowest terms, to which it can be reduced. Thus the assumption can be written in the following way, which is

equivalent to its original statement:

> *Assume that the square root of 2 can be represented as a fraction in lowest terms.*

Let p be the numerator and q the denominator of the fraction which is assumed to exist. That is,

$$\frac{p}{q} = \sqrt{2} \qquad (p \text{ and } q \text{ have no common factors.})$$

and so,

$$\left(\frac{p}{q}\right)^2 = (\sqrt{2})^2 = 2$$

and

$$p^2 = 2q^2.$$

The last equation implies that p^2 is an even number (why?). Therefore, p^2 is the square of an even number. This implies that p^2 is divisible by 4.

But if p^2 is divisible by 4, then q^2 must be an even number (why?). Therefore, q is even, as well.

But if p and q are both even, then they must have a common factor of 2, which contradicts the original assumption. Therefore, the assumption must be false, and so the theorem has been proven. ❖

Activities

The Infinity of the Primes

1. Repeat the numerical example in the proof using different numbers of primes. Explain why you always get a remainder of 1 when you divide m by one (or more) of the primes in your list.

2. If you subtract 1 instead of adding 1 to obtain the number m in the proof, how would the proof be affected?

3. Do some research on the number theory and find some of the unproven conjectures such as the **Goldbach Conjecture** and the **Twin Primes Conjecture.** Explain them in your own words and give numerical illustrations of them.

The "Incommensurability" of the Square Root of 2

1. Prove that the square of an even number is divisible by 4.

2. Prove that the square of an odd number is odd.

3. In the incommensurability proof, there are two points at which the reader is asked to explain why a result follows. Explain why in your own words.

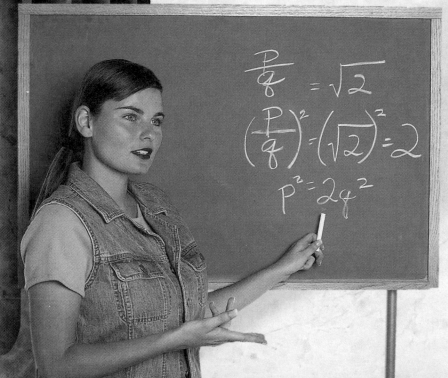

Chapter 12 Review

Vocabulary

argument	644	contrapositive	660	negation	652
binary number system	671	disjunction	651	premise	644
compound statement	650	exclusive or	651	proof by contradiction	666
conclusion	644	inclusive or	651	truth table	650
conjunction	650	input-output table	673		
contradiction	666	logic gate	673		

Key Skills and Exercises

Lesson 12.1

➤ Key Skills

Write a valid conclusion based on premises.

We have two premises:

> If the Memorial Day parade passes the house, then the house is on Main Street.

> The house is not on Main Street.

A valid conclusion would be: *Therefore, the Memorial Day parade does not pass the house.*

This argument is in the form: *If p then q; not q; therefore, not p.*

Spot false arguments.

What is wrong with the following argument?

> If the Memorial Day parade passes the house, then the house is on Main Street.

> The Memorial Day parade does not pass the house.

> Therefore, the house is not on Main Street.

This is a false argument known as denying the antecedent. This fallacy has the form: *If p then q; not p; therefore, not q.*

➤ Exercises

1. Write a valid conclusion for the following premises:

> *If groms are plamous, then they are rute.*
> *Merts are plamous groms.*

2. Is the following argument valid or invalid? Explain.

> *Some butterflies migrate.*
> *The monarch is a butterfly.*
> *Therefore, the monarch migrates.*

Lesson 12.2

> ### Key Skills

Write a conjunction, disjunction, and negation.

You have two statements: (*p*) *We eat muffins*, and (*q*) *We have breakfast at the diner*. The conjunction of these two statements is *p* AND *q*:

> *We eat muffins, and we have breakfast at the diner.*

The truth table for this conjunction is shown above right.

For statements *p* and *q* above, the disjunction is *p* OR *q*:

> *We eat muffins or we have breakfast at the diner.*

The truth table for this disjunction is shown below right.

The negations of the statements above are (~*p*) *We do not eat muffins*, and (~*q*) *We do not have breakfast at the diner.*

p	q	p AND q
T	T	T
T	F	F
F	T	F
F	F	F

p	q	p OR q
T	T	T
T	F	T
F	T	T
F	F	F

> ### Exercises

For Exercises 3-5, create two statements and follow the directions.

3. Write a conjunction.　　　　**4.** Write a disjunction.

5. Write a negation of your conjunction in Exercise 3. Give the truth table for the negation.

Lesson 12.3

> ### Key Skills

Write the converse, inverse, and contrapositive of a statement.

Statement: If Socks is purring, then he is happy.
Converse: If Socks is happy, then he is purring.

Statement: If Socks is purring, then he is happy.
Inverse: If Socks is not purring, then he is not happy.

Statement: If Socks is purring, then he is happy.
Contrapositive: If Socks is not happy, then he is not purring.

> ### Exercises

In Exercises 6–8, use the statement "If a quadrilateral is a trapezoid, then it has two parallel sides." Is each new statement true or false.

6. Write the converse.　　**7.** Write the inverse.　　**8.** Write the contrapositive.

Lesson 12.4

> ### Key Skills

Use indirect reasoning in a proof.

Prove indirectly that $\overline{SQ}$ and $\overline{PT}$ do not bisect each other inside △*RPQ*.

Assume that the segments intersect each other. They are diagonals of *PQTS*, and if diagonals of a quadrilateral bisect each other, then the quadrilateral is a parallelogram. If *PQTS* is a parallelogram, $\overline{PS}$ is parallel to $\overline{QT}$. Thus $\overline{PQ}$ is a transversal, and ∠*RPQ* and ∠*RQP* are supplementary angles. That means m∠*RPQ* + m∠*RQP* = 180°. However, △*RPQ* has three nonzero angles that sum to 180°. This contradicts m∠*RPQ* + m∠*RQP* = 180°, indirectly proving that the segments do not bisect each other.

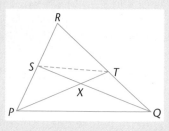

➤ Exercises

9. Prove by indirect reasoning that the bisector of an angle of a scalene triangle is not perpendicular to the opposite side.

Lesson 12.5

➤ Key Skills

Construct input-output tables.

Create an input-output table for this network.

p	q	(NOT p) AND q
1	1	0
1	0	0
0	1	1
0	0	0

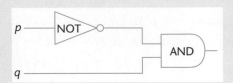

Write a logical expression for a network.

We can create a logical expression that represents the network above. Read from left to right, top to bottom. The left-hand terms, top to bottom, are (p OR q), (NOT r). The right-hand term combines these: (p OR q) AND (NOT r).

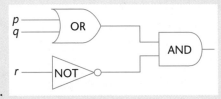

➤ Exercises

10. Write a logical expression for the network.

11. Construct an input-output table for the network.

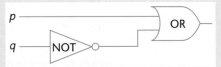

Lesson 12.6

➤ Key Skills

Use coordinate geometry in proofs.

Prove that the angle inscribed in the semicircle measures 90°.

Use the coordinates of the endpoints to find the slopes of $\overline{AB}$ and $\overline{BC}$. If the slopes of $\overline{AB}$ and $\overline{BC}$ are negative reciprocals, the lines are perpendicular and $\angle ABC$ is a right angle.

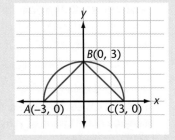

slope of $\overline{AB}$ = (3 − 0)/(0 − 3) = −1
slope of $\overline{BC}$ = (3 − 0)/(0 − (−3)) = 1

The slopes are negative reciprocals, and so inscribed $\angle ABC$ is a right angle.

➤ Exercises

12. Using coordinate geometry, prove that if the diagonals of a parallelogram are equal, then the figure is a rectangle.

Applications

13. Mario, his sister, his daughter, and his son are sprinters. Two facts describe these people: (1) The worst sprinter's twin and the best sprinter are of opposite sex. (2) The worst sprinter and the best sprinter are the same age. Which one of the four is the best sprinter?

14. **Entertainment** The cable TV signal goes into my VCR, which is connected to the TV set. I want to watch cable channel 24. Create a network diagram and a logical expression to describe what I must do before I can select and watch channel 24.

Chapter 12 Assessment

1. Write a valid conclusion for these premises:
   If it was raining when I arrived, my umbrella is wet.
   My umbrella is dry.

2. Write a disjunction for these statements:
   Farah makes a shot.
   Richard blocks the shot.

3. Write a negation of your disjunction from Exercise 2. Give the truth table for the negation.

In Exercises 4–6, use the statement "If an angle is inscribed in a semicircle, then it is a right angle."

4. Write the converse. Is the new statement true or false?
5. Write the inverse. Is the new statement true or false?
6. Write the contrapositive. Is the new statement true or false?

7. Prove by indirect reasoning that a triangle can have at most one obtuse angle.

In Exercises 8–9, refer to the network below.

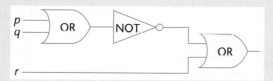

8. Write a logical expression for the network.
9. Construct an input-output table for the network.

10. Prove that the triangle below is isosceles.

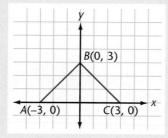

Chapters 1-12   Cumulative Assessment

College Entrance Exam Practice

Quantitative Comparison   Exercises 1–3 consist of two quantities, one in Column A and one in Column B, which you are to compare as follows:

A. The quantity in Column A is greater.

B. The quantity in Column B is greater.

C. The two quantities are equal.

D. The relationship cannot be determined from the information given.

	Column A	Column B	Answers
1.	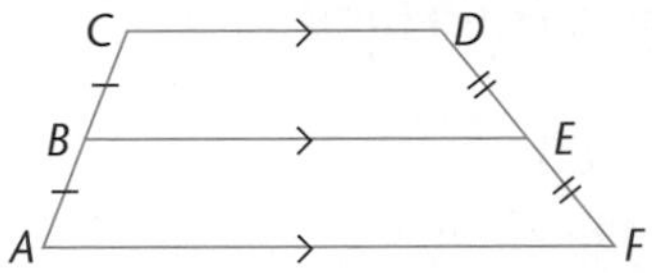 BE	$AF/2$	Ⓐ Ⓑ Ⓒ Ⓓ **[Lesson 3.7]**
2.	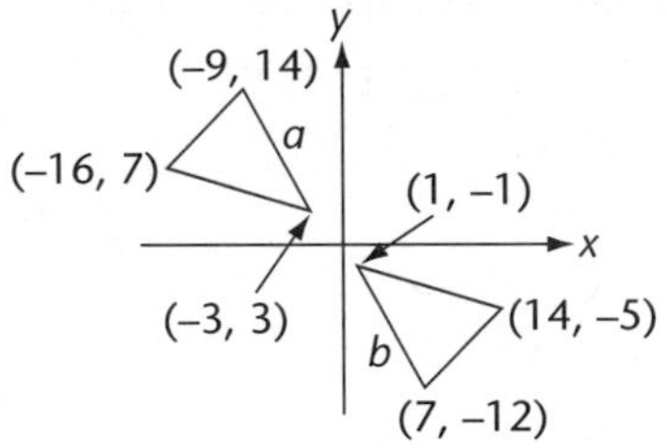 Side a	Side b	Ⓐ Ⓑ Ⓒ Ⓓ **[Lesson 1.7]**
3.	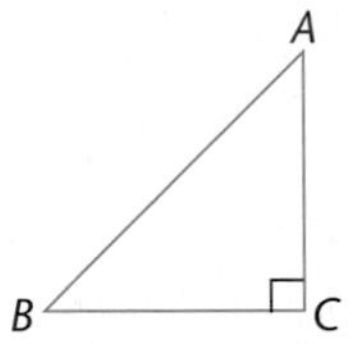 $AB + BC$	$AC + BC$	Ⓐ Ⓑ Ⓒ Ⓓ **[Lesson 5.4]**

4. Which is a definition?   **[Lesson 2.3]**

   a. A monarch is an orange butterfly.

   b. A line segment is the shortest path between two points.

   c. A pile of loose rubble is a hazard.

   d. A rhombus is a parallelogram.

5. Line $\overline{AF}$ contains points $A(13, 15)$ and $F(9, 20)$. Which points define a line parallel to $\overline{AF}$?   **[Lesson 3.8]**

   a. $(0, 0), (25, 16)$ **b.** $(15, 13), (20, 9)$

   c. $(4, -6), (8, -1)$ **d.** $(4, 3), (12, -13)$

6. Which value is largest in $\triangle ABC$?   **[Lessons 5.5, 10.2]**
   a. $\cos A$ **b.** $\sin B$ **c.** $\cos B$ **d.** $\sin C$

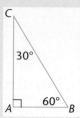

7. Identify the correct expression for length EC.   **[Lesson 9.5]**

   a. $ED \times \dfrac{EB}{EA}$ **b.** $EB \times \dfrac{EA}{ED}$

   c. $EB \times \dfrac{CD}{AB}$ **d.** $EB \times \dfrac{AB}{CD}$

8. Which statement is true in a hyperbolic geometry?   **[Lesson 11.5]**
   a. No lines are parallel.
   b. Two-sided polygons exist.
   c. The sum of the angles of a triangle is less than 180°.
   d. The shortest distance between two points is a great circle.

In Exercises 9–10, refer to the triangle at right.   **[Lesson 5.6]**
   9. Make a conjecture about $\overline{DE}$.
10. Use coordinate geometry to prove your conjecture from Exercise 9.

In Exercises 11–12, use these statements: **[Lessons 12.1, 12.3]**
   a. If the fog has lifted, the boat can leave the harbor.
   b. The boat cannot leave the harbor.

11. Write a valid conclusion.
12. Write a contrapositive of statement (a).
13. Project points A, B, and C onto line n.   **[Lesson 11.6]**
14. Write an equation for a circle with radius 6 and center at $(11, -13)$.   **[Lesson 9.6]**

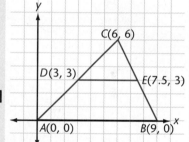

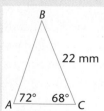

Free-Response Grid   Exercises 15–17 may be answered using a free-response grid commonly used by standardized test services.

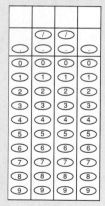

15. I parachute into a taxicab circle of radius 6 that includes the dangerously busy intersection of Fifth and Broadway. If I have an equal chance of landing at any intersection within or on the circle, what are the chances that I'll land at Fifth and Broadway?   **[Lessons 5.7, 11.2]**
16. What is the length of segment $\overline{AB}$?   **[Lessons 10.5, 10.6]**
17. How long is $\overline{FG}$?   **[Lesson 5.6]**

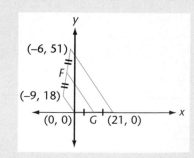

TABLES

Table of Squares, Cubes, Square and Cube Roots

No.	Squares	Cubes	Square Roots	Cube Roots	No.	Squares	Cubes	Square Roots	Cube Roots
1	1	1	1.000	1.000	51	2,601	132,651	7.141	3.708
2	4	8	1.414	1.260	52	2,704	140,608	7.211	3.733
3	9	27	1.732	1.442	53	2,809	148,877	7.280	3.756
4	16	64	2.000	1.587	54	2,916	157,464	7.348	3.780
5	25	125	2.236	1.710	55	3,025	166,375	7.416	3.803
6	36	216	2.449	1.817	56	3,136	175,616	7.483	3.826
7	49	343	2.646	1.913	57	3,249	185,193	7.550	3.849
8	64	512	2.828	2.000	58	3,364	195,112	7.616	3.871
9	81	729	3.000	2.080	59	3,481	205,379	7.681	3.893
10	100	1,000	3.162	2.154	60	3,600	216,000	7.746	3.915
11	121	1,331	3.317	2.224	61	3,721	226,981	7.810	3.936
12	144	1,728	3.464	2.289	62	3,844	238,328	7.874	3.958
13	169	2,197	3.606	2.351	63	3,969	250,047	7.937	3.979
14	196	2,744	3.742	2.410	64	4,096	262,144	8.000	4.000
15	225	3,375	3.873	2.466	65	4,225	274,625	8.062	4.021
16	256	4,096	4.000	2.520	66	4,356	287,496	8.124	4.041
17	289	4,913	4.125	2.571	67	4,489	300,763	8.185	4.062
18	324	5,832	4.243	2.621	68	4,624	314,432	8.246	4.082
19	361	6,859	4.359	2.668	69	4,761	328,509	8.307	4.102
20	400	8,000	4.472	2.714	70	4,900	343,000	8.367	4.121
21	441	9,261	4.583	2.759	71	5,041	357,911	8.426	4.141
22	484	10,648	4.690	2.802	72	5,184	373,248	8.485	4.160
23	529	12,167	4.796	2.844	73	5,329	389,017	8.544	4.179
24	576	13,824	4.899	2.884	74	5,476	405,224	8.602	4.198
25	625	15,625	5.000	2.924	75	5,625	421,875	8.660	4.217
26	676	17,576	5.099	2.962	76	5,776	483,976	8.718	4.236
27	729	19,683	5.196	3.000	77	5,929	456,533	8.775	4.254
28	784	21,952	5.292	3.037	78	6,084	474,552	8.832	4.273
29	841	24,389	5.385	3.072	79	6,241	493,039	8.888	4.291
30	900	27,000	5.477	3.107	80	6,400	512,000	8.944	4.309
31	961	29,791	5.568	3.141	81	6,561	531,441	9.000	4.327
32	1,024	32,768	5.657	3.175	82	6,724	551,368	9.055	4.344
33	1,089	35,937	5.745	3.208	83	6,889	571,787	9.110	4.362
34	1,156	39,304	5.831	3.240	84	7,056	592,704	9.165	4.380
35	1,225	42,875	5.916	3.271	85	7,225	614,125	9.220	4.397
36	1,296	46,656	6.000	3.302	86	7,396	636,056	9.274	4.414
37	1,369	50,653	6.083	3.332	87	7,569	658,503	9.327	4.431
38	1,444	54,872	6.164	3.362	88	7,744	681,472	9.381	4.448
39	1,521	59,319	6.245	3.391	89	7,921	704,969	9.434	4.465
40	1,600	64,000	6.325	3.420	90	8,100	729,000	9.487	4.481
41	1,681	68,921	6.403	3.448	91	8,281	753,571	9.539	4.498
42	1,764	74,088	6.481	3.476	92	8,464	778,688	9.592	4.514
43	1,849	79,507	6.557	3.503	93	8,649	804,357	9.644	4.531
44	1,936	85,184	6.633	3.350	94	8,836	830,584	9.695	4.547
44	2,025	91,125	6.708	3.557	95	9,025	857,375	9.747	4.563
46	2,116	97,336	6.782	3.583	96	9,216	884,736	9.798	4.579
47	2,209	103,823	6.856	3.609	97	9,409	912,673	9.849	4.595
48	2,304	110,592	6.928	3.634	98	9,604	941,192	9.899	4.610
49	2,401	117,649	7.000	3.659	99	9,801	970,299	9.950	4.626
50	2,500	125,000	7.071	3.684	100	10,000	1,000,000	10.000	4.642

Table of Trigonometric Ratios

Angle	sin	cos	tan	Angle	sin	cos	tan
0°	0.0000	1.0000	0.0000	45°	0.7071	0.7071	1.0000
1°	0.0175	0.9998	0.0175	46°	0.7193	0.6947	1.0355
2°	0.0349	0.9994	0.0349	47°	0.7314	0.6820	1.0724
3°	0.0523	0.9986	0.0524	48°	0.7431	0.6691	1.1106
4°	0.0698	0.9976	0.0699	49°	0.7547	0.6561	1.1504
5°	0.0872	0.9962	0.0875	50°	0.7660	0.6428	1.1918
6°	0.1045	0.9945	0.1051	51°	0.7771	0.6293	1.2349
7°	0.1219	0.9925	0.1228	52°	0.7880	0.6157	1.2799
8°	0.1392	0.9903	0.1405	53°	0.7986	0.6018	1.3270
9°	0.1564	0.9877	0.1584	54°	0.8090	0.5878	1.3764
10°	0.1736	0.9848	0.1763	55°	0.8192	0.5736	1.4281
11°	0.1903	0.9816	0.1944	56°	0.8290	0.5592	1.4826
12°	0.2079	0.9781	0.2126	57°	0.8387	0.5446	1.5399
13°	0.2250	0.9744	0.2309	58°	0.8480	0.5299	1.6003
14°	0.2419	0.9703	0.2493	59°	0.8572	0.5150	1.6643
15°	0.2588	0.9659	0.2679	60°	0.8660	0.5000	1.7321
16°	0.2756	0.9613	0.2867	61°	0.8746	0.4848	1.8040
17°	0.2924	0.9563	0.3057	62°	0.8829	0.4695	1.8807
18°	0.3090	0.9511	0.3249	63°	0.8910	0.4540	1.9626
19°	0.3526	0.9455	0.3443	64°	0.8988	0.4384	2.0503
20°	0.3420	0.9397	0.3640	65°	0.9063	0.4226	2.1445
21°	0.3584	0.9336	0.3839	66°	0.9135	0.4067	2.2460
22°	0.3746	0.9272	0.4040	67°	0.9205	0.3907	2.3559
23°	0.3907	0.9205	0.4245	68°	0.9272	0.3746	2.4751
24°	0.4067	0.9135	0.4452	69°	0.9336	0.3584	2.6051
25°	0.3420	0.9063	0.4663	70°	0.9397	0.3420	2.7475
26°	0.4384	0.8988	0.4877	71°	0.9455	0.3256	2.9042
27°	0.4540	0.8910	0.5095	72°	0.9511	0.3090	3.0777
28°	0.4695	0.8829	0.5317	73°	0.9563	0.2924	3.2709
29°	0.4848	0.8746	0.5543	74°	0.9613	0.2756	3.4874
30°	0.5000	0.8660	0.5774	75°	0.9659	0.2588	3.7321
31°	0.5150	0.8572	0.6009	76°	0.9703	0.2419	4.0108
32°	0.5299	0.8480	0.6249	77°	0.9744	0.2250	4.3315
33°	0.5446	0.8387	0.6494	78°	0.9781	0.2079	4.7046
34°	0.5592	0.8290	0.6745	79°	0.9816	0.1908	5.1446
35°	0.5736	0.8192	0.7002	80°	0.9848	0.1736	5.6713
36°	0.5878	0.8090	0.7265	81°	0.9877	0.1564	6.3138
37°	0.6018	0.0786	0.7536	82°	0.9903	0.1392	7.1154
38°	0.6157	0.7880	0.7813	83°	0.9925	0.1219	8.1443
39°	0.6293	0.7771	0.8098	84°	0.9945	0.1045	9.5144
40°	0.6428	0.7660	0.8391	85°	0.9962	0.0872	11.4301
41°	0.6561	0.7547	0.8693	86°	0.9976	0.0698	14.3007
42°	0.6691	0.7431	0.9004	87°	0.9986	0.0523	19.0811
43°	0.6820	0.7314	0.9325	88°	0.9994	0.0349	28.6363
44°	0.6947	0.7193	0.9657	89°	0.9998	0.0175	57.2900
45°	0.7071	0.7171	1.0000	90°	1.0000	0.0000	∞

POSTULATES, THEOREMS, AND DEFINITIONS

Def 1.4.1 **Measure of Segment $\overline{AB}$** Let A and B be points on a number line with coordinates a and b. Then the measure of segment $\overline{AB}$, which is called the length, is $|a - b|$ or $|b - a|$. The measure of segment $\overline{AB}$ is written as mAB or simply $\overline{AB}$. (30)

Post 1.4.2 **Segment Addition Postulate** On segment $\overline{PQ}$, if R is between points P and Q, then $PR + PQ = PQ$. (32)

Def 1.5.1 **Measure of Angle** Suppose the vertex V of $\angle AVB$ is placed in the center point of a half-circle with coordinates from 0 to 180. Let a and b be the coordinates of the points where $\overrightarrow{VA}$ and $\overrightarrow{VB}$ cross the half-circle. Then the measure of $\angle AVB$, written as $m\angle AVB$ is $|a - b|$ or $|b - a|$. (38)

Post 1.5.2 **Angle Addition Postulate** If point S is in the interior of $\angle PQR$, then $m\angle PQS + m\angle SQR = m\angle PQR$. (39)

Def 1.5.3 **Special Angle Sums** If the sum of the measures of two angles is $90°$, then the angles are **complementary**. If the sum of the measures of two angles is $180°$, then the angles are **supplementary**. (39)

Post 2.2.1 **If-Then Transitive Property** Suppose you are given: If A then B and if B then C. You can conclude: If A then C. (76)

Post 2.4.1 **Postulate** Through any two points there is exactly one line, or two points determine exactly one line. (88)

Post 2.4.2 **Postulate** Through any three noncollinear points, there is exactly one plane, or three noncollinear points determine exactly one plane. (88)

Post 2.4.3 **Postulate** If two points are on a plane, then the line containing them is also on the plane. (88)

Post 2.4.4 **Postulate** The intersection of two lines is exactly one point. (88)

Post 2.4.5 **Postulate** The intersection of two planes is exactly one line. (88)

Thm 2.5.1 **Overlapping Segments Theorem** Given a segment with points A, B, C, and D arranged as shown, the following statements are true:

1. If $AB = CD$, then $AC = BD$
2. If $AC = BD$, then $AB = CD$

(94)

Thm 2.5.2 **Overlapping Angles Theorem** Given four rays with common endpoint arranged as shown, the following statements are true:

1. If $m\angle AOB = m\angle COD$, then $m\angle AOC = m\angle BOD$
2. If $m\angle AOC = m\angle BOD$, then $m\angle AOB = m\angle COD$

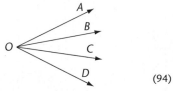

(94)

Post 2.5.3 **Reflexive Property of Equality** For all real numbers a, $a = a$. (94)

Post 2.5.4 **Symmetric Property of Equality** For all real numbers a and b, if $a = b$, then $b = a$. (94)

Post 2.5.5 **Transitive Property of Equality** For all real numbers a, b, and c, if $a = b$ and $b = c$, then $a = c$. (94)

Thm 2.6.1 **Vertical Angles Theorem** All vertical angles have equal measure. (100)

Thm 2.6.2 **Reflection Through Parallel Lines** Reflection twice through a pair of parallel lines is equivalent to a translation by twice the distance between the lines in a direction perpendicular to the lines. (101)

Thm 2.6.3 **Reflection through Intersecting Lines** Reflection twice through a pair of intersecting lines is equivalent to a rotation about the intersection point. The rotation angle measure is double that of the angle formed by the lines. (102)

Def. 3.1.1 **Polygon** A polygon is a closed plane figure formed from three or more segments such that each segment intersects exactly two other segments, one at each endpoint. (116)

Def. 3.1.2 **Reflectional Symmetry** A plane figure has reflectional symmetry if and only if its reflection image through a line coincides with the original figure. The line is called an **axis of symmetry**. (117)

Def. 3.1.3 **Rotational Symmetry** A figure has rotational symmetry if and only if it has at least one rotation image that coincides with the original image. We say that a figure has a **rotation image of n degrees** if a rotation by n degrees about a fixed point results in an image that coincides with the original. Rotation images of $0°$ or multiples of $360°$ do not have rotational symmetry. (118)

Def. 3.3.1 **Transversals**   A transversal is a line, ray, or segment that intersects two or more coplanar lines, rays, or segments, each at a different point.   (132)

Post 3.3.2 **Corresponding Angles Postulate**   If two lines cut by a transversal are parallel, then corresponding angles are congruent.   (134)

Thm 3.3.3 **Alternate Interior Angles Theorem**   If two lines cut by a transversal are parallel, then alternate interior angles are congruent.   (135)

Thm 3.4.1 **Converse of the Corresponding Angles Theorem**   If two lines are cut by a transversal in such a way that corresponding angles are congruent, then the two lines are parallel.   (139)

Thm 3.4.2 **Theorem**   If two lines are cut by a transversal in such a way that consecutive interior angles are supplementary, then the two lines are parallel.   (140)

Thm 3.4.3 **Theorem**   If two lines are perpendicular to the same line, the two lines are parallel to each other.   (141)

Thm 3.4.4 **Theorem**   If two lines are parallel to the same line, then the two lines are parallel to each other.   (142)

Post 3.5.1 **The Parallel Postulate**   Given a line and a point not on the line, there is one and only one line that contains the given point and is parallel to the given line.   (143)

Thm 3.5.2 **Triangle Sum Theorem**   The sum of the measures of the angles of a triangle is 180°.   (144)

Thm 3.5.3 **Exterior Angle Theorem**   The measure of an exterior angle of a triangle is equal to the sum of the remote interior angles.   (147)

Thm 3.8.1 **Slope Formula**   The slope of segment PQ, with endpoints $P(x_1, y_1)$ and $Q(x_2, y_2)$ is the following ratio:

$$\text{Slope} = \frac{y_2 - y_1}{x_2 - x_1}. \quad (161)$$

Thm 3.8.2 **Equal Slope Theorem**   If two nonvertical lines are parallel, then they have the same slope.   (161)

Thm 3.8.3 **Converse of the Equal Slope Theorem**   If two nonvertical lines have the same slope, then they are parallel.   (161)

Thm 3.8.4 **Slopes of Perpendicular Lines**   If two nonvertical lines are perpendicular, then the product of their slopes is −1.   (162)

Thm 3.8.5 **Converse of Slopes of Perpendicular Lines**   If the product of the slopes of two nonvertical lines is −1, then the lines are perpendicular.   (162)

Def 4.1.1 **Congruent Polygons**   Two polygons are congruent if and only if there is a way of setting up a correspondence between their sides and angles, in order, so that

1. all pairs of corresponding angles are congruent, and

2. all pairs of corresponding sides are congruent.   (176)

Post 4.3.1 **SSS(Side-Side-Side) Postulate**   If three sides in one triangle are congruent to three sides in another triangle, then the triangles are congruent.   (186)

Post 4.3.2 **SAS(Side-Angle-Side) Postulate**   If two sides and the angle between them in one triangle are congruent to two sides and the angle between them in another triangle, then the triangles are congruent.   (186)

Post 4.3.3 **ASA(Angle-Side-Angle) Postulate**   If two angles and a side between them in one triangle are congruent to two angles and the side between them in another triangle, then the triangles are congruent.   (186)

Thm 4.3.4 **AAS(Angle-Angle-Side) Theorem**   If two angles and a side that is not between them in one triangle are congruent to the corresponding two angles and the side not between them in another triangle, then the triangles are congruent.   (188)

Thm 4.3.5 **HL (Hypoteneuse Leg) Theorem**   If the hypotenuse and a leg of a right triangle are congruent to the hypotenuse and the corresponding leg in another right triangle, then the two triangles are congruent.   (189)

Thm 4.4.1 **Isosceles Triangle Theorem**   If two sides of a triangle are congruent, then the angles opposite those sides are congruent.   (196)

Thm 4.4.2 **Isosceles Triangle Theorem: Converse**   If two angles of a triangle are congruent, then the sides opposite those angles are congruent.   (196)

Cor 4.4.3 **Corollary**   An equilateral triangle has three angles with measure 60°.   (197)

Cor 4.4.4 **Corollary**   The bisector of the vertex angle of an isosceles triangle is the perpendicular bisector of the base.   (197)

Thm 4.4.5 **Theorem**   The median from the vertex to the base of an isosceles triangle divides the triangle into two congruent triangles.   (197)

Thm 4.4.6 **Theorem**   The bisector of the vertex angle of an isosceles triangle bisects the base.   (198)

Thm 4.4.7   **Theorem**   The opposite sides of a parallelogram are congruent.   (198)

Thm 4.4.8   **Theorem**   The opposite angles of a parallelogram are congruent.   (198)

Thm 4.5.1   **Theorem**   A diagonal of a parallelogram divides the parallelogram into two congruent triangles.   (201)

Thm 4.5.2   **Theorem**   The diagonals of a parallelogram bisect each other   (204)

Thm 4.5.3   **Theorem**   The diagonals and sides of a rhombus form four congruent triangles.   (205)

Thm 4.5.4   **Theorem**   The diagonals of a rhombus are perpendicular.   (205)

Thm 4.5.5   **Theorem**   The diagonals of a rectangle are congruent.   (205)

Thm 4.5.6   **Theorem**   The diagonals of a kite are perpendicular.   (206)

Thm 4.6.1   **Theorem**   If two pairs of opposite sides of a quadrilateral are congruent, then the quadriateral is a parallelogram.   (210)

Thm 4.6.2   **The "Housebuilder" Rectangle Theorem**   If the diagonals of a parallelogram are congruent, then the parallelogram is a rectangle.   (211)

Thm 4.6.3   **Theorem**   If one pair of opposite sides of a quadrilateral are parallel and congruent, then the quadrilateral is a parallelogram.   (211)

Thm 4.6.4   **Theorem**   If the diagonals of a quadrilateral bisect each other, then the quadrilateral is a parallelogram.   (211)

Thm 4.6.5   **Theorem**   If one angle of a parallelogram is a right angle, then the parallelogram is a rectangle. (211)

Thm 4.6.6   **Theorem**   If one pair of adjacent sides of a parallelogram are congruent, then the quadrilateral is a rhombus.   (211)

Thm 4.6.7   **Triangle Midsegment Theorem**   If a segment joins the midpoints of two sides of a triangle, then it is parallel to the third side and its length is one-half the length of the third side.   (212)

Thm 4.7.1   **Congruent Radii Theorem**   In the same circle, or in congruent circles, all radii are congruent.   (213)

Post 4.9.1   **Converse of the Segment Addition Postulate ("Betweenness")**   Given three points P, Q, and R, if $PQ + QR = PR$ then Q is between P and R.   (228)

Post 4.9.2   **Triangle Inequality Postulate**   The sum of the lengths of any two sides of a triangle is larger than the length of the other side.   (229)

Post 5.1.1   **The Area of a Rectangle**   The **area of a rectangle** with base b and height h is $A = bh$.   (245)

Post 5.1.2   **Sum of Areas**   If a figure is composed of nonoverlapping regions A and B, the area of the figure is the sum of the areas of regions A and B.   (245)

Def 5.3.1   **Circle**   A **circle** is the figure that consists of all the points on a plane that are the same distance r from a given point known as the **center** of the circle. The distance r is the **radius** of the circle and $d = 2r$ is the **diameter**.   (261)

Thm 5.4.1   **"Pythagorean" Right-Triangle Theorem**   For any right triangle, the square of the length of the hypotenuse is equal to the sum of the squares of the lengths of the legs.   (269)

Thm 5.4.2   **Converse of the "Pythagorean" Right-Triangle Theorem**   If the square of the length of one side of a triangle equals the sum of the squares of the lengths of the other two sides, then the triangle is a right triangle.   (270)

Thm 5.5.1   **45-45-90 Right-Triangle Theorem**   In any 45-45-90 right triangle, the length of the hypotenuse is $\sqrt{2}$ times the length of a leg.   (276)

Thm 5.5.2   **30-60-90 Right Triangle Theorem**   In any 30-60-90 right triangle, the length of the hypotenuse is two times the length of the shorter leg, and the longer leg is $\sqrt{3}$ times the length of the shorter leg.   (277)

Thm 5.5.3   **Area of a Regular Polygon**   The area of a regular polygon with apothem a and perimeter p is $A = \frac{1}{2}ap$.   (279)

Thm 5.6.1   **Distance Formula**   The distance between two points (x_1, y_1) and (x_2, y_2) is

$$d = \sqrt{(x_2 - x_1)^2 + (y_2 - y_1)^2}.   \text{(284)}$$

Thm 5.6.2   **The Midpoint Formula**   For any two points (x_1, y_1) and (x_2, y_2) in the coordinate plane, the midpoint is given by $\left(\frac{x_1 + x_2}{2}, \frac{y_1 + y_2}{2} \right)$.   (285)

Def 6.2.1   **Definition**   Two planes are parallel if and only if they do not intersect, no matter how far they are extended.   (312)

Def 6.2.2   **Definition**   A line is perpendicular to a plane if and only if it is perpendicular to every line in the plane that intersects it.   (312)

Def 6.2.3 **Definition**   A line is parallel to a plane if and only if it is parallel to a line in the plane.   (313)

Def 6.2.4 **Definition**   The measure of a dihedral angle is the measure of the angle formed by a line in one of the two planes, perpendicular to the intersection of the planes, and a line in the other plane which intersects the first line and is also pependicular to the intersection of the planes.   (313)

Thm 6.6.1 **Sets of Parallel Lines**   In a perspective drawing, all lines that are parallel to each other, but not to the picture plane, will seem to meet at the same point.   (338)

Thm 6.6.2 **Lines Parallel to the Ground**   In a perspective drawing, a line on the plane of the ground will meet the horizon of the drawing if it is not parallel to the picture plane. Any line parallel to this line will meet at the same point on the horizon.   (338)

Thm 7.2.1 **Lateral Area of a Right Prism**   The lateral area L of a right prism with height h and perimeter of a base p is

$$L = hp \quad (360)$$

Thm 7.2.2 **Surface Area of a Right Prism**   The surface area S of a right prism with lateral area L and area of a base B is

$$S = L + 2B \quad (360)$$

Thm 7.2.3 **Cavalieri's Principle**   If two solids have equal heights, and if the cross-sections formed by every plane parallel to the bases of both solids have equal areas, then the two solids have the same volume. (362)

Thm 7.2.4 **Volume of a Prism**   The volume V of a prism with height h and the area of a base B is

$$V = Bh \quad (362)$$

Thm 7.3.1 **Lateral Area of a Right Regular Pyramid**   The lateral area L of a right regular pyramid with slant height l and primeter p of a base is

$$L = \frac{1}{2}lp. \quad (367)$$

Thm 7.3.2 **Surface Area of a Pyramid**   The surface area S of a pyramid with lateral area L and area of base B is

$$S = L + B \quad (367)$$

Thm 7.3.3 **Volume of a Pyramid**   The volume V of a pyramid with area of its base B and altitude h is

$$V = \frac{1}{3}Bh \quad (369)$$

Thm 7.4.1 **Lateral Area of a Right Cylinder**   The lateral area L of a right cylinder with radius r and height h is

$$L = 2\pi rh. \quad (374)$$

Thm 7.4.2 **Surface Area of a Right Cylinder**   The surface area S of a right cylinder with radius r, height h, area of base B, and lateral area L is

$$S = L + 2B \quad \text{or} \quad S = 2\pi rh + 2\pi r^2. \quad (374)$$

Thm 7.4.3 **Volume of a Cylinder**   The volume V of a cylinder with radius r, height h, and area of a base B, is

$$V = Bh \quad \text{or} \quad V = \pi r^2 h. \quad (376)$$

Thm 7.5.1 **Surface Area of a Right Cone**   The surface area S of a right cone with radius of base r, height h, and slant height l is

$$S = L + B \quad \text{or} \quad S = \pi rl + \pi r^2. \quad (381)$$

Thm 7.5.2 **Volume of a Cone**   The volume V of a cone with radius r and height h is

$$V = \frac{1}{3}Bh \quad \text{or} \quad V = \frac{1}{3}\pi r^2 h. \quad (382)$$

Thm 7.6.1 **Volume of a Sphere**   The volume V of a sphere with radius r is

$$V = \frac{4}{3}\pi r^3. \quad (389)$$

Thm 7.6.2 **Surface Area of a Sphere**   The surface area S of a sphere with radius r is

$$S = 4\pi r^2. \quad (390)$$

Def 8.2.1   **Similar Polygons**   Two polygons are similar if and only if there is a way of setting up a correspondence between their vertices so that

1. the corresponding angles are congruent, and
2. the corresponding sides are proportional.   (420)

Thm 8.2.2   **Cross-Multiplication Property**   If $\frac{a}{b} = \frac{c}{d}$ and $b, d \neq 0$, then $ad = bc$.   (422)

Thm 8.2.3   **Reciprocal Property**   If $\frac{a}{b} = \frac{c}{d}$ and $a, b, c, d \neq 0$, then $\frac{b}{a} = \frac{d}{c}$.   (422)

Thm 8.2.4   **Exchange Property**   If $\frac{a}{b} = \frac{c}{d}$ and $a, b, c, d \neq 0$, then $\frac{a}{c} = \frac{b}{d}$.   (422)

Thm 8.2.5   **"Add-One" Property**   If $\frac{a}{b} = \frac{c}{d}$ and $b, d \neq 0$, then $\frac{a+b}{b} = \frac{c+d}{d}$.   (422)

Post 8.4.1   **AA (Angle-Angle) Similarity Postulate**   If two angles of one triangle are equal in measure to two angles of another triangle, then the triangles are similar.   (436)

Post 8.4.2   **SSS (Side-Side-Side) Similarity Postulate**   If the measures of pairs of corresponding sides of two triangles are proportional, then the two triangles are similar.   (436)

Post 8.4.3   **SAS (Side-Angle-Side) Similarity Postulate**   If the measures of two pairs of corresponding sides of two triangles are proportional and the measures of the included angles are equal, then the triangles are similar.   (437)

Thm 8.4.4   **Side-Splitting Theorem**   A line parallel to one side of a triangle divides the other two sides proportionally.   (437)

Cor 8.4.5   **Two-Transversal Proportionality Corollary**   Two or more parallel lines divide two transversals proportionally.   (439)

Thm 8.5.1   **Proportional Altitudes Theorem**   If two triangles are similar, then their corresponding altitudes have the same ratio as the corresponding sides.   (445)

Thm 8.5.2   **Proportional Medians Theorem**   If two triangles are similar, then their corresponding medians have the same ratio as the corresponding sides.   (446)

Thm 8.5.3   **Proportional Angle Bisectors Theorem**   If two triangles are similar then the corresponding angle bisectors are proportional to the corresponding sides.   (450)

Thm 8.5.4   **Proportional Segments Theorem**   The angle bisector of a triangle divides the opposite side into segments proportional to the other two sides of the triangle.   (451)

Def 9.1.1   A **circle** consists of the points in a plane that are equidistant from a given point known as the **center** of the circle. A **radius** is a segment from the center of the circle to a point on the circle. A **chord** is a segment whose endpoints lie on the circle. A **diameter** is a chord that passes through the center of the circle.   (470)

Thm 9.1.2   **Theorem**   If two chords of the same circle are congruent, then their intercepted arcs are congruent.   (474)

Def 9.2.1   A **secant** to a circle is a line that intersects a circle at two points. A **tangent** to a circle is a line that intersects a circle at just one point, known as the **point of tangency.**   (479)

Thm 9.2.2   **Theorem**   A radius perpendicular to a chord bisects the chord.   (479)

Thm 9.2.3   **Tangent Theorem**   If a line is perpendicular to a radius of a circle at its endpoint, then the line is tangent to the circle.   (481)

Thm 9.2.4   **Theorem**   The perpendicular bisector of a chord passes through the center of the circle. (482)

Thm 9.2.5   **The Converse of the Tangent Theorem**   If a line is tangent to a circle, then it is perpendicular to a radius of a circle at a point of tangency.   (483)

Thm 9.3.1   **Inscribed Angle Theorem**   An angle inscribed in an arc has a measure equal to one-half the measure of the intercepted arc.   (487)

Cor 9.3.2   **Inscribed Angle Corollary**   An angle inscribed in a half-circle is a right angle.   (487)

Thm 9.3.3   **Theorem**   If two inscribed angles intercept the same arc, then they have the same measure.   (487)

Thm 9.4.1   **Theorem**   If a tangent and a secant (or a chord) intersect on a circle at the point of tangency, then the measure of the angle formed is one-half the measure of its intercepted arc.   (493)

Thm 9.4.2   **Theorem**   The measure of an angle formed by two secants or chords intersecting in the interior of a circle is one-half of the sum of the measures of the arcs intercepted by the angle and its vertical angle.   (493)

Thm 9.4.3   **Theorem**   The measure of an angle formed by two secants or chords intersecting in the exterior of a circle is one-half of the difference of the measures of the intercepted arcs.   (494)

Thm 9.4.4 **Theorem** The measure of a secant-tangent angle with its vertex outside the circle is one-half the difference of the measures of the intercepted arcs. (496)

Thm 9.4.5 **Theorem** If two tangents intersect in the exterior of a circle, the measure of the angles formed is one-half the difference of the measures of the intercepted arcs. (497)

Thm 9.5.1 **Theorem** If two secants intersect outside of a circle, then the product of the lengths of one secant segment and its external segment equals the product of the lengths of the other secant segment and its external secant segment. (504)

Thm 9.5.2 **Theorem** If a tangent and a secant intersect outside of a circle, then the product of the length of the secant segment and its external segment equals the square of the length of the external secant segment. (504)

Thm 9.5.3 **Theorem** If two chords intersect inside a circle, then the product of the lengths of the segments of one chord equals the product of the divided lengths of the other chord. (505)

Thm 10.5.1 **The Law of Sines** For any triangle ABC: $\frac{\sin A}{a} = \frac{\sin B}{b} = \frac{\sin C}{c}$. (556)

Thm 10.6.1 **The Law of Cosines** For any triangle ABC:

$$a^2 = b^2 + c^2 - 2bc \cos A$$
$$b^2 = a^2 + c^2 - 2ac \cos B$$
$$c^2 = a^2 + b^2 - 2ab \cos C. \text{(564)}$$

Thm 11.4.1 **Jordan's Theorem** Every simple closed curve divides the plane into two distinct regions. (609)

Thm 11.4.2 **Euler's Formula** For any polyhedron with vertices V, edges E, and faces F,

$$V - E + F = 2. \text{(610)}$$

Thm 11.6.1 **Affine Transformation** An affine transformation maps all pre-image points P in a plane to image points P' so that

1. collinear points project to collinear points.
2. straight lines project to straight lines.
3. intersecting lines project to intersecting lines.
4. parallel lines project to parallel lines. (622)

Thm 11.6.2 **Theorem of Pappus** If A,B, and C are three distinct points on one line and A',B', and C' are three distinct points on a second line, then the intersections of $\overline{AB'}$ and $\overline{BA'}$, $\overline{AC'}$ and $\overline{CA'}$, and $\overline{BC'}$ and $\overline{CB'}$ are collinear. (624)

Thm 11.6.3 **Theorem of Desargues** If one triangle is a projection of another triangle, then the intersections of the lines containing the corresponding sides of the two triangles are collinear. (625)

Def 12.4.1 **Proof by Contradiction** To prove **s**, assume ~**s**. Then the following argument form is valid:

If ~**s** then (**t** and ~**t**)

Therefore, **s**. (666)

GLOSSARY

acute triangle A triangle with three acute angles. (555)

adjacent angles Two angles in a plane that share a common vertex and a common side but have no interior points in common. (84)

affine transformation A tranformation in which all pre-image points are mapped to image points so that collinear points, straight lines, intersecting lines, and parallel lines remain as such. (621)

alternate exterior angles Two nonadjacent exterior angles which lie on opposite sides of a transversal. (133)

alternate interior angles Two nonadjacent interior angles which lie on opposite sides of a transversal. (133)

altitude of a cone A segment from the vertex perpendicular to the plane of the base. (379)

altitude of a cylinder A segment joining the two base planes and perpendicular to both. (373)

altitude of a prism A segment joining the two base planes and perpendicular to both. (359)

altitude of a pyramid A segment from the vertex perpendicular to the plane of the base. (366)

altitude of a triangle A line segment from a vertex drawn perpendicular to the line containing the opposite side. (23)

angle A figure formed by two rays that have the same endpoint. (11)

angle bisector A ray that divides an angle into two congruent angles. (18)

annulus The region between two circles which have the same center but different radii. (387)

arc A part of a circle. (472)

area The number of nonoverlapping unit squares that will cover the interior of a figure. (245)

argument A sequence of statements. (644)

axis of cylinder The segment joining the centers of the two bases. (373)

axis of symmetry The line that divides a figure into two symmetrical halves. (117)

Base of isosceles triangle The side opposite the vertex angle. (196)

base angles of isosceles triangle The angles opposite the legs. (196)

bases of a prism Two congruent polygonal faces that lie in parallel planes. (318)

base of a cone The circular face of the cone. (379)

bases of a cylinder The two congruent circular faces that lie in parallel planes. (373)

base of a pyramid The polygonal face that is opposite the vertex. (366)

betweenness Given three points, A, B, and C, if $AB + BC = AC$, then B is between A and C. (228)

binary number system A number system based on the digits 0 and 1. (671)

Cavalieri's Principle If two solids have equal heights, and if the cross-sections formed by every plane parallel to the bases of both solids have equal areas, then the two solids have the same volume. (362)

center of a circle The point inside the circle that is equidistant from all the points on the circle. (261)

center of mass See centroid. (25)

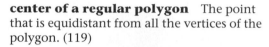

center of a regular polygon The point that is equidistant from all the vertices of the polygon. (119)

central angle of a circle An angle formed by two rays orginating from the center and passing through two points on the circle. (471)

central angle of a polygon An angle formed by two rays orginating from the center and passing through adjacent vertices of the polygon. (119)

centroid The point where the three medians of a triangle intersect. (25)

chord A segment whose endpoints lie on a circle. (470)

circle The set of points in the plane that are equidistant from a given point known as the center of the circle. (261)

circumcenter The point where the three perpendicular bisectors of the sides of a triangle intersect; it is equidistant from the three vertices of the triangle and is the center of the circumscribed circle. (25)

circumference The distance around a circle. (262)

circumscribed circle A circle is circumscribed about a polygon if each vertex of the polygon lies on the circle. (24)

collinear Points that lie on the same line. (11)

complementary angles Two angles whose measures have a sum of 90°. (39)

compound statement A statement formed when two statements are connected by "and" or by "or". (650)

concave polygon A polygon that is not convex. (149)

conclusion The phrase in a conditional statement following the word "then". (72)

conclusion The final statement of an argument. (644)

conditional statement A statement that can be written in the form "If p, then q,"

where p is called the hypothesis, and q is called the conclusion. (72)

cone An object that consists of a circular base and a curved lateral surface which extends from the base to a single point called the vertex. (379)

congruence The relationship between figures having the same shape and same size. (31)

congruent polygons Two polygons are congruent if their vertices can be matched such that corresponding angles are congruent, and corresponding sides are congruent. (176)

conic section The plane curves that can be formed by the intersection of a plane with a right circular cone; they include the circle, ellipse, parabola, and hyperbola. (499)

conjecture An "educated guess" based on observation. (18)

conjunction A compound statement that uses the word "and". (650)

consecutive interior angles Two interior angles which lie on the same side of a transversal. (133)

contraction A dilation where the figure that is transformed is reduced in size. (414)

contradiction A contradiction asserts that a statement and its negation are both true. (666)

contrapositive The statement formed by interchanging the hypothesis and conclusion of a conditional statement and negating both parts. (660)

converse The statement formed by interchanging the hypothesis and conclusion of a conditional statement. (74)

convex polygon A polygon in which any line segment connecting two points of the polygon has no part outside the polygon. (149)

coordinate plane The plane of the x- and y-axes. (51)

corresponding angles Two nonadjacent angles, one interior and one exterior, that lie on the same side of the transversal. (133)

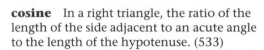

cosine   In a right triangle, the ratio of the length of the side adjacent to an acute angle to the length of the hypotenuse. (533)

cotangent   In a right triangle, the ratio of the length of the side adjacent to an acute angle to the length of the opposite side. (527)

counterexample   An example which proves that a conditional statement is false in that the hypothesis is true but the conclusion is false. (74)

cube   A prism with six square faces. (304)

deductive reasoning   The process of drawing conclusions by using logical reasoning. (73)

diagonal   A segment that joins two nonadjacent vertices of a polygon. (201)

diameter   A chord that passes through the center of a circle. (261)

dihedral angle   An angle formed by the intersection of planes. (313)

dilation   A dilation with center C and scale factor k is a tranformation that maps every point P to a point P' determined as follows

(1) if P is point C, then $P = P'$,
(2) otherwise, P' lies such that
     $CP' = k \bullet CP$, where $k > 0$ and $k \neq 1$. (414)

direction of vector   The component of a vector, usually indicated by an arrow, that indicates orientation. (570)

disjunction   A compound statement that uses the word "or". (651)

Distance Formula   The distance between two points in the plane containing (x_1, y_1) and (x_2, y_2) is

$$d = \sqrt{(x_2 - x_1)^2 + (y_2 - y_1)^2}. \text{ (284)}$$

edge   A segment formed by the intersection of two faces of a polyhedron. (318)

equation of a circle   A circle with its center at (h, k) and a radius of length r has an equation

$$(x - h)^2 + (y - k)^2 = r^2. \text{ (511)}$$

equilateral triangle   A triangle in which all three sides are congruent. (196)

Equivalence Properties of Equality
The reflexive, symmetric, and transitive properties of equality. (94)

equivalence relation   Any relation that satisfies the three properties of reflexivity, symmetry, and transitivity. (95)

even vertices   The vertices of a network that have an even number of paths going to them. (601)

exclusive *or*   Indicating either one or the other, but not both. (651)

exterior angle of a polygon   An angle formed between one side of polygon and the extension of an adjacent side. (147)

external secant segment   The portion of a secant segment that lies outside the circle. (502)

face of a prism   Each flat surface of a prism. (318)

fractal   A structure which is self-similar in that each subdivision has the same structure as the whole. (629)

glide A combination of a translation and a reflection. (49)

Golden Ratio See Golden Rectangle. (586)

Golden Rectangle The rectangle in which the length l and the width w satisfy the proportion $\frac{l}{w} = \frac{w}{l-w}$. The ratio lw is called the golden ratio.

great circle The intersection of a sphere with a plane that passes through the center of the sphere. (145)

head-to-tail method To find the sum of two vectors place the tail of one vector at the head of the other; the vector drawn from the tail of the first to the head of the second represents the vector sum. (572)

heptagon A polygon with seven sides. (117)

hexagon A polygon with six sides. (117)

hyperbolic geometry The geometry of a surface that curves inward like a "saddle". (616)

hypotenuse The side opposite the right angle in a right triangle. (189)

hypothesis The phrase in a conditional statement following the word "if". (72)

identity matrix A square marix in which all the entries on the main diagonal (from upper left to lower right) are one, and all other entries are zero. (550)

image See transformation. (44)

incenter The point where the three angle bisectors of a triangle intersect; it is equidistant from the three sides of the triangle and is the center of the inscribed circle. (25)

inclusive *or* Indicating either one or the other, or both. (651)

indirect proof Proving a conjecture by showing that the opposite of the conjecture is impossible. (665)

inductive reasoning Forming conjectures on the basis of an observed pattern. (99)

input-output table A table which records what happens to the input as it passes through a logic gate. (673)

inscribed angle An angle whose vertex is on the circle and whose sides contain chords of the circle. (484)

inscribed circle A circle is inscribed in a polygon if each side of the polygon is tangent to the circle. (24)

intercepts The points where a line in the coordinate plane crosses the x- and y- axes. (331)

intercepted arc The arc that lies in the interior of an angle inscribed in a circle. (484)

invariant Properties that stay the same regardless of how a figure is deformed. (609)

inverse The statement formed by interchanging the hypothesis and conclusion of a conditional statement. (660)

isometric drawing A drawing on graph paper that has three rather than two sets of parallel lines; therefore, three-dimensional objects can be represented. (304)

isosceles triangle A triangle with at least two congruent sides. (196)

kite   A quadrilateral in which two pairs of adjacent sides are congruent. (259)

knot theory   A branch of topology which investigates a curve formed by looping and interlacing a piece of string and then joining the ends together. (607)

lateral area   The sum of the areas of the lateral faces. (359)

lateral faces   The faces of a prism or pyramid that are not bases. (318)

lateral surface   The curved surface of a cylinder or cone. (373)

legs of a right triangle   The sides adjacent to the right angle. (189)

legs of an isosceles triangle   The congruent sides of an isosceles triangle. (196)

line   An undefined term in geometry, a line is understood to be straight, contain an infinite number of points, extend infinitely in two directions, and have no thickness. (10)

linear pair of angles   Two adjacent angles whose noncommon sides are opposite rays. (40)

logic gate   An electronic circuit that represents "not," "and," or "or". (673)

logical chain   Linking several conditionals together. (75)

logical reasoning   Linking true conditionals together to form a valid conclusion. (72)

logically equivalent   Two logic statements that have the same truth values. (653)

magnitude of vector   The length of the vector arrow. (570)

major arc   The major arc consists of points A and B and all points of the circle in the exterior of central angle *AOB*. (472)

median   A segment from a vertex to the midpoint of the opposite side in a triangle. (23)

midpoint of a segment   The point that divides the segment into two congruent segments. (221)

midsegment of a trapezoid   The segment that connects the midpoints of the nonparallel sides. (155)

midsegment of a triangle   A segment that connects the midpoints of two sides. (155)

minor arc   The minor arc $\overarc{AB}$ of central angle *AOB* consists of all the points on the circle that lie in the interior of the central angle. (472)

Modus Ponens   In logic, a valid argument of the following form:
    If *p*, then *q*.
    p, therefore *q*. (645)

Modus Tollens   In logic, a valid argument of the following form:
    If *p*, then *q*.
    Not *q*, therefore not *p*. (646)

Möbius strip   The one-sided surface formed by taking a long rectangular strip of paper and pasting its two ends together after giving it half a twist. (607)

Monte Carlo Method   A simulation technique used to obtain a probability in an experiment which has *n* number of equally likely outcomes. (291)

negation If *p* is a statement then not *p* is its negation. (652)

nets Flat figures that can be folded to enclose a particular solid figure. (359)

network A collection of points called vertices some of which may be connected by edges. (600)

non-Euclidean geometries A system of geometry in which the Parallel Postulate does not hold. (615)

noncollinear Three or more points not all of which lie on the same line. (11)

number line A line whose points correspond with the set of real numbers. (29)

oblique cone A cone that is not a right cone. (379)

oblique cylinder A cylinder that is not a right cylinder. (373)

oblique prism A prism that is not a right prism. (319)

oblique pyramid A pyramid that is not a right pyramid. (366)

obtuse triangle A triangle which has one obtuse angle. (555)

octagon A polygon with eight sides. (117)

octant One of the eight spaces into which the whole of space is divided by the *x*-, *y*-, and *z*-axes. (327)

odd vertices The vertices of a network that have an odd number of paths going to them. (601)

orthocenter The point where the three lines containing the altitudes of a triangle intersect. (25)

orthogonal Perpendicular to. (617)

orthographic projections A parallel projection with all rays perpendicular to the plane of projection. (306)

parallel lines Two coplanar lines that do not intersect. (17)

parallel planes Two planes that do not intersect. (312)

parallelogram A quadrilateral in which opposite sides are parallel. (124)

pentagon A polygon with five sides. (117)

perimeter The distance around a geometric figure that is contained in a plane. (244)

perpendicular bisector A line that is perpendicular to a segment at its midpoint. (18)

perpendicular lines Two lines that intersect to form a right angle. (16)

plane An undefined term in geometry, a plane is understood to be a flat surface that extends infinitely in all directions. (11)

Platonic Solids The five regular polyhedra which are the tetrahedron, the hexahedron. (cube), the octahedron, the dodecahedron, and the icosahedron. (404)

point An undefined term in geometry, a point can be thought of as a dot that represents a location on a plane or in space. (10)

point of tangency The one point at which a tangent intersects a circle. (478)

polygon A closed plane figure formed by three or more segments such that each segment intersects exactly two other segments, one at each endpoint. (116)

polyhedron A geometric solid with polygons as faces. (404)

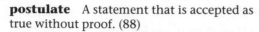

postulate   A statement that is accepted as true without proof. (88)

pre-image   See transformation. (44)

premise   The statements in an argument that come before the conclusion. (644)

prism   A polyhedron that has two parallel faces called bases; all other faces, called lateral faces, are parallelograms formed by joining corresponding vertices of the bases. (317)

probability   A ratio that compares the number of successful outcomes with the total number of possible outcomes. (289)

projective geometry   The study of the properties of figures that do not change under projection. (624)

proof   An organized series of statement used to form a convincing argument that a given statement is true. (66)

proof by contradiction   See indirect proof. (666)

proportional sides   Sides are proportional if all the ratios of corresponding sides of two polygons are the same. (420)

pyramid   A solid figure which consists of one base and lateral faces which are triangles. (366)

Pythagorean triple   A set of three positive integers, a, b, and c such that $a^2 + b^2 = c^2$. (268)

quadrilateral   A polygon with four sides. (124)

radius   A segment which connects the center of a circle with a point on the circle. (261)

ratio   A comparison of two numbers by division. (354)

ray   Consists of an initial point A and all points on that lie on the same side of A as B does. (10)

rectangle   A parallelogram with four right angles. (124)

reflection   A transformation in which a line of reflection acts as a mirror reflecting points to their images. (46)

reflectional symmetry   A plane figure has reflectional symmetry if its reflection image through a line coincides with the original figure. (117)

Reflexive Property of Equality   For all real numbers a, $a = a$. (94)

regular polygon   A polygon which is both equilateral and equiangular. (119)

regular pyramid   A pyramid whose base is a regular polygon. (366)

remote interior angle   An interior angle of a triangle that is not adjacent to a given exterior angle. (147)

resultant vector   The vector that represents the sum of two given vectors. (571)

rhombus   A parallelogram with four congruent sides. (124)

Riemannian geometry   A geometry in which there are no parallel lines, such as on the surface of a sphere. (616)

right angle   An angle with a measure of 90°. (17)

right cone   A cone in which the altitude intersects the base at its center. (379)

right cylinder   A cylinder whose axis is perpendicular to the bases. (373)

right prism   A prism in which all the lateral faces are rectangles. (319)

right pyramid   A pyramid in which the altitude intersects the base at its center. (366)

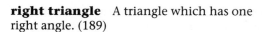

right triangle   A triangle which has one right angle. (189)

right-handed system   The most common arrangement of the three axes in a space coordinate system, as illustrated in the figure. (326)

rigid transformation   A transformation that does not change the size or shape of a figure. (44)

rise and run   The vertical and horizontal distances between two points in the plane. (160)

rotation   The transformation that moves a geometric figure about a point known as the turn center. (45)

rotation matrix   A matrix used to rotate a figure around a given point *(x, y)* through an angle. (550)

rotational symmetry   A figure has rotational symmetry if it has at least one rotation image that coincides with the original image. (118)

scale factor   In a transformation the number used to multiply each coordinate of the pre-image to obtain the coordinates of the image. (412)

secant to a circle   A line that intersects a circle at two points. (478)

secant segment   A portion of a secant with two endpoints. (502)

sector of a circle   A region of a circle bounded by two radii and their intercepted arc. (262)

segment   A portion of a line with two endpoints. (10)

segment bisector   A line that divides a segment into two congruent segments. (18)

self-similarity   A structure where each subdivision has the same structure as the whole. (629)

semicircle   The arc of a circle that is cut off by a diameter. (243)

similar figures   Two figures that have the same shape; if they are polygons, the corresponding angles are congruent and the corresponding sides are proportional. (419)

simple closed curve   A figure that encloses a distinct region of the plane and whose segments or curves do not cross themselves. (608)

sine   In a right triangle, the ratio of the length of the side opposite an acute angle to the length of the hypotenuse. (533)

skew lines   Lines that are not coplanar. (312)

slide arrow   An arrow that shows the direction and motion of points in a translation. (45)

slope   The ratio of the vertical rise to the horizontal run of a segment. (160)

slope formula   The slope of a segment whose endpoints have coordinates (x_1, y_1) and (x_2, y_2) is the ratio $\frac{y_2 - y_1}{x_2 - x_1}$. (161)

solid of revolution   The object formed by rotating a plane figure about an axis in space. (399)

sphere   The set of points in space that are equidistant from a given point known as the center of the sphere. (387)

square   A parallelogram with four congruent sides and four right angles. (124)

supplementary angles   Two angles whose measures have a sum of 180°. (39)

surface area of a prism   The sum of the areas of all the faces of a prism. (359)

Symmetric Property of Equality   For all real numbers *a* and *b*, if $a = b$, then $b = a$. (94)

tangent to a circle   A line that intersects a circle at a single point. (478)

tangent ratio   In a right triangle, the ratio of the length of the side opposite an acute angle to the length of the adjacent side. (525)

tangent segment   A portion of a tangent with two endpoints. (502)

taxicab circle   The set of points in the plane that are at a given taxidistance from a given point known as the center of the circle. (596)

taxicab radius   The distance between the center of a taxicab circle and any point on the taxicab circle. (596)

taxidistance   In taxicab geometry the smallest number of grid units that must be traveled to go from one point to another point. (594)

theorem   A statement that can be proved true. (88)

topology   The geometry that deals with the properties of figures that are unchanged by distortion. (607)

torus   A donut-shaped surface in topology. (609)

transformation   The movement of a figure in the plane from its original position, the *preimage*, by translation, rotation, or reflection to a new position, the *image*. (44)

Transitive Property of Equality   For all real numbers a, b, and c, if $a = b$ and $b = c$, then $a = c$. (94)

translation   A transformation that glides all points in the plane the same distance in the same direction. (44)

transversal   A line, ray, or segment that intersects two or more coplanar lines, rays, or segments, each at a different point. (132)

trapezoid   A quadrilateral with only one pair of parallel sides. (124)

trefoil knot   A knot which has the shape of a plane figure made of congruent arcs of a circle arranged on an equilateral triangle so that the figure is symmetrical about the center of the triangle and the ends of the arcs are on the triangle. (607)

triangle   A polygon with three sides. (23)

triangle rigidity   The property that if the sides of a triangle are fixed, there is only one shape the triangle can have. (180)

triangulation   The process of using known measurements in a triangle to find other measurements indirectly rather than by direct measurement. (186)

trigonometry   The geometry of the triangle. (525)

true conditional   A conditional statement that leads only to a true conclusion. (73)

truth table   A table used to list all possible combinations of truth values for a given statement or compound statement. (650)

turn center   The point about which a figure is rotated. (45)

two-column proof   A proof in which the statements are written in the left-hand column and the reasons in the right-hand column. (68)

unit circle   A circle with its center at the origin of the coordinate plane and with a radius of 1. (542)

vanishing point   In a perspective drawing the point at which the parallel lines seem to meet. (337)

vector   A mathematical quantity that has both direction and magnitude (or length). (570)

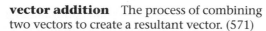

vector addition   The process of combining two vectors to create a resultant vector. (571)

vector sum   The resultant vector created by vector addition. (571)

vertex angle of an isosceles triangle
The angle opposite the base of the triangle. (196)

vertex of a cone   The single point opposite the base of the cone. (379)

vertex of a prism   A point where three or more edges meet. (318)

vertex of a pyramid   The point where the lateral triangular faces meet. (366)

vertical angles   The opposite angles formed when two lines intersect. (99)

volume   The number of nonoverlapping unit cubes that will fill the interior of a solid figure. (360)

A-shaped level, 344–345
Absolute value, 29–30, 283–284
Acute triangle, 555
"Add-one" property of proportion, 422
Addition
    of vectors, 571
    property of equality, 92
Adjacent angles, 84
Adjacent side, 524
Affine transformations, 621–623
Algebra *(See Connections—Algebra)*
Algebraic properties of equality, 92
Alhambra, 114
Alice in Wonderland, 665
Alternate exterior angles, 133
Alternate interior angles, 133
    theorem, 135
Altitude(s)
    of cone, 379
    of cylinder, 373
    of prism, 359
    of pyramid, 366
    of similar triangles, 445
    of trapezoid, 256
    of triangle, 23
Analytic geometry *(See Coordinate geometry)*
AND gate, 673
Angle(s), 11, 60
    addition postulate of, 39
    adjacent, 84
    alternate exterior, 133
    alternate interior, 133
    associated with a circle, 491–495
    base, 196
    between two tangents, 497
    bisector, 18, 23
    complementary, 39
    congruent, 39, 61, 419
    consecutive interior, 133
    constructing bisector of, 214–215
        corresponding, 133–134, 666
    dihedral, 313
    exterior, of a polygon, 147, 151, 170
    inscribed, 484–488

linear pair, 40, 100
measure of, 38, 90
of a polygon, 150, 170, 296–297
of a triangle, 144
remote interior, 147
right, 17, 109
secant-tangent, 492
supplementary, 39, 100
transversals and, 132–135, 169
vertex of, 491–493, 496–497
vertical, 99–100,108
Angle-angle similarity postulate, 436
Angle-angle-side theorem (AAS), 188,
Angle bisectors of similar triangles, 450
Angle-side-angle postulate (ASA), 186
Annulus, 387, 388
    area of, 388
Apothem, 279
Applications
    business and economics
        advertising, 427
        agriculture, 251, 260
        architecture, 5, 136, 365, 424, 559
        construction, 27, 164, 184, 192, 197, 202, 249, 368, 372
        irrigation, 264
        landscaping, 249, 427
        manufacturing, 384
        small business, 385
    language arts
    Communicate, 13, 20, 25, 33, 40, 48, 55, 68, 76, 85, 90, 96, 103, 120, 127, 136, 140, 146, 152, 157, 163, 177, 183, 190, 197, 203, 210, 215, 222, 248, 257, 263, 271, 279, 287, 292, 307, 314, 320, 328, 333, 340, 354, 361, 368, 374, 381, 389, 398, 416, 424, 433, 440, 447, 456, 474, 481, 489, 495, 506, 512, 528, 537, 545, 552, 558, 567, 574, 590, 597, 604, 610, 618, 626, 633, 647, 653, 661, 668, 675, 683
    Eyewitness Math, 80, 130, 252, 323, 392, 500, 562, 656
    life skills
        carpentry, 136, 178, 184, 488

computer programming, 110
consumer awareness, 377
cooking, 358
decorating, 259
drafting, 142, 281
food, 392
house painting, 249
machining, 507
mechanical drawing, 304, 306
model building, 427
navigation, 42, 137, 187, 315, 491, 494
traffic signs, 200
miscellaneous
    area, 256
    buried treasure, 550
    computers, 671
    criminal justice, 78
    distance, 52
    earth satellite, 478
    gasoline storage, 376
    law, 658, 667
    materials handling, 612
    measurment, 277, 428, 441
    product packaging, 356, 359, 364, 386, 458
    pyramids, 527
    solar energy, 250
    stained glass, 490
    statistics, 35, 476
    surveying, 199, 435, 448, 531, 559
    telescope mirror, 562–563
    triangulation, 186
    typography, 436
    water storage, 378
science
    aeronautical engineering, 358
    archaeology, 24, 156
    astronomy, 138, 330, 448, 535, 540, 546, 547
    automobile engineering, 265
    biology, 355, 357
    botany, 354, 357
    cartography, 389, 505
    chemistry, 85
    civil engineering, 280, 443, 475
    communications, 482, 497, 63
    crystals, 319
    earth science, 418
    engineering, 120, 531
    environmental protection, 287
    environmental science, 678

flood control, 159
genetics, 607
geology, 35, 85, 382
ichthyology, 361
marine biology, 426
meteorology, 292
optics, 417
petroleum engineering, 545
physics, 28, 570, 572
physiology, 357, 358
science, 9
seismology, 500–501
space flight, 482
wildlife management, 426, 560
social studies
geography, 123, 145
history, 219, 273, 630
sports and leisure
aquarium, 14
baseball, 124, 206
candy making, 635
chess, 642
fine art, 5, 78, 129, 204, 295, 336, 342, 427, 481
hot air ballooning, 389, 391
knitting, 254
marching band, 517
music, 541
olympics, 516
origami, 19, 22
photography, 323, 422
quilting, 178, 184
recreation, 184, 309, 385, 393, 444, 535
scale models, 386
scuba diving, 42
sky diving, 293
soccer, 570
sports, 266, 271, 392, 458
theatre arts, 255
transportation, 595
travel distance, 558
Arc(s)
degree measure of, 473
intercepted, 484
length of, 473
major, 472
minor, 472
Area, 245
cross-sectional, 454
maximum, 246–247
of rectangle, 245
of circle, 262, 285–286

of isosceles trapezoid, 253
of parallelogram 255–256
of polygons, 278–279, 296–297
of rectangle, 245
of regular hexagon, 278
of regular polygons, 278–279
of trapezoid, 256, 259
of triangle, 254–255, 260
ratios of similar figures, 452
Argument
invalid, 645
valid, 644
Assessment
Chapter Assessment 63, 111, 171, 239, 301, 349, 409, 465, 521, 581, 641, 691
Chapter Review, 60, 108, 168, 234, 298, 346, 406, 462, 518, 578, 638, 688
Cumulative Assessment 112, 240, 350, 466, 582, 692
Ongoing (*See Communicate, Critical Thinking, Try This*)
Axis of cylinder, 373
Axis of symmetry, 117, 168
in triangles, 117–118

Base
of cone, 379
of cylinder, 373
of isosceles triangle, 196
of prism, 318, 359
of pyramid, 366
Base angles of isosceles triangle, 196
Betweenness, 228
postulate, 228
Biconditional, 663
Binary number system, 671
Biographical/historical note
Alhambra, 114
Brunelleschi, 336, 339
Caliban, 106–107
Cantor, Georg, 631
Carroll, Lewis, 665
Chiu Chang, 268–269
DaVinci, 6
Desargues, 625
Dürer, Albrecht, 337

Eiffel Tower, 180
Einstein, Albert, 618
Escher, M.C., 115, 129, 165, 193, 614, 617
Euclid, 88, 92, 200, 468, 614, 685
Euler, 27, 322, 600, 610
Garfield, President, 273
Gauss, Karl Friedrich, 615, 685
Heron, 260
Ibn al Haitham, 412
Jordan, Camille, 609
Königsberg, 600, 603
King Tut's tomb, 394
Lobachevski, 616
Lune of Hippocrates, 243
Ma, Yo Yo, 74
Modrian, Piet, 5
Napoleon, 219
Notre Dame, 468
Pappus, 200, 624
Parthenon, 5
Pascal, 632
Pick, G., 297
Plato, 272
Poincare, Henri, 617
President Garfield, 273
Pythagoras, 268
Pythagoreans, 268
Riemann, G.F.B., 616
Stella, Joseph, 132
Stonehenge, 540
Warhol, Andy, 174
Yucatan Peninsula, 502
Bisector
angle, 18, 23, 214–215
of vertex angle, 197
perpendicular, 18, 23, 197
segment, 18
Brunelleschi, 336, 339

Caliban, 106
Calculator (*See Technology*)
Cantor, Georg, 631
Carroll, Lewis, 665
Cavalieri's Principle, 362
Center
of circle
of dilation, 414
of mass (See Centroid)

of projection, 623
of regular polygon, 119
of rotation, 45
Central angle
  of a circle, 471
  of a polygon, 119
Centroid, 25
Chaos game, 636
Chapter Assessment 63, 111, 171,
  239, 301, 349, 409, 465, 521,
  581, 641, 691
Chapter Projects, 58, 106, 166, 232,
  344, 404, 460, 516, 576, 636,
  685
Chapter Review, 60, 108, 168, 234,
  298, 346, 406, 462, 518, 578,
  638, 688
Chiu Chang, 268–269
Chord(s), 502
  angle between, 493
  product of segments of, 505
Chords and arcs theorem, 474
Circle, 7
  arcs of, 472–473, 484
  area of, 262
  center of, 511–512
  central angle of, 471
  chord of, 470, 502
  chords and arcs of, 474
  circumference of, 262
  circumscribed, 24, 61
  construction of "flowers", 472
  definition of, 261, 470
  diameter of, 261, 262, 470
  equation of, 511
  estimating area of, 285–286
  graph of, 509–512
  great, 145, 616
  inscribed angle of, 484
  inscribed, 24, 61
  intercepted arc of, 484
  major arc of, 472
  minor arc of, 472
  radius of, 261, 470
  secants of, 478, 491–494,
    502–504
  sectors of, 262
  semicircle, 472
  special angles of, 491–495
  tangents to, 478, 480–481, 483,
    491–493, 497, 502–503
  theorems on special angles of,
    493, 494, 496, 497

unit, 542
Circuit, Euler, 605
Circumcenter, 25
Circumference
  of circle, 262
  of earth, 444
Circumscribed circle, 24, 61
Collinear, 11
Collinearity, 228
Common notions (Euclid), 92
Communicate *(See Applications, lan-*
  guage arts)
Compass constructions, 24–26,
  30–31, 95, 213–219, 221, 226,
  262, 429, 430, 431, 471, 472,
  474, 479, 480, 485, 503, 504,
  516, 587, 589, 591, 622
Complementary angles, 39
Compound statement, 650
Computer graphics *(See Technology)*
Concave polygon, 149
Conclusion, 72, 108, 138, 644
Conditional statements, 72–76,
  108–109, 658
  true, 73
Cone
  altitude of, 379
  base of, 379
  lateral surface of, 379
  oblique, 379
  right, 379
  slant height of, 380, 381
  surface area of, 380–381
  vertex of, 379
  volume of, 382
Congruence, 95
  reflexive property of, 176
  transitive property of, 135
Congruent
  angles, 39, 61, 134–135, 419
  parts of a parallelogram, 203
  polygons, 174–176
  radii theorem, 213
  segments, 31, 61, 95, 213
  sides of triangles, 117
  triangles, 53, 180–182, 186–189,
    201–202
Conic sections, 499
Conjectures, 18, 19, 24, 46, 47, 52,
  53, 60, 99, 100, 101, 102, 108,
  118, 119, 125, 126, 132, 133,
  134, 144, 151, 154, 155, 187,
  201, 207–209, 308, 414, 479,

480, 485, 503, 555, 595, 624,
  625
Conjunction, 650
  negation of, 652
Connections *(See Applications,*
  Cultural Connections, Math
  Connections)
Consecutive interior angles,
  133
Construction *(See also Compass con-*
  structions, Paper cutting, Paper
  folding)
  congruent segments, 31, 226
  of "circle flowers"
  of circumscribed circle, 24
  of dilation, 415
  of hexagon, 471
  of inscribed circle, 24
  of reflections, 46–47
  of ruler, 30
  of parallel line through a point,
    139
Contractions, 414
Contradiction, 309, 483
  proof by, 666
Contrapositive, 649, 660
Converse, 74, 82, 109, 138–139,
  161, 162, 196, 209, 228, 270,
  659
Convex polygon, 149
Cooperative learning, 81, 131, 253,
  325, 395, 501, 563, 657
Coordinate geometry, 51–54, 62,
  160–162, 220–222, 327–328,
  365, 396–400, 509–512,
  548–551, 679–681
Coordinate plane, 51
Coordinates, 55, 327
  of midpoint, 285
Corollary, 438
  inscribed angle, 487
  two transversal proportionality,
    439
Corresponding
  angles postulate, 134
  angles, 133, 175, 666
  parts of polygons, 174–176
  sides of congruent triangles
    (CPCTC), 193–195
  sides, 175
Cosine(s), 533, 543
  Law of, 564
Cotangent, 527, 533

Counterexample, 74, 109, 187, 207
Critical thinking, 18, 25, 31, 32, 38,
    39, 45, 46, 53, 54, 75, 76, 83,
    90, 95, 101, 102, 120, 125,
    126, 144, 145, 150, 175, 176,
    181, 188, 194, 202, 209, 213,
    214, 221, 222, 227, 228, 244,
    246, 247, 255, 256, 263, 269,
    275, 286, 307, 314, 319, 326,
    338, 339, 355, 356, 360, 361,
    366, 375, 382, 389, 414, 415,
    420, 421, 432, 437, 438, 439,
    443, 454, 471, 472, 473, 480,
    481, 486, 488, 494, 510, 511,
    525, 533, 542, 550, 551, 565,
    572, 589, 602, 608, 646, 652,
    653, 660, 661, 667, 672, 675,
    680, 681
Cross-multiplication property of
    proportion, 422
Cross-sectional area, 454–455
Cube(s)
    constructing, 49
    drawing, 304
Cultural connections
    Africa, 34, 57, 122, 209, 244, 260,
        353, 370, 439, 444, 449, 468,
        527, 624
    Asia, 8, 19, 22, 43, 252, 266, 267–
        268, 272, 468, 469, 576–577,
        634
    Europe, 130, 268, 322, 600
    South America, 656
Cumulative Assessment 112, 240,
    350, 466, 582, 692
Cuneiform, 576
Curves, simple closed, 608
Cylinder
    altitude of, 373
    axis of, 373
    bases of, 373
    lateral area of, 374
    lateral surface of, 373
    oblique, 373
    right, 373
    surface area of, 374
    volume of, 376

DaVinci, 6
Decagon, 117

Deductive reasoning, 73, 99,
    100, 110
Definitions, writing, 82–83, 109
Desargues, Girard, theorem of, 625
Diagonal(s)
    of a prism, 319
    of a square, 275
    of parallelogram, 201, 204, 680
    of rectangle, 205
Diameter of a circle, 261, 470
Dihedral angle, 313
Dilation, 414–415
    center of, 414
Direction of a vector, 570
Disjunctions, 651
Distance
    between two points, 274
    formula, 283-284
    formula in three-dimensions,
        328
    indiect measure of, 443–444
Dividing segment into equal parts,
    442
Dodecagon, 117
Dodecahedron, 404, 405
Dürer, Albrecht, 337

Edge of prism, 318
Egg Over Alberta, 130
Eiffel Tower, 180
Elements (Euclid), 92
Ellipse, 499
Endpoints of a segment, 10, 60
Equal slope theorem, 161
Equality, algebraic properties of,
    92–93
Equation
    of a circle, 511
    of a line, 331
    of a plane, 331
Equations
    inconsistent, 161
    parametric, 333
    transformation, 548
Equilateral triangle, 73,
    117–118, 196
    corollary, 197
Equivalence properties, 94
Equivalence relation, 95

Equivalence, topological, 608
Einstein, Albert, 618
Escher, M.C., 115, 129, 165, 193,
    614, 617
Euclid, 88, 143, 468, 685
    fifth postulate, 615
    postulates, 614
Euler, 600
    circuit, 605
    lines, 27
Euler's Formula, 322, 610
Exclusive "or", 651
Explorations,
    Discovering Geometry Ideas in a
        Model, 12
    Perpendicular Lines and Parallel
        Lines, 17
    Segment and Angle Bisectors, 18
    Some Special Points in Triangles,
        24
    Constructing Special Circles, 25
    Linear Pairs, 40
    The Reflection of a Point, 46
    Reflections of Segments and
        Triangles, 47
    Translation, 52
    Reflection Through the x- or y-
        Axis, 53
    180° Rotation About the Origin,
        54
    Three Challenges, 67
    Capturing the "Essence of the
        Thing", 83
    Creating Your Own Objects, 84
    The Vertical Angles Conjecture,
        99
    Reflections Through Parallel
        Lines, 101
    Reflections Through Intersecting
        Lines, 102
    Axes of Symmetry in Triangles,
        117
    Rotational Symmetry in Regular
        Polygons, 119
    Parallelograms, 125
    Rhombuses, 125
    Rectangles, 126
    Squares, 126
    Special Angle Relationships, 133
    The Triangle Sum Theorem, 144
    Angle Sums in Polygons, 149
    Exterior Angle Sums in Polygons,
        151

Midsegments of Triangles, 154
Midsegments of Trapezoids, 155
Making the Connection, 156
Polygon Congruence, 174
Triangle Rigidity, 180
Two More Congruence Postulates, 181
Parallelograms, Rectangles, and Rhombuses, 207
Midpoints of Horizontal Segments, 221
Midpoints of General Segments, 221
Translating a Segment, 226
Congruence in Polygon Translations, 227
Betweeness in Translations, 228
Fixed Perimeter, Maximum Area, 246
Fixed Area, Minimum Perimeter, 247
Areas of Triangles, 254
Areas of Parallelograms, 255
Areas of Trapezoids, 256
The Circumference of a Circle, 262
The Area of a Circle, 262
Solving the Puzzle, 267
30-60-90 Right-Triangles, 277
Estimating the Area of a Circle, 285
A "Monte Carlo" Method for Estimating π, 291
Using Isometric Grid Paper, 305
Using Unit Cubes, 305
Volume and Surface Area, 307
Parallel Lines and Planes in Space, 311
Segments and Planes, 312
Measuring Angles Formed by Planes, 313
The Lateral Faces of Prisms, 318
Ratio of Surface Area to Volume, 354
Maximizing Volume, 355
Pyramids and Prisms, 368
Analyzing the Volume of a Cylinder, 375
The Surface Area of a Right Cone, 380
The Surface Area Formula for a Cone, 381
Reflections in Coordinate Space, 396

Reflectional Symmetry in Space, 398
Rotational Symmetry in Space, 399
Transforming a Point Using a Scale Factor, 413
Constructing a Dilation, 415
The Position of Transformed Points, 415
Defining Similar Polygons, 419
Similar Triangles Postulate 1, 429
Similar Triangles Postulate 2, 430
Similar Triangles Postulate 3, 431
Ratios of Areas of Similar Figures, 452
Ratios of Volume, 453
Increasing Height and Volume, 455
Constructing a Regular Hexagon, 471
Constructing a Circle Flower, 472
Chords and Arcs Theorem, 474
Radii Perpendicular to Chords, 479
Radii and Tangents, 480
Intercepted Arcs, 485
Proving the Relationship, 485
Vertex on Circle, Secant-Tangent Angles, 492
Lines Intersect Inside a Circle, 493
Vertex of an Angle Is Outside a Circle, 493
Lines Formed by Tangents, 503
Lines Formed by Secants, 503
Intersection of Chords Inside a Circle, 504
A Familiar Ratio, 524
Graphing the Tangent Ratio, 525
Using the Tangent Ratio, 526
Sines and Cosines, 534
Two Trigonometric "Identities," 541
Redefining the Trigonometric Ratios, 543
Graphing the Trigonometric Ratios, 544
Rotating a Point, 549
Combining Rotations, 551
Law of Sines, 555
The Dimensions of the Golden Rectangle, 587
Seashells, 587

Constructing a Golden Rectangle, 589
Exploring Taxidistances, 595
Taxicab Circles, 596
Networks, 601
Affine Transformations, 622
Two Projective Geometry Theorems, 624
Creating a "Cantor Dust," 631
Pascal and Sierpenski, 632
The Negation of a Conjunction, 652
On-Off Tables, 671
Triangle Midsegment Theorem Revisited, 679
The Diagonals of a Parallelogram, 680
Reflection About the Line $y = x$, 681
Exchange property of proportion, 422, 423
Extensions, 84, 182, 209, 297, 345, 479, 487, 488, 495, 542
Exterior angle(s), 147, 151, 170
   theorem, 147–148
Eyewitness Math, 80, 130, 252, 323, 394, 500, 562, 656

Face of a prism, 318
Fibonnaci sequence, 79, 592
Flexagons, 232
Formal proof, 68, 93, 100 (*See also Proof, two-column*)
Formulas
   area of rectangle, 245
   distance in three-dimensions, 328
   distance, 284
   Euler's, 322
   lateral area of cylinder, 374
   lateral area of prism, 360
   lateral area of pyramid, 367
   midpoint, 285
   slope, 161
   surface area of cone, 381
   surface area of cylinder, 374
   surface area of prism, 360
   surface area of pyramid, 367
   surface area of sphere, 390
   volume of cone, 382

volume of cylinder, 376
volume of prism, 362
volume of pyramid, 369
volume of sphere, 389
writing, 150, 151, 153, 156
Fractal geometry, 585, 629–632
Function, 251

Garfield, President, 273
Gates, logic, 673–674
Gauss, Karl Friedrich, 615, 685
Geodesic dome, 390
Geometry
   coordinate, 51–54, 62, 160–162,
      220–222, 365, 396–400,
      509–512, 548–551, 679–681
   fractal, 585, 629–632
   hyperbolic, 616, 620
   Lobachevskian, 616
   non-Euclidean, 145, 614–618
   projective, 623–625
   Riemannian, 616
   spherical, 616
   taxicab, 594–596
Geometry graphics (See Technology)
Golden ratio, 585, 586, 588
Golden rectangle, 5, 586
   constructing, 589
   dimensions of, 587
Graph
   of a circle, 509–512
   of a plane, 332
Graph paper exercises, 52, 53, 54,
   162, 220–224, 246, 247,
   254–256, 285–286, 413, 534,
   549, 551, 593, 595, 596, 622,
   632, 679, 680, 681
Graph theory, 603
Graphics calculator (See Technology)
Great circle, 145, 616

Hands-on activities (See Models)
**Head-to-tail method of vector
   addition**, 572

Heptagon, 117
Heron's formula, 260
Hexaflexagon, 232
Hexagon, 117, 150, 151, 165
   area of regular, 278
   construction of, 471
Hexagonal prism, 317
Hexahedron, 404
Hexahexaflexagon, 233
Hyperbola, 499
Hyperbolic geometry, 616, 620
Hypotenuse, 189, 269, 275–277,
   524–525, 533–536
Hypotenuse-leg congruence
   postulate (HL), 189
Hypothesis, 72, 138, 269

Ibn al Haitham, 412
Icosahedron, 404
Identities, trigonometric, 536
Identity matrix, 550
If-and-only-if statements, 82, 663
If-then statements, 72–76
If-then transitive property, 648
Image of transformation, 44, 51–54,
   62, 397, 412
Incenter, 25
Inclusive "or", 651
Incommensurability of the square
   root of 2, 686
Inconsistent systems of equations,
   161
Indirect measurement, 428, 441,
   443–444, 46–461
Indirect proof, 483, 665, 667
Indirect reasoning
   law of, 646
Inductive reasoning, 79, 99, 100
Inequalities, 270
Informal proof, 66–67, 101, 108,
   145
Input-output table, 673–674
Inscribed angle, 484–488
   corollary, 487
   measure of, 487
   theorems, 487
Inscribed circle, 24, 61

Intercepted arc, 484
Intercepts, 331, 509–510
Intersecting lines, 12, 102
Invariants, 609, 621
Inverse statement, 660
Islamic art, 8
Isometries, 228
Isosceles trapezoid, area of, 253
Isosceles triangle, 73, 117–118,
   196
   base angles of, 196
   base of, 196
   legs of, 196
   theorem, 196
   vertex angle of, 196, 197

Jordan, Camille, 609
   theorem, 609

King Tut's Tomb, 394
Kite, 16, 205, 259
   construction of, 218
Klein bottle, 613
Knot designs, 8
   theory of, 607
   trefoil, 607
Koch Snowflake, 585, 633

Lateral area
   of prism, 359, 360
   of cone, 379
   of cylinder, 374
   of pyramid, 367
Lateral face of prism, 318
Lateral surface of cylinder, 373
Law of Cosines, 564

proof of, 567
Law of Sines, 556
Legs
    of isosceles triangle, 196
    of right triangle, 189, 269,
        276–277
Length
    of arc, 473
    of segment, 220
Line segment *(See Segment)*
Line(s), 10
    formed by secants, 503
    formed by tangents, 503
    intersecting, 102
    parallel to a plane, 313
    parallel, 17, 101, 132, 135, 139,
        161
    perpendicular to a plane, 312
    perpendicular, 17, 162
    skew, 312
    slope of, 160–161, 414, 415
Linear pair, 40, 100
Lobachevski, 616
Logic gates, 673–674
Logic puzzles, 106–107
Logical chain, 75
Logical conclusions, 73
Logical reasoning, 72–76
Lune, 243

Ma, Yo Yo, 74
Magnitude of a vector, 570
Major arc, 472
Mandalas, 7
Manipulatives *(See Models)*
Math Connections
    algebra, 15, 32, 34, 36, 41, 42, 51,
        67, 68, 69, 70, 93, 100, 122,
        137, 141, 153, 161, 164, 206,
        246, 247, 249, 251, 254, 255,
        256, 257, 259, 262, 266, 269,
        270, 271, 272, 273, 274, 275,
        276, 278, 284, 285, 288, 292,
        331, 332, 333, 334, 354, 367,
        374, 381, 384, 388, 415, 417,
        422, 423, 425, 428, 435, 444,
        445, 446, 447, 455, 459, 473,

476, 477, 479, 482, 483, 489,
    495, 496, 504, 506, 507, 508,
    509, 512, 513, 514, 526, 537,
    548, 557, 565, 567, 568, 573,
    588, 631, 632
    coordinate geometry, 179
    maximum/minimum, 258, 265
    pre-calculus, 288
    probability, 290, 292, 428, 635
    statistics, 35, 476
Matrix
    identity, 550
    rotation, 550
Maximizing volume, 355, 385
Maximum/minimum problems,
    246–247, 258, 264, 265
Mayan calendar, 656–657
Measure
    indirect, 428, 441, 443–444,
        460–461
    of angle, 37–38, 90
    of arc, 473
    of inscribed angle, 487
    of segment, 30
    of special angles associated with
        a circle, 491–497
Median(s)
    of a triangle, 23
    of similar triangles, 446, 449
Menger Sponge, 629
Mercator projection, 623
Midpoint
    formula, 285
    of a segment, 221–222
Midsegment
    of trapezoid, 155, 170
    of triangle, 154, 170, 212, 679
Minor arc, 472
Mirror of transformation, 230
Models, 16–19, 28, 36, 181, 207,
    220–222, 246–247, 253,
    262–263, 291, 313, 318–319,
    325, 364, 368–369, 404–405,
    611, 613
    dot paper, 296–297
    graph paper 52, 53, 54, 162,
        220–224, 246, 247, 254–256,
        285–286, 413, 534, 549, 551,
        593, 595, 596, 622, 632, 679,
        680, 681
    grid paper, 291, 305, 354
    orthographic projections, 306
    paper cutting, 28, 36, 40, 144,

151, 174, 180–182, 253,
    255–256, 262–263, 313, 368,
    404–405, 611, 613
    paper folding, 16–19, 24, 25, 47,
        58–59, 101, 102, 117–118,
        232–233, 313
    paper, tracing, 119
    scale, 461
    unit cubes, 305–306, 354
Modus ponens, 645
Modus tollens, 646
Möbius strip, 585, 607, 611
Mondrian, Piet, 5
**Monte Carlo method for esti-
    mating pi**, 291

N-gon, 117, 150, 151
Negation, 652
Nets, 49, 359, 370, 508
Network, 600
Nim, 65, 71, 98
Nine Chapters on the Mathematical
    Art, 252–253
Non-Euclidean geometries, 145,
    614–618
Nonagon, 117
Noncollinear, 11
NOT gate, 673
Notre Dame, 468
Number line, 29
Number theory, 685

Oblique
    cone, 379
    cylinder, 373
    prism, 319
    pyramid, 366
Obtuse triangle, 555
Octagon, 117
Octahedron, 404
Octants, 327, 397
On-off tables, 671–672
OR gate, 673

Ordered
   pairs, 55
   triples, 327
Origami, 16, 19, 22, 58–59
Orthocenter, 25
Orthogonal, 617
Orthographic projections, 306, 309, 310
Overlapping
   angles theorem, 94
   segments theorem, 94, 195
   triangles, 195

Paper cutting, 28, 36, 40, 144, 151, 174, 180–182, 253, 255–256, 262–263, 313, 368, 404–405, 611, 613
Paper folding, 16–19, 24, 25, 47, 58–59, 101, 102, 117–118, 232–233, 313
Paper, tracing, 119
Pappus, theorem of, 624
Parabola, 499
Paragraph proof, 202, 203
Parallax, 324
Parallel lines, 17, 101, 132, 135, 139, 169
   construction of, 218
   in perspective drawings, 338
   in space, 311–312
   theorem, 141, 142
   to a plane, 313
Parallel planes, 312
Parallel postulate, 143, 615
Parallel vectors, 571
Parallelogram(s), 61, 124, 125, 201–203
   area of, 255–256
   conjectures about, 208
   diagonals of, 211, 680
   that is a rhombus, 211
   vector addition using, 572
Parametric equations, 333
Parthenon, 5, 586
Pascal's triangle, 632
Pentagon, 117, 119, 150 151, 586
   constructing, 591
Pentagonal prism, 317
Percentile, 35

Perimeter, 244
   minimum, 247
Perpendicular bisector, 18, 23, 60, 197,
   construction of, 217
Perpendicular lines, 17, 217
   in space, 311–312
   slopes of, 162
   to a plane, 312
Perspective drawing, 336–339, 342
Pi, 262–263
   estimating, 291
Planes, 11, 12
   graphs of, 332
   parallel, 312
   trace of, 334
Plato, 272
Platonic solids, 404
Plimpton 322, 267–268, 576–577
Poincare, Henri, 617
Point of tangency, 478
Points, 10
Polygon(s), 116
   area of, 296–297
   area of regular, 279
   center of, 119
   central angle of, 119
   concave, 149
   congruent, 174–176
   convex, 149
   naming, 175
   regular, 119, 150
   rotating, 551
   similar, 419–420
   sum of angles of, 150
Polyhedron (polyhedra), 359, 404
   regular, 404
Portfolio Activities, 3, 6–8, 27, 50, 65, 71, 98, 115, 129, 165, 173, 241, 281, 293, 303, 342, 353, 364, 377, 378, 384, 411, 428, 475, 476, 515, 523, 585, 611, 643, 670
Postulate(s), 88, 181, 182
   angle addition, 39
   area of rectangle, 245
   ASA, 186
   betweenness, 228
   corresponding angles, 134
   Euclid's 614, 615
   HL, 189
   list of basic, 88
   parallel, 143, 615

   SAS, 186
   segment addition, 32
   similar triangles, 436, 437
   SSS, 186
   sum of areas, 245
   triangle inequality, 229
Pre-image of transformation, 44, 51–54, 62, 412
Premises, 644
   false, 645
Primes, 685
Prisms, 317–319
   altitude of, 359
   base of, 318
   diagonal of, 319
   edge of, 318
   face of, 318
   hexagonal, 317
   lateral area of, 359, 360
   lateral faces of, 318
   oblique, 319, 362
   pentagonal, 317
   rectangular, 317
   right, 319, 359
   surface area of, 359, 360
   triangular, 317
   vertex of, 318
   volume of, 360, 362
Probability, 289–291, 428
Projection, center of, 623
Projections, 623
   orthographic, 306
Projective geometry, 623–625
   theorems, 624–625
Projective rays, 623
Projects (See Chapter Projects)
Proof, 102
   by contradiction, 666
   formal, 68, 93, 100 (See also Two-column proof)
   indirect, 483, 665, 667
   informal, 66–67, 101, 108, 145
   of "Pythagorean" Right-Triangle Theorem, 269, 273
   need for, 99
   paragraph form, 141, 142, 202, 203
   two-column, 68, 100, 105, 135, 145, 148, 194, 195, 197, 198, 202, 210, 211, 212, 437, 450, 451
Properties
   if-then transitive, 76, 648

of equality, 92–93
of proportions, 422
reflexive of equality, 94
substitution, 93–93, 100
symmetric of equality, 94
transitive of equality, 94
Proportional, 420
altitudes, 445
angle bisectors, 450
medians, 446, 449
segments, 451
sides, 436–438
Proportionality, scale factor of, 421
Proportions, properties of, 422
Protractor, 37, 41, 46, 47, 52, 53, 54,
102, 125, 126, 133, 151, 154,
155, 156, 174, 181, 207, 262,
419, 429, 430, 431, 479, 480,
485, 516, 524, 525, 526, 549,
551, 555, 593
Prove
parallelogram is a rectangle, 211
quadrilateral is a parallelogram,
210, 211
Puzzles, logic, 106, 107
Pyramid
altitude of, 366
base of, 366
lateral area of, 367
lateral faces of, 366
oblique, 366
of Cheops, 370
of Gisa, 353
regular, 366
right, 366
slant height of, 367
surface area of, 367
vertex of, 366
volume of, 369
Pythagoras, 268, 592
"Pythagorean" Right-Triangle
Theorem, 268, 269, 273
using, 270, 274, 275, 276, 277
Pythagorean triples, 268, 272,
576–577

Quadrature, 285
Quadrilateral(s), 117, 124,
168–169

classifying, 124
exterior angles of, 151
kite, 205
sum of angles of, 150

Radius (radii)
congruent, 213
of circle, 261, 470
of inscribed circle, 253
Ratio(s)
golden, 585, 586, 588
of areas of similar figures, 452
of surface area to volume,
354–355
of volumes of similar figures, 453
Raw score, 35
Ray, 10, 60
Reasoning *(See also Critical Thinking)*
deductive, 73, 99, 100, 110
inductive, 79, 99, 100
Reciprocal property of proportion,
422, 423
Rectangle, 124, 126
area of, 245
conjecures about, 208
diagonals of, 205
golden, 586
Rectangular prism, 317
Reductio ad absurdum, 665
Reflection
image through a plane, 397
of a point, 46, 397, 681
of a segment, 47, 62
of a triangle, 47, 53, 101
through intersecting lines, 102
through parallel lines, 101
Reflectional symmetry, 117,
168, 398
**Reflexive property of congru-
ence**, 176
of equality, 94
Regular polygon, 119, 150
Remote interior angles, 147
Resultant vector, 571
Revolutions in space, 399
Rhombus, 124, 125, 204–205
conjectures about, 208
diagonals of, 204–205
Riemann, G.F.B., 616

Right angle, 17, 109
Right cone, 379
Right cylinder, 373
Right prism, 319
Right pyramid, 366
Right triangle, 269
hypotenuse of, 269, 275–277,
524–525, 533–536
legs of, 269, 276–277
special, 275–277
Right-handed system of axes,
326
Rigid tranformation, 44
Rigidity, triangle, 180
Rise and run, 160
Rotating
a point, 549
poygons, 551
Rotation, 45, 54, 62
combining, 551
image, 118, 542
matrices, 550
tessellation, 165
turn center, 45
Rotational symmetry, 118–119,
168, 201, 399
in regular polygons, 119
Ruler, 29

Scale factor
of proportionality, 421
of tranformation, 412–413
Scale models, 461
Scalene triangle, 117–118
Scientific calculator *(See Technology)*
Secant(s), 478
segments of, 502, 504
special angles formed by,
491–495
Secant-tangent
angle, 492
theorem, 493
Sector of a circle, 262
Seed point, 636
Segment(s), 10, 60
addition postulate, 32, 93
bisector, 18
congruent, 31, 61, 95, 213, 220
dividing into equal parts, 442

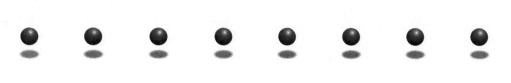

measure of, 30, 220
midpoint of, 221–222
proportional, 451
reflection of, 47
secant, 502, 504
tangent, 502, 503
translating, 226
Self-similarity, 629
Semicircle, 472
Sequences, 69,
  Fibonacci, 79, 592
Seven Bridges of Königsberg, 600,
  603
Side-angle-side postulate (SAS), 186
Side-angle-side similarity postulate,
  437
Side opposite, 524
Side-side-side postulate (SSS), 186
Side-side-side similarity postulate, 436
Side-splitting theorem, 437
Sierpinski Gasket, 585, 632
Similar
  polygons, 419–420
  triangles, 429–432, 436, 437
Similar figures, 410, 419
  ratio of areas of, 452
  ratio of volumes of, 453
Similar triangles, parts of, 445, 446
Simple closed curve, 608
Sine(s), 533, 543
  Law of, 556
Skew lines, 312
Slant height
  of cone, 380, 381
  of pyramid, 367
Slide arrow, 45
Slide rule, 36
Slope(s)
  equal, theorem, 161
  formula, 161
  of a line, 160–161, 170, 414, 415
  of parallel lines, 161
  of perpendicular lines, 162
Solid of revolution, 399
Solids, Platonic, 404
Spatial contradictions, 309
Sphere
  surface area of, 389–390
  volume of, 387–389
Spherical geometry, 616
Spreadsheet, 223
Square, 124, 126
  diagonal of, 275

Star of Pythagoras, 585, 592
Statement, compound, 650
Stonehenge, 540
String figures, 166–167
Substitution property, 93–93, 100
Subtraction property, 100
Sulbastutras, 272
Sum of vectors, 571
Supplementary angles, 39, 100
Surface area, 307
  nets for, 359
  of cone, 380–381
  of cylinder, 374
  of prism, 359, 360
  of pyramid, 367
  of sphere, 390
  ratio to volume of, 354
**Symmetric property of equal-
  ity**, 94
Symmetry, 116
  axis of, 117
  reflectional, 117, 398
  rotational, 118–119, 168, 399

Tables
  input-output, 673–674
  on-off, 671–672
  truth, 650, 651, 652, 653, 654,
    659, 660
Tangent(s), 478
  angle between, 497
  segment, 502, 503
  segments from external point,
    503
  special angles formed by,
    491–495
  theorems, 481, 483
  to a circle, 478
Tangent ratio, 525, 526, 533, 543
Taxicab
  circle, 596
  geometry, 594–596
  radius, 596
Taxidistance, 594–595
Technology
  geometry graphics, 24–26, 99,
    101, 102, 125, 126, 129, 133,
    154, 155, 156, 173, 174, 180,

    181, 214, 224, 246, 247, 288,
    429, 430, 431, 479, 480, 485,
    503, 504, 524, 525, 526, 555,
    593, 624, 679, 680
  graphics calculator, 224, 246,
    247, 354, 355, 385, 509, 551,
    552, 553, 554, 637
  scientific calculator, 149, 267,
    285, 413, 452, 453, 524, 526,
    529, 530, 534, 536, 538, 539,
    541, 543, 549, 555, 587
  spreadsheet, 223, 246, 247, 260,
    285, 288, 354
Tessellation, 115, 150, 165
  rotation, 165
  translation, 129
Tetrahedron, 404
Theorems, 88, 141, 142, 197–198,
    210–212, 482, 487, 493, 494,
    497, 504, 505
  30,60,90 right triangle, 277
  45-45-90 right triangle, 276
  AAS, 188
  alternate interior angles, 135
  chords and arcs, 474
  congruent radii, 213
  converse of corresponding
    angles, 139
  converse of "Pythagorean" Right-
    Triangle Theorem, 270
  converse of tangent, 483
  diagonals of a parallelogram,
    204, 680
  diagonals of rectangle, 205
  diagonals of rhombus, 204, 205
    equal slope, 161
  exterior angle, 147
  "housebuilder" rectangle, 211
  inscribed angle, 487
  isosceles triangle, 196
  Jordan's, 609
  of Desargues, 625
  of Pappus, 624
  overlapping angles, 94
  overlapping segments, 94, 195
  proportional altitudes, 445
  proportional angle bisectors, 450
  proportional medians, 446, 449
  proportional segments, 451
  Pythagorean, 269, 273
  side-splitting, 437
  slopes of perpendicular lines,
    162

supplements of congruent
  angles, 105
tangent, 481
triangle midsegment, 680
triangle sum, 144–145
vertical angles, 100
Three-dimensional coordinates,
  326–328, 396–400
Topology, 603, 607
Topological equivalence, 608
Torus, 609
Trace of a plane, 334
Transformation(s)
  affine, 621–623
  dilations, 414–415
  equations of, 548
  image of, 44, 51–54
  mirror of, 230
  pre-image of, 44, 51–54
  reflections, 46–47, 53, 62, 101
  rigid, 44, 228
  rotations, 45, 54, 62, 549
  scale factor of, 412–413
  slide arrow of, 45
  translations, 44, 62, 554
  turn center of, 45
Transit, 90
Transitive property
  if-then, 76, 648
  of congruence, 135
  of equality, 94
Translating
  a circle, 514
  a polygon, 227
  a segment, 226
Translations, 44–45, 52, 62
Transversal, 132, 134–135, 169
  divided proportionally, 439
Trapezoid, 124
  altitude of, 256
  area of, 253, 256, 259
  midsegment of, 155, 170, 259
Traveling salesman problem, 80–81
Trefoil knot, 607
Triangle(s)
  acute, 555
  altitude of, 23
  area of, 254–255, 260
  congruent, 53, 180–182,
    186–189, 201–202
  constructing congruent, 214, 216

equilateral, 73, 117–118, 196
exterior angles of, 151
hypotenuse of, 269
inequality postulate for, 229
isosceles, 73, 117–118, 196
median of, 23
midsegment of, 154, 170, 212, 679
obtuse, 555
overlapping, 195
reflection of, 47, 53
right, 269
rigidity of, 180–181
similar, 429–432, 436, 437
special right, 275–277
sum of angles of, 144, 150, 169
using law of cosines to solve,
  565–566
using law of sines to solve,
  556–558
Triangular prism, 317
Triangulation, 186
Trigonometry, 525, 522–577
Trigonometric ratios, 533, 542–543
  graphs of, 544
  on a calculator, 541
Trigonometric identities, 536
Tripod, 90
True conditional, 73
Truth table, 650, 651, 652, 653, 654,
  659, 660
Try This, 30, 47, 74, 89, 94, 95, 102,
  134, 189, 196, 203, 209, 228,
  262, 263, 278, 279, 285, 290,
  328, 400, 421, 430, 431, 432,
  454, 512, 527, 545, 551, 572,
  596, 645, 646, 661
Turn center of rotation, 45
Two-column proof, 68, 100, 105,
  135, 145, 148, 194, 195, 197,
  198, 202, 210, 211, 212, 437,
  450, 451
Two transversal proportionality
  corollary, 439

Vanishing point, 337
Vector(s), 570
  addition of, 571, 572
  direction, 570
  magnitude of, 570
  resultant, 571
  sum of, 571
Venn diagrams, 72–74, 83, 126
Vertex
  of cone, 366
  of prism, 318
  of pyramid, 366
  on a circle, 492
  outside a circle, 493
**Vertex angle of isosceles trian-
  gle**, 196
Vertical angles, 99–100, 108, 135
Vertices
  even, 601
  odd, 601
Volume, 307
  Cavalieri's Principle of, 362
  maximizing, 355
  of cone, 382
  of cylinder, 376
  of prism, 360, 362
  of pyramid, 369
  of sphere, 387
  ratio of to surface area, 354
  ratios of in similar figures, 453

Warhol, Andy, 174

Unit circle, 542
Unit length, 31

Yin-Yang symbol, 266, 469
Yucatan Peninsula, 502

CREDITS

PHOTOS

Abbreviations used: (t) top, (c) center, (b) bottom, (l) left, (r) right, (bckgd) background, (bdr) border.
FRONT COVER: (l), Julie Vanderchmitt/AllSport, France; (c), Homer Smith © 1994 Art Matrix; (r), David Ladd Nelson. TABLE OF CONTENTS: Page vi(tr), Dennis Fagan/HRW Photo; vi(cr), Michael Thompson/COMSTOCK; vi(cr), Denis Valentine/The Stock Market; vi(cr), COMSTOCK; vi(bl), Dennis Fagan/HRW Photo; vi(br), Coco McCoy/Rainbow; vii(tr), Gary A. Conner/PhotoEdit; vii(bl), Tomas Pantin/HRW Photo; vii(br), Alex Bartell/Science Photo Library/PhotoResearchers, Inc.; viii(tr), George Whiteley/Photo Researchers Inc.; viii(cl), Guy Sauvage/Photo Researchers, Inc.; ix(tl), Bob Daemmrich/The Image Works; ix(cr), Tony Freeman/PhotoEdit; ix(br), Bob Daemmrich/Stock, Boston; x(tl), Dennis Fagan/HRW Photo; x(cr), R. Sydney/The Image Works; x(br), Okoniewski/The Image Works; xi(tl), Tomas Pantin/HRW Photo; xi(cr), xi(bl), Dennis Fagan/HRW Photo; xi(br), H. L. Romberg. CHAPTER ONE: Page: 2-3(t), COMSTOCK; 2(c), COMSTOCK; 2-3(b), Scott Van Osdol/HRW Photo; 4(t) (c), Georgia Tblisi/Sovfoto/Eastphoto; 4-8(bckgd), Scott Van Osdol/HRW Photo; 5(tl), Lisa R. Glass/Still Life Stock; 5(cr), Giraudon/Art Resource; 5(bl), Judy Canty/Stock, Boston; 6(tl), Planet Art Collection; 7(c), Dennis Fagan/HRW Photo; 7(br), Russell Dian/Monkmeyer Press Photo; 8(cl), Macduff Everton/The Image Works; 8(cr), Addison Geary/Stock, Boston; 9(bckgd), (tr), (bl), Fermilab Visual Media Services; 10-11(bckgd), Magrath Photography/Science/Photo Library/Photo Researchers Inc.; 12(bl), Myrleen Ferguson Cate/PhotoEdit; 13(tr), Lowell Georgia/Photo Researchers, Inc.; 14(tr), Thomas J. Magno/Still Life Stock, Inc.; 15(br), 16(t), (br), 17(all), Dennis Fagan/HRW Photo; 18(tl), Mug Shots/The Stock Market; 18(bl), Dennis Fagan/HRW Photo; 19(cl), (bl), (br), Dennis Fagan/HRW Photo; 20(br), Michael Newman/PhotoEdit; 21(cr), 22(cr), (br), Dennis Fagan/HRW Photo; 23(tr), Dassault-Breguet/Science Photo Library/Photo Researchers, Inc.; 24(br), Rogers Fund, 1918/The Metropolitan Museum of Art; 25(br), Scott Van Osdol/HRW Photo; 28(tr), Dennis Fagan/HRW Photo; 28(br), 29(tr), Dennis Fagan/HRW Photo; 29 D.P. Wilson/Science Source/Photo Researchers, Inc.; 30(br), COMSTOCK; 33(tr), Scott Van Osdol/HRW Photo; 34(br), Courtesy of Mathematics in the Time of the Pharaohs, by Gillings; 35(br), Courtesy of USGS; 36(cr); 36(br), Dennis Fagan/HRW Photo; 37(tr), Joe Towers/The Stock Market; 37(tl), Ron Sherman/Stock, Boston; 37(br), (bckgd), Dennis Fagan/HRW Photo; 40(br), Wellzenbach/The Stock Market; 42(bckgd), Mark M. Lawrence/The Stock Market; 43(br), Babylonian Collection (YBC 7289)/Yale University; 44(bl), David Ulmer/Stock, Boston; 44(br), Lisa R. Glass/Still Life Stock, Inc.; 44(tr), F. Gohier/Photo Researchers, Inc.; 44(bl), Thomas Zimmerman/FPG International; 45(br), Alex Bartell/Science Photo Library/PhotoResearchers, Inc.; 46(tr), 48(cl), (cr), Dennis Fagan/HRW Photo; 48(bckgd), Stephen Kaufman/Tony Stone Images; 49(cr), Paul Steel/The Stock Market; 49(br), Rod Planck/Photo Researchers, Inc.; 50(tr), Courtesy of Quarto Publishing, England; 50(br), Frank Siteman/Stock, Boston; 51(t), (bckgd), Vee Sawyer/MertzStock; 57(cr), Building in Egypt, by Dieter Arnold; 58(all), 59(all), Dennis Fagan/HRW Photo. CHAPTER TWO: Page 64-65(Bckgd), Chris Noble/Tony Stone Images; 65(br), (tl), 66(tr), 67(tr), 68(br), 69(tr), 71(br), Dennis Fagan/HRW Photo; 72(tl), Michael Thompson/COMSTOCK; 72(t), (c), Denis Valentine/The Stock Market; 72(tr), COMSTOCK; 73(tl), Gerald Gusthall/FPG International; 74(tl), Courtesy of Karen Kuehn/Sony Music; 76(br), NASA/The Image Works; 77(tl), Andy Sacks/Tony Stone Images; 78(bl), Tate Gallery, London/Art Resource; 78(br), Art Resources, N.Y.; 79(b), Joseph Sachs/Still Life Stock, Inc.; 80-81(bckgd), George Diebold/The Stock Market; 82(tr), Dennis Fagan/HRW Photo; 87(br), Michael Newman/PhotoEdit; 88(tr), Llewellyn/Uniphoto Picture Agency; 90(tr), Coco McCoy/Rainbow; 92(tr),Vee Sawyer/MertzStock; 95(cr), Ernest Manewal/FPG International; 96(tr), Randy Duchaine/The Stock Market; 97(br), Michael Newman/PhotoEdit; 98(br), Dennis Fagan/HRW Photo; 99(bckgd), (br), 105(cr), Dennis Fagan/HRW Photo; 107(br), Courtesy of Broderbund Software; 107(br), Dennis Fagan/HRW Photo; 107(bl), Scott Van Osdol/HRW Photo; 106-107(b), (t), Sam Dudgeon/HRW Photo. CHAPTER THREE: Page 114-115(bckgd), Photo Edit; 114(t), Greg Hursley; 115(tl), COMSTOCK; 115(br), Murray Alcosser/The Image Bank; 116(t), Coco McCoy/Rainbow; 118(tr), Dennis Fagan/HRW Photo; 118(bl), IPA/The Image Works; 119(bl), Color Box 1992/FPG International; 120(cl), Steve Elmore/The Stock Market; 120(br), © 1995 M.C. Escher/Cordon Art–Baarn–Holland. All rights reserved; 123(b), (bckgd), Vee Sawyer/MertzStock; 124(t), Jerry Wachter/Photo Researchers, Inc.; 124(bl), © Jim Richardson; 127(tr), Dennis Fagan/HRW Photo; 128(bl), Michael Newman/PhotoEdit; 128(tl), Michael Newman/PhotoEdit; 128(b), (c), VanBucher/Photo Researchers, Inc.; 128(tr), Scott Camazine/Photo Researchers, Inc.; 128(br), Chromosohn/The Stock Market; 129(tr), © 1995 M.C. Escher/Cordon Art–Baarn–Holland. All rights reserved; 130-131(bckgd), Annette Del Zoppo; 130(tl), (bl), (b) Annette Del Zoppo; 131(br), Annette Del Zoppo; 130-131(c), Robin White/Fotolex Associates; 132(bckgd), Robert Rathe/Scock, Boston; 132(tr), Joseph Stella, Bridge, 1936. Oil on canvas; 50-1/8 x 30-1/8 (127.3 x 76.5 cm). San Francisco Museum of Modern Art, WPA /Federal Arts Project allocation to the San Francisco Museum of Art 3760.43; 134(bl), Robert Frerck/Tony Stone Images; 135(t), (bckgd), Gabe Palmer/Mug Shots/The Stock Market; 136(tr), Peter Pearson/Tony Stone Images; 138(tr), Courtesy of NASA; 140(tr), 142(tr), Dennis Fagan/HRW Photo; 143(bl), Tony Stone Images; 143(t), (bckgd), Blair Seitz/Photo Researchers, Inc.; 143(br), Crandall/The Image Works; 144(tr), Dennis Fagan/HRW Photo; 145(bl), Tony Craddock/Science Photo Library/Photo Researchers, Inc.; 146(tr), Dennis Fagan/HRW Photo; 149(tr), Tom Sanders/The Stock Market; 152(tr), Brownie Harris/The Stock Market; 153(br), © Jim Richardson; 154(bckgd), Larry Nelson/Stock, Boston; 155(tr), Bob Daemmrich/Stock, Boston; 155(tl), Harvey Lloyd/The Stock Market; 157(cr), HRW Photo/Tomas Pantin; 158(bckgd), Lester Lefkowitz/Tony Stone Images; 159(tr), Chris Rogers/Rainbow; 160-361(bckgd), Joseph Nettis/Stock, Boston; 160(tr), Hank Morgan/Rainbow; 162(bckgd), EuniceHarris/Photo Researchers, Inc.; 164(tr), Steve Weber/Stock, Boston; 165(br), Tomas Pantin/HRW Photo; 165 TessellMania!; 165(tr), © 1995 M.C. Escher/Cordon Art–Baarn–Holland. All rights reserved; 166(tr), (bl), Dennis Fagan/HRW Photo; 166(b), (c), (br), 167(all), Dennis Fagan/HRW Photo. CHAPTER FOUR: Page 172(cr), Mark Joseph/Tony Stone Images; 172-173(t), McConnell, McNamara & Company; 173(c), Ron Kimball; 172-173(b), Art Wolfe/Tony Stone Images; 174(t), Andy Warhol, 100 Cans. 1962; Synthetic polymer paint and silkscreen, ink on canvas. © The Andy Warhol Foundation. Collection of Albright-Knox Art Gallery. Gift of Semour Knox; 176(bl), Richard Payne; 178(br), Gary A. Conner/PhotoEdit; 179(br), Dennis Fagan/HRW Photo; 180(t), G. Zimbel/Monkmeyer; 180-181(bckgd), Lee Snider/The Image Works; 183(tr), Tomas Pantin/HRW Photo; 184(tl), M. Antman/The Image Works; 184-185(bckgd), Cary Wolinsky/Stock, Boston; 186(bl), Sullivan/Texastock; 186(tr), Daniel Brody/Stock, Boston; 192(b), Landform Slides; 193(tr), (bl), © 1995 M.C. Escher/Cordon Art–Baarn–Holland. All rights reserved; 197(br), Tomas Pantin/HRW Photo; 200(tl), (tr), Michael Newman/PhotoEdit; 201 HRW Photo; 203(cr), Dennis Fagan/HRW Photo; 206(b), Dennis O'Clair/Tony Stone Images; 207(tr), Scott Van Osdol/HRW Photo; 208(tl), Dennis Fagan/HRW Photo; 210(br), 211(tr), Dennis Fagan/HRW Photo; 213(tr), Scott Van Osdol/HRW Photo; 216(t) (bckgd), Michael Newman/PhotoEdit; 220(tr), Tomas Pantin/HRW Photo; 222(tl), 224(br), Dennis Fagan/HRW Photo; 226(tr), Jean Marc Barey/Photo Researchers, Inc.; 232-233(bckgd), Phototone; 232-233(all) Scott Van Osdol/HRW Photo; 238(bl), Luis Villota/ The Stock Market. CHAPTER FIVE: Page 242-243(bckgd), Scott Van Osdol/HRW Photo; 242(c), NASA; 243(c), Richard Stockton; 244(bckgd), Kazuyoshi Nomachi/Photo Researchers, Inc.; 245(tr), Kazuyoshi Nomachi/Photo Researchers, Inc.; 246(cr), J. Barry O'Rorke/The Stock Market; 247(cr), Frank Siteman/Stock, Boston; 248(tl), Steve Elmore/The Stock Market; 248(tr), Michael Rosenfeld/Tony Stone Images; 249(b), Carroll Seghers/Photo Researchers, Inc.; 249 John Running/Stock, Boston; 250(t), Conklin/Monkmeyer Press Photo; 252-253(bckgd), Boden/Ledingham/MASTERFILE; 254(tr), Jonathan Nourok/PhotoEdit; 254(tl), (br), Vee Sawyer/Mertzstock; 256(bl), Robert Rathe/Stock, Boston; 257(tr), Tomas Pantin/HRW Photo; 259(tl), Gary A. Connor/PhotoEdit; 260(tr), Gary Irving/Tony Stone Images; 261(tr), NASA; 263(bl), Dennis Fagan/HRW Photo; 264(tr), Gary Braasoh; 264(b), © Jim Richardson; 265(tr), Tony Stone Images; 266(tr), Bob Daemmrich/The Image Works; 267(c) , Columbia University Rare Book and Manuscript Library. Gift of George A. Plimpton; 267(bckgd), Vee Sawyer/Mertzstock; 267(r), Art Resource; 268(bl), Vee Sawyer/Mertzstock; 269(br), G. Gardner/The Image Works; 271(cr), Dennis Fagan/HRW Photo; 272(br), Frank Siteman/Monkmeyer Press Photo; 275(tr), Tomas Pantin/HRW Photo; 276-277(bckgd), Bill Binzen/Rainbow; 279(br), Tomas Pantin/HRW Photo; 280(br), Vee Sawyer/Mertzstock; 281(br), Dennis Fagan/HRW Photo; 283(tr), L. Kolvoord/The Image Works; 284(b), (bckgd), SIU/Photo Researchers, Inc.; 285(br), 286(bl), Dennis Fagan/HRW Photo; 287(b), (c), Edward Pieratt/Stock, Boston; 291(bl), Dennis Fagan/HRW Photo; 293(tr), Guy Sauvage/Photo Researchers, Inc.; 294(t), David M. Grossman/Photo Researchers, Inc.; 294(bckgd), (bl),Vee Sawyer/Mertzstock; 295(c), H. L. Romberg; 297(b), Scott Van Osdol/HRW Photo. CHAPTER SIX: Page 304(tr), K. Kai/FujiFotos/The Image Works; 305(bl), Dennis Fagan/HRW Photo; 309(tl), H. L. Romberg; 309(bl), Scott Van Osdol/HRW Photo; 311(tr), George Whiteley/Photo Researchers Inc.; 311(bl), Charles M. Falco/Photo Researchers, Inc.; 312(cr), 313(tr), Dennis Fagan/HRW Photo; 315(b), Bair/Monkmeyer Press Photo; 316(tr), Sam Dudgeon/HRW Photo; 316(cr), (bl), (br), Dennis Fagan/HRW Photo; 317(tr), © Jim Richardson; 318(tl), Sam Dudgeon/HRW Photo; 318(bl), 320(br), 322(cr), Dennis Fagan/HRW Photo; 322(br), David Parker/Science Photo Library/Photo Researchers, Inc.; 323(bckgd), 324(bckgd), © Chuck Solomon; 325(tl), Dennis Fagan/HRW Photo; 326(tr), Frank Rossotto/The Stock Market; 326(bl), Dennis Fagan/HRW Photo; 330(bckgd), David A. Hardy/Science Photo Library/Photo Researchers, Inc.; 331(tr), Click Chicago/Tony Stone Images; 336(tr), "Grandes Chroniques de France", Castres, Musee Goya/Giraudon/Art Resource; 336(c), "Veduci di citta ideale"/Art Resource; 336(bc), Dennis Fagan/HRW Photo; 336(bl), Sam Dudgeon/HRW Photo; 337(tl), Fotomas Index; 337(cr), Donovan Reese/Tony Stone Images; 337(bl), Dennis Fagan/HRW Photo; 337(bl), Cindy Verheyden/HRW Photo; 338(br), H. L. Romberg; 339(tl), Stan Osolinski/Tony Stone Images; 339(cl), Brunelleschi, "Interno verso l'altare S. Lorenzo"/Art Resource; 339(b), Kathy Bushue/Tony Stone Images; 343(b), Scott Van Osdol/HRW Photo; 344(tl), Building in Egypt, Pharaonic Stone Masonry ; 344(bl), Michelle Bridwell/Frontera Fotos; 344-345(bckgd), Doug Armand/Tony Stone Images; 345(tr), Dennis Fagan/HRW Photo. CHAPTER SEVEN: Page 352-353(bckgd), Fabrizio Bensch REUTERS/Bettman; 352(c), Lutz Schmidt REUTERS/Bettman; 353(bl), Kurgan-Lisnet/Liaison International; 354(tl), Phillip Bailey/The Stock Market; 354(tr), Paul Conklin/Monkmeyer Press Photo; 355(tr), Jany Sauvanet/Photo Researchers, Inc.; 355(cr), W. K. Almond/Stock, Boston; 357(tr), Courtesy of NASA; 357(cr), Fred McConnaughey/Photo Researchers, Inc.; 357(br), G. W. Ott/The Stock Market; 358(tr), Tom Tracy/The Stock Market; 358(br), Sam Dudgeon/HRW Photo; 359(tr), Scott Van Osdol/HRW Photo; 361(tr), Mark Stouffer/Animals Animals;

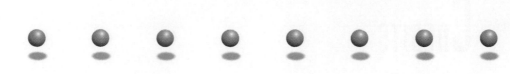

362(tr), Scott Van Osdol/HRW Photo; 363(tr), Tomas Pantin/HRW Photo; 363(br), B. Mahoney/The Image Works; 365(tr), Mark Antman/The Image Works; 366(t), 368(bl), 369(cl), Dennis Fagan/HRW Photo; 370(tl), Will and Deni McIntyre/Photo Researchers, Inc.; 372(t), Spencer Jones/FPG International; 373(t), Tony Freeman/PhotoEdit; 377(br), Tomas Pantin/HRW Photo; 378(tr), David Wells/The Image Works; 378(br), Dennis Fagan/HRW Photo; 379(tr), Tony Stone Images; 381(b), Bilderberg/The Stock Market; 382(bl), Courtesy of US GS; 386(br), Peter Menzel/Stock, Boston; 386(br), Dean Abramson/Stock, Boston; 387(tr), Stephen Agricola/Stock, Boston; 389(cl), Trask/Stock Imagery, Inc; 389(bl), Imtek Imagineering/MASTERFILE; 390(tl), Matthew McVay/Tony Stone Images; 391(tr), Bob Daemmrich/The Image Works; 391(br), Tomas Pantin/HRW Photo; 392(c), (tr), Dennis Fagan/HRW Photo; 394(bckgd), Scott Van Osdol/HRW Photo, 394, 395 BPL; 396(tr), Frank Rossotto/The Stock Market; 396(bl), Dennis Fagan/HRW Photo; 398(bcl), Fay Torresyap/Stock, Boston; 398(collage),Vee Sawyer/MertzStock; 399(t), Biophoto Associates/Science Source/Photo; Researchers, Inc.; 399(tr), (tl), (bcr), Patrice Ceisel/Stock, Boston; 399(bl), (tcl), (tcl), (bckgd), Vee Sawyer/MertzStock; 401(bl), Bob Daemmrich/Stock, Boston; 402(bl), Amy Etra/PhotoEdit; 403(tl), Pun Nio/HRW Photo; 403(bl), (br), Dennis Fagan/HRW Photo; 404-405(bckgd), Richard Pasley/Stock, Boston; 405(tr), (br), (bl), Scott Van Osdol/HRW Photo. CHAPTER EIGHT: Page 410-411(bckgd), Ron Kimball; 412(tr), Michelle Bridwell/HRW Photo; 413(tl), Tomas Pantin/HRW Photo; 414(br), Oscar Burriel/Latin Stock/Science Photo; 414(bl), Oscar Burriel/Latin Stock/Science Photo; 416(tr), Felicia Martinez/PhotoEdit; 418(cr), World Perspectives; 419(tr), Scott Van Osdol/HRW Photo; 419(bl), Dennis Fagan/HRW Photo; 421(cl), Ed Michaels/Photo Researchers, Inc.; 422(bl), Tony Freeman/PhotoEdit; 424(tl), Donald Dietz/Stock, Boston; 426 Courtesy of Texas Highway Department; 426 Courtesy of Texas Parks and Wildlife; 429(tr), Stephen Dunn/ALLSPORT USA; 429(tl), Al Bello/ALLSPORT USA; 433(tr), Tomas Pantin/HRW Photo; 435(br), Michael Newman/PhotoEdit; 436(t), Vee Sawyer/MertzStock; 441(bckgd), Ken Reid/FPG International; 441(bl), Tomas Pantin/HRW Photo; 443(tr), Michael Collier/Stock, Boston; 447(cl), Peter Miller/Photo Researchers, Inc.; 447(cr), Dick Dietrich/FPG International; 448(br), Daemmrich/UNIPHOTO Picture Agency; 449(cr), Robert Caputo/Stock, Boston; 452(tr), Tim Barnwell/Stock, Boston; 454(bl), Dennis Fagan/HRW Photo; 455(tl), Kevin Schafer/Tony Stone Images; 456(cr), Dennis Fagan/HRW Photo; 459(tr), Vic Bider/PhotoEdit; 460(bckgd), (cl), 461(b), Scott Van Osdol/HRW Photo. CHAPTER NINE: Page 468-469(bckgd), COM-STOCK; 468(c), 469(c), COMSTOCK; 470(tr), Scott Van Osdol/HRW Photo; 471(cl), Dennis Fagan/HRW Photo; 472(br), LeDuc/Monkmeyer Press Photo; 474(br), Scott Van Osdol/HRW Photo; 475(bl), LeDuc/Monkmeyer Press Photo; 476(br), Dennis Fagan/HRW Photo; 478(bckgd), Courtesy of NASA; 482(t), (b), Tony Stone Images; 484(t), 488(br), Dennis Fagan/HRW Photo; 489(tr), Tomas Pantin/HRW Photo; 490(tr), Okoniewski/The Image Works; 491(t), John Henley/The Stock Market; 497(b), Tony Stone Images; 500(tl), Steven Starr/Stock, Boston; 500-501(bckgd), Spencer Grant/Monkmeyer Press Photo; 501(bl), Krafft/Explorer/Photo Researchers inc.; 502(t), Clark James Mishler/Westlight; 502(bl), Peter Menzel/Stock, Boston; 507(bckgd), Peter Saloutos/Tony Stone Images; 507(t), Peter Saloutos/Tony Stone Images; 507(br), Vee Sawyer/MertzStock; 509(b), Tomas Pantin/HRW Photo; 510(cl), 512(br), Dennis Fagan/HRW Photo; 514(tr), Randy Duchaine/The Stock Market; 507(br), Vee Sawyer/MertzStock; 509(b), Tomas Pantin/HRW Photo; 516(bl), Dennis Fagan/HRW Photo; 516(bl), Bob Daemmrich/Stock, Boston; 517(tl), Mark Green/Tony Stone Images; 517(br), Alan Schein/The Stock Market. CHAPTER TEN: Page 522-523 COMSTOCK; 523(br), Eric Crossan; 522(b) left, Zigy Kaluzny/Tony Stone Images; 522-523(bckgd), Pete Saloutos/The Stock Market; 524(tr), (bckgd), Paul Steel/The Stock Market; 524(l), Vee Sawyer/MertzStock; 525(br), Dennis Fagan/HRW Photo; 527(bckgd), Carolyn Brown/Photo Researchers, Inc.; 529(br), Jeff Greenberg/PhotoEdit; 531(t), Martin Frick/M&PPRY; 531(b), Michael Collier/Stock, Boston; 533(t), Ken Straiton/The Stock Market; 536(bl), Dennis Fagan/HRW Photo; 537(br), R. Sydney/The Image Works; 538(r), Alain Evrard/Photo Researchers, Inc.; 539(cr), Amy Etra/PhotoEdit; 540(bckgd),

Geoff Dore/Tony Stone Images; 541(tl), Eastcott & Momatiuk/The Image Works; 541(br), Mark Antman/The Image Works; 543(bckgd), Vandystadt/Photo Researchers, Inc.; 545(br), Bob Daemmrich/The Image Works; 546(br), Dennis Fagan/HRW Photo; 547(bckgd), NASA/Photo Researchers, Inc.; 548(tr), NASA/Photo Researchers, Inc.; 549(cr), Dennis Fagan/HRW Photo; 555(tr), 558(tl), Earth Satellite Corporation/Science Photo Library/Photo Researchers, Inc.; 559(cr), Bill Bachman/Photo Researchers, Inc.; 560(t) (bckgd), Kathy Bushue/Tony Stone Images; 561(br), HRW Photo; 562(c), NASA/Science Photo Library/Photo Researchers, Inc.; 563(br), NASA/Frank Rossotto/ The Stock Market; 562-563(bckgd), Digital Art; 564(c), Scott Van Osdol/HRW Photo; 564(bckgd), Robert Rathe/Stock, Boston; 564(bcdgd), (b), Vee Sawyer/MertzStock; 570(t), David Young Wolff/PhotoEdit; 570(br), Tony Freeman/PhotoEdit; 571(tl), Tom Tracy/The Stock Market; 571(tr), Ed Pritchard/Tony Stone Images; 573(bckgd), Bob Abraham/The Stock Market; 574-575(bckgd), George Disario/The Stock Market; 576-577(bckgd), COM-STOCK; 576(tr), Columbia University Rare Book and A. Plimpton. CHAPTER ELEVEN: Page 584-585(bckgd) Gregory Sams/Science Photo Library/Photo Researchers,Inc; 584(bl), Lisa R. Glass/Still Life Stock; 584(cl), Michelle Bridwell 584(br),Scott Van Osdol/HRW Photo; 585(tl), Judy Canty/Stock, Boston; 586(cl), Lisa R. Glass/Still Life Stock; 586(cr), Judy Canty/Stock, Boston; 586(bl),Photri Inc.; 586(bckgd),Planet Art Collection; 588(bckgd), Jim Erickson/The Stock Market; 590(bckgd), Scott Van Osdol/HRW Photo; 593(br), Rue/MonkMeyer Press Photo; 594(t) Ed Pritchard/The Stock Market; 594(br), Robert Brenner/PhotoEdit; 596(tr), Robert Brenner/PhotoEdit; 597(tr), Chad Ellers/Tony Stone Worldwide; 598 (tl), Susan McCartney/Photo Researchers, Inc.; 599(b), Robert Estall/Tony Stone Worldwide; 599(bl), Mike Mazzaschi/Stock, Boston; 600(tr) Sovfoto/Eastfoto; 604(tr), Scott Van Osdol/HRW Photo; 605(br), Spencer Grant/Stock, Boston; 606(br), Dennis Fagan/HRW Photo; 607(tl), Tony Stone Worldwide; 607(tr),Michael Burgess/The Stock Market; 611(br), Dennis Fagan/HRW Photo 612(tr), Tomas Pantin/HRW Photo; 614(t) © 1995 M.C. Escher/Cordon Art–Baarn–Holland. All rights reserved.; 615(cl), Wayne Scherr/Photo Researchers, Inc.; 618(br), The Image Works; 619(tr), Dennis Fagan/HRW Photo; 621(tr), Don Hay; 621(tr), Scott Van Osdol/HRW Photo; 625(br), 628(b) Dennis Fagan/HRW Photo; 629(bckgd), Scott Van Osdol/HRW Photo; 629(bl),Fractal Geometry of Nature by Mandelbrot/Freeman; 630(tl), Worldsat International, Inc./Science Photo Library/Photo Researchers, Inc.; 630(c), Courtesy of NASA; 630(b), Sam Dudgeon/HRW Photo; 631(tr), Tomas Pantin/HRW Photo; 632(bckgd), Stephen Johnson/Tony Stone Worldwide; 634 Przemyslaw Prusinkiewicz; 636(br), Seattle Times/Gamma-Liaison International; 637(tr), Sam Dudgeon/HRW Photo. CHAPTER TWELVE: Page 642-643(bckgd), Scott Van Osdol/HRW Photo; 644(tr), Joe Arcure; 645(bl), Zefa-Bach/The Stock Market; 647(tl), John Lund/Tony Stone Worldwide; 647(br), Scott Van Osdol/HRW Photo; 648(tr), Bob Daemmrich/The Image Works; 650(t), Tony Arruza/Tony Stone Worldwide; 651(bckgd), Ronnie Kaufman/The Stock Market; 652(bckgd), The Kobal Collection; 653(br), Bob Daemmrich/Tony Stone Worldwide; 654(cr), Gary Lefever/Grant Heilman; 656(bckgd), Graham French/MASTERFILE ; 658(t), Shooting Star International Photo Agency; 661(br), Vee Sawyer/MertzStock; 663(br), Frank Siteman/Stock, Boston; 664(b), Esbin-Anderson/The Image Works; 665(t), Vee Sawyer/MertzStock; 666(tl), H.L. Romberg; 667(b), Paramount Pictures/The Kobal Collection; 668(tr), Magritte,"Ceci n'est pas une pipe", 1929. Oil on canvas, 60x81cm. © ARS, NY. Los Angeles County Museum of Art, Los Angeles, CA/Giraudon/Art Resource; 668-669(bckgd), Eric Neurath/Stock, Boston; 671(tr), Courtesy of Motorola; 672(t), Tony Freeman/PhotoEdit; 672(c), Spencer Jones/FPG International; 672(b),David Woods/The Stock Market; 673-674(bckgd),SB Photography/Tony Stone Worldwide; 675(br), Tomas Pantin/HRW Photo; 676(br), David Parker/Science Museum/Science Photo Library/Photo Researchers Inc.; 677(br), David Bassett/Tony Stone Worldwide; 678(br), J. B. Diderich/Contact Press Images/The Stock Market; 679(br), Photri/The Stock Market; 681(bl), 682-683(bckgd)Dennis Fagan/HRW Photo; 685(br), 687(bl), Scott Van Osdol/HRW Photo; 685-687 Letraset Phototone.

ILLUSTRATIONS/DESIGN

Abbreviations used: (t) top, (c) center, (b) bottom, (l) left, (r) right, (bckgd) background, (bdr) border.

Brooks, Janet pages 252-253, 394-395
Cericola, Anthony pages 443, 447, 448 (c), 530, 531, 535, 559, 609, 613
Claunch, David/Liason pages 172-173, 242-243, 468-469, 562-563, 656-657
Design Island 584-585
Davis, Will pages 385 (t), 417, 550, 608, 656-657
Effler, Jim pages 315, 323, 427, 540, 566
Farrell, Russell pages 289
Fischer, David pages X, 54, 219, 368, 385, 395, 417, 490, 494, 604, 633
Kell, Leslie pages 500-501, 516-517, 576-577, 636-637, 685-687
Lee, Kanokwalee pages 232-233, 296-297, 404-405-460-461
Maryland Carto Graphics pages 426, 444, 505, 568, 623

Nigro, Lisa pages costume 419
Obershan, Mike pages 2-8, 46, 58-59,106-107, 130-131, 302-303, 323-325, 329 (t), 330 (c)(r), 336 (b), 337 (b)(l), 344-345, 410-411, 468-469, 642-643
Pembroke, Richard pages 428
Randazzo, Tony pages 302-303
Sawyer, Vee pages 35, 51, 52, 55, 267, 507, 524, 600, 650 Collages: 6, 37, 44, 92, 123, 128, 143, 244, 245, 249, 254, 267, 294, 354, 398, 399, 434, 436, 500, 502, 507, 509, 524, 564, 586, 647, 650, 661, 665
Scrofani, Joe pages 75, 199, 670
Sullivan, Alicia pages 57, 64-65, 80-81, 114-115, 166-167, 172-173, 252-253, 468-469, 522-523, 656-657
Szetela, Chris page 603
Tenniel, Sir John pages 665

SELECTED ANSWERS
Practice & Apply and Look Back

Chapter 1

Lesson 1.1

7. $\overline{AB}$, $\overline{BC}$, $\overline{AC}$ **9.** $\angle A$: $\overrightarrow{AB}$, $\overrightarrow{AC}$; $\angle B$: $\overrightarrow{BA}$, $\overrightarrow{BC}$; $\angle C$: $\overrightarrow{CA}$, $\overrightarrow{CB}$ **11.** line **13.** plane **15.** point
17. False. Lines go on without bound in both directions. **19.** False. Suppose two lines intersect and another line is perpendicular to the lines at the intersection point. The third line does not lie in the same plane as the first two.
21. True. For example, the planes containing the front wall, a side wall and the floor of a classroom intersect at a point, the corner of the room. **23.** False. Any 3 points are contained in a unique plane, so a point not on the plane cannot be coplanar with them.
25. False **27.** True **29.** True **31.** True
33. True **35.**

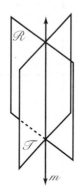

37. 1 **39.** 6 **41.** 10 ; $\frac{n(n-1)}{2}$ angles for n points. **43.** 22 **45.** $-81 + 30 = 51$
47. $|3| = 3$ **49.** $|-14| = 14$

Lesson 1.2

5. A triangle **7.** right triangle **9.** Such a triangle would have two equal sides. **11.** Lines l and m are perpendicular. The bisectors of angles formed by intersecting lines are perpendicular. **13.** The square diagonals are perpendicular bisectors of each other and are equal in length. **15.** plane **17.** point **19.** segment

$\overline{MN}$, $\overline{NM}$ **21.** $\overrightarrow{PS}$ **23.** A **25.** -26 **27.** 46
29. 46

Lesson 1.3

7. E is between F and D unless all three points coincide. **9.** The sides of the triangle must all have the same length.

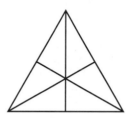

11. outside **13.** It divides the circle into two equal parts. **15.** About 23 in. **17.** The distance from the vertex to the centroid is $\frac{2}{3}$ of the distance from the vertex to the midpoint opposite the vertex. **19.** The orthocenter, centroid and circumcenter of any triangle are on the same line. **21.** positive **23.** negative, positive
25. negative **27.** negative, positive

Lesson 1.4

7. $AB = |-3 - -1| = |-1 - -3| = 2$,
$AC = |-3 - 3| = |3 - -3| = 6$, $BC = |-1 - 3| = |3 - -1| = 4$ **9.** Congruent segment pairs are: $\overline{AF}$ and $\overline{ED}$, $\overline{FG}$ and $\overline{EG}$, and $\overline{GB}$ and $\overline{GC}$.
11. $AB = 85$

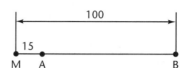

13. $x = 5$ **15.** $XY = 51$ **17.** 35 miles **19.** N
21. S **23.** S **25.** $\frac{1}{8}$, $\frac{1}{16}$, $\frac{1}{32}$, Sample answers:
$\frac{4}{11}$, $\frac{1}{5}$, $\frac{2}{7}$ **27.** An earthquake with magnitude 3 is 100 times more energetic than an earthquake with magnitude 1. An earthquake of magnitude 8 has 10,000,000 times the energy of an earthquake with magnitude 1. **29.** circumcenter, can be outside the triangle **31.** centroid, can't be outside the triangle **33.** circumcenter

35. A right angle can be inscribed in a half circle.

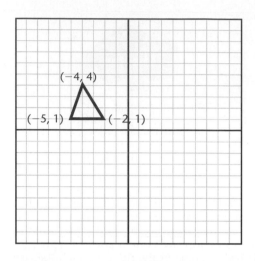

11. A 180° rotation about the origin.

Lesson 1.5

7. 46° **9.** 119° **11.** 73° **13.** ∠BVO and ∠MVT : 20°, ∠BVM, ∠OVT and ∠TVR : 45°, ∠BVT and ∠MVR : 65° **15.** 60°, The 12 numbers on a clock divide the circle into 12 angles, each of measure 30°. **17.** 114° **19.** 18° **21.** 15° **23.** $x = 8.75$; 73.75° **25.** 27.25° **27.** 90° **29.** 270° **31.** 315° **33.** 67.5 **35.** 45°, 135°, 225°, 315°; The diver must always swim the same distance in each direction. **37.** 20° **39.** One gradian is smaller, since one gradian is $\frac{9}{10}$ of a degree. **41.** Lines n and m **43.** A, B, C **45.** |−6 − 10| = |10 − −6| = 16

Lesson 1.6

9. Shift it by 2 cm in the same direction that A moved. **11.** 90° **13.** line l. **15.** Each image point is the same distance from the reflection line, but on the opposite side of the line.
17.

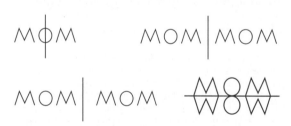

19. B **21.** Translate $ABCD$ one side unit up, one unit down, and one unit to the left. Then slide the square twice to the right by one unit.
23. Rotate $ABCD$ around each of its vertices 90° counterclockwise. Then rotate the image of $ABCD$ at its right by 90° counterclockwise around the point E at the upper right. **25.** Collinear
27. Yes. Through any two points there is exactly one line. **29.** Yes. The line containing the intersection of the front wall and the ceiling of a classroom and the line containing the intersection of a side wall and the floor of the classroom are noncoplanar lines. **31.** 23 **33.** 13°

Lesson 1.7

5. $(x, y) \rightarrow (x, y - 32)$ **7.** $(x, y) \rightarrow (x - 8, y)$
9. $(x, y) \rightarrow (-x, y)$

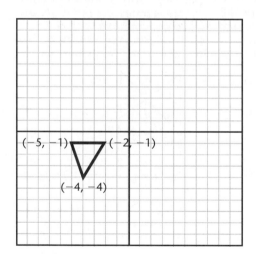

13. translation 7 to the right
15. translation left 6 and up 7
17. translation down 4
19. rotation of 180° about the origin
21. translation up 2
23. Sample table

x	y
0	0
2	2
5	5

25. 45°

27.

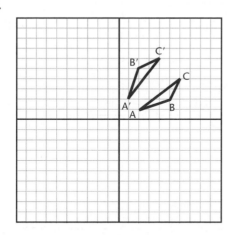

29. $(x, y) \to (y, x)$ **31.** Find the point where the angle bisectors meet. **33.** Find the point where the altitudes meet. **35.** 270° **37.** 225° **39.** A figure slides a given distance in a given direction. **41.** A figure is transformed to its mirror image through a line.

Lesson 2.1

7. Smaller, because 1 is successively divided by bigger numbers. **9.** 1
11.

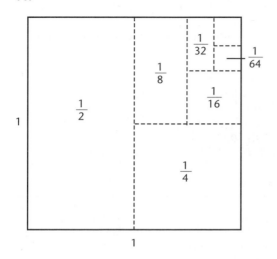

13. Column A; It takes a long time. **15.** 2, 5, 8, 11, . . . ; 3, 6, 9, 12, . . . **17.** C, because 999 is a multiple of 3. **19.** If $n = 5, \frac{n(n + 1)}{2} = \frac{5 \cdot 6}{2} = 15$ and $1 + 2 + 3 + 4 + 5 = 15$, If $n = 6, \frac{n(n + 1)}{2} = \frac{6 \cdot 7}{2} = 21$ and $1 + 2 + 3 + 4 + 5 + 6 = 21$
21. $n(n + 1)$ **23.** It suggests relating the sum to the number of dots in the constructed triangle, and the pattern is the same for any size of the triangle, and hence n. **25.** 6, $\overline{GR}$, $\overline{GE}$, $\overline{GP}$, $\overline{RE}$, $\overline{RP}$, $\overline{EP}$ **27.** $x = 21\frac{5}{7}$, m$\angle KIY = 65\frac{1}{7}°$, m$\angle TIY = 14\frac{6}{7}°$

Lesson 2.2

7. Hypothesis: a person lives in Ohio. Conclusion: the person lives in the United States **9.** If a person lives in the United States, then the person lives in Ohio. False. A person who lives in Indiana is a counterexample. **11.** Socrates is mortal.

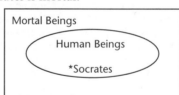

13. If a person is an independent farmer, then the person is disappearing. That man is an independent farmer. Therefore, that man is disappearing. The statement does not mean that farmers are vanishing visually, only that many cannot continue as farmers. Also, the statement does not refer to individual independent farmers, but to independent farmers as a group.
15. If m$\angle C = 90°$, then triangle ABC is a right triangle. True. If triangle ABC is a right triangle, then m$\angle C = 90°$. False, triangle ABC may have m$\angle A = 90°$. **17.** If druskers leer, then homblers fawn. If homblers fawn, then quompies pawn. If quompies pawn, then rhomples gleer. If druskers leer, then rhomples gleer. No, as long as the reasoning chain is valid. **19.** If a person has no hair, then that person will not carry a comb. If a person does not carry a comb, then the person cannot drop a comb. Therefore, if a person has no hair, then the person cannot drop a comb. Since the criminal dropped a comb, the suspect is innocent. **21.** The triangle is obtuse. **23.** If the triangle is obtuse, then m$\angle A$ + m$\angle B < 90°$. **25.** Answers will vary. **27.** Two lines or segments are perpendicular if they form right angles when they meet. Fold

one endpoint of the segment onto the other endpoint. The crease is the perpendicular bisector. **29.** single point. **31.** single point.
33.

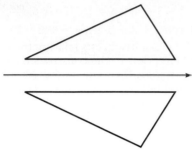

Lesson 2.3

5. ...the person is over 12 years of age. ...the person is a teenager. ...the person is over 12 years of age. Not a definition since the converse is false. A person who is 21 years old is over 12 but is not a teenager. **7.** If a number is 0, then the number is an integer between −1 and 1. If a number is an integer between −1 and 1, then the number is 0. A number is 0 if and only if the number is an integer between −1 and 1. A definition, since both statement and converse are true. **9.** If an instrument is a sitar, then it is a lutelike instrument of India. If an instrument is a lutelike instrument of India, then the instrument is a sitar. An instrument is a sitar if and only if it is a lutelike instrument of India. Not a definition, since the converse is false. There are other lutelike instruments of India — the vena, for example. **11.** b, d **13.** b, c **15.** Student definitions should specify the object as uniquely as possible.
17.

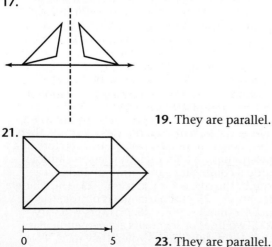

19. They are parallel.

21.

23. They are parallel.

Lesson 2.4

5. any 3 points in plane R **7.** line l **9.** exactly 1
11. Z, F, O **13.** 5, 9 **15.** 8
17.

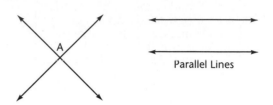

Parallel Lines

19.

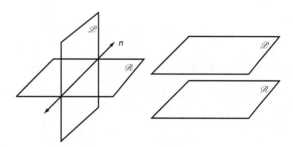

21.

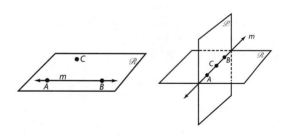

23. The midpoint of a segment is equidistant from the endpoints of the segment, and the angle bisector is the set of points equidistant from the sides of an angle.

Lesson 2.5

5. Transitive Property of Equality **7.** UT and TA **9.** AH and TA **11.** $UT = AH$, Given, $UT + TA = AH + TA$, Addition Property Of Equality **13.** Overlapping Segments Theorem **15.** $\angle PLA$ and $\angle ALS$ **17.** m$\angle SLC$ and $\angle ALS$ **19.** Given m$\angle PLA$ = m$\angle SLC$, m$\angle PLA$ + m$\angle ALS$ = m$\angle SLC$ + m$\angle ALS$ **21.** Overlapping Angles Theorem **23.** $AB = GH$ **25.** Sample answer : If m$\angle BAC$ = m$\angle EDF$ and m$\angle EDF$ = m$\angle HGI$, then m$\angle BAC$ = m$\angle HGI$.

27.

Statements	Reasons
1. $m\angle PLS = m\angle ALC$,	1. Given
2. $m\angle PLA + m\angle ALS = m\angle PLS$, and $m\angle SLC + m\angle ALS = m\angle ALC$	2. Angle Addition Postulate
3. $m\angle PLA + m\angle ALS = m\angle SLC + m\angle ALS$	3. Substitution Property of Equality
4. $m\angle PLA + m\angle ALS - m\angle ALS = m\angle SLC + m\angle ALS - m\angle ALS$, and so $m\angle PLA = m\angle SLC$	4. Subtraction Property of Equality

29.

Statements	Reasons
1. $m\angle BAC + m\angle ACB = 90°$, $m\angle DCE + m\angle DEC = 90°$, and $m\angle ACB = m\angle DCE$	1. Given
2. $m\angle BAC + m\angle ACB = m\angle DCE + m\angle DEC$	2. Substitution Property of Equality
3. $m\angle BAC + m\angle ACB = m\angle ACB + m\angle DEC$	3. Substitution Property of Equality
4. $m\angle BAC + m\angle ACB - m\angle ACB = m\angle ACB + m\angle DEC - m\angle ACB$, and so $m\angle BAC = m\angle DEC$	4. Subtraction Property of Equality

31. $x = 5$; $FN = 26$; $FU = UN = 13$
33. Example. If it is raining, then it is cloudy.
Converse: If it is cloudy, then it is raining.

Lesson 2.6

5. 8 and 6; 5 and 7; 4 and 2; 1 and 3   **7.** $\angle LEP$
9. $\angle PMQ$ and $\angle SMF$ measure 150°, $\angle PMS$ and $\angle QMF$ measure 30°   **11.** $m\angle ABC = 68°$
13. $m\angle ABC = \frac{180°}{2} = 90°$ **15.** $x = 7$
17. $x = 11\frac{1}{16}$ **19.** 20 cm   **21.** 5 cm
23. Yes

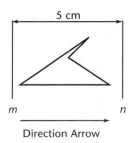
5 cm

m   *n*

Direction Arrow

25. a. 60°, b. 100°, c. $2y°$ **27.** Given   **29.** bisector
31. If it is Saturday, then I am in a good mood.

Lesson 3.1

7. 1 reflectional   **9.** 1 reflectional
11.   **13.**

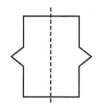

15.

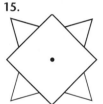

17. 11 and 13 have 180° rotational symmetry
19.

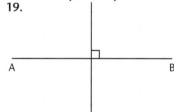

A   B

21. It is the perpendicular bisector.   **23.** $\overline{BC}$, $\overline{BA}$, $\angle CBA$
25. Divide the angle into two 40° angles by drawing the bisector.   **27.** Draw the angle bisector.   **29.** reflection through $\overline{AC}$ **31.** reflection through $\overline{BD}$ **33.** rotation of 180° clockwise around X **35.** 4 symmetry axes; Rotational symmetries of 45°, 90°, 135°, 180°, 225°, 270°, 315°, 360°   **37.** No symmetry axes; Rotational symmetries of 120°, 240°, 360
39. $x = 1$ **41.** $x = -2$

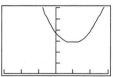   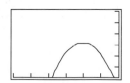

43. $x = 0$

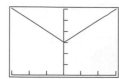

45. O **47.** $\overleftrightarrow{AB}$ **49.** $85°$

Lesson 3.2

5. False. A rectangle is a special parallelogram.
7. True. All squares are parallelograms. **9.** False.
A square is a rhombus that is also a rectangle.
11. $70°$ **13.** 6 **15.** $20°$
17.

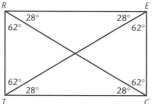

19.

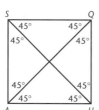

21. 6 **23.** isosceles triangle **25.** $x = 30$; $m\angle D = m\angle A = m\angle I = m\angle M = 90°$ **27.** If a device is a hypsometer, then it is used to measure atmospheric pressure and thus determine height above sea level. If a device is used to measure atmospheric pressure and thus determine height above sea level, then it is a hypsometer. It is not a definition because the converse is false. Devices can be used to determine atmospheric pressure and height above sea level that are not hypsometers. An example is an altimeter, which has a different operating mechanism than the hypsometer. **29.** If a creature is a whale, then it is a mammal. If a creature is a mammal, then it is a whale. It is not a definition, because the converse is false. Some mammals are not whales. **31.** Rotational and reflectional **33.** Rotational and reflectional

Lesson 3.3

5. p and q; also p and k **7.** $\angle 3$ and $\angle 8$; $\angle 4$ and $\angle 7$ **9.** $\angle 3$ and $\angle 7$; $\angle 4$ and $\angle 8$ **11.** It is a transversal to them.

13. It is a transversal to them. They are corresponding angles. **15.** No. They are not formed from the same transversal. **17.** $m\angle 2 = m\angle 3 = m\angle 4 = 45°$ **19.** $m\angle 3 = 30°$ **21.** $m\angle 5 = 150°$
23. $m\angle 8 = 150°$ **25.** $50°$ **27.** $25°$ **29.** $25°$
31.

Given: Line l ∥ line m
          and transversal p.
Prove: $\angle 1 \cong \angle 2$

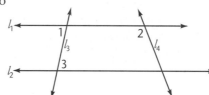

Statements	Reasons
1. Line l ∥ line m, and line p is a transversal	1. Given
2. $\angle 1 \cong \angle 3$	2. ∥s⇒Corr. $\angle$s $\cong$
3. $\angle 3 \cong \angle 2$	3. Vertical $\angle$s $\cong$
4. $\angle 1 \cong \angle 2$	4. Transitive Prop of Cong

33. Perpendicular
35. $m\angle 1 = 30°$; $m\angle 2 = 150°$

Lesson 3.4

5. definition of supplementary **7.** definition of supplementary **9.** Subtraction Property of Equality **11.** l_1 and l_2. If two lines are cut by a transversal in such a way that alternate interior angles are congruent, then the two lines are parallel.
13. No

15. congruent alternate interior angles imply parallel lines **17.** supplementary consecutive interior angles imply parallel lines **19.** $12°$
21. $12°$ **23.** $168°$ **25.** $12°$ **27.** transversal
29. m or p **31.** corresponding **33.** $\angle 3$ **35.** the Transitive Property of Congruence **37.** by using the Corresponding Angle Theorem converse.
39. Hypothesis: The figure is a rectangle. Conclusion: The figure is a parallelogram. **41.** If a figure is a parallelogram, then it is a rectangle.

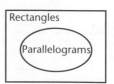

Lesson 3.5

7. 50° **9.** The sum of the other two angles is 90°, so each must be less than 90°, or acute. **11.** 60° **13.** 50° **15.** 50° **17.** 70° **19.** 60° **23.** …the sum of the measures of its remote interior angles **25.** $\angle BCD$ and $\angle BCA$ form a linear pair. **27.** Transitive Property of Equality **29.** intersection **31.** linear **33.** lls⇒Corr. $\angle$s $\cong$ **35.** or Substitution $\cong$ Corr. $\angle$s ⇒lls

Lesson 3.6

5. $x = 78°$; $y = 102°$; $z = 127°$ **7.** 108° **9.** 45° each **11.** 105° **13.** Quadrilateral, 90°; Pentagon , 72° ; Hexagon, 60° **15.** $\frac{360}{n}$ **17.** 12; 360° **19.** 6; 720° **21.** 15; 2340° **23.** 72° **25.** 144° **27.** two pairs; vertical angles **29.** For any line there is one and only one line passing through a point not on the line that is parallel to the given line.

Lesson 3.7

7. $DE = 25$ **9.** $AB = 80$ **11.** $EF = 67.5$; $GH = 75$; $IJ = 82.5$ **13.** The successive differences between the segment lengths is the same, 7.5. **15.** Let the bases have lengths b_1 and b_2. Then the length of each successive parallel segment decreases by $\frac{b_1 - b_2}{n}$, if b_1 is the longer base. **17.** Perimeter of the outer triangle is 46. Perimeter of the inner triangle is 23. Perimeter of the inner triangle is half the perimeter of the outer triangle. **19.** Rectangle, because opposite sides are parallel and adjacent sides meet at right angles.

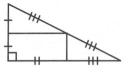

21. Square, because opposite sides are parallel, all sides are congruent and adjacent sides meet at right angles.

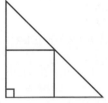

23. 380 m **25.** Yes, since 400 m < 418 m **27.** False **29.** True **31.** 70°

Lesson 3.8

7. $\frac{3}{2}$ **9.** Rectangle: slope of $\overline{CR} = 0$; slope of $\overline{RA}$ is undefined; slope of $\overline{AB} = 0$; slope of $\overline{BC}$, is undefined; Vertical lines have undefined slopes and horizontal lines have slopes = 0. Because vertical lines and horizontal lines are perpendicular, $CRAB$ is a rectangle. **11.** Rectangle slope of $\overline{RA} = \frac{-4}{3}$; slope of $\overline{AI} = \frac{3}{4}$; slope of $\overline{IN} = \frac{-4}{3}$; slope of $\overline{NR} = \frac{3}{4}$; Since both pairs of opposite sides have the same slope, both pairs of opposite sides are parallel and the figure is a parallelogram. Also, adjacent sides are reciprocals and therefore are perpendicular. Therefore $RAIN$ is a rectangle. **13.** One diagonal segment has a slope of 7. The other diagonal segment has a slope of $-\frac{1}{7}$. Since $(7)(-\frac{1}{7}) = -1$, the diagonals are perpendicular. **15.** Answers will vary. **17.** The pitch of the roof is 1.0, so the house is in violation of the building codes. **19.** $y = \frac{-1}{3} x + 7$ **21.** $m\angle 2 = 35°$ **23.** $m\angle FBC = 35°$ **25.** The Parallel Postulate: Given a line and a point not on the line, there is one and only one line that contains the given point and is parallel to the given line. **27.** 108° **29.** 90°

Chapter 4

Lesson 4.1

7. $\angle N \cong \angle W$

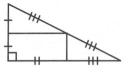

9. a. $\overline{WX} \cong \overline{NT}$, b. $\overline{ET} \cong \overline{YX}$, c. $\overline{UY} \cong \overline{KE}$, **11.** $\overline{AC} \cong \overline{DF}$, $\angle A \cong \angle D$, $\overline{BC} \cong \overline{EF}$, $\angle B \cong \angle E$ **13.** $\overline{MN} \cong \overline{RS}$, $\angle LMN \cong \angle QRS$, $\overline{NO} \cong \overline{ST}$, $\angle MNO \cong \angle RST$, $\overline{OP} \cong \overline{TU}$, $\angle NOP \cong \angle STU$, $\overline{PL} \cong \overline{UQ}$, $\angle OPL \cong \angle TUQ$ **15.** Transitive Property of Congruence **17.** Yes. For every angle in $QRST$ there is a corresponding congruent angle in $WXYZ$ and for every side of $QRST$ there is a corresponding congruent side in $WXYZ$. **19.** Yes. Rotation is a rigid

motion that preserves congruence. **21.** 8 cm
23. Yes. All corresponding sides and angles are congruent. **25.** If parallel lines are cut by a transversal, then alternate interior angles are congruent. **27.** True **29.** True

Lesson 4.2

7. Yes; $\triangle DFR \cong \triangle XWY$; Angle−Side−Angle **9.** Yes; $\triangle FDH \cong \triangle FGH$; Angle−Side−Angle **11.** No. The congruent angles are not between the two congruent pairs of corresponding sides.
13. The diagonal board breaks the shelf rectangle into two triangles, both of which are rigid.
15. Yes. $\overline{AO} \cong \overline{DO}$ and $\overline{BO} \cong \overline{CO}$ are shown. $\overline{AB} \cong \overline{DC}$ because the sides of a regular hexagon are congruent, Also, $\angle BOA \cong \angle COD$ because central angles in a regular hexagon are congruent. So $\triangle BOA \cong \triangle COD$ by either side-side-side or side-angle-side. **17.** Yes **19.** No **21.** a. $\overline{PO} \cong \overline{TO}$, b. $\angle POH \cong \angle TON$ because vertical angles are congruent. c. $\angle P$ because $\|$ s $\Rightarrow$ Alternate interior $\angle$s $\cong$, d. $\triangle POH \cong \triangle TON$ by angle-side-angle
23. $RB + BT = RT$ by the Segment Addition Postulate **25.** The diagonals of a rhombus are perpendicular, and bisect the rhombus angles as well as each other. **27.** $\angle I$ **29.** Possible

Lesson 4.3

7. $\triangle WEB \cong \triangle ARF$; ASA **9.** $\triangle ZMI \cong \triangle BQA$; AAS
11. $\overline{UW}$ and $\overline{UV}$; $\overline{EF}$ and $\overline{EG}$ **13.** $\overline{AE}$; $\overline{RS}$ **15.** No, at least one pair of congruent corresponding sides is needed. **17.** $\angle TKA$, hypotenuse $\overline{TA}$; $\angle LKA$, hypotenuse $\overline{LA}$ **19.** Reflexive Property of Congruence **21.** $\angle A \cong \angle D$, $\overline{AF} \cong \overline{CD}$, and $\angle BFA \cong \angle ECD$ are given. Setting up the correspondence, then $\triangle AFB \cong \triangle DCE$ by ASA.
23.   **25.**

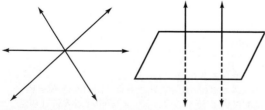

27. Choose answers from: $\angle 1$ and $\angle 2$; $\angle 3$ and $\angle 4$; $\angle 5$ and $\angle 6$; $\angle 7$ and $\angle 8$; $\angle 1$ and $\angle 8$; $\angle 2$ and $\angle 7$; $\angle 3$ and $\angle 6$; $\angle 4$ and $\angle 5$; $\angle 1$ and $\angle 4$; $\angle 2$ and $\angle 3$; $\angle 5$ and $\angle 8$; $\angle 6$ and $\angle 7$. **29.** The two interior angles adjacent to the congruent exterior angles are also congruent.

Lesson 4.4

7. SAS **9.** bisector **11.** Draw the altitude. Then the corresponding base angles are congruent and have measure of 90°. The corresponding sides are congruent by definition of an isosceles triangle. The triangles share an altitude leg, which implies the triangles are congruent by HL. **13.** opposite sides
15.

Statements	Reasons
1. $XY \parallel ZW$, $YZ \parallel XW$	1. Given,
2. $\overline{YW} \cong \overline{YW}$	2. Reflexive Property of Congruence
3. $\angle ZYW \cong \angle XWY$	3. $\parallel$s $\Rightarrow$ Alt. int $\angle$s $\cong$
4. $\overline{YZ} \cong \overline{XW}$	4. Opp sides of parl are $\cong$
5. $\triangle XWY \cong \triangle ZYW$	5. SAS
6. $\angle X \cong \angle Z$	6. CPCTC

By drawing diagonal $\overline{XZ}$ and repeating the argument, $\angle XYZ \cong \angle XWZ$. completing the proof.
17. congruent **19.** The vertex angle bisector divides the isosceles triangle into two congruent triangles. The two angles formed by the base and the bisector are congruent by CPCTC. They are also a linear pair, which means their measures are $\frac{180}{2} = 90°$. Thus the bisector is perpendicular to the base.
21.

Statements	Reasons
1. $\triangle ABC \cong \triangle DEC$	1. Given
2. $\angle BAC \cong \angle EDC$	2. CPCTC
3. $AB \parallel DE$	3. Alt. Int $\angle$s $\cong \Rightarrow \parallel$ s

23.

Statements	Reasons
1. $\angle A \cong \angle D$, $\overline{AB} \cong \overline{DE}$ $\overline{AF} \cong \overline{DC}$	1. Given
2. $\overline{AC} \cong \overline{DF}$	2. Overlapping Segment Thm
3. $\triangle ABC \cong \triangle DEF$	3. SAS
4. $\angle B \cong \angle E$	4. CPCTC

25. $\triangle XYZ \cong \triangle XBT$ by SAS using the given sides and the vertical angles. Then $\overline{BT} \cong \overline{YZ}$ by CPCTC. **27.** All its angles are 60° **29.** $\overline{AC} \cong \overline{DB}$; Overlapping Segment Theorem **31.** $\angle ABE \cong \angle CBD$; Overlapping Angles Theorem **33.** 78°
35. $\frac{180-x}{2}$

Lesson 4.5

5. the definition of a parallelogram **7.** opposite
9. ASA **11.** $\overline{AB} \parallel \overline{DC}$ by definition of parallelogram, so $\angle BDC$ and $\angle ABD$ are congruent alternate interior angles. Also, $\angle ACD$ and $\angle CAB$ are congruent alternate interior angles. $\overline{AB} \cong \overline{DC}$ because opposite sides of a parallelogram are congruent. $\triangle ABE \cong \triangle CDE$ by ASA, and $\overline{AE} \cong \overline{CE}$ since CPCTC. Therefore, point E is the midpoint of $\overline{AC}$ by definition of midpoint and $\overline{BD}$ bisects $\overline{AC}$ at point E.
13. The diagonals of a parallelogram bisect each other. **15.** Definition of rhombus. **17.** Diagonals of a parallelogram bisect each other. **19.** SSS
21. they are triangles formed by sides and diagonals of a rhombus **23.** 180° **25.** right **27.** right **29.** $\overline{RU} \cong \overline{ST}$ and $\overline{RS} \cong \overline{UT}$ because opposite sides of a parallelogram are congruent. By the definition of a rectangle, $\angle URS$ and $\angle STU$ are congruent since they are both right angles. Therefore $\triangle URS \cong \triangle STU$ by SAS, so $\overline{RT} \cong \overline{SU}$ because CPCTC. **31.** $\overline{KI} \cong \overline{KE}$ and $\overline{TE} \cong \overline{TI}$ is given. Therefore $\angle KEA \cong \angle KIA$ and $\angle TEA \cong \angle TIA$ because base angles of an isosceles triangle are congruent. Thus m$\angle E$ = m$\angle KEA$ + m$\angle TEA$ and m$\angle KIA$ + m$\angle TIA$ = m$\angle I$. Therefore, m$\angle E$ = m$\angle I$ by angle addition. Then $\triangle KET \cong \triangle KIT$ by SAS, so that $\angle EKA \cong \angle IKA$ because CPCTC. Next, $\triangle KEA \cong \triangle KIA$ by ASA, and $\angle KAE \cong \angle KAI$ by CPCTC. But the last two angles form a linear pair, which means they must each measure 90°. Finally, $\overline{KT} \perp \overline{EI}$ by the definition of perpendicular lines. **33.** The triangles may be the same shape, but not necessarily the same size. **35.** HL **37.** $x = 20, y = 20$

Lesson 4.6

3. Given **5.** SSS **7.** If alt. int $\angle$s $\cong \Rightarrow \parallel$ s
9. Given **11.** Reflexive Property of $\cong$
13. CPCTC **15.** Congruent supplementary angles are right angles. **17.** Rectangle definition **19.** Given: Quadrilateral $ABCD$ with diagonals $\overline{AC}$ and $\overline{BD}$ meeting at E, $\overline{DE} \cong \overline{BE}$, $\overline{AE} \cong \overline{CE}$

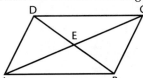

Prove: $ABCD$ is a parallelogram. Quadrilateral $ABCD$ with diagonals $\overline{AC}$ and $\overline{BD}$ meeting at E is given. It is also given that the diagonals bisect each other, so $\overline{DE} \cong \overline{BE}$ and $\overline{AE} \cong \overline{CE}$. $\angle DEC \cong \angle AEB$ and $\angle DEA \cong \angle BEC$ because they are vertical angles. Thus $\triangle DEC \cong \triangle BEA$ and $\triangle DEA \cong \triangle BEC$ by SAS. $\angle DCA \cong \angle BAC$ and $\angle DAC \cong \angle BCA$ by CPCTC. Because

alternate interior angles are congruent, $\overline{AB} \parallel \overline{DC}$ and $\overline{AD} \parallel \overline{CB}$. Therefore $ABCD$ is a parallelogram by definition.
21.
Given: Parallelogram $ABCD$; $\overline{AB} \cong \overline{BC}$
Prove: $ABCD$ is a rhombus

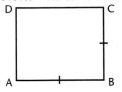

Statements		Reasons
1. Parallelogram $ABCD$, $\overline{AB} \cong \overline{BC}$		1. Given,
2. $\overline{AB} \cong \overline{CD}$, $\overline{BC} \cong \overline{DA}$		2. Opposite sides of a parallelogram are $\cong$,
3. $\overline{AB} \cong \overline{BC} \cong \overline{CD} \cong \overline{DA}$		3. Transitive Prop of $\cong$,
4. $ABCD$ is a rhombus		4. Rhombus definition

23. SAS **25.** $\overline{AC}$ is $\parallel$ to $\overline{FB}$ **27.** If opp side pair of a quadrilateral are parallel and $\cong$, it is a parallelogram **29.** Substitution **31.** They are supplementary.

Lesson 4.7

7. Draw a circle with center D. Mark the points where the circle intersects the sides of the angle as Y and X, then draw congruent circles with centers X and Y. Mark the point Z in the interior of the angle where the circles meet. Draw $\overline{DZ}$, the bisector of $\angle EDF$. **9.** $\overline{RS} \cong \overline{RQ} \cong \overline{BC} \cong \overline{BA}$ and $\overline{QS} \cong \overline{AC}$ because they are radii of congruent circles. $\triangle QRS \cong \triangle ABC$ by SSS. So $\angle ABC \cong \angle QRS$ by CPCTC. **11.** they are radii of congruent circles **13.** $\overline{AC} \perp \overline{BD}$ **15.** the diagonals of a parallelogram bisect each other **17.** $\overline{AD}$ **19.** $\overline{BC}$ **21.** Diagonals of a kite are perpendicular.
23.

Statements	Reasons
1. Line l and point M not on l	1. Given
2. $\overline{PR} \cong \overline{PT} \cong \overline{MN} \cong \overline{MO}$	2. Congruent Radii Theorem
3. $\overline{RT} \cong \overline{NO}$	3. Congruent Radii Theorem
4. $\triangle RPT \cong \triangle NMO$	4. SSS
5. $\angle RPT \cong \angle MNO$	5. CPCTC
6. line $u \parallel$ line l	6. Corresponding $\angle$s $\cong \Rightarrow \parallel$ s

25. Quadrilateral, parallelogram, rhombus.
27. Quadrilateral, parallelogram

Lesson 4.8

7. No; $PQ = 3$, $RS = 7$ **9.** Yes; $TO = OP$ by CPCTC   **11.** (4,1)   **13.** (–5,–1)
15.

Statements	Reasons
1. Points $L(-5,0)$, $C(0,3)$, $O(0,0)$, $R(5,0)$	**1.** Given
2. $LO = OR = 5$	**2.** Definition of length of horizontal segments
3. $OC = OC$	**3.** Reflexive Prop of Equality
4. $m\angle COL = m\angle COR = 90°$	**4.** Rt $\angle$s are between vertical and horizontal segments
5. $\triangle COL \cong \triangle COR$	**5.** SAS
6. $CL = CR (CL \cong CR)$	**6.** CPCTC
7. $\triangle LCR$ is isosceles;	**7.** Isosceles $\triangle$ definition

17. $X(2,5)$

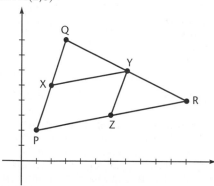

19. Trapezoid;The midsegment is parallel to the third side but half the length. So one pair of opposite sides is parallel, but not the other.
21. Parallelogram; Both midsegments parallel to a side of the triangle. Since both pairs of opposite sides are parallel, $PXYZ$ is a parallelogram.
23. $PR = 2(XY)$ **25.** Graph two lines with the same slope in the form $y = m_1x + a$ and $y = m_1x + b$ and two lines with the same slope but steeper than the slope of the other two lines in the form $y = m_2x + c$ and $y = m_2x + d$. For example: $y = 2x + 5$, $y = 2x - 8$, $y = -3x + 2$, $y = -3x - 4$ **27.** Graph two lines in the form $y = x + a$ and $y = x - a$ and two lines in the form $y = -x + a$ and $y = -x - a$. For example: $y = x + 8$, $y = x - 8$, $y = -x + 8$, $y = -x - 8$
29. The graph is a parallelogram. The slope of

$QU = 1$; slope of $\overline{UA}$ is undefined ; slope of $\overline{AD}$ = 1; slope of $\overline{DQ}$, is undefined. Both pairs of opposite sides have the same slope and are therefore parallel.
31. The graph is a trapezoid. The slope of $\overline{QU} = \frac{4}{3}$; slope of $\overline{UA} = \frac{1}{4}$; slope of $\overline{AD} = -3$; slope of $\overline{DQ} = \frac{1}{4}$. Only one pair of opposite sides have the same slope and are therefore parallel.
33. The graph is a rectangle. The slope of $\overline{QU} = 1$; slope of $\overline{UA} = -1$; slope of $\overline{AD} = 1$; slope of $\overline{DQ}$, $= -1$. Slopes of adjacent sides are negative reciprocals and therefore perpendicular.
35.

Statements	Reasons
1. Parallelogram $PQRS$, $\overline{QS}$, $\overline{PR}$, $\overline{XY}$ intersect at Z	**1.** Given
2. $\overline{RZ} \cong \overline{PZ}$	**2.** Diagonals of a parallelogram bisect each other
3. $\overline{QR} \parallel \overline{SP}$	**3.** Parallelogram definition
4. $\angle RXY \cong \angle PYX$	**4.** $\parallel$ s $\Rightarrow$ Alt int $\angle$s $\cong$
5. $\angle XZR \cong \angle PZY$	**5.** Vertical $\angle$s $\cong$
6. $\triangle XZR \cong \triangle YZP$	**6.** ASA
7. $\overline{XZ} \cong \overline{YZ}$	**7.** CPCTC

Lesson 4.9

5. They must all march the same distance along the given slide arrow.   **7.** $AB + BC > AC$
9. greater than the length of the third side.
11. Constructions will vary.   **13.** Constructions will vary.   **15.** $\angle BPA \cong \angle B'PA'$; $\angle APA' \cong \angle BPB' = 40°$ by construction. $\angle APA'$ and $\angle BPB'$ overlap at $\angle BPA'$. By the Overlapping Angle Theorem $\angle BPA \cong \angle B'PA'$.   **17.** To construct a rotation of $n°$ counterclockwise where $n > 180°$, construct a rotation of $360° - n°$ clockwise.   **19.** $\overline{AA'} \perp \overline{XY}$; $\overline{BB'} \perp \overline{XY}$; Every segment connecting a point and its preimage is cut at right angles by the mirror of the reflection.

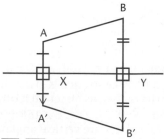

21. $\overline{AA'} \parallel \overline{BB'}$; $\overline{AA'}$, and $\overline{BB'}$, are both perpendicular to line l.   **23.** Parallelograms.

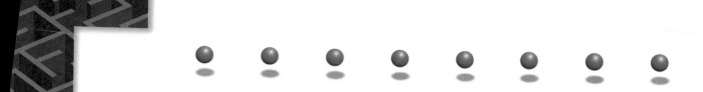

Lesson 6.6

19.

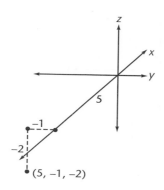

21.

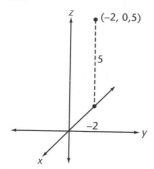

23. 17.35   **25.** $(\frac{-1}{2}, 5, \frac{9}{2})$

$\frac{\text{Surface area}}{\text{Volume}} = 1.5$ **11.** The flat-
maximizes its surface area and
s the amount of oxygen it can
e volume of water the cactus can
ed while the surface area
ater can evaporate and heat
minimized. When water is
s can store the maximum

amount and retain it more easily.   **15.** For an
$n \times 1 \times 1$ prism, the ratio approaches 4. For an
$n \times n \times n$ prism, the ratio is inversely proportional to n.   **17.** The ratio of surface area to volume is smaller in a larger roast, so there is
relatively little surface area through which heat
can enter. There is also more meat to heat in a
larger roast.   **19.** The engine that pushes a larger
amount of air not as quickly is both quieter and
more fuel efficient. It is quieter because the differences between the speeds of the air coming
out of the engine and the air outside the engine
are not as great. It is more efficient because it
delivers more power per weight of the plane for
a given fuel consumption.   **21.** $1 \times 9 = 9$ m^2; 2
$\times 8 = 16$ m^2; $3 \times 7 = 21$ m^2; $4 \times 6 = 24$ m^2; 5
$\times 5 = 25$ m^2 **23.** 25π units^2 **25.** 68 cm^2

Lesson 7.2

15. 35 cm^3 **17.** 391 cm^3 **19.** 520 units^3 **21.**
294 yd^2 **23.** 1995.32 cm^3 **25.** $S = 197.87$ in^2,
$V = 103.18$ in^3 **27.** If doubled: The surface area
is multiplied by $2^2 = 4$ and the volume is multiplied by $2^3 = 8$. If tripled: The surface area is
multiplied by $3^2 = 9$ and the volume is multiplied by $3^3 = 27$.   **29.** Power flakes: $S = 158$ in^2,
$V = 120$ in^3; Nutri Flakes: $S = 232$ in^2,
$V = 160$ in^3 **31.** $2\sqrt{5}$ **33.** 175   **35.** 5 in
37. 15.59 in

Lesson 7.3

9.

11. 120 cm^2 **13.** 211.24 cm^2 **15.** 100 units^3
17. 168 m^3 **19.** $SA = 864$ cm^2, $V = 1296$ cm^3
21. 1125 m^3 **23.** 39°   **25.** 30 units^2

Lesson 7.4

9. 477.52 units^2 **11.** 753.98 units^3 **13.** 785.40
cm^2 **15.** 157.08 m^3 **17.** 37.70 mm   **19.** Cylinder
x: $L = 113.10$; $S = 169.65$; Cylinder y: $L = $
113.10; $S = 339.29$; The lateral areas are the

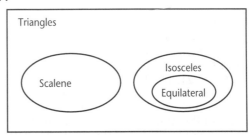

Lines m and n are parallel to line l. $\overline{AA'} \parallel \overline{BB'}$ from Exercise 21. Since both pairs of opposite sides are parallel, $ACYX$ and $A'C'YX$ are parallelograms. **25.** $BC = B'C'$. $AX = CY$ and $A'X = C'Y$ because opposite sides of a parallelogram are congruent. Since $AX = A'X$, $CY = C'Y$ by the Substitution Property of Equality. Because $BY = B'Y$ and $CY = C'Y$, $BC = B'C'$ by the Subtraction Property of Equality. **27.** $AB = A'B'$ by CPCTC. **29.** Constructions will vary. **31.** Label the angle to be copied $\angle ABC$. Draw a line and mark a point on the the line Y to be the vertex of the copy. Draw a circle with center B and a congruent circle with center Y. Mark the points where the circle intersects $\angle ABC$ as D and E. Mark the point where the circle intersects line l as Z. Set your compass equal to DE. Make an arc with center Z and radius DE. Mark the point where the circle with center Y intersects the arc with center Z as X. Draw $\overline{YX}$. $\angle ABC \cong \angle XYZ$.

Area of $ABCD = \frac{1}{2}(AC)(XD) + \frac{1}{2}(AC)(XB) = \frac{1}{2}(AC)(XD + XB) = \frac{1}{2}(AC)(BD)$.
27.

Triangles

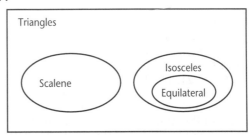

29. $x^2 + 2xy + y^2$

Lesson 5.3

5. $C = 37.7$ in; $A = 113.1$ in^2 **7.** $C = 22.0$ in; $A = 38.5$ in^2 **9.** 1963.5 m^2 **11.** Area of a piz with $d = 12$ in: 36π in^2; Area of a pizza wit' $d = 18$ in: 81π in^2; The area of an 18 in p' more than twice the area of a 12 in piz 18 in pizza should feed six people. **1** **15.** The inside tires do not have to they are not moving as fast. **17.** tions **19.** 172.79 in^2 **21.** 7.07 $A = 199$ m^2, Square: $A = 156$ area is larger. **25.** 254.47 ir **29.** 17.5 cm^2

Lesson 5.1

5. 52 in. **7.** 26 in. **9.** 42 in. **11.** 24 in. **13.** 16 in.^{2} **15.** Segment Addition Postulate; Overlapping Segments Theorem **17.** base $= 35x$; height $= 5x$; Area $= 175x^2$ **19.** 66 **21.** 690 **23.** maximum area: 2500 ft^2 **25.** $b = 50 - h$ or $h = 50 - b$, Linear function; b and h must both be positive and less than 50. **27.** 120° **29.** $(n - 2)180°$ **31.** -2

Lesson 5.2

5. 75 units^2 **7.** 28 units^2 **9.** 70 units^2 **11.** 270 units^2 **13.** 108 units^2 **15.** 375 units^2 **17.** 20 cm **19.** Triangle a: $A = \frac{1}{2}(8)(6.93) = 27.72$ units^2, Triangle: b: $A = \frac{1}{2}(6)(8) = 24$ units^2, Triangle a has the larger area. **21.** Parallelogram a: $A = (6)(7) = 42$ units^2, Parallelogram b. $A = (5)(7) = 35$ units^2, Parallelogram a has the larger area. **23.** 1.5 units^2 **25.** Area of $\triangle ACB = \frac{1}{2}(AC)(XB)$. Area of $\triangle ACD = \frac{1}{2}(AC)(XD)$.

Lesson 5.4

5. 5 **7.** $\sqrt{7445}$ 9 **13.** $12^2 > 5^2 +$ Right **17.** 3^2 5,12,13; 7,2ℓ

Let $a = m$ $+ \left(\frac{m^2 - }{2}\right)$ $= \frac{m}{}$

Lesson 7.1

7. 2 **9.** $s =$ worm's shap thus maximiz absorb. **13.** T hold is maximi through which can be absorbed i available, the cact

740

Selected Answers

sa. units defined

m∠*EID* = 36° **19.** congruent sides; golden ratio
21. $\frac{610}{377}$ = 1.618037135; $\frac{377}{233}$ = 1.618025751; $\frac{233}{144}$ = 1.618055556; $\frac{144}{89}$ = 1.617977528; $\frac{89}{55}$ = 1.618181818; $\frac{55}{34}$ = 1.617647059; $\frac{34}{21}$ = 1.619047619; $\frac{21}{13}$ = 1.615384615 ; $\frac{13}{8}$ = 1.625; $\frac{8}{5}$ = 1.6; $\frac{5}{3}$ = 1.666666667; $\frac{3}{2}$ = 1.5; $\frac{2}{1}$ = 2; $\frac{1}{1}$ = 1; The number converges on the golden ratio.

Lesson 11.2
5.

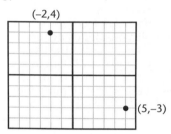

AB = | −2 − 5 | + | 4 − -3 | = 14; Find the distance between the x-coordinates and add this to the distance between the y-coordinates.
11.

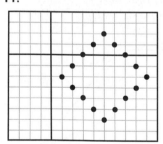

17. π = $\frac{\text{Circumference}}{\text{diameter}}$ **19.** Yes. The distance from a point on the circle to another point on the circle on a line through the center of the circle is twice the distance from a point on the circle to the center of the circle. **21.** $\frac{C}{d}$ = 4 is the taxicab equivalent of π. **23.** Points on the perpendicular bisector of a segment are equidistant from the endpoints of the segment in both Euclidean and Taxicab geometry. However, in Taxicab geometry, the perpendicular bisector is not "straight" and the points on the perpendicular bisector are not "connected". **25.** $8\frac{4}{7}$ units

27. 11.22 units

Lesson 11.3
7. Yes. There are 2 odd vertices.
9. Sample answer:

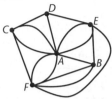

11. Start at *D* and end at *E*. Or, start at *E* and end at *D*.
12. Sample answer:

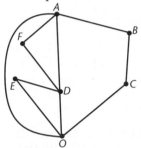

14. Sample answer:

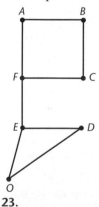

23.

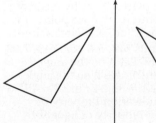

25. If a person lives in California, then the person lives in the USA. **27.** If a parallelogram has

four congruent sides and four congruent angles, then it is a square.

Lesson 11.4

7. Yes. They are both simple closed curves.
9. Assume the top and bottom faces of the original cube are *ABCD* and *EFGH*.

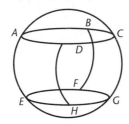

11. $V = 8$, $E = 12$, $F = 6$; $8 - 12 + 6 = 2$; Yes
13. The path traverses what would have been both sides of the original strip of paper, but the path never goes over the edge of the strip. The path finishes where it began. The strip has only one side. **15.** Sample answer: Only one side would have to be cleaned and maintained. Workers would not have to go "under" the conveyor belt to clean or repair it. The belt would wear evenly—one side would not wear more quickly than the other. **17.** The strip that is at the top of your finger when you start is at the bottom of your finger when you finish going around the nested strips. The "ceiling" becomes the "floor." **19.** The result is a twisted strip of paper that has three loops. **21.** Both angles measure 60°. **23.** 115.61 units

Lesson 11.5

7. Consider polar opposite sphere points a single "point." Two distinct points are then poles that are oriented differently. For any two "points," draw the great circles through these points. The two great circles will then intersect in two polar opposite places. Thus by identifying polar opposite points as a single point, it is true that any two "lines," or great circles, intersect in a single point. Also, points on two distinct poles have only one great circle passing through them, so that two sphere points have only one line passing through them. **9.** Each of the angles of a 2-gon could measure nearly 180°, so the sum of the measures of the angles of a 2-gon < 360°. **11.** The sum of the measures of the angles of a quadrilateral is greater than 360°. The quadrilateral can be divided into two triangles with a diagonal. The sum of the

measures of the angles of each triangle is greater than 180°, so the sum of the measures of the angles of the quadrilateral must be greater than 360°. **13.** Since distance measure keeps shrinking as you come closer to the edge of the circle, you would never come to the edge of the circle when traveling along a line. So the line would effectively have infinite length because you never approach the end of the line. **15.** The lines that contain the radii are perpendicular. **17.** An infinite number.; An infinite number.; An infinite number.
19.

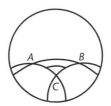

21. The length of the midsegment appears to be close to half the length of the third side of the triangle.
23. Quad *ABCD*. **25.**

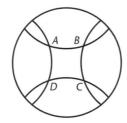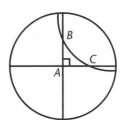

27. Answers will vary. **29.** Sample answers: a donut, a life preserver, a key

Lesson 11.6

5.

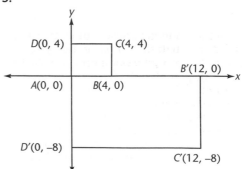

9. N; $\overrightarrow{NJ}$, $\overrightarrow{NK}$, $\overrightarrow{NL}$ **11.** C; H; K **13.** B; D or E

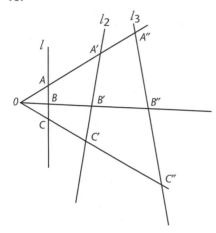

Lines *m* and *n* are parallel to line *l*. $\overline{AA'} \parallel \overline{BB'}$ from Exercise 21. Since both pairs of opposite sides are parallel, *ACYX* and *A'C'YX* are parallelograms. **25.** $BC = B'C'$. $AX = CY$ and $A'X = C'Y$ because opposite sides of a parallelogram are congruent. Since $AX = A'X$, $CY = C'Y$ by the Substitution Property of Equality. Because $BY = B'Y$ and $CY = C'Y$, $BC = B'C'$ by the Subtraction Property of Equality. **27.** $AB = A'B'$ by CPCTC. **29.** Constructions will vary. **31.** Label the angle to be copied $\angle ABC$. Draw a line and mark a point on the the line Y to be the vertex of the copy. Draw a circle with center *B* and a congruent circle with center Y. Mark the points where the circle intersects $\angle ABC$ as *D* and *E*. Mark the point where the circle intersects line *l* as *Z*. Set your compass equal to *DE*. Make an arc with center *Z* and radius *DE*. Mark the point where the circle with center *Y* intersects the arc with center *Z* as *X*. Draw $\overline{YX}$. $\angle ABC \cong \angle XYZ$.

Lesson 5.1

5. 52 in. **7.** 26 in. **9.** 42 in. **11.** 24 in. **13.** 16 in.^{2} **15.** Segment Addition Postulate; Overlapping Segments Theorem **17.** base = $35x$; height = $5x$; Area = $175x^2$ **19.** 66 **21.** 690 **23.** maximum area: 2500 ft^2 **25.** $b = 50 - h$ or $h = 50 - b$, Linear function; b and h must both be positive and less than 50. **27.** 120° **29.** $(n - 2)180°$ **31.** –2

Lesson 5.2

5. 75 units^2 **7.** 28 units^2 **9.** 70 units^2 **11.** 270 units^2 **13.** 108 units^2 **15.** 375 units^2 **17.** 20 cm **19.** Triangle a: $A = \frac{1}{2}(8)(6.93) = 27.72$ units^2, Triangle: b: $A = \frac{1}{2}(6)(8) = 24$ units^2, Triangle a has the larger area. **21.** Parallelogram a: $A = (6)(7) = 42$ units^2, Parallelogram b. $A = (5)(7) = 35$ units^2, Parallelogram a has the larger area. **23.** 1.5 units^2 **25.** Area of $\triangle ACB = \frac{1}{2}(AC)(XB)$. Area of $\triangle ACD = \frac{1}{2}(AC)(XD)$.

Area of $ABCD = \frac{1}{2}(AC)(XD) + \frac{1}{2}(AC)(XB) = \frac{1}{2}(AC)(XD + XB) = \frac{1}{2}(AC)(BD)$.
27.

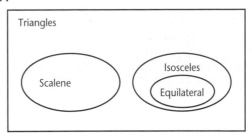

29. $x^2 + 2xy + y^2$

Lesson 5.3

5. $C = 37.7$ in; $A = 113.1$ in^2 **7.** $C = 22.0$ in; $A = 38.5$ in^2 **9.** 1963.5 m^2 **11.** Area of a pizza with d = 12 in: 36π in^2; Area of a pizza with d = 18 in: 81π in^2; The area of an 18 in pizza is more than twice the area of a 12 in pizza, so the 18 in pizza should feed six people. **13.** 78.5 ft **15.** The inside tires do not have to go as far, so they are not moving as fast. **17.** 10.44 revolutions **19.** 172.79 in^2 **21.** 7.07 m^2 **23.** Circular: $A = 199$ m^2, Square: $A = 156.25$ m^2, Circular area is larger. **25.** 254.47 in^2 **27.** 16.5 cm^2 **29.** 17.5 cm^2

Lesson 5.4

5. 5 **7.** $\sqrt{7445}$ **9.** $4\sqrt{130}$ **11.** 128 cm^2 **13.** $12^2 > 5^2 + 9^2$; Obtuse **15.** $25^2 = 24^2 + 7^2$; Right **17.** $3^2 + 4^2 = 5^2$; Right **19.** 3,4,5; 5,12,13; 7,24,25; 9,40,41; 11,60, 61; 13,84,85

Let $a = m$, $b = \frac{m^2 - 1}{2}$ and $c = \frac{m^2 + 1}{2}$, $a^2 + b^2 = m^2 + \left(\frac{m^2 - 1}{2}\right)^2 = m^2 + \frac{m^4 - 2m^2 + 1}{4} = \frac{4m^2}{4} + \frac{m^4 - 2m^2 + 1}{4} = \frac{m^4 + 2m^2 + 1}{4}$, $c^2 = \left(\frac{m^2 + 1}{2}\right)^2 = \frac{m^4 + 2m^2 + 1}{4}$,

$a^2 + b^2 = c^2$, so any triple generated will work. **21.** The area of the constructed square is $BR^2 = AR^2 + AB^2$, and AR^2 is the area of EFGH and AB^2 is the area of ABCD. **23.** 43.6 ft. **25.** Both squares have the same area: $(a+b)^2$. In the first square, after the triangles are removed, a square with side *c* and area c^2 remains. In the second square, after the triangles are removed, two squares with sides a and b and total are $a^2 + b^2$ remain. Since the areas of the triangles are the same in both squares: $c^2 = a^2 + b^2$. **27.** 18.7 units^2 **29.** 30.5 units^2 **31.** True. A rhombus is defined to be a parallelogram with four congru-

ent sides.   **33.** True. The sum of the areas of the angles of a quadrilateral is $(4 - 2)(180) = 360°$.   **35.** $2\sqrt{3}$

Lesson 5.5

7. 25.25 units²   **9.** $y = 6\sqrt{3}$; $z = 12$ **11.** $x = 7$; $y = 7\sqrt{3}$ **13.** $q = 6$; $r = 6\sqrt{2}$ **15.** $p = 4\sqrt{2}$; $r = 8$ **17.** $h = 3.4\sqrt{3}$; $g = 6.8$ **19.** $k = 8.5$; $h = 8.5\sqrt{3}$ **21.** 113.14 ft.   **23.** 16 cm²
25. $a = 5.2$ cm; $A = 93.6$ cm²   **27.** $P = 54$ cm; $A = 140.3$ cm²   **29.** 15.2 cm²   **31.** π units each
33. Area of the lune = (Area of the isosceles right triangle) + 2 (Area of the semicircle with diameter equal to a leg of the triangle) – (Area of the semicircle with diameter equal to the hypotenuse of the triangle); $A = 4 + 2\pi - 2\pi = 4$ units²   **35.** Lunes could be constructed on regular pentagons or hexagons, for example.   **37.** They are parallel. Corresponding angles are congruent.   **39.** $A =$ 17.89 m², $P = 20$ m   **41.** 9 ft.   **43.** $A = \frac{1}{2}$ ab

Lesson 5.6

5. 5.39   **7.** 5.39   **9.** 6.08 units   **11.** $BD = \sqrt{17}$, $BC = \sqrt{20}$, $CD = 3$. Since no two pairs of sides have the same length, the triangle is not isosceles.   **13.** $EJ = 9.01$ units; $JH = 6.26$ units; $EH = 10.51$ units   **15.** Sample answer shown: $AB = |5 - 0| = 5, AC = |12 - 0| = 12,$ $BC = \sqrt{(5 - 0)^2 + (0 - 12)^2} = \sqrt{5^2 + (-12)^2} = \sqrt{169} = 13$

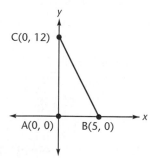

17. Sample answer when rectangles of width 1 are used is as follows.
From the left: $(1)(2) + (1)(3) = 5$ (low estimate)
From the right: $(1)(6) + (1)(3) = 9$ (high estimate)
Average $= \frac{5+9}{2} = 7$ **19.** Slope of $\overline{WX} = \frac{-6}{8} = \frac{-3}{4}$; Slope of $\overline{XY} = \frac{4}{3}$; Slope of $\overline{YZ} = \frac{6}{-8} = \frac{-3}{4}$; Slope of $\overline{ZW} = \frac{4}{3}$; Since the slopes of the opposite sides are equal, the opposite sides are parallel and the figure is a parallelogram. Since the slopes of consecutive sides are negative reciprocals, the

sides are perpendicular and the figure is a rectangle.   **21.** Perimeter of square: 60 cm Perimeter of circle: 53.17 cm. Square has the greater perimeter.   **23.** 9.18

Lesson 5.7

5. $\frac{1}{2}$ **7.** $\frac{3}{16}$ **9.** 75%   **11.** $66\frac{2}{3}$%   **13.** $\frac{1}{2}$ or 0.5   **15.** $\frac{4}{5}$ or 0.8   **17.** 0.089   **19.** 0.717   **21.** $\frac{2}{9}$
23. Sample answer: Let the board be circular, with $r_{outer} = \sqrt{2}r_{inner}$ **25.** 0.44 Depending on the skill of the dart thrower, areas on the outer part of the board are less likely for the dart to land on.   **27.** 15 in²   **29.** 27 ft²   **31.** 804.57 mm²

Chapter

Lesson 6.1

17. 10   **19.** No   **21.** 14 units   **25.** 5 cubes   **29.** Answers will vary.   **33.** 24.61 m²   **35.** 78.54 m²
37. 170.29 yds

Lesson 6.2

7. $\overline{KO}$; $\overline{JN}$; $\overline{HK}$; $\overline{IJ}$ are skew   **9.** $\overline{IJ}$; $\overline{HK}$; $\overline{MN}$; $\overline{LO}$ are perpendicular   **11.** True. If the three lines are coplanar, then they will all be parallel. If the lines are not coplanar, then the first and second lines are contained in a plane parallel to the third line, and hence will never intersect.
13. False. Consider the front wall and a side wall of your classroom. Both are perpendicular to the ceiling, but they are not parallel to each other.   **15.** True. There is a line in the plane that contains both the points where the two perpendicular lines intersect the plane. So both lines will be perpendicular to that line and therefore parallel.   **17.** Skew lines   **19.** Acute. A plane will not take off straight up or upside down.
21. Insert in $\overline{CD}$ **23.** The dihedral angle formed by insert $\overline{AB}$ is larger than the angle formed by insert $\overline{CD}$.

Lesson 6.3

9. $\overline{AD}$ and $\overline{CF}$ **13.** $\angle ADE$; $\angle ADF$; $\angle DAB$; $\angle DAC$; $\angle BED$; $\angle BEF$; $\angle EBA$; $\angle EBC$; $\angle CFD$; $\angle CFE$; $\angle FCA$;

$\angle FCB$ **15.** $\overline{JN}$; $\overline{IM}$; $\overline{HL}$ **17.** Parallelogram
19. $\angle KNJ$; $\angle JGK$; $\angle LMI$; $\angle IHL$ **25.** 22 faces; 40
vertices; 60 edges **27.** 11.20 **29.** 18.03 ft
31. False. The diagonals will only bisect the corner angles of a rectangle if the rectangle is a
square. **35.** (0,0) **37.** (3.5,2)

Lesson 6.4

13. Top-back-right **15.** Bottom-front-right
17.

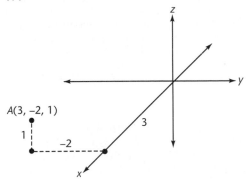

19.

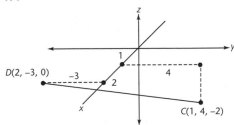

21. 5.83 **23.** 10.25 **25.** $F(0,10,7)$ **27.** $C(6,10,0)$
29. (2,2,2) **31.** $(1,\frac{-1}{2},\frac{1}{2})$ **33.** $(\frac{1}{2},\frac{5}{2},\frac{-1}{2})$ **35.** 8.60
37. (2,–5)

Lesson 6.5

5.

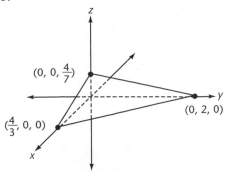

9.

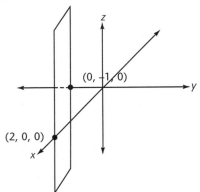

13.

t	x	y	z
1	$\frac{2}{3}$	1	0
2	$\frac{4}{3}$	2	–1

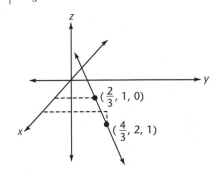

15. $x + 3y = 7$ **19.** toward the southwest
23.

Statements	Reasons
1. $\overline{DA} \cong \overline{RT}$ and $\overline{DT} \cong \overline{AR}$	**1.** Given,
2. $DART$ is a parallelogram	**2.** If both pairs of opposite sides of a quadrilateral are congruent, then it is a parallelogram.

Lesson 6.6

19.

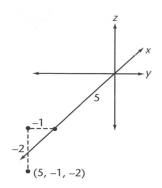

21.

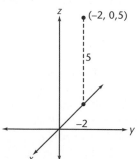

23. 17.35   **25.** $(\frac{-1}{2}, 5, \frac{9}{2})$

Lesson 7.1

7. 2   **9.** s = 4; $\frac{\text{Surface area}}{\text{Volume}}$ = 1.5   **11.** The flat-worm's shape maximizes its surface area and thus maximizes the amount of oxygen it can absorb.   **13.** The volume of water the cactus can hold is maximized while the surface area through which water can evaporate and heat can be absorbed is minimized. When water is available, the cactus can store the maximum

amount and retain it more easily.   **15.** For an n × 1 × 1 prism, the ratio approaches 4. For an n × n × n prism, the ratio is inversely proportional to n.   **17.** The ratio of surface area to volume is smaller in a larger roast, so there is relatively little surface area through which heat can enter. There is also more meat to heat in a larger roast.   **19.** The engine that pushes a larger amount of air not as quickly is both quieter and more fuel efficient. It is quieter because the differences between the speeds of the air coming out of the engine and the air outside the engine are not as great. It is more efficient because it delivers more power per weight of the plane for a given fuel consumption.   **21.** 1 × 9 = 9 m²; 2 × 8 = 16 m²; 3 × 7 = 21 m²; 4 × 6 = 24 m²; 5 × 5 = 25 m²   **23.** 25π units²   **25.** 68 cm²

Lesson 7.2

15. 35 cm³   **17.** 391 cm³   **19.** 520 units³   **21.** 294 yd²   **23.** 1995.32 cm³   **25.** S = 197.87 in², V = 103.18 in³   **27.** If doubled: The surface area is multiplied by 2² = 4 and the volume is multiplied by 2³ = 8. If tripled: The surface area is multiplied by 3² = 9 and the volume is multiplied by 3³ = 27.   **29.** Power flakes: S = 158 in², V = 120 in³; Nutri Flakes: S = 232 in², V = 160 in³   **31.** 2√5   **33.** 175   **35.** 5 in   **37.** 15.59 in

Lesson 7.3

9.

11. 120 cm²   **13.** 211.24 cm²   **15.** 100 units³   **17.** 168 m³   **19.** SA = 864 cm², V = 1296 cm³   **21.** 1125 m³   **23.** 39°   **25.** 30 units²

Lesson 7.4

9. 477.52 units²   **11.** 753.98 units³   **13.** 785.40 cm²   **15.** 157.08 m³   **17.** 37.70 mm   **19.** Cylinder x: L = 113.10; S = 169.65; Cylinder y: L = 113.10; S = 339.29; The lateral areas are the

same. Cylinder y has greater surface area.   **21.** 3141.59 mm³ or 3.14 cm³   **23.** The volume of a cylinder depends on both the radius and the height of the cylinder. Both dimensions must be taken into consideration when finding volume.   **25.** About 3118 ft³   **27.** $S = 230.53$ units²; $V = 144$ units³   **29.** $S = 276$ ft²; $V = 280$ ft³   **31.** If two solids have equal heights, and if the cross-sections formed by every plane parallel to the bases of both solids have equal areas, then the two solids have the same volume.

Lesson 7.5

7. 301.59 m²   **9.** 6.46 in   **11.** 54.45 in³
13. 297.35 cm²   **15.** $S = 24.53$ cm²; $A = 19.63$ cm²   **17.** Cone; $SA = 3522.07$ units²
19. Straight angle: $A = 56.55$ cm²; 120° angle: $A = 37.70$ cm²   **21.** Approximately 116.31 cm³
23. $\frac{2}{\pi}$ **25.** 41.04   **27.** 89.91   **29.** 153.56
31. $V = \frac{1}{3}\pi(x^2)(\sqrt{100 - x^2})$ **33.** V ≈ 403.1 units²   **35.** $(2a,0)$

Lesson 7.6

7. $S = 615.75$ cm², $V = 1436.76$ cm³   **9.** $S = 1017.85$ cm², $V = 3053.54$ cm³   **11.** $V = 42.4$ in³   **13.** $V = 22.4$ in³   **15.** 13.09 in³
17. Cube: 600 units², Sphere: 483.1 units²; the cube has a larger surface area.   **19.** 904.78 in³
21. 2827.43 cm²   **23.** 420 ft³   **25.** 3534.29 m³

Lesson 7.7

9. front-right-bottom; $(6,5,-8)$ **13.** back-right-bottom; $(-4,1,-1)$ **21.** 314.16 units²
23. 785.40 units³   **25.** 67.02 units³
27. $SA = 456.63$ units², $V = 445.48$ units³
29. $V = (8)(5)(4) = 160$ m³

Lesson 8.1

5. 2   **9.** $\left(\frac{-2}{3}, 2\right)$ **15.** The pinhole is the center of dilation.   **17.** Light travels in straight lines. The light rays containing the points of the pre-image and the image cross at the pinhole of the camera.   **23.** $y = \frac{5}{2} x$ **25.** $y = \frac{1}{4} x + 4$ **27.** Line

containing $A(5,9)$ and $A'(10,18)$ has equation $y = \frac{9}{5} x$. $A''(25,45)$ is on the line because $45 = \frac{9}{5} (25)$.   **29.** Line containing $C\left(\frac{2}{3}, \frac{1}{5}\right)$ and $C'\left(\frac{8}{3}, \frac{4}{5}\right)$ has equation $y = \frac{3}{10} x$. $C''\left(2, \frac{3}{5}\right)$ is on the line because $\frac{3}{5} = \frac{3}{10} (2)$.   **37.** $A = 4\pi\left(\frac{40000}{2\pi}\right)^2 \approx$ 509,295,817 km²

Lesson 8.2

9. $\frac{2}{5} = \frac{4}{10}$ **11.** $\frac{6+9}{9} = \frac{2+3}{3}$ **13.** $x = 12$
15. $x = 14$ **17.** $x = \frac{43}{4} = 10\frac{3}{4}$ **19.** $GH = 15$
21. False; $\frac{2}{3} = \frac{4}{6}$ but $\frac{2+5}{3+5} \neq \frac{4+5}{6+5}$ **23.** True : $xs = yr$, $xn = ym \Rightarrow xy + xs + xn = xy + yr + ym$
31. 200 in. by 300 in. or 16 ft 8 in by 25 ft   **33.** $\frac{1}{288}$ **41.** Anthony can find the distance from where he is standing to point A, which is equal to the distance across the river.

Lesson 8.3.

5. Yes; AA-Similarity   **7.** Yes; SSS-Similarity
9. $\frac{2}{3}, \frac{2}{3}$ **11.** No; The ratios between pairs of corresponding sides will not be equal.   **13.** $\Delta RST \sim \Delta XYZ$; AA-Similarity   **15.** $YZ = 22$ units; $XZ = 27.5$ units   **21.** d = $\sqrt{(8-2)^2 + (3+6)^2} = \sqrt{6^2 + 9^2} = \sqrt{117} \approx 10.82$ **23.** \$15; Fine for every mile over the speed limit.   **25.** $y = \frac{2}{3}x + 6$

Lesson 8.4

7. $\Delta ABC \sim \Delta EDC$; AA-Similarity; Scale factor = 2 or $\frac{1}{2}$ **9.** $x = 9.6$

11. $x = 2.4$ **13.** The AA Similarity Postulate
15. By the Two-Transversal Proportionality Corollary, the parallel segments divide $\overline{AC}$ and $\overline{AB}$ proportionally. By construction, the proportionality ratio is 1. Therefore, $\overline{AB}$ is cut into congruent segments.   **19.** Perimeter of $\Delta ABC = 29.31$ cm   **21.** Perimeter of $\Delta DEF = 14.66$
23. $BD = 5.85$ cm

Lesson 8.5

9. $x = 14.4$ units   **11.** $x = 704$ m   **15.** $m\angle BCA = m\angle B'C'A = 90°$; $m\angle BAC = m\angle B'AC'$ because vertical angles have equal measures; $\Delta ABC \sim \Delta AB'C'$ by AA-Similarity.

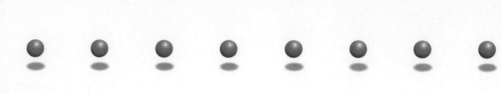

17. The lens should be half the distance from the object to the image.
23. Median definition **25.** $BD + DC = BC$; $YM + MZ = YZ$
27. Substitution and fraction simplification
31. $\frac{AD}{XM} = \frac{BA}{YX}$ **35.** Angle bisector definition
39. m$\angle DAC$ = m$\angle MXZ$ **43.** $\frac{AD}{XM} = \frac{AC}{XZ} = \frac{AB}{XY} = \frac{BC}{YZ}$ **47.** $\angle 3 \cong \angle 2$ **51.** Isosceles Triangle Theorem
57. Sometimes

Lesson 8.6

7. The 16-in pizza. If the radius of a pizza is doubled, the area of the top is multiplied by 4. The 16-in pizza is approximately 4 times as big as the 8-in pizza, but it doesn't cost 4 times as much. **9.** $\left(\frac{7}{5}\right)^2 = \frac{7^2}{5^2} = \frac{49}{25}$ **11.** $\frac{5}{7}$ **13.** $\sqrt{\frac{16}{25}} = \frac{4}{5}$
15. $\left(\frac{4}{5}\right)^3 = \frac{4^3}{5^3} = \frac{64}{125}$ **17.** $\left(\frac{10}{3}\right)^2 = \frac{10^2}{3^2} = \frac{100}{9}$
19. Area: $\frac{x}{50} = \frac{25}{16} \Rightarrow 16x = 1250 \Rightarrow x =$ 78.125m^2 **21.** $\frac{7}{9}$ **23.** $\left(\frac{7}{9}\right)^2 = \frac{7^2}{9^2} = \frac{49}{81}$ **25.** $\left(\frac{12}{13}\right)^3 = \frac{12^3}{13^3} = \frac{1728}{2197}$ **27.** $\frac{16}{25}$ **29.** $\frac{4}{5}$ **31.** Perimeter: $\frac{x}{42} = \frac{4}{6} \Rightarrow 6x = 168 \Rightarrow x = 28$ m, Area: $\frac{x}{96} = \frac{16}{36} \Rightarrow 36x = 1536 \Rightarrow x = 42\frac{2}{3}$m^2 **33.** $x = 686$ cm^3
35. About 2.8 times. **37.** $x = 4.5$ cm^3 **39.** the diameters are in ratio $\frac{201}{50}$, and the volumes are in ratio $\left(\frac{201}{50}\right)^3$. **41.** $400 \times 1.5 \times 1.75 = 1050$ m^2
43. 1420 tons **45.** 24 sides **47.** b and c

Lesson 9.1

7. $\overline{PA}$ or $\overline{PB}$ or $\overline{PC}$ **9.** $\overline{AC}$ **15.** 126° **17.** 116°
19. 166° **21.** 90 ft **25.** 36° **27.** 36° **29.** If central angles of a circle are congruent, their chords are congruent. **31.** SSS congruence. **33.** Yes; AA-Similarity **35.** Yes; AA-Similarity **37.** $SA \approx$ 166 in^2, $V = 140$ in^3

Lesson 9.2

5. $\overline{WZ}$ **7.** $XW = WZ \approx 2.24$ units **11.** 1124 mi

13. 1.73 **15.** 5.39 **19.** $P = 58\frac{2}{3}$ ft, $A = 156\frac{4}{9}$ ft^2
21. $S = 12{,}465.84$ ft^2

Lesson 9.3

7. m$\angle C = 34°$; m$\angle D = 34°$ **9.** m$\angle B = 43.5°$; m$\angle A = 43.5°$ **11.** 29° **13.** 61° **15.** 50° **17.** 80°
19. 200° **21.** 80° **23.** m$\angle A = 40°$; m$\angle B = 80°$
25. 20 units.

Lesson 9.4

5. 75° **7.** 85° **9.** 85° **11.** 40° **19.** 100°
25. Acute secant-tangent angle, vertex on circle; m$\angle AVC = \frac{1}{2}$m$\widehat{AV}$ **27.** Chord-chord angle, vertex inside circle; m$\angle AVC = \frac{1}{2}\left(\text{m}\widehat{AC} + \text{m}\widehat{BD}\right)$
29. Secant-tangent angle, vertex outside circle; m$\angle AVC = \frac{1}{2}\left(\text{m}\widehat{BC} - \text{m}\widehat{AC}\right)$ **31.** 105°

Lesson 9.5

7. 5 units **9.** 3 cm **13.** $\angle BPV$ **15.** $CD = 3$
19. 12.5 units **21.** 7.24 or 2.76 **27.** 60°
29. 50° **31.** No . There are no 90° angles.

Lesson 9.6

13. Center $= (0,0)$: $r = \sqrt{101}$ units **15.** $x^2 + y^2 = 13$ **17.** Center $= (6,0)$; $r = 3$ units
19. Center $= (-5,2)$; $r = 4$ units **21.** Center $= (0,0)$; $r = 6$ units **27.** $(x - 2)^2 + (y - 1)^2 = 9$
29. $(x - 5)^2 + (y - 3)^2 = 9$ **33.** $(3,0)$, $(-3,0)$, $(0,9)$, $(0,-1)$ **37.** $(x + 4)^2 + (y - 2)^2 = 9$ **41.** $(x - 2)^2 + (y - 3)^2 = 36$ **43.** $y = \frac{3}{4}x + 12.5$

Lesson 10.1

7. $\tan A = \frac{5}{2}$ or 2.5 **9.** $\tan A = 2.4$ or $\frac{12}{5}$
11. $\tan A = 1$ **13.** $\tan C = 1$ **15.** $\tan C = \frac{5}{3}$ or 1.667 **17.** $\tan C = 1$ **19.** About 56° $-57°$
21. 2.3559 **23.** 19.0811 **25.** 1.0724
27. 0.3839 **29.** 25° **31.** 67° **33.** As the angles get larger, the tangent values get larger. As the angles get smaller, the tangent values get

smaller. **35.** 37.25 units **37.** 22.57 units
39. 49.40° **41.** 10.2°, 14° **43.** 1119.62 feet
45. $V = 1099.56$ units^3, $SA = 596.90$ units^2
47. $V = 65.45$ units^3, $SA = 78.54$ units^2
49. $MQ = 7.42$ units $= QN$

Lesson 10.2

7. $\frac{12}{13}$ **9.** $\frac{5}{13}$ **11.** $\frac{12}{5}$ **13.** 67.38° **15.** $\angle Y$ **17.** $\angle X$
19. $\angle Y$ **21.** 0.3090 **23.** 0.8387 **25.** 17°
27. 56° **29.** 68° **31.** $\sin^{-1}(\sin \theta) = \theta$. If you find the sine of an angle and the inverse sine of the result, you return to the original angle.
33. $\cos^{-1}(\cos q) = q$. If you find the cosine of an angle and the inverse cosine of the result, you return to the original angle. **35.** $\sin A = \cos B$
37. $\cos B = \sin(90° - m\angle B)$, $\sin B = \cos(90° - m\angle B)$, $\cos A = \sin(90° - \angle A)$ **39.** Area = 19.28 units^2 **41.** $Y = 7\sqrt{3}$; $Z = 14$ **43.** $X = \frac{14\sqrt{3}}{3}$; $Z = \frac{28\sqrt{3}}{3}$ **45.** $q = 1$; $r = \sqrt{2}$ **47.** 108° **49.** 21.6°

Lesson 10.3

9. $\sin 110° = 0.94$; $\cos 110° = -0.34$; $\tan 110° = -2.75$ **11.** $\sin 450° = 1$; $\cos 450° = 0$; $\tan 450°$ is undefined **13.** $\cos 157° = -0.9205$
15. $\sin 480° = 0.8660$ **17.** 139°; 221°
19. P'(0.7071, 0.7071) **21.** P'(0.3090, 0.9511)
23. P'(−0.3420, −0.9397) **25.** P'(0.7660, 0.6428)
27. a **29.** b **31.** $(\cos A)^2$ **33.** $\sin A = \cos(90° - m\angle A)$

Lesson 10.4

5. (5,4) **7.** (−1,−4) **9.** (0,3)
11. (−1.278,11.107) **13.** (−1.981,−17.919)
15. (−3.163,−1.731) **17.** 180° **19.** 0° or 360°
21. (−5.7994,−.6058) **23.** (−2.2321,0.1340)
25. $\begin{bmatrix} 0 & 1 \\ -1 & 0 \end{bmatrix}$ **27.** $\begin{bmatrix} .5 & -.866 \\ .866 & .5 \end{bmatrix}$ **29.** A''(0,2), B''(−4,4), C''(−2,5) **39.** 69° **43.** −0.12; −0.99; 0.12 **47.** 0; 1; 0

Lesson 10.5

7. 8.02 cm **9.** 2.02 units **11.** 5.64 cm **13.** 7.93 cm **15.** 3.62 miles **17.** Let A be the angle opposite Oak Street and angle B be the angle opposite 3rd Ave. $m\angle A = 56.24°$; $m\angle B = 48.76°$
19. $a \sin B$ **21.** Multiplication Property of Equality **23.** Division Property of Equality
25. $\frac{h_2}{c}$ **27.** $a \sin C$ **29.** Multiplication Property of Equality **31.** $\frac{\sin A}{a} = \frac{\sin C}{c}$

33. Transitive Property of Equality **35.** $\frac{\sin A}{a} = \frac{1}{c}$; $\frac{\sin B}{b} = \frac{1}{c}$; $\frac{\sin C}{c} = \frac{1}{c}$ **37.** $SA = 2827.4$ m^2; $V = 14{,}137.2$ m^3; If the radius were tripled, the surface area would be multiplied by $3^2 = 9$ and the volume would be multiplied by $3^3 = 27$.
39. $x = 4$ **41.** 60°

Lesson 10.6

5. 9.53 units **9.** 70.99 units **13.** a. 24.88 km, b. 42.25° **15.** The triangle is obtuse. **17.** By the "Pythagorean" Right Triangle Theorem, $a^2 = h^2 + (c_1 + c)^2 = h^2 + c_1^2 + 2c_1c + c^2$. **18.** From right angle trigonometry, $h = b \sin (180 - A) = b \sin A$. Also, $c_1 = b \cos(180 - A) = -b \cos A$.
19. Substituting, $a^2 = (b\sin A)^2 + \cos A)^2 = (-b\cos A)(c) + c^2$. Simplifying and using the identity $(\sin A)^2 + (\cos A)^2 = 1$ gives the Law of Cosines: $a^2 = b^2 + c^2 - 2bc \cos A$ **23.** length = $6\frac{2}{3}$ cm; width = $33\frac{1}{3}$ cm **25.** No, the congruent angles are not between the corresponding sides in both triangles.

Lesson 10.7

15. The angles of the parallelogram are $180° - 60° - 25° = 95°$ and $180° - 95° = 85°$
17. 20.74 units **19.** $|a| = \sqrt{89} = |b|$ The vectors have the same magnitude, but they are not equal because they are not parallel.
21. 1.65 mph

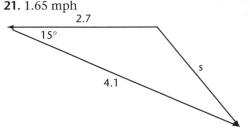

23. 287.92 units **25.** 7.53 units **27.** 3.74 units

Chapter

Lesson 11.1

11. $\frac{a}{b} \approx 1.6$ **13.** $m\angle D = 108°$ **15.** $\frac{a}{b} \approx 1.618033989$ **17.** $m\angle DEI = m\angle EDI = 72°$;

m∠*EID* = 36° **19.** congruent sides; golden ratio
21. $\frac{610}{377}$ = 1.618037135; $\frac{377}{233}$ = 1.618025751;
$\frac{233}{144}$ = 1.618055556; $\frac{144}{89}$ = 1.617977528; $\frac{89}{55}$ =
1.618181818; $\frac{55}{34}$ = 1.617647059; $\frac{34}{21}$ =
1.619047619; $\frac{21}{13}$= 1.615384615 ; $\frac{13}{8}$ = 1.625;
$\frac{8}{5}$ = 1.6; $\frac{5}{3}$ = 1.666666667; $\frac{3}{2}$ = 1.5; $\frac{2}{1}$ = 2; $\frac{1}{1}$ =
1; The number converges on the golden ratio.

Lesson 11.2

5.

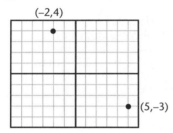

AB = | −2 − 5 | + | 4 − -3 | = 14; Find the distance between the x-coordinates and add this to the distance between the y-coordinates.
11.

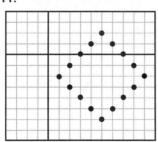

17. π = $\frac{\text{Circumference}}{\text{diameter}}$ **19.** Yes. The distance from
a point on the circle to another point on the circle on a line through the center of the circle is twice the distance from a point on the circle to the center of the circle. **21.** $\frac{C}{d}$ = 4 is the taxicab equivalent of π. **23.** Points on the perpendicular bisector of a segment are equidistant from the endpoints of the segment in both Euclidean and Taxicab geometry. However, in Taxicab geometry, the perpendicular bisector is not "straight" and the points on the perpendicular bisector are not "connected". **25.** 8$\frac{4}{7}$ units

27. 11.22 units

Lesson 11.3

7. Yes. There are 2 odd vertices.
9. Sample answer:

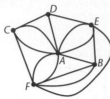

11. Start at *D* and end at *E*. Or, start at *E* and end at *D*.
12. Sample answer:

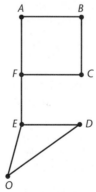

14. Sample answer:

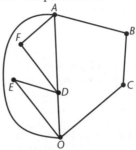

23.

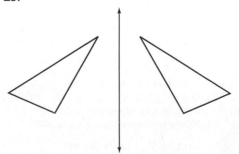

25. If a person lives in California, then the person lives in the USA. **27.** If a parallelogram has

four congruent sides and four congruent angles, then it is a square.

Lesson 11.4

7. Yes. They are both simple closed curves.
9. Assume the top and bottom faces of the original cube are *ABCD* and *EFGH*.

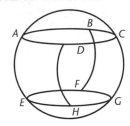

11. $V = 8$, $E = 12$, $F = 6$; $8 - 12 + 6 = 2$; Yes
13. The path traverses what would have been both sides of the original strip of paper, but the path never goes over the edge of the strip. The path finishes where it began. The strip has only one side. **15.** Sample answer: Only one side would have to be cleaned and maintained. Workers would not have to go "under" the conveyor belt to clean or repair it. The belt would wear evenly—one side would not wear more quickly than the other. **17.** The strip that is at the top of your finger when you start is at the bottom of your finger when you finish going around the nested strips. The "ceiling" becomes the "floor." **19.** The result is a twisted strip of paper that has three loops. **21.** Both angles measure 60°. **23.** 115.61 units

Lesson 11.5

7. Consider polar opposite sphere points a single "point." Two distinct points are then poles that are oriented differently. For any two "points," draw the great circles through these points. The two great circles will then intersect in two polar opposite places. Thus by identifying polar opposite points as a single point, it is true that any two "lines," or great circles, intersect in a single point. Also, points on two distinct poles have only one great circle passing through them, so that two sphere points have only one line passing through them. **9.** Each of the angles of a 2-gon could measure nearly 180°, so the sum of the measures of the angles of a 2-gon < 360°. **11.** The sum of the measures of the angles of a quadrilateral is greater than 360°. The quadrilateral can be divided into two triangles with a diagonal. The sum of the

measures of the angles of each triangle is greater than 180°, so the sum of the measures of the angles of the quadrilateral must be greater than 360°. **13.** Since distance measure keeps shrinking as you come closer to the edge of the circle, you would never come to the edge of the circle when traveling along a line. So the line would effectively have infinite length because you never approach the end of the line. **15.** The lines that contain the radii are perpendicular. **17.** An infinite number.; An infinite number.; An infinite number.
19.

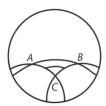

21. The length of the midsegment appears to be close to half the length of the third side of the triangle.
23. Quad *ABCD*. **25.**

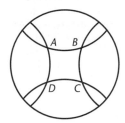 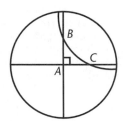

27. Answers will vary. **29.** Sample answers: a donut, a life preserver, a key

Lesson 11.6
5.

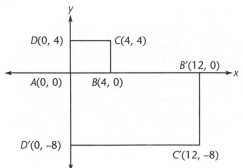

9. N; $\overrightarrow{NJ}$, $\overrightarrow{NK}$, $\overrightarrow{NL}$ **11.** C; H; K **13.** B; D or E

16.

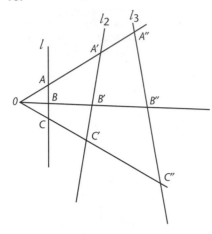

19. 4   **21.** 6   **23.** None

Lesson 11.7

6. Since the depth of self-similarity is endless, it would be impossible to measure around all the "edges" of the coastline. There would always be another similar shape to measure around.
10. The perimeter is increasing at an increasing rate with each iteration. The perimeter will approach infinity with continued iterations.
15. The entire plane would be filled.   **17.** 14.76 units   **19.** 7 units   **21.** 14.25 units

Lesson 12.1

7. The team did not win on Saturday.   **9.** conclusion: b **11.** if n, then r; if n, then q; if m, then r; r, q, m   **13. a.** The conclusion is valid because the argument is in the form *modus ponens*. **b.** False. **c.** The conclusion is true. The diagonals of a square are congruent. **d.** The conclusion is false. The diagonals of a parallelogram are not always congruent.   **15.** Yes; SSS   **17.** Yes; SAS   **19.** 110°

Lesson 12.2

9. A carrot is a vegetable and Florida is a state. True   **11.** The sum of the measures of the angles of a triangle is 180° and two points determine a line. True   **13.** Triangles are circles or squares are parallelograms. True   **15.** The figure is not a rectangle.   **17.** Rain does not make the road slippery.   **19.** $\triangle ABC$ in not isosceles or $\triangle ABC$ is scalene.   **21.** $\triangle ABC$ is isosceles and $\triangle ABC$ has two equal angles.   **23.** $\angle 1$ and $\angle 2$ are not acute angles OR $\angle 1$ is not an acute angle or $\angle 2$ is not an acute angle.   **25.** $\angle 1$ and $\angle 2$ are acute angles and $\angle 1$ and $\angle 2$ are not adjacent angles.
27. This weekend we will go camping and hiking. This weekend it will rain and we will cancel the trip.
29.

p	q	r	p OR q	$(p$ OR $q)$ OR r
T	T	T	T	T
T	T	F	T	T
T	F	T	T	T
T	F	F	T	T
F	T	T	T	T
F	T	F	T	T
F	F	T	F	T
F	F	F	F	F

p OR q OR r is false when all three statements are false.   **31.** No. 4 is not proportional to 15.
33. Yes; SAS-Similarity   **35.** $\frac{4}{5}$ **37.** $\frac{3}{5}$

Lesson 12.3

7. Conditional is true by the Multiplication Property of Equality. Converse: If $a^2 = b^2$, then a = b. False. $(-2)^2 = (2)^2$; Inverse: If a ≠ b, then $a^2 \neq b^2$; False. $(-2)^2 = (2)^2$; Contrapositive: If $a^2 \neq b^2$, then a ≠ b. True. If the conditional is true, the contrapositive is also true.   **11.** If ~q, then ~p ; If ~(~p), then ~(~q); If p, then q. The contrapositive of the contrapositive of a statement is the same as the original statement.   **17.** If a point is on the perpendicular bisector of a segment, then the point is equidistant from the endpoints of the segment.
19. If you do your mathematics homework every night, then your mathematics grade will improve.
21. b is the same form as the given statement; c is the contrapositive of the given statement.   **25.** 3.53 units

Lesson 12.4

9. No. The proof does not start by assuming the opposite of the statement to be proved.
11. $\angle 1 \not\equiv \angle 2$ **13.** vertical angles   **17.** not vertical angles   **21.** ...perpendicular lines form congruent adjacent angles.   **23.** ...congruent angles have equal measures.   **25.** ...l not perpendicular to m.

27.

Statements	Reasons
1. $\overline{CT} \not\cong \overline{BK}$	1. Given
2. Assume $\overline{BC}$ and $\overline{KT}$ bisect each other.	2. Negate proof statement
3. $\overline{CW} \cong \overline{BW}$, $\overline{TW} \cong \overline{KW}$	3. Definition of bisection
4. $\angle CWT \cong \angle BWK$	4. Vertical $\angle$s $\cong$
5. $\triangle CWT \cong \triangle BWK$	5. SAS
6. $\overline{CT} \cong \overline{BK}$	6. CPCTC

The last statement is a contradiction of the given statement that $\overline{CT} \not\cong \overline{BK}$. The assumption must therefore be false. So the conclusion is $\overline{BC}$ and $\overline{KT}$ do not bisect each other. **31.** Assume that the temperature is below 32°F. If the temperature is below 32°F, then the water on the sidewalk would freeze. But that contradicts the fact that the water on the sidewalk is not frozen. The assumption must therefore be false and the temperature must be 32°F or higher. **35.** 4 **39.** Sample answers: cube, pencil

Lesson 12.5

9. 0, 1 **11.** 0, 1 **15.** NOT(p OR q) = (NOT p) AND (NOT q) **19.** 0, 0, 1, 1
23. 1, 0, 1, 1
25.

27. p AND (q OR r)

p	q	r	(q OR r)	p AND (q OR r)
1	1	1	1	1
1	1	0	1	1
1	0	1	1	1
1	0	0	0	0
0	1	1	1	0
0	1	0	1	0
0	0	1	1	0
0	0	0	0	0

31. AND gate ; Power Button AND Play Button

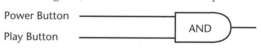

Power	Play	CD Player
1	1	1
1	0	0
0	1	0
0	0	0

35. False ; For example: $\frac{1}{2} = \frac{2}{4}$ but $\frac{1}{2} \neq \frac{1+2+3}{2+4+3}$.

Lesson 12.6

5 a. $C(7,3)$, b. $C(q,p)$ **7** a. $C(10,5)$, b. $C(p+r,q)$
11. $M = (1,3)$; $S = (10,3)$ **13.** $MS = \frac{AD + BC}{2}$;
The length of the median of a trapezoid is equal to the average of the lengths of its bases.
15. $M = (p,q)$; $S = (r + s,q)$ **17.** Slope of $\overline{AD} = \frac{0 - 0}{0 - 2s} = 0$; Slope of $\overline{BC} = \frac{2q - 2q}{2r - 2p} = 0$;
Slope of $\overline{MS} = \frac{q - q}{r + s - p} = 0$. Since the slopes of the bases and the midsegment are equal, they are parallel. **19.** $M = (3,2)$ **25.** 30 cm³
27. 137.44 in³ **29.** 301.59 cm²